IMPETUS AND EQUIPOISE IN THE LIFE-STRATEGIES OF REASON

Logos and Life
Book 4

ANALECTA HUSSERLIANA

THE YEARBOOK OF PHENOMENOLOGICAL RESEARCH

VOLUME LXX

Editor-in-Chief:

ANNA-TERESA TYMIENIECKA

The World Institute for Advanced Phenomenological Research and Learning

Belmont, Massachusetts

For sequel volumes see the end of this volume.

IMPETUS AND EQUIPOISE IN THE LIFE-STRATEGIES OF REASON

Logos and Life
Book 4

by

ANNA-TERESA TYMIENIECKA

The World Phenomenological Institute

Published under the auspices of
The World Institute for Advanced Phenomenological Research and Learning
A-T. Tymieniecka, President

SPRINGER SCIENCE+BUSINESS MEDIA, B.V.

Library of Congress Cataloging-in-Publication Data is available

ISBN 978-0-7923-6731-4 ISBN 978-94-010-0946-1 (eBook)
DOI 10.1007/978-94-010-0946-1

Printed on acid-free paper

To the memory of my beloved sister

Romana Zaremba/Tymieniecka Janicka

— who instilled in me the love of learning

TABLE OF CONTENTS

PART ONE

FROM THE ELUSIVE PRIMEVAL LOGOS TO THE OPEN-ENDED GREAT PLAN OF LIFE

CHAPTER ONE
THE PRIMEVAL LOGOS

CHAPTER TWO
LIFE AND NON-LIFE

CHAPTER THREE
LIFE IN ITS SPECIFICS

TYING POINT ONE

THE MANIFESTATION OF LIFE THROUGH THE NATURE-LIFE COMPLEX AND ITS RADIUS

Nature-Life

TYING POINT TWO

ANTICIPATING THE MANIFESTATION OF THE LOGOS OF LIFE

CHAPTER ONE
METAPHYSICS OF MANIFESTATION
Logos in the Individualization of Life, Sociability, and Culture

CHAPTER TWO
SPONTANEITY, CONSTRUCTIVE DYNAMISM, AND CIPHERING IN THE HUMAN CONDITION

PART THREE

MANIFESTATION AND DIFFERENTIATION

CHAPTER ONE
THE SURGING MANIFESTATION OF LIFE

CHAPTER TWO
THE STRATEGIES OF DIFFERENTIATION AND HARMONY IN THE SELF-INDIVIDUALIZING LIFE PROCESS

CHAPTER THREE
ONTOPOIETIC DIVERSITY AND THE UNITY OF APPERCEPTION

CHAPTER TWO
THE GREAT PLAN OF LIFE, THE ESOTERIC PASSION OF THE MIND

CHAPTER TWO
KNOWLEDGE AND COGNITION IN THE SELF-INDIVIDUALIZING PROGRESS OF LIFE

CHAPTER THREE
THE CREATIVE RISE OF THE HUMAN SPIRIT

CHAPTER THREE
THE RELEASE OF SUBLIMINAL YEARNINGS

PART FIVE

THE PROMETHEAN DIRECTION OF THE LOGOS OF LIFE IN QUEST OF ACCOMPLISHMENT

The Dialectic of Embodiment and Freedom

CHAPTER ONE
THE HUMAN SELF IN THE COMMUNAL FABRIC

CHAPTER TWO
FROM HUSSERL'S FORMULATION OF THE SOUL-BODY PROBLEM TO THE DIFFERENTIATION OF FACULTIES

PART SIX

THE STRATEGIES OF IMPETUS/EQUIPOISE IN COMMUNAL SHARING-IN-LIFE

CHAPTER ONE
THE FULGURATION OF THE LOGOS IN THE "OVERT" STRATEGIES OF THE EXISTENTIAL INTERACTION
The Communal Significance of Life

CHAPTER TWO
THE DIALECTIC JUNCTION IN THE LOGOIC STRATEGIES: MORAL LAW VS. COMMITMENT

CHAPTER FOUR
MORAL AND CIVIC VIRTUE AS THE BEDROCK OF THE MANIFEST GAME OF LIFE, THE CORNERSTONE OF DYNAMIC SOCIAL EQUIPOISE

TYING POINT SIX

THE GOLDEN MEASURE:
TOWARD A NEW ENLIGHTENMENT

THE META-ONTOPOIETIC CLOSURE

ACKNOWLEDGMENTS

In presenting this work, the fruit of more than two decades of reflection, a few words are in order. Although this project has been prompted and guided by key intuitions of its incipient phase, its progress and their elaboration have proceeded slowly and in fragments. These fragments have in turn been guiding and nourishing our programs in the World Phenomenology Institute, and consequently have constituted the systematic nucleus of our collective work as it has appeared in the Analecta Husserliana book series. Throughout I kept receiving feedback from the reflection and insights of my collaborators, who have both checked and enriched my thought. I want to acknowledge my debt to the multifarious discourse carried on in the Institute's work and to express my deep gratitude to all the colleagues who sustained with me this *discourse fleuve* over three decades.

Several of the fragments that find their proper place in this treatise were previously published in Analecta Husserliana volumes. We owe thanks to Kluwer Academic Publishers for permission to include these here; and also to Professor Marcelo Sánchez Sorondo for permission to reprint the fragment "The Virtues and the Passions of the Soul," that appeared in Silvano Allasia and Marcelo Sánchez Sorondo (eds.), *Ragione pratica, libertà, normativita,* Collana "Dialogo di Filosofia" 8 (Rome: Herder; Università Lateranense, 1991), pp. 291-332 as well as to Professor Robert Cohen for permission to reprint "Knowledge and Cognition in the Self-Individualizing Process of Life," which appeared in Ioanna Kuçuradi (ed.), *The Concept of Knowledge: the Ankara seminar* (Dordrecht, Boston: Kluwer Academic Publishers, 1995), pp. 197-217.

Jeffrey Hurlburt and Louis Houthakker, my assistants, deserve thanks for their work preparing the manuscript.

Above all I owe a debt of warm gratitude to my valued collaborator Robert Wise Jr. for preparing this work for publication and for his careful editing of my work over the years.

Anna-Teresa Tymieniecka

IN PRAISE OF PHILOSOPHY

Among our intellectual endeavors philosophy merits the highest praise. In essence it amounts to a theoretical view on reality, the human being, the aspirations and limitations of human endeavor. Pointing the way to this intellectual theorizing is the "philosophical streak" of the entire human mind at work, which streak when traced proves to be pursuing its path in all of the the mind's engagements with reality from birth to death. This unique pursuit among all of the propensities of human beings, this need to understand all, to "philosophize," crystalizes in the mind's innumerable confrontations with reality upon its acquiring full consciousness and self-awareness. These confrontations are mediated by innumerable tasks as well as the capacity with which the mind is endowed. We may say that this philosophical urge surreptitiously presides already as we differentiate all things, as we cognize-constitute them as ideas, values. Then in a reverse turn of the mind in its synthetic swing, discernment becomes experience as all is brought together in the crystalization of the Human Condition within the individual's experiential modalities.

But although present from the start, it is in philosophy that the radical recasting of orientation finds expression. Having thus far pursued its constructive as well as cognitive work in the cause of the propagation of life, the Logos arrives at a point that Husserl called in relation to transcendental philosophy the "heroism of the mind" and turns toward the discovery and scrutiny of its own workings.

Indeed, the Logos turns toward its own working already in scientific exploration and reflection. Logos unfolds its inquisitive working within the scientific field, partly scrutinizing the constitutive work accomplished with the establishment of the manifestation of the world, nature, life, and partly inventively catching it again in the logoic cognitive/inventive modalities. However, philosophy involves a radical turn for the inquisitive logos of the human mind, a self-conscious shifting of its vision in its quest after 'truth.' In this quest are mobilized all the numerous streaks of the logos of the mind — the Dionysian logos — proposing innumerable furrows in pursuit of which the path toward 'truth' may proceed. Just what is the unique role of the philosophical quest? As I just said, in the mind's reverse swing, its philosophical streak crystalizes into the quest, essential to humanhood, to understand all, that is, to rise above all that is singular, specific, and concrete that it has confronted in its differentiating swing and to find meaning in it all. In this quest, philosophy becomes the hound of all knowledge, the conductor of the inquiries of a mind seeking the universal truth of things, persons, the world. In striving to discover a unifying principle the mind projects unifying links forming a "great plan" of life. This quest is not an abstract pursuit of universal forms; discrimination, valuation, judgement enter into all its findings. Philosophy proceeds from the inmost selfhood of the human being in progress and so brings all to the forge of experience.

To philosophize is not only to project a view of the world and life and so obtain life's bearings amid the All, but is also to offer guidelines for human conduct, to distinguish right and wrong, good and evil.

Rising like a sun above the struggles of life, but enlightening them all from its height, philosophy provides rest for the anxious mind, comforts the heart in its predicaments, and consoles amid the vicissitudes of life. Its rays throw a sublime light of beauty on an individual life.

Most of all, in its quest to understand all, philosophy bestows on life its specifically human significance. In its final thrust toward the meaning of life, philosophy encounters the drive to transcend, which takes one in the direction of faith and the religious quest. Here philosophy plays an indispensable role as it helps the believer to clarify revelation. Thus, even truth that is not immanent receives an essential frame of reference by which it may establish its claims immanently, receiving interpretation that incorporates it into life.

Thus, philosophy is key to pursuing understanding of the meaning of all human endeavor. Philosophizing is the uniquely human aspect of man's self-interpretation-in-existence.

In its swings in opposing directions, now differentiating and now synthesizing, philosophy is part and parcel of the entire human enterprise. It is called on to provide a perspective on the All, the thread uniting what would otherwise be a topsy-turvy conglomeration. Through Western history philosophers of various orientations have undertaken this task. In our day thinkers in all fields, in the physical sciences, the human sciences, literature and art, sensing our total disorientation, are desparately searching nature and the cosmos for such a key to human endeavor. The natural sciences and human cultural creation are now profound enough to lay bare the foundations of our lives and beckon philosophy to enter the arena.

INTRODUCTION

ANTICIPATION IN OUR QUEST FOR THE LOGOS IN LIFE AND THE LIFE OF THE LOGOS

Every attempt in history at what is generally called 'metaphysics' has been proposed upon its author's conviction as the ultimate vision of things. So is the present work. What can be more final, more definitive as the vision of the All than the exfoliation of the historical progress of the Logos itself from its genesis and through its logistics, yielding the story of the Logos? — the Logos from which unfold all of rationality's matters, beings, events, processes, generative schemas, and origins. Here we follow all the strings that lead to coming to be and passing away. After having laid down a philosophical groundwork in phenomenology/philosophy of life in the first three books of my treatise, *Logos and Life*, and in the other works from which this inquiry sets out, I have embarked upon this culminating elucidation of the reason of all reasons — relational, seminal, generative, transformative, inventive, creative... — from the perspective of the gigantic game that we may pursue in the Logos' manifestation in reality, in life. It took me almost three decades of painstaking effort to pursue this elucidation of the inner workings of the Logos per se as it manifests itself in its accomplishments.

The focus upon logos as the key to the ultimate questions of philosophy is by no means new. From the days of Heraclitus and the Stoics, logos has been evoked throughout the history of Western philosophy by numerous thinkers, including Edmund Husserl. And yet the way in which we have proceeded in the present inquiry is new. The traditional approach to the logos has been purely speculative. No doubt our ascent to the logos as the last principle of order is also speculative. Nevertheless, we are not postulating this rational principle as the answer to the many questions that the world and the cosmos and the enigmatic nature of the human being bring forth, as an answer that hides all concrete answers, jumping on the wings of the human mind as it flies to its limit. On the contrary, our mode of 'speculation' is to conjecture step by step what led to the logos' being posited within the gigantic schema of dynamic reality. In the conjecture itself is contained an essential element of the foothold of its surging, a logoic instance, as well as anticipation of the answer point of logoic destination.

A WORD ABOUT THE PRESENT WORK

We will follow the dynamic, ever flexible network of the strategies of the logos, tracing it to its intrinsic nature as "the reason of all reasons." The greatest category of all, to which even the logos is subject, emerges as "constructivity."

It was by pursuing the constructive entanglements of life's spontaneities, dynamisms, forces in their radiating relevancies for the system of life and cosmic forces in which the logos flourishes that we first, in Book 1 of our *Logos and Life,* assessed the work of logos at its peak: the human creative act. We then, in Books 2 and 3, pursued the work of logos within the human sphere and reached its diminishment and passage into anti-logos. Subsequently, and this is the great move, in a reverse turn pursuing the inner workings of life, we reached the station of the emergence of life, at which the logoic constructive work of the logos may be traced to its very initiation, to its incipient forces and the settings prompting the entire strategic engine at work.[1] We may pursue as well this unfurling of the strategic powers of logos amid emergence, differentiation, dynamic plays of opportunities, in its magnificent achievement: the manifestation of the reality of life within the relevancies of the All.

Within the vision of this gigantic game of life played by the logos, nothing may be lost. All the questions that philosophers used to ask will reappear, albeit in a different form. May we claim the final philosophical truth of this vision? Would it need to be legitimated in an apodictic fashion? How could we find a legitimation that is an instance prior to the logos itself? Does not its 'truth' reside in its very intuition? Encompassing the *entire* reality could it be challenged or its evidences denied, put into question, doubted, on the basis of any fragmentary evidence?

The ultimate vision of the real in the strategies of the logos culminates in its manifestation. Yet some discussion of these preliminary issues will be given.

Seen in the mirror of the present day defiance of rationality, rejection of the truth value of statements, of a stabilizing order in the world, things and beings, and that of the reductionistic modeling of the mind, the workings of bios and the human spirit, in terms of the artificial mechanism of the computer, and in view of the novum whereby communication on the earth's surface can be doubled, perfectioned, nay, replaced by unlimited cybernetic space wherein we human beings can establish our mental, emotional and imaginary existence, such an attempt at retrieving the relative recurrence, stability, order of the old earth and the vital dimensions of life — human life — may appear not only an anachronistic effort but even folly.

We may retort that although humanity may be momentarily blinded by the turmoil created by ever new inventions and need time to appropriately appreciate, critically evaluate, and healthily absorb it all, nevertheless our entire earthly existence relies on the truth value of statements and facts, whether we believe in them or not. Although we stand in awe before ever new views on the nature of the earth, its place in the cosmos, and the many foreseeable consequences following from these views, in all human memory earth has remained a solid — albeit changeable — rock upon which humanity has been established. Although life on this earth undergoes transformations and we are terrified by the possible developments that future generations may face, nevertheless as far back as humanity can remember the basic cycles of development, the essential modalities of its progress and continuity amid contingent fluctuation have remained well in place.

In brief, meteors might fall, regions get cooler or warmer, oceans may sweep fertile islands and shores, and cataclysms like volcanoes and earthquakes may destroy flourishing cities, but the intrinsic order of the cosmos in its expansion remains the cornerstone of its dynamisms. As long as the intrinsic blueprint, its conduits, its reservoirs of synergies and forces remain in place, there is the same ordering power at work.

It might also seem folly to venture an inquiry into the All in view of the enormous development of human knowledge, the expansion of individual consciousness, the increased demands made upon judgement by new worries and ideas. The present endeavor is possible in its concrete speculative form because it draws essentially on these above-mentioned developments, owing to which there come more clearly to light the articulation of the order pervading and maintaining the All as well as its particular knots revealing the inward hidden working of the logos.

The enormous expansion of knowledge has led the practitioners of the various hard sciences to recognize the necessity of revising their inveterate assumptions and has permitted them to approach the humanities to ask after hidden links between the sciences and the humanities. This opens them to a confrontation with the pursuit of the work of the logos throughout All there is. It has been a painstaking effort to scrutinize the innumerable ways and paths of the logos in the concrete, following how its workings emerge and pinpointing the accomplishments of its fulgurating rationalities, its dynamisms, its seminal wealth, its expanding vigor matched by constructive ingeniousness from the first principles of the primeval logos: *impetus and equipoise.* We will pursue the strategies of logos in the ways it takes to establish its immense manifestation in life, its existential context.

BY THE SAME AUTHOR

Essence et Existence. Etude à propos de la philosophie de Roman Ingarden et de Nicolai Hartmann. Paris: Editions Montaigne, Aubier, 1957.

Phenomenology and Science in Contemporary European Thought. New York: Farrar, Straus and Cudahy, 1962.

Leibniz' Cosmological Synthesis. Assen: Royal Van Gorcum/Humanities Press, 1964.

Why Is There Something Rather Than Nothing? Prolegomena to the Phenomenology of Cosmic Creation. Assen: Royal van Gorcum/Humanities Press, 1966.

Eros et Logos. Louvain: Nauwelaerts, 1972.

Logos and Life, Book 1: *Creative Experience and the Critique of Reason,* Analecta Husserliana 24. Dordrecht: Kluwer Academic Publishers, 1988.

Logos and Life, Book 2: *The Three Movements of the Soul,* Analecta Husserliana 25. Dordrecht: Kluwer Academic Publishers, 1988.

Logos and Life, Book 3: *The Passions of the Soul and the Elements in the Ontopoiesis of Culture. The Life Significance of Literature.* Dordrecht: Kluwer Academic Publishers, 1990.

Monographs in the Analecta Husserliana book series:

"Beyond Ingarden's Idealism-Realism Controversy with Husserl: the New Contextual Phase of Phenomenology," in Anna-Teresa Tymieniecka (ed.), *Ingardeniana,* Analecta Husserliana 4. Dordrecht: D. Reidel, 1976. Pp. 241-418.

"The Moral Sense: A Discourse on the Phenomenological Foundation of the Social World and of Ethics," in Anna-Teresa Tymieniecka (ed.), Analecta Husserliana 15. Dordrecht: D. Reidel, 1983. Pp. 3-78.

"The Moral Sense and the Human Person within the Fabric of Communal Life," in Anna-Teresa Tymieniecka (ed.), *The Moral Sense in the Communal Significance of Life,* Analecta Husserliana 20. Dordrecht: Kluwer Academic Publishers, 1986. Pp. 3-100.

"Tractatus Brevis: First Principles of the Metaphysics of Life Charting the Human Condition; Man's Creative Act and the Origin of Rationalities," in Anna-Teresa Tymieniecka (ed.), *The Phenomenology of Man and of the Human Condition,* Part II: *The Meeting Point between Occidental and Oriental Philosophies,* Analecta Husserliana 21. Dordrecht: Kluwer Academic Publishers, 1986. Pp. 3-73.

FOREGROUND

FOLLOWING THE LOGOS THROUGH THE LABYRINTH OF LIFE

I. THE "FIRSTNESS" OF THE ONTOPOIETIC METAPHYSICS OF LIFE

In the course of history and its culmination in the present, we have learned several important lessons about the philosophical procedure. First of all, we have seen that the semantic system of language cannot determine what a given philosophical inquiry aims at, since any new insight or fresh flight of creative imagination would disrupt the semantic logic of the system in the effort to reach further areas. The system assumed at the start of an inquiry has inevitably to be transformed or replaced by another that emerges.

Nor can the givenness of life, which in its manifestation extends over multiple spheres of significance, be brought adequately to disclosure and clarified in understanding by any one procedure in philosophical inquiry.

Lastly and most of all, and this is strictly connected to these two previous states of affairs, we became aware that the vast sweep of the significant modalities entering into and interplaying in the vertiginous outburst of unfolding forces in the ongoing gigantic play of the *manifestation of beingness* and all the fragments, sequences, segments of complete constructive processes subject to disruption by unforeseeable conditions and influences could never be adequately estimated and grasped by strict rational rules and logical interconnections, by the schematization that each and every language of rational discourse employs. No one single so-called "philosophical method" or approach applies to the numerous modalities of the real.

It should be granted to postmodernism that no one approach in philosophy may claim absolute validity over others. We will, however, add to that a twist: in order to get at the "bottom of things," at the source of "truth," that is, at the spring of reason that carries on the great streaming edifice of life, we have to take into consideration insights from any of them that fall within our purview.

And how obvious this is with our particular subject matter: LIFE! The vertiginous wealth of emerging, unfolding, transformation, degeneration — the vast spectrum of the dynamic fluctuations — have to find their own place within the philosophical discourse. That is to say, we cannot try to squeeze the riches of life's innumerable signs into a perfectionist effort at organizing them in the stereotypical language of so-called "scholarly" discourse that would ape science but be merely pseudo-scientific. Here we will, as in the three preceding books, employ all manner of style, all metaphors, all ways of proceeding: intellectual, poetic, pragmatic if only they serve the aim of sketching the infinitely rich profusion of the logos in life's manifestation and perfectly fit the purpose insofar as they coincide with our discourse's central constructive direction. Furthermore, if the exposition of reflection on such an all-embracing subject as life were to follow a sequential "therefore" order of writing, it would inevitably meet a dead end and force observations from numerous streamlets into one narrow bed. Such an approach forces oversimplification since this discourse by nature follows a direct line of argument while life streams in all directions and will at any point refract its modalities and their apparatus into innumerable rays that flow concurrently onward.

Phenomenology properly understood — terminology aside — did not intend to invent a method, evidentiary or intuitive, but to devise an impartial way of clarifying the ways in which humans cognize/constitute their reality. In the new phenomenology, expanded to take in all the ways of life per se, what is at stake is not strictly human reality, specifically human centered focusing operations, but all modes of human functioning, all human involvement in the orbit of life.

'Givenness' remains the direct objective, but in this expansion of vistas we mean by 'givenness' not givenness as it responds to the cognitive processes of the human mind, but the inward givenness of the life progress common to all living beings as such; 'givenness' is then more specifically understood as the modality of a role fulfilled within the constructive progress. It is only in a direct, immediate insight into the constructivism of life and its coincidence with our own creative constructivism that we may expect to disentangle and grasp life's patterns.

No exposition following the classical rules of organization may convey this flood-like concatenation of processes that is life. To grasp life's patterning, *all should be presented at once in one cross section of an image.* But then from what angle could such a cross section, if at all feasible, be made since life's very patterning is ever in development?

Hence the unorthodox articulation of our presentation. We do not start at random and develop our conception in an arbitrary, topsy-turvy way, however. To the contrary, we attempt to follow a strategic trajectory. Nevertheless, at each junction at which we make a new point, we have to retrace some steps taken in making the old point, to give it its natural context. Hence, numerous what may be called 'repetitions' of what we have already exfoliated are indispensable; these substantiate the new steps we are taking. I hope only that patient readers will proceed and profit from this amplification that retraces steps.

It is not the concern of our presentation, then, to call upon or insist upon the mind's usual projections of consistencies and logical interconnections, as is usually the case with homogenous scholarly writing. Yet I believe I do justice to the intrinsic character of scholarship, even if, as in the previous three books of this treatise, I follow the 'truthfulness' of the real by presenting experiences of it in the 'languages' in which the givens of reality surge. An attempt will be also made to place the fragmentary disclosures of the 'hidden logos' in its manifestations within the context of the logoic concatenations proper to the circuit in which each partakes.

However, and this is the novelty and the clue to this our "spontaneous" presentation, instead of advancing by adding new segments in the usual "linear" way, our argument will evolve around its center. This center is movable and displaces itself, but it holds the vital strings of beingness, in which the logoic work of the manifestation of life — and hence of reality, of the All — converges: the self-individualizing constructive advance of life in its ontopoietic design of self-individualizing progress. It is from this center that flow the rays of becoming, from whence flow other rays, down to the primeval origins and upwards to free proliferation that follows the differentiating rules of reason. That is to say, *we will first seek to grasp the Logos in its constructive impetus and then in the unfolding of its vehicle, the self-individualization of life.* The latter receives the impact of the impetus as the measure of *a constructive equipoise.* There lies the centralizing factor of the logos of life. It lurks through the innumerable concatenations of life's dynamic spread posing a real challenge to the human mind. We hope in taking up this challenge to avoid the snarls of relativism that have sneaked into philosophy and the biases of tradition as well. We will keep our inquiry from going wayward at the mercy of the winds on an uncharted sea precisely by moving always around the center — and by seeking to give an "all at once" presentation.

II. TYING THE NODAL POINTS: THE SEQUENCE OF DISCOURSE VERSUS THE "ALL-AT-ONCE" EMERGENCE OF THE MANIFESTATION OF LIFE

Philosophical discourse by nature follows the temporal flow of the human reflective mind in its exposition. Still, out of concern for continuity, it strives to smooth out the route the mind takes in its work of presenting reality, so that its becoming and its features are presented in a sequence of arguments and so that there is continuity among the descriptive fragments. The exposition of a ripe conception bridges the gaps in the fragmentary temporal work of the mind and strives for a linear continuity, arranging fragments in a sequence in which piece by piece, layer by layer, a presentation is built up. However, the manifestation of the life of the logos, which it is our objective to present, has to appear "all-at-once," even though the dynamic logoic work runs through various phases and in a great variety of dimensions. In the manifestation of life, in its emergence, the "manifesting" work is not to be fulfilled in a linear sequence or a continuity, but in all its generic as well as functional operations, with all the logoic strategies in place, all the connective strings properly tied in a specific vibration, in a spontaneous effort that has to "happen" all-at-once. Therefore, in order to bring forth but an approximate account of this incommensurable game of life, we have to go both ways: first, for the sake of discursive understanding, we will introduce descriptively at the proper place the notions that our exploratory work arrived at *last,* namely, the various elements of the game intuited in creative abstraction; secondly, on achieving a certain momentum in this discursive exfoliation, we will at a 'nodal point' take hold of the tying strings of the formative work of the logos "all-at-once."

Grasping the tying knot "naturally" calls for moving in a twofold direction: the actual stage of the manifestation of the logos draws on a radius of generic and formative links that, though I will have already brought them out in the descriptive phase of exposition, have to be called on again, that is, I have to bring them back *in retrospect* to sustain the knot to be tied. In bringing forth the knot binding the strings of the logoic momentum, we draw in again the elements of the radius on which the momentum itself draws. But, in this radius we have also to anticipate the strings and the elements that have not yet been exfoliated descriptively, that is, we move *prospectively* toward the logoic strategies sustaining the strings being tied together and belonging to the knot's full radius.

In retrospect, we will go back to numerous points sustaining the spectacle of the game of forces. We will as well, in a reverse move, advance to the gist of the manifestation of life that we aim at, grasping its most salient as well as life-significant formative knots, which projects our presentation into realms of exploration beyond those we have dealt with descriptively and which will follow.

This intrinsic movement of the exposition, reminiscent of the French popular maxim: "reculer pour mieux sauter," so contrary to the usual style of reflection, but to a great extent in accord with the dynamism of intellectual exploration, may not only sharpen our inquisitive attention to the marvel of the play of the forces of life, but also lead to a penetrating grasp of the enigma of life behind the curtain of the marvelous spectacle of the world, nature, human being.

That is to say, faithful to our logoic principle that each significant move of life simultaneously stems from a radius of conjoined operating forces and by its emergence circumscribes the new radius of a constructive conjoined operational schema, with each radius drawing on a circumambient respectively selective play of forces, we will in the linear segments of our discourse so situate such points of the coming together of the main lines of logoic unfolding that they will simultaneously reveal the radius within which they bring about a crucially significant knot out of which a specific novel constructive logoic factor emerges even as they simultaneously galvanize and make subservient to its specific aims the forces of the radius around it. This radius of forces is at the same time a constructive coming together toward an achievement and a rebound that throws out lines of operations imbued already by a significant factor making it operate.

It has to be pointed out from the start that our investigation of the logos of life, being philosophical, is most deeply embedded in the traditional philosophical conundrum of issues, approaches, arguments etc. with their various bents and assumptions. It is within the treatment of the nodal point that the pertinent issues of traditional philosophy will come into their own either in reference, or discussion, or underlying assumptions. Thus, without entering into polemics over the history of philosophy, through the spread of the radius of the nodal points we will encircle the main constructive arteries of life accounting for its specific achievements, as we throw charting buoys on the immeasurable ocean of life's constructive turmoil within the cosmos.

While not following the traditional outlines of discourse, this presentation does not mean to be a literary effort, certainly not an exercise in fiction. On the contrary, it offers philosophical reflection and makes statements claiming the validity of truth rooted in the real and worked out directly from it.

Hence it is in order to engage now in a succinct discussion of our reflection's source and procedures.

III. THE ONTOPOIESIS OF LIFE. SIGNIFICANT COINCIDENCES BETWEEN AND AMONG VARIOUS FIELDS OF COGNITION, AND THE "FIRSTNESS" OF THE METAPOIESIS OF LIFE

(a) Having originated together with what subsequently separated itself out as 'science,' metaphysics shares with science its essential quest, inquisitiveness, and commitment to ultimate pursuits. We could venture to say that it has often been — as it is somewhat today — in competition with science over priority of place.

Yet we should bring it out strongly that as today's science has expanded its horizons, approaches, evaluative principles, it has abandoned its classic "arrogance" in claiming certainty. As I have pointed out elsewhere, it is humbly recognizing that its certitudes and paradigms change, that its questing and its cognitive and evaluative criteria as well err. Thus, today's science recognizes the validity of other than hard scientific pursuits and leaves space for the humanities' playing a role in the appreciation of reality.

By the same stroke science in seeking to unravel the nature of reality has opened itself to a new era of a dialogue with metaphysics. The competition over priority, over "first place" within the orbit of cognition is taking place on a new platform.

But we may say that the metaphysical quest needs to undergo an analogous transformation. As a matter of fact, inspired by scientific discoveries, metaphysics recognizes that neither speculative universality nor a focus on the most universal — mysterious, elusive — idea of Being may give us the appropriate access to the reality of the All. This brings out, in contrast, the crucial need for concreteness and penetration into the hidden recesses of reality-in-becoming, what science itself is pursuing in its way, as being indispensable for metaphysics' specific aims. In the last stage of preparing this book for publication, I reviewed contemporary biological research in order to check on my strictly metaphysical intuitions in empirical science and found such an amazing agreement with the essential ontopoietic intuitions of my so long maintained ideas that I consider this support of science to be an important factor in legitimating onto-metaphysical insights.

There should be no surprise at this unexpected agreement between metaphysics and science; after all, such cross-pollination goes back not only to their historical origin but to what caused it, their common source, that is, the creative experience that natural science and my creative phenomenology of life share. May the present metaphysics of the Logos contribute to a further cross-pollination with natural science and become a platform that offers a firm, valid basis for the philosophical quest after truth so lacking today. In what do science and metaphysics come more closely together than in my own focus, the mysterious events of life?

(b) The ontopoietic unfolding of the logos of life manifests itself in the spectacle of the All, cosmos, world, nature, life, the works of human spirit.... We live within its network and speculate without noticing it. Yet the philosophical query dives below the spectacle to find its genesis and underpinnings.

There is this, the quest after the *universal* order of the All, the order underlying the spectacle and carrying its cogent manifesto, which quest is our quest after the unfolding of the logos of life. This quest meets the quests of the seekers of knowledge in the various scientific pursuits, so called. What is remarkable is that there occur most significant coincidences between these two types of questing and the findings of the phenomenology/philosophy of life. I submit that this is owing to the way in which we approach our objective and pursue our quest. It is opportune, then, to give some attention to our procedures.

Let us begin with the contrasting ways of science and philosophy. First of all, each natural science is already itself differentiated according to the cross section it makes into the workings of nature, its key focus. While anatomy cuts through all the arteries of the organism and focuses on the structural patterning of its organs, epidemiology will do the same taking excursions through particular functions. Cardiology, neurology, ophthalmology, etc. single out a system or a specific organ and focus on its inner workings using specific approaches, instruments, chemical substances, etc. that are particularly adapted to addressing its problems. Thus, in medical science itself the cross sections made within each specific perspective on the functioning organism yield specialized sets of insights; these are compartmentalized and isolated, requiring afterwards the delineation of the functional links between them. Other natural sciences like botany, zoology, anthropology — each again subdivided into numerous specializations — isolate life's infinitely intertwined processes. Where life's large networks comprise all types of living being in tight ecological conundra, the sciences segment the types from each other and, cutting through their existential ties, work on finding the particular construction of each using particularly adapted methods. Even the new sciences of the environment, ecology, which are concerned precisely with those existential ties, have a particular narrowness of focus, emphasizing a specific perspective on the immense game of life.

Cosmology, and physics, focus on the great laws of the cosmos and the real world, its constructive features and regulation, without descending into the course of life, while chemistry seeks in all of the world's components the foundations and regulatory principles of composition and transformation.

Although all of the sciences enter, at least vicariously, through the creative/inventive channel into the innermost workings of nature, nevertheless they first install in its genuine transformational progress their own particularized machinery for dissecting, recombining, interpreting and, lastly, grasping in a discursive, intelligible mold. There is already in the very investigation of the workings of nature the lenses through which they are seen — different lenses for each perspective. Through these lenses prepared in various perspectives, an entire "station" of investigation, an entire "scaffold" is installed within the womb of nature, allowing each investigator to probe, register, and lastly grasp segments of its workings within a pattern of discourse.

At this last stage there comes sharply to light the researcher him/herself either as a man/robot or then as a human being endowed with specific equipage by the workings of life: the constitutive expressive system of forms and their meaningfulness proper to the human significance of life. Although, owing to the influence of C. S. Peirce, researchers are now aware of their participatory role in the discovery process and take that into account in the formulation of its progress and results, this process is carried out throughout with reference to the forms of that meaningfulness. All is referred to the latest phase of life-significance accomplished by human beings. The results of research are grasped interpretively within mankind's already established patterns of meaningfulness.

A twofold conclusion follows. *First,* the natural as well as physical sciences make only cross sectional excursions into being on their own account, putting a dent in it, attempting to stop it for a while, cutting away at some segment — each science using its own equipment — of the vast expanse of life's network and treating that segment of Life as its own, mindless of the reach of that network. Each grasps its cross section with its own sense of rational interconnectedness, its own rationale, while the bulwark of the vastly expanding flux of Life that carries them incessantly onward does not emerge into the light. The great dynamics ever being transformed in process, intertwining all and mixing the streams of these workings, is not even intuitively present in these sciences.

Second, the creative function particular to the human agent that is the nuclear physicist, the molecular biologist, etc. enters into the workings of this great game in which he or she plays a role — a role of centralizing and harnessing, and tying into knots of articulation the segments of its flowing forces, energies, transformations, etc. But that function is partly vitiated at the root so far as the yield that it could offer given the significance of human agency is concerned. Specifically human transcendental meaningfulness is introduced in intellectually stabilized/sclerosed forms into the work of the scientist from start to finish. These cover over its creative source, the vibrant point of interchange between the vital significance of the logos of life and the virtualities of human cognition This vibrant source of sense is obscured by what is later, the transcendental meaningfulness established by human intellect.

We should not forget, however, that technology is the direct application of scientific discoveries in which the focus is not on theorizing itself but on the immediate transformative virtualities of the discoveries, on their virtual impact on the workings of nature. As such, it, indeed, keeps close to the universal springs and proceedings of the life/world process. Thus, technical inquiries into points of juncture between the investigation and its concrete object can, if not bypass, at least disregard to some degree the barriers to interpretation by allowing the human mind to vicariously get into the Life processes themselves.

IV. THE CIPHERING OF THE INNER WORKINGS OF LIFE AND THE FULL MANIFESTATION OF THE LOGOS OF LIFE

In introducing the notion of the 'manifestation of the Logos,' we owe to the reader used to the popular semiotic approach in interpreting meanings some explanation of what is to be understood by 'manifestation.' It could be assumed that 'manifestation' is connatural with the emergence, appearance of a sign. Following such a perspective, we may say that the manifestation of the Logos through life and its encompassing cosmic system comprises systems of signs through which we may discover the workings and the generic nature of rationality, reason, the Logos.

These inner workings may be discovered by the philosophizing mind by entering into its very own logoic manifestation first; then into the circuits of bios, zoe, cosmos progressively/regressively encircling it; and lastly into the inner workings of the 'primeval' logos, dynamic, self-constitutive, and self-directed. These inner workings are themselves constitutive of the cosmic laws that find their most specific extension in the significance of life, its origination, constructive becoming, and overarching continuity.

Thus cosmos, nature, bios, my discovering their laws within my own living operating frame are systems of 'signs' in which the primeval logos is differentiated, distributing its meaningful rays in networks of rational interconnections modulating their kinds, projecting its very own syntax, its interpretive levels and modalities. If we envisage these signs as being only on the "surface" of what is already present there and in terms of how it appears to our merely cognitive faculties, which are already formed and sclerosed, we only skim their effects without reaching their inner springs. But on entering into the very workings of the Logos through creative acts that participate in them, we discover the very language of the logos in action along with the new set of semantic categories that it reveals.

Already C. S. Peirce — militating against the strictly cognitive Cartesian approach to reality by which the philosopher attempts to become a dispassionate, detached, objective spectator — set the rule that, on the contrary, we may gain access to reality only if the observation of philosopher/agent takes into account the fact of that observation. This level of the philosophical approach to reality he called 'firstness.'

Peirce thus proposed what he believed to be three most universal categories that derive from the philosopher's own congenital immersion in the world of life. Since without a reflecting being there would be no concern with anything, *the reflecting agent, who is self-organized and self-regulating and "independent of anything else," is the primary reality of philosophizing, its firstness.* That congenital enmeshing of the singular agent is relative to interaction with and reaction to something else, that is, with and to others within the context of the world indicates that *interaction is the second level of reality,* a 'mediation' by which independent beings are "brought into relation."[1]

We could not agree more with the crucial call for envisaging the inquiring philosopher as a participant in that into which he inquires; this is, indeed, the way to avoid the undue "dualisms" perpetuated by philosophers from the time of Antiquity up till now. However, with our direct concern to enter into the circuits of the vertiginous differentiation of the logos as the very springs of primogenital becoming, we must either be able to reach these springs or to renounce such a quest altogether. That means that our approach has to be more specific, particular (and more demanding) than Peirce's. We substitute Peirce's approach with our phenomenology of the creative act. As a matter of fact, the descent through the creative act of the human agent conducting the inquiry into the ontopoietic workings of life not only involves "taking into account the participation of the inquirer in the inquiry" but endows this inquiry with an "immediacy" of an unprecedented type. This is not the immediacy of the 'givenness of the data of consciousness to itself.' Neither is it the more deeply elaborated immediacy of the self-destructuring conscious life to its carrier and performer, which was Husserl's focus. It is the living agent partaking in full with its own creative powers in the processes of life in which its own life-structuring and fashioning at all circuits of becoming takes place. In this partaking in the deepest workings of life, the creative act reveals most importantly the origins and fashioning of the participating functions of the observing agent in the processes that pertain to the preconscious modality of its participation as well as those of reflective observation as such.

There has been since the beginning of Western thought the interplay between philosophical intuitive speculation and reflection about nature and the inquisitive investigation of natural science; seeking to approach their subject matter-in-the-process-of-change, aware philosophy and science aware or not may descend creatively into the common source seeking its reasons and regularities: the workings of life in which they themselves as inquiring minds participate.

Yet they take different routes in interpreting these workings. The scientific mind aims at clarity and simplicity by isolating specific categories of these workings, specific circuits (e.g., anatomical, botanical, sociological, psychological, etc.) and probe with optimally matched intellectual/experimental tools the specific nature of each as if they were working apart from each other. The philosophers, ontologists or metaphysicians, meanwhile, are attempting to take into consideration all the main lines of life as they approach them through their creative lenses.

Though scientist and philosopher are isolated from each other in concrete practice, they both have to face questions that wait like the minotaur at the end of the labyrinth of life.

'Firstness,' to use C. S. Peirce's expression, albeit in an entirely transformed way — i.e , to mean the philosophizing inquirer's descent into the creative act that he himself performs disseminating the innumerable rays of his/her inward life performance and the tentacles it throws out — reveals that life itself manifests the inner workings of the logos with its strategies for incarnating, embodying, concretizing in a moving spectacle.

Thus the "creative function of the human condition" is within this perspective not only 'first' but the very condition of the discovery category.

Furthermore, participation in the self-constructive, self-individualizing becoming of life itself — the second category in the order of discovery — is the "ontopoiesis of life" revealed in the creative plunge.

Furthermore, this latter reveals itself in its very own "inner working" to be an inward-outward oriented "poietic entelechial principle." That means that ontopoiesis proceeds from its innermost in the self-construction of the living individual-in-progress by the most intimate exchange with the outer circuits of the world and life, while it builds up its "inward" means of living. Thus the inner/outer orientation of the ontopoietic entelechial principle itself is the crucial self-individualizing principle that directs an individualizing life's autonomy, distinctiveness, and relative independence within the network of its interdependencies.

In this perspective we may consider the life performing functions of a living being as the immediate 'interpretation' of the signs that the logos manifests through life. Thus a 'vital significance' is spontaneously being established, and the living beings performing it are the interpreters themselves. With societal interdependence and interaction, there emerge the signs, signals, etc. with which ordinary semiosis is concerned.

It is then at several circuits of life's complexity that the "script" — as it is fashionable nowadays to say — of the logoic manifestation is laid down by the very functional progress of individualizing life that I have in this perspective also called "self-interpretation in existence." Differentiated through life's complex stages, the specifically human significance of life is being synthesized. The logos of life is manifested through all of these stages up to the appearance of its central processing factor, the human being, the creator who with his own functional powers refashions the preceding factors and creates new forms of experience of life as the cultivation of the logos. The human being as creator transforms, through ever newly invented logoic strategies, the vital-gregarious-societal significance of the animal logoic interchange into its own specific intelligible script having signs that assume the novel function of representational meaningfulness. In this creative, inventive swing of the human mind, the entire workings of the logos in its life manifestation are recorded ready to be retrieved in a uniquely human deciphering.

Thus, the legitimation found for scientific investigation lies not in its so-called "methods" but in the indications that their application to the life processes has truly reached the concrete processes of life itself.

Yet the gist of the great game of the logos remains hidden out of sight in spite of the enormous progress of science and technique. It is left to hypothesizing, and that borders already on philosophical speculation. But these hypotheses do not for the reasons discussed above enter into the heart of the matter.

Reaching the heart of the matter requires, in fact, a creative approach, the full-fledged philosophic entrance into the game.

And so when we raise the question of the "firstness of metaphysics of life," we mean first its privileged role with respect to all the cognitive enterprises that are to be seen in the perspective of the "genealogy of cognition." It is indeed from the first principles of the system of life that there proceed all the branches of reality that then become the focus of the cognitive quest. We mean, second, that the metaphysics — the ontopoiesis — of life subtends and gives a moving and yet reliable, for a time, framework for inquiring into reality. In this respect it may serve as the system of reference for the progress of the sciences hard and soft. Thirdly, and most profoundly, we grasp the primordial role of the metaphysics of life when we ask the crucial and most enigmatic question: What are the first principles, the "sufficient" and indispensable reasons that may account for the capacity of the human mind to bring forth and exercise its prerogatives of constituting/presenting a human world with roots in the immeasurable cosmos, that is, of seeking the very existential principles of the possibilities of both the human being and the cosmos? Is there a place to be assigned to such an inquiry?

Challenging all supposition we affirm that it comes absolutely first. Phenomenology of life stands in radical contrast to the prevailing opinions. The questions we raise are prime, before all the inquiries that we have to enter into subsequently. What are the very first reasons for the existence of something rather than nothing? Unraveling the origin of life, we find that first reasons manifest the nature of all reason; whether we call it order or logos, reasons are self-explanatory; they are not subject to further questioning after "origins." No reason may come before reason. The metaphysics of the manifestation of the logos comes first in the absolute sense, whereas metaphysics of the human creative act, of the human condition provides access to it.

There is no sequence of inquiry that conditions it: the knower and the known are enmeshed in the entire dynamic set of ramified conditions by reason of reason.

V. THE PRIMAL CIPHERING OF THE LOGOS OF LIFE

Sketching the groundwork of our procedure legitimates the validity of its results. We will now enter into the knot of methodological considerations that have evolved around the method of interpretation that has gained such authority in our times by sharply contrasting with it what I call primal "creative ciphering."

Indeed, in radical contrast to the approach that science takes to access the workings of nature, the creative approach of phenomenology is guided by the game that the mind plays in the creative process through which the artist, poet, musician invents, initiates, and carries out the creative constructivism of an exemplary work. Instead of seeking to understand the particular message, significance, or content of this exemplary work by dismantling the semantic levels of its structure — a structure that is itself semantic and represents the last phases of the creative process — and instead of being satisfied with seeking to understand this message in its human significance in/with a particular, specific, and unprecedented object of conveyance, which significance the message contains through the sedimentations of other messages conveyed in texts constructed by the human mind in the progress of culture, and thus moving within the self-encircled dimensions of the human significance of life as the human mind has unfolded it from its primitive simple phase of development until now, the creative descent of the phenomenological quest seeks to see these messages in a process of constructive digging into that phase at which the Human Condition with its virtualities lies at the borderline of the vitally significant processes of nature as they emerge from their vital source, a source that never loses its revitalizing role, its role of basic depository and carrier of all that is human as well, or its pervasive presence in whatever the human mind may experience. That is to say, in our phenomenology of the creative function, this function differentiates itself resolutely from the cognitive function — albeit while continuing naturally to employ it vicariously — and proceeds to cipher the real with a set of creative intuitions, insights, intimations, etc.

In this way creative phenomenology enters spontaneously into the workings of life in its circuits, in which circuits living human beingness performs its unfolding and obtains its sustenance. Thus, we do not proceed by deciphering already ciphered scripts. On the contrary, we dwell in the ciphering itself. From ciphering's own subliminal forge, in which the creative mind originates and unfolds while ciphering its works, the mind naturally delves into the workings of life progress, which are themselves the ciphering of the Logos of life. It is then through the ciphering of the logos itself, its innumerable trajectories in life's progress through innumerable concurrent streamlets at various stages of unfolding, each one implicated in all the others, that the creative function of the human condition reveals the logos in its own manifestation.

At the basis of this strict differentiation between what is in general called 'interpretation' and what I call 'ciphering,' there lies the prior distinction between the 'cognitive' approach and 'creative' approach. In the first approach, we seek to discriminate the elements of intelligibility deposited in the manifestation of the real by the constitutive human mind seeking to grasp the specifically human significance of life. In the second approach, we aim at finding an access to the ways of reality's making, its 'poiesis,' reaching beyond the confines of the mind.

And this implies that interpretation proceeding by the methods directly available in the expressive modality of the text's constitutive cognitive form (e.g., focusing on what is seen as the delimited distinctiveness of the text) stands in contradiction to the creative approach, which seeks the clues for interpretation in the modes of poiesis itself. Interpreting the text by attempting to find the sense indicated by the "surface" signifying strata in the "deeper substrata" of meaning stands in contrast to descending with the creative intuition of the creative process down to the ciphering processes out of which the deciphering should proceed. Lastly, where in standard theories of interpretation the meaning of the text is assumed to be this intelligible expressive matrix spun by the *last* circuit in which the logos of life manifests itself, namely, in the constitutive expressive circuit of the specifically human significance of life, in our creative interpretation we substitute for the current constitutive phase of significance a complete descent into the initial circuits of the primeval logos, seeking thereby to grasp the ways and means in which its progressive unfolding in all the spheres of the cosmos, bios, psyche, etc. is being ciphered. We seek to attune our specifically human means of deciphering and grasping in intelligible discourse to that ciphering.

Let us now focus on creative phenomenology and its radical novum. Our foregoing interpretation of the scientific endeavor is itself already made in the perspective of the creative phenomenology in which it is life as such that is taken as the originary soil of the human venture into the inner workings of reason through science. In fact, we situate the scientific endeavor within the processes of life itself as installing a scaffold therein, this in contrast to those philosophers who assume philosophy's final stance to be the Husserlian/Heideggerian concept of the lifeworld. Cornelis van Peursen, for one, considers the lifeworld to be the "ground of invention."[2]

We also resolutely turn upside down the entire perspective maintained by Merleau-Ponty, in which the lifeworld is the final point of reference. *It is life and not the world that offers the platform for the scientific investigation that takes off from the lifeworld but aims to install itself in life's workings.* Beginning the inquiry into the first principles on the basis of the already differentiated is at the root of the universalizing tendency. Hence we have the conception of being as the culminating point of Western tradition, seen as applying to all particulars in a 'univalent' as well as 'universal' way.

With Leibnizean metaphysics we see a shift of perspective: Leibniz is seeking the first "ciphers" (universal characteristics) in which the universe and the world with its monadic beings is ciphered — ciphered with the same ciphers throughout, including the human mind. But Leibniz did not strike the proper path for approaching this; his all-embracing cosmological syntheses accommodates all the levels of creation, but this creation is only speculatively assumed/projected; it lacks the vertebrae of life itself, which carries all on. Even more, it does not do justice to the innermost nature of the dynamism so strongly emphasized by Leibniz, so that this abstractly devised "synthesis" assumes a stationary character. Lastly and foremostly, it does not penetrate the key source of intuitions, which only the creative approach to life and to its cosmic circuit may open.

VI. THE CREATIVE VISION AND CIPHERING

Indeed, what the creative channel of the path of discovery implies is an unprecedented opening. Where the interpretive path of inquiry goes deep into the sedimentation of meaning, we seek the simplest meanings as originary factors, and on isolating them, we are left with a worldview reduced to elementary particles. With the creative descent along the logos' ciphering course through life, we proceed from the more complex to the least complex, simplest moments of living beings, and then from the more complex to the least complex moves and laws of the primeval logos itself. Yet in our registering of the ciphering, we begin with the beginning: with registering the differentiation of the logos itself.

Second, we do not rest with our intuitions of the set of the simplest moves or seminal particles; on the contrary, the creative process prompts this initial move to expand into the great vision of its unfolding.

Here is a vision, not of a stationary accomplishment, neither of an abstract universalized one, nor of a concreteness generalized to a system of creation, but a vision of the initially open, unsoundable, spontaneous bursting forth, at the one pole, and of the profuse, indeed, innumerable, unlimited radiating streams of dynamic/constructive forces, energies, binding, fusing, synergetically coalescing myriads of materials and forms with their ungraspable wealth of vibrating propensities into in a constructive flux onward and onward into the equally unforeseeable, illuminating infinite, the ever receding horizon, on the other.

This creative vision of the ciphering of the primeval logos of unknowable origins projecting itself through the game of the cosmos and specializing in that of life does not stop at any abstract being. The focus of its registering philosophically — metaphysically — is the making of "beingness" as such: how beingness as such occurs, how it comes to "be." That is to say, we are moving metaphysics into the ontopoiesis of life.

The creative deciphering of the signs deposited by the unfolding logos is, thus, oriented not directly by a limited focus on particular circuits of life's itinerary enclosed by a particular field of scientific investigation that imposes on it the instruments and intellectual scaffold installed by the scientific investigator, but by a vision of the logos in its immense radiation of constantly in transformation processes going onward through infinitely open dimensions that it opens even while bursting forth. That is to say, from the outset the deciphering is not interpreted or "scripted" in terms of the differentiation of frameworks, division into structural sectors, particular patternings but distinguishes the ties that unite all these to the principles of the self-intending surge of the logoic primeval bursting forth; here are tied the streamlets of forces bringing forth generative synergies that project vast networks of ever rejuvenating cycles of life in an infinitely open expanse, such and nothing else.

At this juncture there emerges the metaphysical quest of the human mind with its unique prerogative. In the spontaneous surge of philosophical marveling, the quest to penetrate the facade of the world and enter into its deeper strata, to grasp its strings and to reach its source, we have a quest that cannot be satisfied by any fragmentary answer, that cannot be stopped at any clear-cut perspective, a quest that tends toward the encompassing All, enters into the creative descent and into the vertiginous dynamism of the logos in the unfolding of life, into the life ciphering of the primeval logos. It does not stop at any juncture; it does not thematize or "interpret"; to follow the ciphering of the logos of life through its meandering workings is to emerge with a metaphysical vision of "beingness."

VII. THE SELF-LEGITIMATION OF THE CREATIVE APPROACH

Given the spontaneous tendency of the speculative mind to seek a point of vantage from which the all-embracing intuition could be obtained, the cognitive mode of the mind is led to a focus on the static, stationary circuits of the artifacts of the lifeworld that the human mind itself establishes. The inquiry remained stuck there. To penetrate the dynamisms of becoming calls for a profound plunge into those dynamisms. Can there be a more penetrating immediacy of intuition that yields more originary insights into the processes creating either the "texts" of human intelligible meaningfulness — we register, we record — or those forming meaningfulness itself-in-the-making? In the creative process, we witness, as the agent who performs the process, the fusions, structurings, ever moving selections of streaming images, of emotional surges and volitional moves, etc. in our own manifestation of the logos, the origination of specifically human meaningfulness, which crystalizes a most intimate texture of intertwinings within the operations of the agent who registers it. Furthermore, by pursuing this *filum Ariadne* that penetrates into the labyrinths of the streaming vital stirrings, albeit in a silenced mode, the creative intuition offers this primogenital evidence that is neither the eidetic intuition we vicariously employ as we pass from the phase of registering to that of forming an intelligible discourse, nor the intuition that Husserl employs in the genetic step by step descent into the unfolding of the structures and acts of transcendental consciousness as well as those of the lifeworld. Those intuitions always remain intuitions fashioned in accord with a level of analysis. And so the level of transcendental consciousness or constitutive consciousness — constitutive of only the last phase of the manifestation of the logos of life — by its presentational, expressive constructivism silences all else: it places a stationary roof over life, covers its nervous arteries. (Merleau-Ponty attempted to enter the arteries of the body, but did not reach its dynamic, living arteries.)

The creative approach offers a legitimation of meaning at its source and entails meeting the demands of immediacy of evidence in a more essential form, the gathering of evidence from encountering the operations of the life ciphering of the logos and from encountering the streamlet of the same process as it crystalizes itself within the agent, who ciphers the logos in his/her mind.

Why do we consider our inquiry emphasizing the creative to be phenomenological? It is so different from both the eidetic ontology of the "regions of reality" and the 'transcendental' phenomenology of the structures of consciousness, as well as its variant, the 'genetic phenomenology' of this same consciousness in its constitution of itself and of the lifeworld, even putting Merleau-Ponty's approach and the results of his phenomenology of the body into question. In what way are we still phenomenologists, in contrast, say, to the current "postmodern" trends in philosophy?

The answer may be formulated in five points. First, the creative inquiry into the ciphering of the logos through its manifestation of life follows "immediate" intuition as it occurs in inquiry into the human creative process first, and into life's unfolding, second. Legitimation through the creative immediacy of the ciphering of the logos is in keeping with the Husserlian postulate even as it offers a much more adequate means by which to justify its validity; entering into the inner workings of nature is furthermore in no explicit need of intellectual/judgmental apodicticity. The insights obtained through the creative evidence gathering are always concurrently checked by other insights and find final "proof" of their validity in their technical application, that is, they are confirmed by the responsiveness of the processes of nature itself.

The enumerated approaches of classical phenomenology seem in this interpretation to postpone the presentation of validating evidence further and further while relying merely upon the surface cognitive evidences of the semantic and cognitive/constitutive meanings by which experiences are grasped. In genetic phenomenology there is no question of the "immediacy" of experience nor the of "bodily presence" of objects presented in the text, nor of any other convincingly "first-hand" evidence.

If the call would be made for the validation of the genetic approach's procedures, interpretation would call for the unraveling of the deeper sedimentations of meaning, with none of them then proving to be definitely "obvious" because those that have been "excavated" from archaic texts have to be interpreted hypothetically with reference to present-day meaningfulness, since their originary meanings have vanished into the obscurity of temporal regress. The circles that interpretation takes still have only one fixed point, precisely that which transcendental consciousness has lately formed.

VIII. THE SELF-LEGITIMATION OF OUR INQUIRY INTO THE POIESIS OF LIFE

In the preceding section we laid the ground for the final argument. Philosophy, in aiming with greater or lesser emphasis at a — if not "absolutely" clear, at least "correct" — grasping of givenness with respect to the matter at hand, the obtainable data, the state of affairs, etc., seeks to support its findings by legitimating them in an explanation of the "method" by which they have been obtained. Our insights into the ontopoiesis of life — into the subterranean strategies of the logos in its manifestation — go so far counter to the established traditional approaches and views that a legitimation of their accuracy is necessary.

Our procedure in gathering data is certainly tributary to the intellectual apparatus and its ways of conceptualizing. But our way of reaching these data and the types of insight that we employ subvert the ways that traditional philosophy took in preparing "material" for conceptualization.

First of all, the unwarranted assumption made since Descartes that by focusing on the cognitive-conceptualizing procedures of consciousness we may reach the "real" has been lifted. Once we reach the authentic springs of the personal human life enactment in which cognitive processes are embedded through the "discovery" of the primordial role played in it by the creative function of the human being, we abandon that well-traveled track — reserving for it its own very important but narrow role — and we follow the line of the creative act of the human being. That leads us into the very current of life wherein the play of the vital forces of the human universe of existence is wrought.

We enter there with intuitions of a different kind than those by which we objectify/constitute reality. Those are left behind. The conscious apparatus that registers the pulsations, moves, processes, transformational motivation, etc. of life's poiesis registers them on multiple plates of its own enactment, which it carries on in a self-identifying fashion. We cannot retrieve these moves and operational segments in their cycles, their tasks, their confrontations with obstacles, and their concluding perfections and satisfactions if we rely straight upon on the conscious tendency to draw out of our functional proceeding a clear-cut perspective;

in focusing on that achievement, we leave the entire ontopoietic enactment in oblivion; we move on the surface.

However, the creative act offers us an access to the forge in which all these otherwise mute and inconspicuous operations of life proceed so that we may retrieve them in their fulfillment directly without constitutive screens and eidetic intermediaries. In this sense the intuition of the forge of life with its individualizing enactment not only touches its object in its absolute immediacy — because it is in its own act — but also conveys it "alive" in a direct intuitive insight. It remains then only to be attentive to the ways in which we will ultimately bring these insights into the conceptual form of a theory. (For more on "living" access to the ontopoiesis of life and its legitimation, see Part IV, Chapter 1 on the intellective act.)

The greatest constructive accomplishment of a living being as such — an accomplishment indispensable for its sustenance in life — seems to be precisely to not disintegrate amid the conflicting forces, tendencies, and to not disperse energies, to not lose them in the chaos of ultimate oblivion, but to find oblique and most intricate ways to pull dispersed and disjointed strings together, to invent within the chaos of forces that it cannot encompass a constructive step forward so that step by tortuous and unforeseeable and yet constructive step a course of subsistence will delineate itself. This is the great and admirable secret of life that we intend to enter into, the secret that allows the constructive logos to achieve, amid excesses, chaotic conflicts between forces, and grand failures, ever renewed constructive accomplishment in a living world.

We might become enlightened by present day probing into the recesses of this struggle for life, probing that attempts to show how 'decentered' human beings' living functions are in their experience, volition, action. But it is the particular greatness of human beings that their enactment of a life route is not strictly determined as that of an animal is, but has a most complex motivating situation with an intrinsic variety of functional complexes that often stand at odds in their particular tendencies while seeking a way out of dilemmas so that a route out of the crossroads is found. The human being — whose impulses often go against his better judgement and whose knowledge of the factors indispensable for a favorable outcome is always hazy and incomplete — often stands at odds with himself; but he still will work out for himself a cohesive course of interactions that serve to sustain his beingness.

It is this striking outcome of coherence that led the first philosophers dealing with living beings to focus on the central forces sustaining it: inward agency, consciousness, self. We may in deeper probings into diversified sectors of the functioning of life focus alternatively on the discrepancies between desire and will, between emotions, tendencies and motivations, etc. We may come in this way to doubt the 'centrality' of the human ego and of the self. However, in spite of such speculative displacements, the striking evidence of life's coherence points ultimately to a centralization that keeps everything from being dissipated. It is this eventual coming together that is the great accomplishment of the logos: dispersion and coming together are just two basic contrary tensions of the logos upon which life is suspended. One of the ways to reach the great vision of Beingness is in becoming.

Following that thread, we traverse intuitively the segments of life's expanse while not fixing our focus on any of its sectors, nor on a few of them together. We are, rather, carried away by life's whirlwind while never approaching the "all," since the totality is never to be encompassed, or even accomplished, its further unfolding being ever postponed to receding "futures."

The route of the Logos might be finite, yet its course of seminal love will never be "complete" because when the lights of life are extinguished, the anti-logos takes over.

It is beingness, suspended upon this gigantic game like a fly in a spider web, that emerges from our human pursuit, itself the most amazing outcome of this play of forces as a station in their processing.

We go our way breezily unknowing of this vision that constitutes the sphere of our human life until its discovery bedazzles us.

PART ONE

FROM THE ELUSIVE PRIMEVAL LOGOS
TO THE OPEN-ENDED GREAT PLAN OF LIFE

CHAPTER ONE

THE PRIMEVAL LOGOS

I. WONDERMENT

While philosophy in general emerges from a quest to understand, metaphysics springs from wonderment. What could be a more striking focus of wonderment than life?

Life! Even as we are engaged in a constant effort to keep ourselves afloat, we believe it an exciting adventure. From our earliest childhood, we live projecting our progress either along the tracks pretraced and prepared for us by our parents, elders, school, and society, or along paths traced by ourselves following our own lights. We live convinced that we have the means, instruments, provisions available to further our wishes and satisfy our needs, that shelters are available in the world to house us. We trust the human community, in the midst of which we expect to find our rightful place, and a cozy one at that; it is our assumption that our life course is meant to be a satisfactory or even blissful one, that it will be punctuated by the 'usual' great events of coming of age, gaining the confidence of the community, acquiring the maturity to start oneself a family, etc.

Unless we "turn pessimistic" — unduly so and in a way harmful to our interests — we hold to this fundamental belief in the regularity of life. We have confidence in those of our prospects that can be foreseen, in the essential stability and security of our existence.

Nothing seems to shake this confidence strongly enough to vanquish it. Wars among nations as well as fratricidal civil turmoil tear humanity apart, taking millions of lives, destroying the institutions that we trust to safeguard our well-being and safety; we see our plans dissipate into thin air owing to changed situations, our projects dwarfed by illness, injury, misfortune, defeat in competition, malicious wrongs. Some people "lose heart," despair, and give up, but the generality of humanity recovers after every cataclysm that destroys their homes, swallows their life securities, deprives them of work and long-term prospects, shatters their hopes. We pick up worn out tools and just the same start all over again. The impetus of life, the unquenchable enthusiasm for living makes us forget the destruction, misfortune, and the fragility of life. We build homes again on the slopes of volcanoes, homestead on tracts subject to periodic severe drought — as if our confidence in life could defy all. Each renewed impetus of life brings about faith in reestablished reason, in stability. With the reestablishment of some equipoise between efforts and attainments, merits and rewards, expectations and their fulfillment, the promise of life appears to be stronger than reality. Even as we are devoured by terminal diseases, we keep scheming projects and hope to realize them.

Is there a greater cause for wonder than that of an intelligent being — one that is capable of observing, seeking reasons, drawing conclusions, planning future events, and directing the course of things — ignoring flagrant facts and proceeding to build a home on a floodplain, amid moving sand, or under a volcano?

Between the impetus for life and the drive to make it secure, life itself originates and unfolds.

If we dare the waves and fire to do their worst, is life anything more than a soap bubble blown from nowhere, floating by for an instant, and bursting in another? And yet what a marvelous soap bubble it is! How can life be? Why, and how, does it continue?

This mystery leads us into the labyrinth of the Logos of Life, which blew it our way, keeps it afloat through all its roving, and brings to light the reasons for its glorious manifestation.

II. TOWARD A METAPOIESIS/METAPHYSICS OF BEINGNESS — ASKING AFTER THE "LOGOS OF SOMETHING"

Human beings are bound to ask questions, first of all, for practical reasons, that is, to orient themselves within their world of life, their context of particular life conditions. Given the intricacies of our involvement with other beings, there are innumerable questions that we have to put and if possible answer properly with respect to our daily routines. Then there are questions concerning our actions, attitudes, reasons for choice and for acceptance of the unavoidable.

We ask further questions about our vocation in life, about embracing or rejecting an ideology, the very aim of our personal pursuits, our fulfillment in destiny or the reasons for its abortive course.... We ask finally — and cannot stop ourselves from raising these questions — about the meaning of our life and of life as such."What do I accomplish at all?" "What is my role in the universal schema of things?" "What finally is the sense of it all?"

We naturally have to bring all our queries to that which is. And philosophy has for ages been distinguishing what-there-is under the aspect of its essence, entity, thing, being, process.... In the distilling of those aspects of what-there-is that distinguish one entity from another, an essence or a being from other beings, we laid down a groundwork of the world order, an ontological groundwork from which we may move to raise questions about that order conditioned by the starting level from which the directives of speculation are drawn.

Asking for "what" a thing, being, event *is,* we single it out from its place within the entire conundrum of other things and beings; we sever its innermost ties with them. We also lift it from the network of causes, reasons, conditions, and the foundational ties of its origin. We make it stand over against all that makes it what it is. We forego the crucial point of explanation, its origin. It is from that point that proceed all the strings of its causes, reasons, conditions, and not only those limited to its singled out nature, but to that very nature considered as an outcome only seemingly standing on its own, but actually something suspended within a web of ties and their conditions. It is the web of the All that accounts for a being's being not simply "something," but this very beingness. When we ask the question of the origin of something, a uniquely revealing question, we are asking after the logos of something.

Asking after this logos, do we mean the rationality of that beingness' essential organization? The specificity of its type? The ways of its generation? Is the focus here always on its structure or "nature" "in itself"?

On the contrary, when we want to "understand something," we have to focus simultaneously on its distinctiveness at the level of phenomenal manifestation from other things, beings, events, etc. and to exfoliate its nature in terms of its most intricate involvement with them all. We do so by probing this involvement at different circuits: the genetic-differentiating modes, the genetic-dynamic compossibilities, the even more encompassing and singularized compossibility patterns. Then we advance in our exfoliation toward the universal system of life and toward the relevancies of its possible patterning in the genetic coming together of materials, mixing, falling out, intergenerating, undergoing influences from the outside and corrosion from within, igniting the spark of life and prompting its advance. But we could not finish this exfoliation of the existential reasons for "something" without tracing genetic progress back to its relevancies in the ordering patterns of the cosmos, its dynamic laws, its own processes.... It is only within this entire schema of "reasons" that we reach the full exfoliation of "something": its exfoliation not in being but in "what it means to be," its exfoliation in its logos. In short, for something, anything to be understood, it has to be exfoliated within the All.

But what about the logos as this infinitely varying system establishing beingness in its multifarious shapes and course, in its spectacular accomplishments and the vicissitudes of failure? I propose to exfoliate the singular beingness in all its existential circuits tracing it back to the origin of its very beingness as such, pursuing the rays of the logos of life. But what about the Logos itself? In this inquiry it appears that all the innumerable "reasons," all the logoic moves, are constructive/destructive; they aim at an advance, but in testing their ground, they have sometimes to retreat or fail altogether. It appears that the motor of becoming is promoted by the inward rationalities themselves, that the logos, reason, is what carries the becoming itself. With its vertiginous versatility in transformation, its irresistible power to expand in every "possible" manner, the logos as the motor, the vehicle, and the constructive power is in its myriad ways carrying the great work of the establishment of reality from its incipient instant to its completion in the manifestation of life, the universe, the world, the human being.

However, going so far, we have also to recognize that it is the Logos, as the reservoir of all the incipient, germinal instances of becoming that is not only their architect, their builder, their engineer, but the very force, dynamic spring, germinal source of the All.

Enough said about the logos *quo ad* creation, the logos with respect to universal becoming, the logos as the reason of all reasons. But can we ask about the logos per se? About the reason of all reasons in its intrinsic dynamism? In its intrinsic nature? Is it in this very nature already a dynamic force that prompts itself and sets the course of unfolding together with setting the timeless instant of its surging force, a timeless instant that having set the clock puts it in motion? The timeless instant that initiates the temporal unfolding of reality, time as such?

Or does it need "outside" prompting to surge with its force? What is its initial dynamism and power? What is its initial spontaneity?

These questions may escape the human mind focusing on the origin of reality. But the universal rules of constructivism of the real may indicate to us some of the crucial intrinsic laws of the reason of all reasons, of the logos itself.

III. THE PRIMEVAL LOGOS

Does the Logos emerge from itself or from without? Did it precede its emergence in life at its origins or was it Life that prompted (and prompts) its activation? These questions may already be the fruit of a discursive form of the logos itself; the "self" and the "other," the "prior" and the "posterior," "to activate" and "to be activated" are already differentiations that proceed from life and its progress. Thus it is futile to ask these questions before we investigate how Logos reveals itself through life.

The exuberant Logos of Life seeks with incomparable persistence in any space, on any occasion, by any mode opportune for doing its work, whether through a plant sprouting in a barely visible crevice in a wall, through a tree making space for itself in a stone wall, through a bush acclimating itself to live in arid soil or heavy clay, through a bug or worm arriving as if from nowhere as soon as picked fruit ripens or an animal dies, or in the way in which each new flower sown seems to provoke new weeds to sprout.

Profligate, at first, with an immeasurable abundance of kinds, species, adaptations, the logos of life is next ready to annihilate, to wipe all from the face of the earth through cataclysms! In the same way, after having brought to fruition an unfolding human community, society, nation, it pushes them into homicidal conflicts over what do not seem to amount to valid reasons for exterminating beings, communities, societies, nations.

Human culture (in which the logos of life attains its highest peak) decays at the very height of its unfolding. Abundance, wealth, opulence, yet another point of pride of the flourishing logos of life, will diminish for no apparent reason. Nations at their height crumble, while others climb from rags to riches. In this seeming turmoil without any order, the Logos of Life whirls and reigns. Has it any reins with which it may be held but that are lost to the guide?

And yet the Logos of Life as capricious, willful, prodigal, and profligate as it is in showering its gifts while throwing them up in generation, growth, corruption, is not haphazard. It would not be "reason" if it were. There is an intricate "logic" in the patterning, the innumerable networks involved in the grand game of life we partake of and enjoy.

Does the logos come from itself or does it come with the outburst of life? Indeed, its life strategies as they are revealed through the manifestation of life show that all the spheres are harmonized. The principles of the primeval logos are revealed in its project. The proto-strategies of the Logos are seen when it brings out its generic "sense of sense"; those of the primeval logos are seen as it brings forth the "reason of reason," the "meaning of meaning" in a series of opposites, contraries, antitheses, triads, schisms, etc.

Indeed, to establish rationality, the "reason of reason," it is necessary that what we call in the language of logic 'quality' first be established, accounted for. To establish the sense of "light" a ring of fulgurating moments must be followed; "light" is knowable only in opposition to the "dark" with innumerable gradations being experienced between "light" and "dark" as one goes from light to darkness, and between dark and light as one goes toward the light. Each quality is in turn experienced in a fulguration of crossings with other qualitative opposites, those of open and closed, of concealed and obvious, of wet and dry, of sound and silence.... Each pair of opposites stands in necessary juxtaposition as well as complementarity, and no quality means anything outside of being an opposite; each element of the pair needs the other for only in contrast can it come into its own sense. It is sense itself that emerges from this play, that is, the sense of "relation" emerges as well. Each element needs to be appreciated in all its degrees ranging towards its opposite in order for its proportions to be rightly guessed. Each needs to be with its qualitative opposite. *This means giving sense to relation. Permutation and combination is the great game.*

Differentiating and comparing, joining and separating, mixing and distilling, what would they mean had there not been specific complementary elements engendering their very sense? Forces, energies with their propensities for entering into the intergenerative process focus the strategy of the Logos.

Springing forth from itself, the Logos is driven by an impetus to advance, to unroll further and further. It is not enough to project all the principles of a blueprint, to provide the building material and the tools, to lay down the "logic" without endowing it with the operational sense it calls for. We have thus to distinguish in the primeval sense unfolding of the Logos its *compositional* and *operational* sides; each is indispensable for the sense of the other.

Would we attribute the first stage of the manifestation of the Logos to the cosmos? But already as from a springboard the next stage is being triggered. Specific laws of the solar system prompt operationally the further unrolling of the logos. The planet earth, without parallel in human ken, appears to be the focus of the constructive expansion of reason within the so-called 'biosphere' — an incubator and sustenance system of unprecedented extravagance, that is, a realm unforeseen in any of the previously projected rings of meaningfulness, something beyond differentiating, compositional, and operational sense, the realm of LIFE.

The manifestation of the Logos through life does not stop with the great gift, the abundance of produce, all that is spread out for innumerable lives to enjoy and share; it can take pride in advance of construction, accomplishment, projecting immeasurable networks of generation, growth, passing away, and beginning all over again.

This network of the constructive progress of life is stretched — like a fabric being woven on a loom — upon several spheres of relevance that are interrelevant, that motivate each other and are indispensable to each other's meaningfulness, sense. Strangely enough, there is in the 'cosmic sphere' or 'pre-life sphere,' the 'biosphere,' and the 'sphere of self-individualizing life,' a shared pattern; all gravitate toward the progress of life's accomplishments.

IV. IMPETUS AND EQUIPOISE — THE FIRST PRINCIPLE GOVERNING THE LOGOS

When in our pursuit of the unfolding logos we reach the territory of its laying the groundwork for life, the biocosmic circuits of its constructive, intergenerative and seminal unfolding, we pause. Its impetus overwhelms us at that point. We are arrested by wonder. The unfolding of the constructive forces of the logos seems like the headlong rush of the waters of a desert freshet, which make their own way and then vanish leaving an empty bed as the supply of water dries up. But the current of the advancing logos does not proceed from an outside source; it flows from within. Neither does it willy-nilly send its waves in various directions. Its continuity is not one with its waves; on the contrary, it rushes onward by its innermost propulsion throwing hooks, projecting trajectories, and constructing vehicles for its constructive routes. It springs forth from within, and each move reaching out is harnessed while in motion by a constructive measure, a harmonizing, structuring, generating, seed sowing measure of equipoise. Indeed the continuity of the splitting or diverging is that of its own projection forward of its innermost impetus, which invigorates by every step of its actualization by virtue of the measures of equipoise it installs in each.

Since construction is its spring, the logos goes ever "onward"; it implies the necessity of discrete continuity (disruption and recurrence), and its spacing and scanning implies the necessity of a measure. Ultimately, *the impetus of becoming implies the necessity of equipoise.*

Indeed, the scale upon which the possibility of All hangs holds in the balance the original impetus, on the one side, and the equipoise of the generative/constructive/creative logos, on the other.

The balance of impetus and equipoise is the innermost law of the logos, its First Principle. This is the first principle of becoming and beingness, the first principle of the ontopoiesis of life. When we ponder the progress of life, from its initial outburst through its unfolding, we see a tremendous impetus sending infinite streamlets through life's arteries — the forces that drive the tree to yield sap and the flower to blossom — an impetus that once in motion reinvigorates itself at each step and diversifies its proficiencies in ever new radiations. Yet life does not drive blindly or aimlessly; its constructive trajectory justifies its direction; the spacing and timing of its constructive accomplishments necessitate that. Even as the impetus drives onward, its very condition implies the necessity of pausing, of retaining. Indeed, while this impetus is life's motor, equipoise is the basis of progress; in the manner of a fisherman, logos keeps hold of the line when it throws out a hook to make a new catch. Life's constructive stages must yield an account, must be assayed for their significance; in order to build, to install progressively novel circuits of sense, both the impetus and its effects have to be balanced.

To accomplish constructive progress beyond the already effectuated significant moments, their very sense, their "reasons" have to be retained.

In every circuit of the unfolding logos there is, first, advance (which retains directedness) as the impetus pushes onward toward an accomplishment. This accomplishment on its completion leaves its essential significance "behind" to be retained as a depository, a reservoir for "future reference." Memory and knowledge are the conjoined main retaining reverse movements of the logos, the main measures of equipoise taken by the forward pushing impetus. As we will show, memory and knowledge pervade, albeit in different modalities, all the avenues of the logos in its course. The central measure of the impetus/equipoise balance is carried on conjointly by what we shall call 'memory' and 'knowledge,' which take numerous forms.

V. MEASURES OF EQUIPOISE IN THE ONTOPOIESIS OF LIFE

1. *Impetus and Equipoise in the Unity-of-Everything-There-Is-Alive*

Universal reason, logos, throughout its unfolding radiates its riches by splitting unities and opposing their elements in order to unite them into a novel source of sense; sense diversifies into discrepant streaks in order to provoke the harmonization that yields original significance. The impetus of life does not spring forth at random, careless and unconcerned; always its impulse and uncommitted thrust enters into an equipoise accomplished through a harmonizing principle.

Even in its incipient phase, the progress of individualizing life is *an unfolding of the entelechial principle with its prefolded order*; this belongs to the synergy of the generation of life.

The life-impetus on bursting forth is bridled as its otherwise wild force encounters the rules of the entelechial code deposited in its synergy as its indispensable counterweight. These direct it to a constructive outlet, giving it constructive propensities and direction; only they may appropriately respond to another measure of logoic constructive equipoise, the preestablished conditions of the life system that the unfolding impulses and means have to meet. The entelechial sequence that is the guideline for the self-individualization of life remains the backbone of life's equipoise, its fundamental ordering tool.

Yet as we have been pointing out all along, the self-individualization of life in its diversified moves, its constructive spontaneities, does not occur in a vacuum. The inward/outward orientation of its operations, into which the singular moves of the life-impetus split, is the second major measure of equipoise by means of which the logos of life in all this dissociating prompts the construction of an individual within the system of life — first of all, in order to apply itself to uniting with all others. The significance of the "individual," in fact, lies in its being a counterpart to a system of others; it is by differentiation from them that an individual as such originates (assumes a sense).

Directly related to this issue is another essential ontopoietic principle, interest in the spacing/scanning of individual life progress, what Leibniz called the principle of economy. The impetus of life has, in fact, to adjust to the limited arena reserved for life's unfolding in accord with biocosmic conditions with their respective relevancies. Remarkable to behold, the measures of impetus/equipoise are, as it were, especially fitted for compliance with the necessity of constraints on life's spread. The unity-of-everything-there-is-alive is grounded in the intrinsic existentially and vitally significant linkage among entire systems of beingness that are mutually indispensable for each other's vital existence. Their selective plenitude (or compensatory filling-in), being in need of constant maintenance, there is only a limited space and duration that may be allotted to each as a result of an infinitely complex "negotiation" over their spacing and scanning between their entelechial exigencies and those of other living beings within the common life context. We may assume that there is also present the principle of plenitude, which is the law of the seminal logos that guarantees that living beings with internal existential/vital interdependencies come about "together" in order that they may sustain themselves reciprocally in their life courses, thus establishing the actual equipoise of the entire life context.

We cannot fail to discern yet another law of the logos, that of the intrinsic/generic principle of evolutionary transformation.

To begin with, the logos of life projects its own inner principle of the self-continuation and propagation of individual life: the principle of generation through splitting or through the union then of two beings that contrast with each other but that aim at equipoise and so harmonize — in a harmonization that holds the promise of continuing generation.

However, this establishment of fecundity as a foothold for future generations is connected with the deposition of generic features that are transformed in each new composition — a deposition that is the basis for evolutionary transformations in future offspring.

Running through and together with these measures for equipoise in the ontopoiesis of life is the cyclic principle of individualization-in-existence. The spacing/scanning phases of life's progress concurrently project constructive progress and hold it back from transgressing any measure. Growth antecedently installs the basis for the promptings that favor the individual's flourishing even as it retains appropriate directives for its decay.

The cycles rooted in fecundity retain the groundwork for new beginnings but leave the scene to let each new life occupy a vacant placing/scanning locus within the unity of living beings.

The cyclic principle of life is the regulative axis for the equipoise of life's becoming and the main executive principle of the economy of existential distribution of individualizing opportunities.

The attainment, fruition, to which the growth of life leads and in which its cyclic unfolding culminates with the passing of the torch of life to another generation is the ultimate principle of measurement and valuation, a principle that saves the transient individual from insignificance and oblivion.

2. *Circumscribing the Imaginable*

As the first, primitive notion of the intrinsic nature of the logos, 'impetus' is beyond definition. Yet we may attempt to describe it. By impetus, I understand a bursting forth, but not a haphazard one. But how can we define force otherwise than by its being a measure in motion? That is, by its impact upon? Yet the bursting forth of the logos of life is neither contained in itself nor does it fall upon anything in order to affect it. The outburst of the logos of life is an outburst of force carrying with itself a synergetic content; it initiates.

When I use the term 'synergy,' I mean "force." By the term 'logos,' I intend the most universal rationality or reason, the "sense of sense." However, the, sense of sense or logos is not to be understood as an abstraction. Abstraction is the fruit of human intellect; logos is a sense-promoting factor and thus the real key to the beingness of whatever exists or could exist. As being's innermost structural blueprint for distributing the lines of becoming, logos does not identify with any modality of existence that crystalizes them. In this sense logos is more real than so-called reality itself. Synergy is a spontaneous force, and logos unfolds gigantic rational schemas through its synergies. A synergy consists of a force comprising an impetus activating appropriate elements at hand.

It is, more, an intergenerative or generative power; logoic synergies infuse the modalities of sense. Most of the sense-bestowing proceedings of logos remain in the dark, inaccessible to the human mind. Yet from the manifestation of its work, we may infer at least five main proficiencies of the logoic synergies: constructing, structuring, generating/intergenerating, qualitatively molding, and the sowing of new kinds of sense.

As force can be assessed only by its impact upon, by the motion it initiates bringing a body from the state of inertia or repose into motion, or energizing a motion already at work, so the impetus that is the first and primal move of nature may be assessed only by its ontopoietic counterpart: the equipoise of its "effect." Indeed, the impetus can be characterized as a bursting forth of logoic force inasmuch as it does not burn out its capital and vanish like fireworks, but carries a synergetic force/endowment. So it is from the outlook of the complementary recipient. There is a generative correspondence between that force and the elements of the recipient, which have propensities to receive its affection (*Affizierbarkeit*). In short, this impetus is not a haphazard carrier of force or energy. On the contrary, its essential nature is to prompt some constructive event, development, an action that is a productive response directed further. This prompting force is directed to prospective constructive responses and on being received acquires its qualitative and specific instrumental nature. In the response it finds its determination: its equipoise. It is in this equipoise that lies the secret life of the ever advancing flux of the logos. Having prompted an effective response, having been, as it were, duly received, the prompting impetus finds equipoise, its measure. And yet this matching effect, the prompting impetus and the equipoise that follows from the challenge, is never completed. It belongs to the infinite constructive propensities of the primeval logos to ignite one of just several responses, reach one or several measures of equipoise from among the infinite reservoir of possibilities that it foreinstalls itself. But the ignited action does not stop there. It may well be that from the vertiginous gamut of promptings a portion of them spend themselves in failure, yet those that catch fire prompt further constructive events according to their initial as well as intergenerative propensities. We could speak of a chain reaction sparked by the prompting force of the impetus and going through all further unfoldings in such a way that its initial force is carried through all their segments.

Its germinal constructive baggage is never exhausted. Its prompting forces may in circumstances unfavorable to combustion shrink and seemingly vanish, but in fact, they stand in waiting for another occasion. This inexhaustible dynamic/generative nature of the logoic drives itself in such a way that no instance of its constructive achievement, no move, no process, no accomplishment is ever final. But in this restless drive onwards, each move fulfilling a constructive task is a measure of equipoise, a halt in the achievement. This halt from which a further step may be plotted and endeavored belongs to the logoic strategy of life.

The strategizing logos as it pauses in its relentless drive onwards takes a particular form within the human Apollonian logos. A human beings puts all his acumen and and the whole reservoir of his energies into the fulfillment of his multifarious plans. On reaching the completion of a plan, he fulfills his dream, his longing, actualizes his will. All of the emotions that were engaged in the fulfillment process, all the actions, decisions, deliberations, hesitations, doubts are synthesized in this achievement. This is not the "achievement" of a singular task or goal, for this fulfillment involves the entire person, for whom it is what we call an "accomplishment." Its becoming a specific station in the life course, this halt in all the dynamic streaks of the creative logos, this measure of equipoise achieved escapes the otherwise linear succession of chronos. Therefore we call this moment one of kairos. Kairic temporality is that in which the logoic prompting complex of the creative spirit finds an accomplishment within an entire spread of a human person.

A most striking methodologically significant point has to be made here: it is in this mobile territory that the human creative self finds itself an access and conductor through the meanders of life, centering all the synergies, reactivities and active outpourings in the wholeness of an autonomous individualizing life. It is in the ramified functionalities of the agent directing the progress that the human creative mind recognizes its own existential outlay and the analogical arteries through which the rhythmic exchange of forces is pumped; my own beingness recognizes itself in this vibrant complex of logoic life strategy, its integral element.

Second, but by no means less significant in the specification of life is the capacity of the individualizing life, nay, the inward orientation of its growth, to produce other living beings through its equally simultaneous conjunction of inward and outward factors, a culmination that is worked out through drawing on the entire living spread of the individual, reproductive apparatus.

It is by retracing the steps of the logos of life in its impetus/equipoise strategies that we will pursue its manifestation in the grand spectacle of life/beingness.

VI. THE INWARD VIBRATION WITHIN THE PRIMEVAL LOGOS

What is the nature of this innermost conjunction of these two primal principles of the primeval logos such that they impel its prompting course onward? Prying with creative intuition step by step from the manifested reality, through its genesis and up to these first reasons, we conjecture that the impetus-equipoise conjunction resides in an inward tension of their distinctive and yet inextricable determination, one implying dynamically the other: there is no impetus not constructively received in an instance of relative rest, an instance assessing it, seizing it, completing it. In this completion resides the determination of the prompting impact. On the other side, equipoise is not a state of inertia, of no development, of nothing happening. It receives from the impetus its determination and its very role in ascertaining their constructive completion, which means repose for the striving impulse, a relative halt in the drive of the logos projecting its constructive rays.

The nature of both impetus and equipoise is dynamic, a dynamism in action striving to effectuate its tendencies and a dynamism in repose, completing and marking its effects.

As such, they remain timelessly in vibrating tension. The marvelous ingeniosity of their mutual motivations, adjustments, conditionings, stimulations lays down all the basic distinctions within the construction of the All. Thereby it escapes any thematization by the human mind. We may, however, witness it in the manifestation of their constructive interplay

As we probe the meanders of reality through the access to it given by the human creative act, through its reach into the generative processes of becoming, we may allude to, evoke these specific modes of the logoic impetus and the tentative rays it throws toward a constructive move that, if the conditions allow it, find completion and equipoise in a station of repose in the onrush of the great flux of universal life progress.

The innermost dialectic dynamic play of these two essential intrinsic principles of the primeval logos from which All appears to have surged defines the very nature of its heroic role as the ultimate generative factor and conductor of the entire game of forces. How could we ever imagine any principle of the All-in-becoming to be a static, abstract principle?

Given the nature of these complementary factors of impetus and equipoise, which hold the key to the prodigal game that the logos carries on, the question of "When did it begin?" has no sense. It is with its unfolding that time and temporality unfolded. The primeval logos holds the ultimate mystery within itself.

The tension between impetus and equipoise is a tension between the seminal forces of the primal logos. In fact, the impetus of the logos could not be an "empty" drive; the impetus of the logoic forces is prompted by forces that are themselves pregnant with seminal logoic dispositions, that is, germinal constructive "materials" to fulfill the primal nature of the logos. This latter consists in its innermost constructive vocation.

The dynamism that the forces in their rush carry is nourished and stimulated by the inward thread of movement. This motion, which is the innermost conductor of the logoic game, partakes of the tension just described and is determined qualitatively, teleogically, and directionally by the unceasing attraction/repulsion play of the seminal materials of the logoic forces. This dynamism of motion expresses in concrete constructive projects the impetus of the logos, while the phase of the entelechial completion of the intended task brings in the modus of the stasis proper to measured equipoise.

The motion and the stasis of the logoic dynamism at work in the play of attraction and repulsion run the entire grand system of the logos: they preside over the great circular, elliptic, and other motions of the heavenly bodies and the stasis into which they revert, and they preside as well over the generation and unfolding of individualizing life. While life establishes all over its significance, it projects the great schema of the All.

So if we seek the main categories of this grand project, we have to begin with impetus/equipoise, seminal material, the grand schema of the All, dynamism, inertia, movement/stasis, attraction/repulsion, all having at their center the constructive ontopoietic design of self-individualizing life.

CHAPTER TWO

LIFE AND NON-LIFE

I. THE INITIAL STEPS OF THE LOGOS IN LAYING DOWN ITS MAIN STRATEGIES

1. *The Constructive Spheres of the Logos: Concurrence, Excess, Waste*

In its spontaneous swing, the logos springs forth in far ranging differentiations of each strain. The constructive/intergenerative proficiencies of a few among these multifarious corresponding differentiations coincide with each other. It is only through their constructive coincidence that they manifest themselves; the aborted formations remain in oblivion, the random strains do not enter into the immense workings of reason but remain at the fringes. The coincidence of the constructively/intergeneratively relevant strains of the sprouting logos accounts for its formative nucleus. And a nucleus it is because in gushing forth from within, reason in its spontaneous outpouring, it does not calculate its effects; its ever renewed wealth is prodigious. Were it not for a detectable formative outline that the coincidences seems to fall into, it would appear as if each plant sprouting from this wealth of seeds pollinates and fruits as a result of pure hazard. And yet, we observe the contours of recurring regulation in other strains of the logoic manifestation of living nature, such that allow for a coincidence and subtend it. Subtending the strains of the logos, they make a coincidence possible without depriving it of the game of fortuitous encounter.

Thus in its drive toward proliferation beyond freedom and determination the logoic mechanisms, generative synergies, etc. always work in profusion, mindless of waste. Aiming to accommodate in some measure the designs of constructive networks, they work in excess of what could be useful, i.e., accommodated. Excess and waste necessarily accompanies the constructive rule of coincidence on the immeasurable field of the exuberant logos.

In fact, the convergence of the surging synergies, their fragments, segments of articulation, operations, etc. is a primary arche-principle underlying differentiation. This is differentiation of the "same" as well as the coming together of its elements into novel formations, clusters, joints, articulations, agglomerates, groups, etc. out of the spontaneous and incalculable spread of forms. However, from the innumerable rays of the spontaneous impetus of differentiation, those that do not find an outlet in coinciding with others are the waste of excessive prodigality. Thus excess and waste refer in their significance to the constructive principles that subtend the coinciding laws of constructive relevancies. But waste in some respects does not mean the extinguishing of the prompting logoic strivings, of the propensities toward a constructive event. They do not vanish from the game of life; they are merely "enfolded." (This, I believe, is what Leibniz had in mind in his conception of nature.)

From the criterion of mere "fitness,"[1] the criterion for the organic selection of energies in nourishment, in the inner/outer oriented organic traffic of exchange, the living being moves in ever further expanding circles to outward interchange and interactive exchange between individuals and among them in gregarious groups. In this outward orientation ever and ever expanding with the unfolding functioning of the living individual, the "communicative" links among organic visceral processes are transformed into symbiotic and further along, interindividual, communication. From the seemingly self-reposing region of inward bound functioning where our body operates in a univocally regulated and orchestrated fashion, in a mute and repetitive fashion, we pass to the outbound flexibility of an orchestration that is oriented by needs, inclinations, tendencies, taste, demands, and the availability of the resources to be found in singular individuals and their groups, who are multireferentially dependent upon the vital, fittedness-regulated operations, and their links. These links are not prepared progressively segment by segment within the exclusive and undisturbed inner workings of the living being; they are to be measured, estimated, negotiated, and agreed in each situation and that for the mutual satisfaction of often diverging individual interests.[2]

They occur in communication beyond the enclosure of the viscerally determined individual domain of the bodily self-communicative links that transgress the mute vital isolation of inward bound individualizing existence. The individual, in the entire selfhood gathered in affect, is mute. It lacks a voice!

2. *The Initial Propulsion and Its Receptacle*

In pursuit of life's becoming through its innumerable fluctuation in forms, conductors of transmutational forces and synergies, we cannot fail to ask what the beginning phase of this formative progress was. *What was the initial propulsion* by which the spontaneity has burst forth? And what was the protogenital recipient of the primeval spontaneities, that could initiate the first move of life? We may acknowledge first, that there was in that event an extraordinary stroke of good luck, hazard, coincidence having such slim probability of occurrence that no "planning" may be suspected to have preceded it. We may also assume that if we pursue the progress of complexity in the forms of life backwards, each prepared by the preceding, by all the sedimentations, there is left the spontaneity of the synergies that account for not only the simplest, but also the "first" in the genealogical line of chance encounter in the prelife, inorganic moments (crystal forming synergies, for example) in which the first ties of forces occurred forming a receptacle for the spontaneities, such a receptacle as would initiate life's outward/inward growth by ingesting the spark of life. Granted that this initial move is owing to an unpredictable hazard of immense significance, to a chance occurrence out of myriads of conditions, the surging of life manifests itself as a vertiginous *strategy* of the logos to project this unlikely harmony of constructivity amid seemingly unacceptable conditions, the play of hazard. And yet it has occurred! The manifestation of flexible and yet relatively stable operations, networks of interdependencies, patterns of transmutations, systems of revelevancies to the various circuits of the articulation manifest in the great outburst of the existing cosmos, all this is a revelation of the *strategies* of the logos itself.

Scientific speculation on the origin of life abounds in theories, some pondering the complexity of the first organism and the complexity of its environment, others hypothesizing that the first organism was simple (e.g., RNA, or clays). The crucial question is to find an element such that manifests the basic characteristics of life: the capacity for self-reproduction and self-guided agency. (See N. H. Horowitz *Proceedings of the National Academy of Sciences* 31 [1945], pp. 153 quoted in Hyman Hartman, "Speculations on the Origin and Evolution of Metabolism," *Journal of Molecular Evolution* 4 [1975], pp. 359-370). The self-guidance of life's progress relies on a metabolism that is growth oriented and self-sustenance oriented, one that in an exchange processes the resources of the environment.

3. *Life on the Move: The Janus-faced Emergence of Individualizing Life*

We intuitively associate life with motility. We seek life in a moving body instead of an inert one.

Janus-like life emerges two-faced: it manifests itself by movement from "within," movement that is already a "kairic" moment accomplishing the subterranean work of a complex of processional operations that coincide in an inwardness out of which the kairic move of a self-propelling beingness surges. Yet this move is of a unique mode: it is carried by and carries a "vital force," a force that is, so to speak, no longer a neutral outburst or an anonymous and indifferent carrier of energy, but is concentrated in a complex of operations of a specific assortment, with specific proficiencies and enactment, ways, and orientation, which force concurrently draws on and is spread upon a complex of environmental forces and proceedings such as enriches and invigorates this central agency. In short, this vital force is thoroughly "qualified," singularized in, and directed toward the promotion of this central complex. As such, it is what Leibniz called "vis viva." Being "personalized" this vital force manifests itself in its enactment, and so we have its manifestation as a directional vector of enactment, a course of life — the course of the ontopoiesis of life through its self-individualization, which promotes as well as manifests a living factor unprecedented in the inorganic realm, the animus.

The primal specifically life carrying moves reside in what in the eighteenth century was called 'reactivity' (*Erregbarkeit*). This is the propensity of the body to be "affected" by inner or outer forces, a propensity activated by the first logoic moves toward life. Yet the manifestation of life as such that is carried by the former consists in such an elementary phenomenon that it cannot be more closely described or defined than by the term 'animus,' for like "breath" it carries with itself an irruption of ungraspable force, one that brings with it freedom from the coercion of inert bonds.

Unraveling the ways of the logos as it projects itself through life, in particular, at its basic laying down of springboards and rails, we will discuss both of the faces that make up the one and the same coin of the self-individualizing surging forth.

Indeed, the ignition of life sets in motion the process that differentiates life and non-life, and determines relevancy to the life process, the degree of its closeness or remoteness to life and its propensities. Specifically characteristic of growth is its setting up the degree of engagement in life processes, and lastly but not least, it sets into motion the complexes of functions that in beings more advanced in complexity constitute what are called "organs," the simplest function being the processing of "material" for the benefit of growth. The living being stays alive, animated through a process of growth. To grow requires self-nurturance. These two central functions determine nature. However, we forget that nature and life are like two sides of a coin, in that it is life that brings with itself nature, or it is with respect to life that nature itself emerges. Life characterizes itself by a specific type of expansion and advance: this or any other type of growth initiates a specific kind of expansion.

All variety of elementary "material" is being processed by these functions, nourishing the being in progress; this maintains it in animation, keeps it alive

Advancing by these two constant functions of growth and nurturance, by its own innermost central agency nourishing itself from the surrounding forces, elements, exchanges, the living being engages and processes the elementary data of the cosmos available through the biosphere. Thus it mediates between non-life and life — transforming one into the other by its very engaging in the workings of the life process, which thus acts as a sieve.

Although the animus suffuses the functional outfit and transforms it into a living beingness, it does not hold any functional apparatus, schema, web of its own. It manifests life in contrast to "inanimate" constructions such as stone formations, crystals, etc. but is not constitutive of life; it results from an intrinsic constitution. This intrinsic functional constitution *in concreto* is its indispensable foundation. This constitution of intrinsic functioning resides in its forces processing core: the most intimately integrated indispensable functions of taking in nourishment, digestion/transformation, and absorption. They constitute the principal device through which logos differentiates itself as a life originating and sustaining system.

II. A PRELIMINARY ASSESSMENT OF THE FULL ROLE OF REASON, THE MANIFESTATION OF THE EDIFICE OF LIFE

Is not the immeasurably extended edifice of life, which holds together in the midst of its intrinsic turmoil, a flagrant manifestation of reason? Expanding in myriad modalities, radiating whole series of rationalities, the Logos of Life appears to those who pierce through the maze of "matter," of "phenomena," of the experiential labyrinths and contortions of the human mind, to be the crucial factor in the gigantic play of forces that in its constructive plans finds an outlet in the relatively minute but innovative projects of Life and Nature. To counteract "postmodern" tendencies, we propose to exfoliate it. *The manifestation of the works of the Logos in the edifice of life constitutes one area of the metaphysics that the phenomenology of life and of the Human Condition indicates.* Its exploration has been prepared for by our previous investigations, which invite the next step.

With its emphasis on the role of the human intellect, especially on so-called mathematical rationality in its theory and practice, the modern era has seen the unprecedented development of human civilization. However, this unfolding of the potentialities offered by translucent intellectual reason appears to have found a limit to exploitation. In the last few decades, reason, mind, rationality seem to have reached a breaking point in their hitherto undisputed sovereignty.

The emergence of innumerable novel human rationalities in scientific theory and in its practical applications has distorted communication between the sciences and has outstripped the hitherto dominant attitudes of human beings toward life and death, the human condition, the human situation on earth, and the prospects of cosmic survival and has further given rise to bewilderment and puzzlement, provoking several types of vitally significant questions.

The first series of questions continues the traditional investigation of reason and can be considered as still belonging to modernity — questions concerning the factual verifiability of the data offered by the various manifestations of human cognition, questions scrutinizing the criteria of cognition, favoring some manifestations over others, and attempting to reduce one to the other, etc.

The second series of questions are those raised in the numerous attempts at a critique of reason — the human faculty of reason being understood as the source of all rationality.

Finally, in the last few decades, there has emerged radical doubt about reason, rationality, the human mind, culminating in the absolute rejection of the plausibility of reason. Although this latest, "postmodern" attitude toward reason stands on weak and unsure legs, it has inspired various types of relativism and enjoys increasing attention.

Those who eagerly lend their ear to easy sophisms do not see how the negation of all rationality leads their reflection into a narrow tunnel, one dark and without exit. It is no wonder that those who enter it are as if blindfolded and miss the absurdity of their situation. This absurdity is blindness to the presence of life within and around us; it is blindness to living reality, to the world — as if both life and the world did not mean articulating, ordering rationality. Can we in the fullness of philosophical experience suspend — be it for a fraction of an instant — the living moment? This instant is an integral element of a gigantic schema for the articulation of life: time means becoming, change, coming to be, growth, fulfillment in fruition, passing away, all spread through events, experiences, processes, durations, that is, through the patterning of innumerable rationalities.

The wholesale negation of basic principles, and with them of all the seriousness of the philosophical enterprise, nay, the abandonment of its vocation that we witness today in "postmodernism" — leaving the human being without a foothold in reality and without a compass — provokes us to continue the investigation into reason that we carried out in our inquiry into the ontopoiesis of life.[3]

Indeed, in the self-individualization of beingness we have found the rational skeleton of life's progress; it appears as the collector, and in turn distributor, of proficient energies, forces, synergetic powers to be shunted into the constructive mechanisms and designs of life's progress; here is a springboard for the philosophical investigation of the Logos of Life in practice. *We were able to find access to this central agency of the Logos as it proceeds from the cosmos and reaches through Nature into the palpitating heart of life and the human mind, because in seeking a starting point for philosophical investigation we bypassed the so much discussed and disputed subject of human cognition and gave absolute priority to the creative act of the human being, thus reaching the Human Condition.* This latter, an incubator gathering all the natural forces for specifically human individualization-in-existence (which means the origination and processing of the human mind), offers us an incomparable vantage point from which to see, at one hand, the grand panorama of the prodigious rationalities of Nature spread before us, of the system of life, of the cosmic laws manifested in the exuberance and the plethora of living beings, creatures. At the other hand, we see, as from the center, the stream of specifically human rationalities gushing forth from each human being and spreading into the common pools of the various cultures, vibrating in an ever renewed superabundance of forms, colors, feelings, reflections, plans, projects, and the struggles that give birth to them and inform them.

Having thus far opened up and brought to light the inner workings of the Logos of Life, we may rise to the vision of its manifestation in the edifice of life. It is the Logos that manifests itself through the workings of life, informing synergies and proficiencies, projecting articulations, purposive joints, processional links, while devising relevancies within the forces, strivings, and "substances," projecting designs, trajectories, organisms, and techniques for the ever renewed and renewing progress of the grandiose game of life. It is the Logos of Life that coordinates this game and its rules with the framework of cosmic forces, at the one extreme, and the orchestration of the human eruption into the biosphere, at the other.

This gigantic game being played by the Logos of Life is neither inferred nor deduced, neither sensed nor imagined, neither theoretically hypothesized nor practically ascertained. It is manifested in a vision that goes beyond the facts of experience and thought; it dwells in all of them, it draws on all of them, it is carried by all the practices of life. Comprising them all, it manifests their intrinsic life of reason.

We will at this stage of our inquiry into reason and rationality survey them in their various sectors.

The individualization of life manifests the prodigal ingenuity of reason, of the Logos, whose fulgurations disperse, diversify infinitely, playfully, and at random, and in doing so fosters conditions for coming together, crystalizing, harmonizing, constructing.

III. THE ONTOPOIETIC DELINEATION OF SELF-INDIVIDUALIZATION-IN-EXISTENCE AND THE SURGING OF THE SPARK OF LIFE

1. *The Animus Bursting Forth As the Manifestation of the Logos of Life*

The manifestation of the Logos begins in a most dazzling fashion with the emergence of the animus. I have previously outlined the self-individualization of life as it emerges from the pre-life phase of alertness toward constructive union with an already formed entelechial principle of inward-outward orientated articulation that follows a constructive path of unfolding from within.

Although this assortment of appropriate elements is "available" and standing in readiness for functional articulation, yet would it come together without an initial "impetus"? Does this latter ignite the spark of aliveness while prompting these elements to coalesce, or does the spark surge from the very falling together of the appropriate mix of elements? We will attempt to answer this question.

The logos of life, which surges along myriad lines with the triggering of the individualization of beingness, manifests itself in the innumerable ways of ordering, articulating, fusing, etc. Two of its main conduits leading to innumerable networks of ramification that account for the progressive development of living beingness into the entire unity-of-everything-there-is-alive subtended by the system of life with its ideal rules and direction are the entelechial order of life's unfolding and the scanning of constructive progress. However, these two devices while being flexible and infinitely diverse in their concrete schemas of ordering are far from being "neutral," "universal," "lifeless," "insensible," "logical" schemas of articulation or rationales. We may apply such terms to artificial intelligence, but they would be inappropriately applied to what is not alien matter but belongs, on the contrary, to an animate zone of beingness. Therein the entelechial order unfolds; likewise, it is the animated nucleus of beingness that scans life's functioning.

The constructive design of the entelechy is not a mere formal blueprint. It is above all a set of selective virtualities — forces and energies endowed with propensities toward intergenerative fusing as well as toward entering into fusions with appropriate elements such that a pattern of growth will be spontaneously outlined by their release. It is from this time conditioned, constructive project that there spring forth constructive means, constructive postulates: inner/outer, and present/past/future. In other words, it is the inner postulate of growth that brings forth what we call the spatiotemporal schema of life.

Is there a way to dissociate the constructive expansion — the result of fusing, coming together, intergenerating, the mutual prompting of the protosynergies that bring life out of the "happening" or qualitative change in which they result? Are these two orders not mere ordering devices surging and unfolding together? But to get back to our point, there is an extraordinary way in which the self-individualizing progress of unfolding life manifests the logos: life as life, life in its emergence, let us emphasize again, is not merely an articulated line of construction, but, on the contrary, this rationale of self-individualization is "animated." Only the spark of "animation" can set the functional apparatus of individualization into motion. As a matter of fact, we may venture to surmise that it is the animating propulsion that in an "instant" articulates the otherwise disparate elements into a functional schema. It is this "spark" that adroitly spurs them into taking cohesive complementary roles and opens up a unique vital dimension of individualizing beingness as its very own zone, a zone of vital exchanges, through tentacles that the individualizing inwardness throws out and through the sensitive openness of its innermost to influences from without. This zone is not a neutral mechanism operating in accord with some or other rules; it processes life in virtue of this pulsating, sensitive spark/propulsion for which we may legitimately retain the traditional term 'animus.'

In short, the crowning point — and the crucial decisive factor of the emerging living beingness through which the logos of life manifests itself and spreads into further spheres spinning vaster and more complex circuits — is the animus. It is the animus that makes the living being "alive"; it is toward its emergence and through its surging that the functional segments of the individualizing apparatus come together; it is the crowning point of the manifestation of the logos of life, its full glory.

The manifestation of the animus as the exponent of individualizing beingness, its manifestation in its various complex paths is striking first in its harmonizing (unifying) power. The animus in itself is unthinkable; it manifests itself through the apparent unity of the articulated, cooperating, confluent, merging segments of operations that, whether simultaneously or in succession, in conglomeration or in selective circuits, provide telic orientation. But this orientation delineates itself through inner-outer pulls, the strife of opposed tendencies, pulls toward differentiation and new coalescence, impulses at variance with each other. In brief, this turmoil comes together in an indissociable harmonious play of attracting/repelling forces within the animus.

We cannot attribute any substance/accident/property/feature to the animus; given the unique all-"animating," indissociable "glow of life" in which "animated" beingness basks and becomes a cogent "self," and given that the animus is connaturally ingrown into everything from which it surges "on its own," the animus cannot be assigned any definition.

However, first, even prior to our apprehending this, we are struck by the fact that in its primogenital harmonizing role, the animus is involved in all the dynamic small streamlets of life that it harmonizes. We are, secondly, struck that in carrying this out it truly encompasses them all in a unique manner, as if it were molding their wild diversity, their otherwise violent interactions by matching them up with each other and all others, an operation through which proportions are settled, correspondences are fixed, etc. We may surmise that the territory of individualizing life is circumscribed by virtue of the animus of each particular living beingness' coming about and unfolding. Thus, the animus being partaken of by all the functional lines of living beingness is undefinable and conceptually ungraspable — that is to say, it eludes objectification — and is by no means an epiphenomenon of the functional schema promoting the entire network of functional operations unfolding its most specific and concrete polyphonic beingness.

Third, the animus is uniquely singular, just as is the beingness that it animates. In fact, its concreteness is grounded in polymorphic and polyphonic operations, energies, and forces in operation. In other words, the animus encompassing the operative, animated zone that determines the identity, the nucleic identity, of each singular entity spreads through innumerable tentacles and forays into the varied sectors of the zone of life where it encounters other living beingnesses, so that it encompasses the entire nucleus. Moreover, it carries on the innermost entelechial agency in control of the entire individual life process.

Recapitulating, we must insist that the animus is not to be identified with the entelechial principle — although it prompts its implementation. Nor is it to be identified with the entelechial schema — although it carries on the interrelational, interchanging, harmonizing task of its concrete actualization. Lastly, it is not to be identified with the already harmonized discrete unity of the actual life forces carrying on the constructive work — although it personifies them all, in fact. The animus is the key to the combining of all the rationalities of the logos into an ever expanding gigantic act — the spectacle, symphony, drama that is life.

Fourth, in its full harmonizing role assuming all sensibilities of conflicting forces, requirements, needs, safeguarding of interests, etc., that which accounts for all the modalities of the animated zone, of the individualizing "pulp" of life, we call, in accord with philosophical tradition, 'animus.' We thus rediscover upon a novel route of inquiry, namely, this probe into the manifestation of the logos of life, a truth about living beingness propounded by Aristotle and Leibniz, that to be a living beingness means to be animated.

Fifth, animus inheres in the modalities of the functional circuits of life and the varied complexities and fullness of the *sensibilia* of the individualizing life in its constructive evolutionary progress. We may distinguish as many modes of animus as there are kinds of living beings, from the simplest to the highest, from the vegetal, through the simplest animalia, to the most complex, the human soul. The soul of the living being stands for the reacting, sensitive, sentient, emotive factors of life's becoming. The animus, grounded in life's complexities, harmonizes them into a self-controlled, self-existent, living individual.

As a matter of fact, surging with the life promoting selection of forces and their respective protogenerative propensities, the animus alerts the latter to "sensitivity" with respect to all that is needed for a specific beingness to unfold its course and maintain it. It is through the thus awakened selection of sensibilia, which with the formation of organs become more specified and complex, that individuals, on the one hand, build their very own nuclei of beingness, as their very own life spaces within the system of life and, on the other, do so by throwing hooks out from themselves thus tying themselves into the unity of life. Through sensibilia "alert to," "attentive to," the soul differentiates, separates, brings together, fuses with or rejects, and forms a network of harmonized diversity neutralizing elements that are often contrary to each other, or even opposed to each other in their own right.

Indeed, even at the lowest level of complexity and consequently of self-control, there is in the animus a doubly oriented pragmatic commitment; first, the animus pervades all the functional and operational links and joints, allowing — promoting — their coordinated movement. In this it remains committed to the task at hand from one performance to another. But, second, it lifts itself above this focus on details toward the attainment of an overview; it moves from the lowest level of mere reactivity to negative phenomena (present already in bacteria) to an overall alertness to danger, to seeking to evade it, to protecting against it. The animus grows in the wealth of its rays as living beingnesses become more complex. Yet in the animalia the animus remains committed to making mere survival (vital) evaluations. It is in the human spirit that the animus finds its full expansion.

Sixth, in a parallel manner, the animus develops different types of modalities. Of these, three modalities are most strikingly instrumental in the manifestation of the logos — the intellectual, the aesthetic, and the moral. We will come to discuss these later.

In the constant effort of the self-individualizing beingness to delineate its own unfolding course while differentiating itself from the connatural elements in the life stream, we may distinguish *grosso modo* four major moments of synthetic pulsation and orchestration, on the one side, and functioning and operating, on the other. These four moments in the orchestration of life are: contrast, differentiation, discrimination, and harmonization. These are the elementary protogenerative ways in which the living operations bring together cohesively the proficiencies of the forces that operate on the basis of the sensibilia, on the one hand, and those that operate according to the genetic code, on the other. Should we then not recognize in the animus and in its multifarious profile a sentient mode of the Logos, the incipient mode of the Logos of Life?

2. *The Individualization of the Logos in Life Reveals the Rationale of the Cosmos*

The central constructive factor of the manifestation of reason in the ontopoiesis of life is evidently the entelechial principle with its encoded assortment of initiating devices, of synergies indicating the phases, articulations, qualitative transformations. These synergies are not mere constructive devices; they flow from virtual constructive propensities, and these forces once activated take their course. Through them the entelechial principle works out its constructive design. They present us with a concrete virtual nucleus of forces with activating propensities for the actual unfolding of the living individual. The entelechial principle's constructive role is in giving directions, but not in an abstract general way. On the contrary, it is its drawing them from the intrinsic virtual propensities of the synergies it gathers into its magazine that is crucial for the unfolding of life. On the one side, it is organized according to the rules and laws of bios per se; it implements and reveals the laws, regulations, and virtualities of the laws projected by the Logos as a strategy for the unfolding of life. On the other side, however, through the individualizing constructive work that the entelechial principle and outline promotes, it also makes manifest the forces, virtualities, laws of the gigantic cosmic system from which life emerged and in the play of which life is immersed and upon which it remains suspended. Let us point out the striking features of this manifestation.

To begin with, and on the one hand, the individualizing constructivism progresses as much through the encounter between virtualities at the nucleus of life, which contain in limbo constructive propensities and which indicate the indispensable requirements for their unfolding, as it does by drawing on circumambient energies, forces, with their specific constructive aptitudes and requirements. On the other hand, the entelechial nucleus, though it is as simple as it could be, already manifests a virtual essential congeniality and not only with the natural/societal situation within which it occurs, e.g., cogenerative propensities that might find in the circumambient situation appropriately relevant elements and factors allowing it to develop its virtualities in the life situation in which they occur. Above all, it already manifests the essential requirements for life to occur and unfold in the order of the cosmos, with its substantial generative dependencies. The culmination is a network of interdependencies between the requirements of individual life, which must change, move, make space for itself, and scan the situation, and the great structural laws of the cosmic order. The individuality of living beings means directly a sudden expansion in motility as well as in extension. In this, the smallest living creature manifests not only the entire life system, the whole of the biosphere, but beyond that the cosmic ordering laws and structure upon which the biosphere is suspended. The extension of living beings is not unlike that of heavenly bodies. However, it manifests a specific set of cosmic laws. Beyond the sphere of its accustomed gravitational pull, a "weightless" body loses life's main orientational proficiencies — posture, directed mobility, ability to guide oneself using the landmarks that the ground provides, etc. The law of gravitation manifests the cosmic logos' provision of particular conditions for life. Hence, each living beingness manifests in its volume/size the laws and ordering of cosmic forces. Life emerged, unfolded, and unfolds within their specifications. This set of relevancies to the cosmic order points beyond, toward more universal structural laws, toward the specification of vaster schemas of rationality and ordering beyond the Logos of Life.

The manifestation of the rationalities of the cosmic order occurs in all phases of individualizing life as it is transcribed in different keys. To mention one instance, the directionality of cosmic spacing and scanning is manifested also in the order of the circuits of the psyche with its "lower levels" and the "higher" circuits of conscious life and the intellect. The gravitational basis that the motility of the body finds in the ground/soil is transcribed into psychic "attachments" to familiar places, people, animals, things, etc. The cycle of the seasons in which the biosphere sets the stage for natural life finds its counterpart in the cycle of conception, generation, birth, growth, decay, and death.

In still different keys the structural laws according to which the volume or size of each living being is conceived manifest in a microcosm the relevance it carries within the macrocosm, within the great cosmic design.

The entire system of the organism shows the ingeniousness with which the Logos of Life has projected its further expansion and specification of the grand macrocosmic design. But what about the logos of life's project of specifically implementing the great laws of the macrocosm when it comes to the autonomous circuits invented by the human mind, those of society and culture?

3. *Life's Constructive Route on the Spacing/Scanning Axis: Proportion, Measure, Valuation*

To recapitulate: the progress of life's individualization through its entelechial synergies not only proceeds from the flexible blueprint its constructiveness follows, not only brings with itself the synergetic forces and spontaneities that initiate and pursue its main line, but with these synergies with their virtual propensities awaiting activation — and in line with the requirements of the foreseen constructive advance — it postulates and brings about the articulations that advance the establishment of living beingness. This, as we have pointed out above, occurs by a spacing "expanding" of individual beingness in the unity-of-everything-there-is-alive, that is, by entering into this unity and occupying a "place," which amounts, however, to performing a set of articulated operations and delineating a segment of the constructive unfolding, some "at once," some "concurrently," some in "continuity," some in "succession," etc. In short, what we in philosophy, in hypostasizing, hylecizing, or waxing transcendental, usually call "time" or "space" — what we seem in practical life to consider as being essentially "measurable" as well as a "measuring stick" — is in reality the "spacing" and "scanning" of life itself.

Spacing/scanning — as articulations of life's progress — are connatural, complementary in the sense that the one can unfold and thus acquire sense only in concurrence with the others' doing so. We say "life spaces and times itself," that is, individualizing life spaces itself constructively and scans its constructive unfolding. However, we should not, as is the habit of philosophers, hastily surmise our way too much toward a beautiful conclusion of the argument. The spacing and scanning of life's progress from within does not mean that we "transcendentalize" or "vitalize" the "spatiotemporal axis" as it is called, ignoring its actual or possible role in the cosmic system as such. As we have argued above, the regulations of the constructivism of life manifest relevant laws in the cosmos. (We cannot ignore the rotation of stars, the falling of meteors, the expansion of the universe, or theorized transformations in stellar systems.)

But even in view of this relevance, assuming that there is a spacing/scanning axis in the laws of the universe, the spacing and scanning of life so uniquely fit the constructive progress of individualizing life, so uniquely fit transformability and flexibility, and are so specifically qualified in their performance with respect to the entelechial design of each individualizing life and its differentiating moves toward other individuals in progress within the unity-of-everything-there-is-alive, that life's spacing/scanning is evidenced by these moves themselves, they being connatural with their own virtual conditions.

To recapitulate, we cannot simplify things either by assuming that the "spatio/temporal axis" is "something" in the cosmic system waiting for life to surge and expand. Nor can we identify this axis with transcendental constitutive conditions having no rapport with the rest. The cosmic laws that also deal with forces, energies, synergies, masses, and life's rationalities are certainly relevant. There is as well a spacing/scanning axis at the center of the transcendental conscious life of humans that is relevant to this life. We have already given above the main hints at the spacing of life. It is now time to turn to its co-essential core, conscious life's concurrence with life's scanning. Indeed, the "movements" of life's unfolding "single out" even as they "bring together and tie," and even as they install a "space" for a living being (the moves being relevant to the telic aim and being "proportionate" first to the entelechial schema and then to the synergies brought into play) and so encroach upon and tie into the vital strings of some one and ultimately innumerable other living spacings, other living beingnesses. In this way, new factors "proportioning" the outcome enter into play. For a beingness to differentiate itself in an individualizing course means for it to launch innumerable tentacles for growth in other congenial and alien life areas of being. Thus life from its innermost projects itself as spacing and scanning augments its voluminosity/size with respect to the above-mentioned factors, an adjustment that means proportioning and measuring.

If we abstractly grasp becoming in its quality of continuity as "temporality," then we will identify it with protogenital articulations in the synergetic surging of life. Second, the profile of becoming will appear in the intergenerative play of forces and energies in the progress of the individualization of life. Third, the scanning of the individual life's progress will assume necessarily its synchronization with that of all other individuals within its existential radius, extending proportionally toward the entire unity-of-everything-there-is-alive in its various phases of unfolding. Fourth, we will identify becoming with the continuities of the interactive individual's unfolding and growth. Having sketched in an anticipatory way the outline of our investigation, let us turn now to its substantiation.

IV. THE LOGOS PROJECTING ITS RAILS OF UNFOLDING: MEASURE, ORDER, TIMING, SPACING

1. *The Process, Propulsion, Move That Introduces Order and Its Primogenital Measure*

"Process" is a concept used in various contexts and consequently it covers different "procedures." However, whether we talk about the process in the so-called "inorganic" circuit (e.g., the procedure through which minerals acquire a specific geometric form, or cosmic energies that fuse into natural "substances"), or about the processes that carry on the transformative steps of vital growth, or even if we call a series of human activities (in the cultivation of the soil, in manufacturing useful objects, in the life of the psyche, in educative endeavors, etc.) a "process," process is a sequence of transformational operations arranged in a line of successive advance, the performance of which leads to an end result. The end result may be just a step on the way to further sequences, but the process as such is a constructive advance. We may say, using an Aristotelian term, that it marks advance by its intrinsic efficacy.

First of all process brings about an "expansion" (or better, is itself this expansion), scanning the steps of the "advance" signifying this move is an "advance" as well as "spacing" by that advance a territory in which to spread. Expansion is the groundwork of what we call "temporality" and "spatiality" — the extension that we abstractly plot on either "perpendicular" or "horizontal" lines of life, along which individualizing becoming proceeds, these lines being the rails on which its interrelated occurrences, clashes of forces, productive exchanges take place.

Leaving aside the question of necessity and contingency, the metaphysical question of whether the logos of life unfolds according to a preestablished plan or whether it springs forth spontaneously and makes its constructive way as it goes, by trial and error, it remains the case that trial and error carry in themselves an intrinsic search "for"...and that the junctions that they favor, favor certain rational intrinsically laid down links.

In this way, the paths of individualizing life as a center attracting blind energies and forces lie between opposition and necessity; in passing through innumerable intermediary phases, in a way, together, they become bound to each other. That is to say, the process is a conductor of motion and its vector. As such, it is also the vehicle of what we abstractly consider "temporality"; it also lays the groundwork for the spacing-map of life's territory. In fact, not only does it indicate the direction of this groundwork, but — what is essential for the ontopoiesis of life — it establishes all the contexts of its proceeding in the inward-outward operative conjunction. Relative to growth, the process draws on environmental forces, energies, supplies of relevant substances in order to work them up into the line of "inwardness" in which the individualizing growth of a living beingness takes place.[4]

The way in which the phases of a process differentiate themselves from each other is indicative of the basic cosmological/natural/intellectual order . Each phase performs distinctive operations that draw on environmental/circumambient fields and these may be favorable/unfavorable in their impact to fostering and working out its efficient assignment; this "assignment" lies in the preceding phase, which, therefore, is not disjoined from its successor. On the contrary it is amplified, adumbrated, efficiently reaffirmed in its line, which with each phase defines itself more clearly. Simultaneously, the actual phase just in performance already anticipates the next one, into which it passes. Thus, when we talk about distinctive phases of the same process we already make an abstraction; this is to introduce an alien fixing of matters into this translacing continuity of constructive operations.

a. *The Process' Intrinsic Projection of Itself through Scanning and Spacing*

To begin with, the process lives in its phases, and in a particular way in only one of them, that of actual performance, that is, the manifested present. It appears as the always changeable stronghold of a translacing continuity in which one phase progressively disappears and another progressively readies itself to come into its own. We introduce the distinctive point of an "instant" into this flowing progress and thereby see the actual appear as the present, vanishing into the past and inaugurating the future — which line of translacing unfolding we envisage as "temporality" in its three phases, which correspond to spatiality's vectors of center and sides. Thus temporality emerges simultaneously with measuring: it is the human mind that introduces the fixating moment by which we distinguish in the flux of the process its phases. But if we cut into this differentiation of the phases that pass into each other, we see that these different fragments delineate the profile of the same efficacy. We see here that the intuition of Heraclitus that we cannot enter into the same river twice refers precisely to this protophenomenological situation in which the logos of life projects its unfolding tracks and finds its substantiation.

The river with its flowing waters remains the same because the waters that make it a river are in a flux of which each segment continues the past and flows into the future in a translacing continuity that makes the sameness of the river, even as each of its segments vanishes at each "insight" into the further regions of the future flowing onward and new waters arrive at each instant. We enter it while it is "the same" and "other" because of the translacing continuity of its intrinsic design: the "same" flux is changing, the sameness of the river is jeopardized.

b. *The Process As the Groundwork for Order, the Vehicle of Becoming*

As our preceding analysis showed, the process brings about temporality and measuring, and so temporality introduces us directly into the process. In short, order and ordering has its genesis in primogenital logoic moves.

The basic primogenital order of the logos can be identified by the human intellect by applying the foundational principles of classical logic. There is the principle of identity: what is is; the principle of non-contradiction: what is not, is not, or what is cannot simultaneously be non-being; and lastly, the principle of the "excluded middle": something is, or is not, there being no third possibility.

These laws are also the laws of intelligibility, that is, of the way in which the human mind in apperception "identifies" and differentiates the reality it participates in.

In fact, the laws established by the mind as those of formal logic are nothing but simplifying abstractions made on the basis of a "fixation" achieved by the abstract "instant" that cuts through the complex of becoming. Within the reality of a being's becoming actualized in a process that gives it its "profile" or "design," there is no clear cut distinction between the phases of the object's becoming. On the contrary, all individualizing beingness, each and every being is in an incessant flux of becoming and is what it is and what it is not. In each present phase of its becoming it carries the past phase, on which it draws, and foreinstalls the future phase.

The "object" of becoming defines itself "in passing." This will become clearer when we proceed below to the precise analysis of identity vs. disparity in the self-individualizing process.

Also, for the same reason — that of the spontaneous force running through the progress of becoming without cease, unfurling its constructive phases, and establishing for the logos of life its base in the manifest world — the ordering mind's fruit of the second principle of formal logic, that of non-contradiction, is an abstraction, for the past phase brings into the present its results and thus is imminently present there, while in the protopoietic moment the future phase is already prefigured in the present phase, which prepares it.

Each thing is either A or non-A as the principle of the excluded middle is embedded in the primary manifestation of the logos in which "everything" is simultaneously "itself" and also otherness. In the fore-installed future phase of becoming, the identity that is never absolutely sharply determined within the translacing continuity of the becoming process is in itself already manifesting the future developments that might transform its shape. The substantiation of this analysis will be given in the following study.

A likely misunderstanding has to be precluded from the start: *That within the flux of process, coming to be means being A and non-A at once does not mean that just any non-A is being envisaged.*

V. PROCESS AND ITS SEQUENCE — OPERATION AND ITS COMPLEX

After we have got an intuitive grasp of the main factor in the logoic strategy in projecting a constructive line of manifestation, which is the process, we may now, moving from this complex constructive device move to its simpler moments.

The phases in which the process outlines its sequence of progress are nothing in themselves; they are sequences of more elementary constructive moments: operations. Operations are the elementary carriers of transformational motion, of exchange of energies, of transmutation of forces. Their logoic significance is to bring about a constructive/destructive change. Operations do not manifest a clear nexus among the factors they involve. We might through the simplifying bent of the abstracting mind introduce here the notion of a causal cause-to-effect relation between an initial situation before the onset of an operation and the outcome of the operation, but it is erroneous to seek such a one-to-one nexus in the primogenital status of the operation itself, which might involve numerous factors at once, e.g., a fusion of factors, or a coalescence, or a resulting decomposition into several moments, or an absorption into the environmental fabric.

It is characteristic of an operation, in the first place, to effect some passage between forces otherwise fluid or disjoined, be they active or inert; it releases in some the activation of propensities vis-à-vis the propensities of others and thus cuts into virtual resources that are otherwise dormant and inert. This release and activation of virtual resources is the crucial strategy of the logos, its dynamic upsurge relying on it.

Second, an operation is seemingly restricted as to the reach of the factors among which it works, and yet it circumscribes closer and further circles of its dynamis. On the one hand, it does not surge of itself, nor does it finish with its performance and then vanish without a trace. On the other hand, not only is it the outcome of a situation with certain circumambient conditions, but in turn, it enters with its performance into this very orbit and introduces some change, or it emerges within an already dynamically charged fabric, or performs a moment within a sequence of other operations it is constructively linked with, performing a special function.

Thus operation is the first initiator of change, transformation, transfusion, etc. within a virtually charged milieu, and it is conditioned in its surging and in its nature by the situation: the dynamic charge and the endowment of virtualities of the milieu. It is also conditioned by its particular composition and its particular state of readiness to be activated, the assortment of available propensities, etc. By its performance, it introduces some transformation into the milieu. In short, an operation encompasses the realm of its relevancies as its closer orbit and a further circuit of conditions as its further orbit.

The process is thus an "advancing sequence" marking a "progress." This progress occurs when each phase marking just one step of the constructive line passes into the next step, which continues it, as does the "following" one. Each process in confluence with a conundrum of other processes marks a constructive advance until all of them are accomplished and the process is terminated in its last phase. In this way, the process forms a groundwork, a moving, dynamic groundwork upon which surge instances of intellective rationality whether those of the cosmos, nature, of the psychic life, or of sociability.

However, as we have already mentioned, it does not initiate its surging on its own.

1. *Operation, Propulsion, Move: the Inward Move/Result, Outward Resources*

Indeed, to this surging's closest circuit belong the factors of its initiation within the conditions of the circumambient/environmental orbit.

It is owing to the composition of this environment that a particular operation and no other will be initiated, as well as to its charge of dynamic propensities, and lastly to its specific "dynamic" state, such that it may sustain a "propulsion" toward a first move. We may distinguish these two last moments, a propulsion and the first move, only after we have intuitively isolated the operative factor at work. In fact, since the motoric initiative of a process does not issue from within itself, the propulsion is indispensable to making it happen. With trial and error, the mere chance meeting among forces on an occasion propitious for the initiation of their operative transformation, with one move resulting from other, the propulsion springs into action and takes on a propitious directedness that seeks fertile ground and sets forth the moves of an operational procedure. It is this operational procedure that carries on the transitory, transforming, absorbing, or fusing effect that first displays the distinction between effective operational inwardness and outward effect.

CHAPTER THREE

LIFE IN ITS SPECIFICS

I. DISCOVERING LIFE IN ITS SPECIFICS — "WITHIN AND WITHOUT"

It is from within that we have to grasp the manifestation of life its logoic arteries and their articulations. No wonder that traditional philosophy by relying on the sclerosed universal features offered by cognition and its analysis barred the way to the revelation of the nervous system of all philosophy, *life*. The revelation of life came through following the royal road opened by the creative act of the human being, a path along which objectifying cognition plays only a subsidiary role in the presentifying of that which the creative act working in the individualizing operations of the human being brings forth. That is intuitive work of the excavating type at its best. Following the very path of life enactment, we do not concentrate on its epistemological aspects where cognition per se is subdued, but follow intuitively the meanders of our own evolutionary progress. Intuition of becoming, the most penetrating type of intuition, does not focus on "objectifying," "presentifying" as its main task, but focuses rather on identification with the very operations of becoming as they plot their station of progress and so retrieves them; it then sinks deeper into the operational arteries by conjectural inference, identifying further, more remote arteries of life, their operational stations with their transformational knots, and follows the line of constructive results, transformational networks spreading into the circumambient/environmental circuits with their forces, energies, synergies, and peculiar propensities in interplay.

The logos of life manifests itself through the self-individualization that occurs from within that which then becomes life's circumambient surroundings, its environment. The self-individualization of life consists in the interchange, and its outward resources surge. Indeed, it is in the split that occurs in the operational performance between life's propulsion and the resources that subtend it, on the one hand, and its efficacious transformational/constructive moves and their results, on the other hand, that the crucial strategy of the logos takes place: the inward/outward split and complementariness of constructivity.

This strategy develops in extraordinary ways and bears fruits at all phases of the evolutionary individualizing of life.

It is already at the level of simplest operations that the impetus of the logos establishes a measure of equipoise that, though it be infinitesimal, provides a springboard for those to come. Within the constructive context of that equipoise, we may now describe the "move."

The move otherwise unqualified, unsituated, undirected, aimless, etc. nevertheless brings in with itself an — even altogether vague — direction and marks an occurrence that breaks "into" the circumambient sphere that it enters in a prototypic instance of scanning and spacing. One could say that as it breaks "into" it marks the sphere "into" which it breaks from "without" until it turns that "without" into the "within" of its own performance.

The process of becoming drawing on the "outward" environment and working out its "inwardness" is not one of what we abstractly call a "causal" chain of clear-cut "events"; on the contrary, each of its operations proceeds within a "complex" of data, energies, forces, with their propensities, tendencies, etc. that circumscribes the range of its possible constructive line and possible deviations from it. Thus, already embedded in the situation of Sameness A is the possibility of the appropriate non-A. (The apple does not fall too far from the apple tree!) The sameness of A and its alterity as a significant unit of relation are codetermined within the complex of becoming in which they are embedded.

Alterity is situated in the entire complex of the situation of a sameness and within a range that the sameness typifies. Alterity to A does not mean total and nondirectional dispersion. Such dispersion occurs within a series of changes in the operational circumambient complex and external influxes as well as influences. Since the progress of life is not itself following any preestablished plan of the logos, but, on the contrary, the logos projects itself spontaneously and capriciously ahead, the sameness/alterity tension, its complementariness within the complex of circumstances of the becoming process, does form a skeleton of order for any ordering but is itself by no means absolute like the posited laws of formal logic derived from it.

As a matter of fact, the leitmotif of differentiation within the otherwise haphazard, unforeseeable, directionless flux of forces and energies is the self-individualizing life. It is with respect to the becoming process of this self-individualization that order — that is, direction, constructive sequencing, articulations of efficient operations, result oriented operations, continuations, continuities and significant discontinuities, telic motives and their accomplishments — surges and crystalizes among the aimless and chance driven, lawless elements.

The self-individualizing process is a painstaking operational differentiation of its own innerness; the conductor of life forms the circumambient chaos and competing individualizing endeavors. This is an effort to find a sameness that attracts, captures, and directs factors allowing progress in constructive becoming, and identifying their relevance and operative allegiance. It is in this constant effort at establishing and maintaining an identity among other self-individualizing beings that the translacing continuity of the processional line of becoming assumes a unique crystalization.

We cannot do justice to the essential and absolutely valid features of beingness — or Being — otherwise than by considering it within the dynamic flux as well as tracing the routes of its differentiation by means of its very own operations, its own self-constructivism and the twofold inward-outward orientation of its operations. In short, it is this ontopoietic model of the way in which the logos of life crystalizes its life projection that gives us the essential ontic profile of beingness.

1. *The Process-like Translacing Continuity/Discontinuity of the Ontopoietic Becoming; the Self-Identifying Quest As a Conductor*

Before we differentiate and isolate the intrinsic operational schema of ontopoietic becoming with its inward/outward operative or functional system, with its interplay of tensions between the two extremes of prescribed direction, on the one hand, and flexibility of selection, on the other, with its compromise between the two seemingly opposed extremes of necessity and contingency, a tension that plays a crucial role in the progress of life, we have to sketch first the phase-like continuity of the poiesis of life. The constant effort to which all the inward/outward oriented operations of life contribute and on which they all more nearly or more distantly converge is that of self-identification.

This bringing to a "center" of all and each of the operational functions throughout the entire progress of becoming in the phases of its being carried out is a further guarantee of an efficacious continuity throughout the discontinuities in the innumerable operations that seem as to work "on their own" and vanish the moment their energies are spent in a result, or for some other reason, the first being the phase-like nature of the process as such.

What is specific and indispensable for ontopoietic becoming is precisely the translacing continuity of the phases, each of them being partly present in the phases before and after so that there is a pipeline of efficacy; the outline of a constructive profile embraces them all, unfolding. However, there is the intrinsic accomplishment of the poiesis as such to be accounted for. The ontopoietic process stands out from all other processes (except the human creative process, which we have analyzed elsewhere) by its formation of an individual life. The processes that define its phases are not only portents of efficacy, they carry on "growth," "maturation," "fruition," "decline," and "extinction," themselves peculiar cycles of individual becoming, of the beingness that it unfolds.

It is the shaping up of this beingness, its self-individualizing course in phases of formation that is the objective of the series of ontopoietic phases carried on by the vital processes.

Having such an objective, these processes proceed in an unfailing, uninterrupted quest for the singular. It is this and no other individual with its own identity that is in process, and that pursues its own unique course by its own, unique enactment.

Its phases of unfolding carry from one phase into another the results and acquisitions of the functions already performed and prepare those to come. There is in each of these phases an intrinsic quest for this isolation of the objective for the sake of which it occurs, for which it prepares the material, the proper means, the proper stage for the further phases of functioning, which then rely upon the previous phase. Thus the "identity" of a living beingness is the constantly repeated effect of its existential genesis at its very core.

Disparity, its opposite, then, is defined by what lies apart from this identifying genetic process, what does not participate in its line of unfolding and hence is "foreign" to the objective. And yet it is in this tension of opposites that that identification occurs.

2. *Life's Categories and Logoic Specifics*

Animus is the final touch of the manifestation of life per se in contrast to what is not life (rock, water, air, metals, etc.). It is only in a secondary instance an operational force. In the first instance the operational power resides with the functional intrinsic agency. It is most specifically the essential manifestation of life's self-individualizing unfolding, which, as has been pointed out repeatedly on various occasions, is outward/inward oriented, and it is precisely given that inward/outward tension between the two seemingly opposite and yet basically complementary directions that unfolding life progressively puts together the functional set up of its self-orienting and self-measuring agency. This is the coincidence of what we could call "categories of the unfolding logos: equipoise, measure, orientation," which in the emergence of the individualizing life acquire an unprecedented character in the differentiation of the logos. While with inanimate formations like rock, soil, rain, the sun, the moon, meteors, etc. it appears that the equipoise and measure of the formation is partly accomplished by partial intrinsic propensities for interchange and transformation proper to the various singular segments of substances coming together as a result of the universal situation in the cosmos and/or in their immediate surroundings, with the emergence of the smallest living unit, the logos of life acquires an overall radiation and on the rebound gathers direction and role: direction from "within," unfolding precisely at this most sensitive, fluctuating moment of inward/outward tension. We could say that it is at the heart of this tension as to direction that the logos of life sets off.

It is to this intrinsic overall concentration of life's functional filaments vibrating at the tender points of the neuralgic tension that the three categories of equipoise, measurement, and orientation find in individualizing their life specific modalities and coordinates. Before we enter into sketching the specific modes of this functional setup of individualizing life — and it is through individualization that we may discover all the springs that subtend it, all the functional lines that make it possible and actualize it, and all the points of the strategy of the logos that establish it and make it work — we will first point out that it is the specific functional translation of the basic categories of life (measurement, equipoise, and orientation) that accounts for the four crucial universal specifics of life: motility, the processing of nutrition (metabolism), a capacity for self-repair, and reproduction.

We could investigate one by one these four specifics of life to see how they distinguish life from non-life, individual living beings from other "substantial" complexes.

Beginning by self-organization, we may point out that other complexes, e.g., crystals and other minerals, have the capacity, first, to absorb only such amounts of other substances as allow them to achieve a differentiating formation around a nucleus of their own and then to stop that absorption when their peculiar complex of crystalization has taken place. However, this self-formation with its very own nucleus once accomplished does not install any directional process. It remains as long as outward conditions allow it, then it disintegrates at random.

There are different kinds of mobility to be observed in the inanimate spheres, falling rocks breaking weaker materials, planets and stars changing position, volcanoes erupting, rivers flowing, etc. And yet, all the mobility to be observed in inanimate spheres is a result of circumstantial conditions, and although in some cases more than in others, like the eruption of volcanoes, for example, this mobility may display the power of a nucleus of forces that gather momentum and issue in a systematically organized fashion, these forces are "blind" and unregulated.

The eruption of fire and lava is a circumstantial synchronization of a number of singular operational sequences; it occurs "at random" whenever it happens that these come together. But the motility of a living being results from a consistent set of operational sequences themselves grounded in a complete functional network that is operating according to a set of intrinsic rules that endow that being with motives as well as propulsions toward pursuing them. It is in itself a self-organizing system of constructive motility with an initial impetus that sets it to work. We may call it a self-propelling system.

With the specific of motility we encroach already on the specific virtuality of self-repair. Natural/cosmic formations that result from random synchronization of phenomena offer in their "substantiality" a resistance to the impacts of external objects and forces; this resistance is according to the measure of their phenomenal substantial composition, the consistency of their particles and their own substantial resilience. Their resistance is altogether passive and inert. Under appropriate pressure, they decompose or disperse.

In contrast, the living individual delineates its unfolding course in a constant active resistance to external, "territorial" pressure, and to internal pressures as well. This manifests a twofold specific of life: self-correction/self-repair.

It is in virtue of its intrinsic functional and operative systems, that is, in virtue of its inward motility that the living individual delineates its course in correcting its inopportune moves, propensities-in-unfolding, operational orientations. Here resistance takes the form of resilience through adaptation and reorientation. Resilience, as opposed to the merely passive resistance offered by substantial composition, is a virtuality residing in the inward motility of the individualizing living being. The crucial virtuality along this line, however, goes beyond adaptive capacities for correction of tendencies, moves, attempts, and — on our level — decisions. Correction goes hand in hand with the crucial specific of life of self-repair.

Owing to the properties of their substances, nonliving substantial formations have a transformational capacity to react to encounters, but this capacity is very limited in regards to the territorial expansion of the complex and is also restricted to some singular effects. And then the occurrence of the encounter is random. In radical contrast, the inward functional system of living beings goes beyond the capacity to endure. Its resilience is particularly striking in lower forms of life and at the embryonic stage of the life of more advanced forms of life, when whole organs may be regenerated.[1] Even the mature individual has a limited capacity to remake destroyed operational circuits, to redesign functional paths, to restore the damaged complex of motoric proficiency, and lastly to reincorporate it into the overall functional schema. The virtuality of self-repair constitutes one of the greatest assets of individualizing life.

II. LIFE'S PROCESSING THE CONTINUITY OF ITS INDIVIDUALIZING ADVANCE WHILE WORKING THROUGH THE INTERACTION WITH EXTERNAL FORCES

With the unraveling of the inward capacities of life, we have reached now the culminating specifics of life, that which makes it stand out from everything nonliving: first, there is the inwardly established universal specific of the transformational metabolic processing of ingested elements and then the capacity to reproduce.

Here the virus represents a transition for it can only replicate itself within a living cell of an appropriate kind, outside of which it is inert. But the next simplest manifestation of life, the one-celled animal has the complete set of life's specifics. Within the nucleus of the cell is its DNA molecule having all the genetic material that it passes on when it divides. The devices of this virtual material define the species. How essential to the logoic system of individualizing life this material is becomes apparent when the nucleus of a cell is removed. It then dies. This indicates that generic replication is the intrinsic rationale of generation as such and a condition of life itself. In higher forms of life, though, which reproduce sexually, removal of the reproductive organs does not mean the death of the individual.

Reproduction is, as we have already mentioned, a key device ordering life's progress from generation to generation. As primed by the initial impetus of life each individual contributes to the common gene pool the effects of its interplay with the forces it encounters giving particular shape to the ontopoietic trajectory through the sedimentations of the process of adaptation.

And so a virtual capacity to advance in complexity is actualized. Replication by means of hereditary virtualities progressively brings into effect ever more specific devices for handling changing environmental conditions. This is at first imperceptible, yet if conditions are favorable for long-term translacing continuity there occurs a generic advance in complexity, there is installed an inward/outward oriented functional pattern defining a new type. At the present stage we have a series of hierarchically ordered types of living beings ranging from viruses and one-celled animals up to human beings, who enter into the workings of nature with their own proficiencies for deciphering its signs.[2]

This particularization or speciation enriches the organs of the enactment of life, and that enactment projects further and further circles of sensitivity, emotivity, interactivity and accounts for greater and greater "awareness" and "participation" — sentient, emotional, sympathetic, interactive — and lastly, human creative entry into their very workings and their intelligible objectified representation. Hence, our creative act, at the climax of this speciation and the accompanying expansion of complexity may reach life's workings themselves in their depths.

Here we see again the logoic strategy in projecting its constructive devices amid a tension of opposites: in discrete continuity progress in speciation moves from the extreme of universality/generality/diffusion to the other extreme of specification, concreteness. At the stage attained by the human being we may ask whether there is still in us a generic virtual opening toward further constructive unfolding, or whether with the human creative entry into the constructive processes of life, which can now extend to the very processes of generation and manipulate heredity, human inventiveness will not direct the unfolding of the further phases of the constructive advance of life.

Nevertheless, as much as we may now inquire into the origination of life through the study of its processes, life's initial impetus remains a mystery.

III. EVOLUTION: OVERARCHING LIFE FROM WITHIN

Since theoretical scientific research has in its descent into the workings of nature proven itself pragmatically in technology, we in phenomenology may draw on creative evidence — and science is to a great degree descending creatively into the workings of life — and admit the insight gained in genetics into the generic unfolding of the complexity of beingness through the transmission of specific features from individuals to their successors. Therefore, in this hypothetical genealogical descent into the very origins of the progress of life, we must ask after the very first, simplest self-individualizing beingness, and its originary, primogenital virtual endowment. Since, on the one hand, the quasi-individual that is the virus is already diversified in virtual specifics such that different kinds seek out and are at home in different types of cells, the question of the initial set of virtualities for unfolding is pushed beyond that point. If we assume with genetics that a) it is the content of the nucleus of the cell that contains the virtual elements for the generative specifics' unfolding and being transmitted from one generation to another, and b) that these elements of the progress of life consist of nucleic acids (DNA and RNA), that is, the so-called nucleotides that are components of what we consider to be biological molecules, then there is no clear borderline, if any, between life and otherwise inanimate chemical material whose virtuality consists merely in entering into various compositions that manifest special propensities (e.g., coenzymes with the addition of phosphates may serve as messengers regulating the activities of the living cell); and then question of the initial dynamic structuring of the initial impetus of life is posed on the level of a stratified inorganic chemical composition. In this composition, then, resides the secret of life's origin.

The crucial strategy of life's continuation lies in coupling an individual's procreation with a translacing ontopoietic strategy that insures the "sameness" of "offspring" and their bearers in such a way that the continuity of life's projected progress is simultaneously a flexible but univocal ordering of life's establishment. This ordering implemented by a concrete contingent system of operations carries an overarching ontopoietic device: the deciphering of a program that is implanted in the plan of generation as such and that it follows.

We are discovering here, indeed, *two of these overarching ontopoietic devices* that are "incorporated" into the system of individual procreative growth from its very initiation. These are: a) *the entelechial design* of the rhythm of unfolding of the cycle of growth in its four phases. It may be flexible in concrete implementation with respect to its scanning and spacing, it may be delayed in its phases and aborted in one or other way, nevertheless it is blindly implemented even as circumambient exchange is allowed. The entelechial design carried within and reaching without in its embodiment in the vital/animal bios brings into the emergent fabric of universal life progress a basic ordering with relevance to the cosmic laws and the natural resources available. The second ontopoietic device is of a more specific and yet crucially strategic significance: it is b) *the device by which procreation reaches over the particular individualizing progress so that it amounts to the "replication" of the type of the self-creating individual.* That does not surprise or astonish us. To the contrary, it is what we always expect, that a ewe, for instance, of regular empirical necessity will deliver offspring that are sheep.

In speaking of the entelechial, we are using a term that has associations with Aristotle and not only with him but with Leibniz, and in modern philosophy with Hedwig Conrad-Martius, and in contemporary science with the biological conception of the living organism of Hans Driesch. It is this last that calls for some comment. Driesch criticized the mechanical conception of the organism for its being incapable of accounting for the "potentialities" of the cells by which an organism builds itself from within by its own forces, and which participate in the overall "harmony of the organism in complying with its various needs for adaptation or even for the restitution of damaged or dysfunctional parts." This manifests an inward "capacity" (*Vermögen*) This inward capacity Driesch sees as a totality that is unchangeable (*Unveränderlich*) and perduring (*Verharrlich*). He writes "Aber die verschiedenen Züge oder Seiten oder 'Eigenschaften' der Entelechie können sich in verrschiedenen Zuständlichkeiten befinden, im Zustand der Latenz oder Ruhe und im Zustand der Aktivität oder Betätigung."[3] It is obvious that the entelechial principle within our conception of the organism also plays the role of maintaining the inward perdurance of the individual identity and autonomy amid the inward/outward traffic as well as changes in nourishment and transformations in the segments of the entire organic apparatus. Furthermore our concept of the entelechial principle, like Driesch's entelechy, is supposed to account for the harmony of organic functioning, as well as for the continuity of species. However, in our conception of the constantly unfolding organism, the harmonies maintain merely concrete ad hoc functioning and may then rely on the organization of functional segments or organs in their self-projected operational functional network; but in the first place the self-projection of the developmental line of this inward/outward functioning in its cyclic phases and in its maintenance of "typical" as well as "individual" coherence delineates an itinerary of becoming such that the steps of the appropriate advance are "measured" from within, being indispensable, because the outline's inward adjustments in functioning are apportioned.

In other words, the "entelechy," or better, "entelechial principle," cannot be a "substance," a concrete part and parcel of the functioning; it is there in the most intimate fashion but rather as a principle of regulation, a principle that is itself adjustable and to a certain point flexible and changeable.

To bring together our preceding and forthcoming reflections, let us emphasize that *the differentiating/compositional logos and operational/intergenerative rationality appear to prompt the emergence of the animus through the sentient logos of life.* Everything seems to acquire full meaningfulness only with the outburst of life through which the Logos enters with a firm foot into the limelight, nay, throws its own light into the darkness, projecting a scene for the spectacle of life within the reach of its rays.

The vital animated Logos, the intellective triadic-noetic Logos, the communicative Dionysian (feeling/sharing) Logos, the Promethean (inventive) Logos, the course of all of these we will follow here all the way to their manifestation in the Grand Vision of Beingness.

IV. THE COINCIDING OF LOGOIC CONSTRUCTIVE LINES

If beyond the enjoyment of living and the struggle to keep ourselves afloat in the life stream, we may still approach life at some "distance" philosophically in order to maintain our determination to steer our own course in the distracting, captivating flux, it is because through the human creative act we enter into the very becoming of the life process — the forging of its significant lines — and we may intuit the basic instrument of its ordering, the self-individualization of living beingness. I have presented a panoramic view of it in some detail elsewhere. I will here simply review some of the most significant moments of the performance of this wondrous means by which the Logos expands itself through life.

The self-individualizing process through which the Logos differentiates itself in innumerable ways in the life stream begins with the emergence of the biosphere from the pre-life realm, with a "move" from an inward-outward oriented station toward the relentless passing flux. We have seen an intrinsic entelechial principle become articulated into a consistent outline of functions that carries and directs the constructive unfolding of life along an existential route on which the living-beingness-in-becoming keeps itself afloat and pursues its constructive advancing course.[4]

We will now follow succinctly this delineation of the individualization of life, which manifests the prodigal ingenuity of reason.

V. THE SELF-INDIVIDUALIZATION OF LIFE AND ITS LOGOIC DEVICES

1. *Physis and Bios. Generative Processes — the Seminal Resources of the Logos and the Emergence of the "Entelechial Principle"*

Is there any sharp distinction possible to be drawn between physis and bios? Is not the set of processes, their relevant resources, and their proportionate setup of intrinsic conditions directly relevant to the conditions of life? That is, had physis not set up the proportions of climatic variations as well as the mineral content of soils, distinguished various degrees of solidity and fluidity as well as the varying reactions of solids and liquids to temperature, had physis not held in reservoir germinal/seminal elements from which the generative processes could unfold in appropriate environments, there would have been no advent of life. We may judge this from the example of our earth, once an inorganic planet from which emerged organic life. How was it possible given moisture's extreme sensitivity to temperature that it was from this very element that a spark of life was released?

2. *Life over against Non-Life in the Ontopoietic Perspective*

Life stands out from non-living constructive processes by its intrinsic telic principle of the reproduction of its forms, which generate other individualizing lines of becoming.

This is owing to the ontopoietic intrinsic operational system that projects itself in appropriate circumstances one way or another and yet maintains a set of operational mechanisms. So that we may lay down clearly some characteristics that absolutely distinguish life from non-life, let it be said that there is no demarcation line between them; on the contrary life draws, feeds itself, existentially relies upon, and ultimately has its origin in non-life.

The antecedent characteristic of the ontopoietic intrinsic life vehicle is its inward-outward exchange of vital resources. All others to be distinguished are rooted in and follow from this one. *First among these is the metabolic exchange system,* which works out of an inwardly directed line of operation that constitutes a point of reference for the distinctive selfhood that the individualizing progress draws on and is directed by. This selfhood operates by the self-enactment of all functions; they are all the inward vehicle of the individualizing course that draws on the outward and adjusts to it.

The second characteristic of life, self-correction, emerges owing to this inward direction and inward operation. Indeed, there is never any fixed range of virtual operations to be maintained regardless of whether external conditions are propitious for their deployment. Depending on whether external circumstances affect the course of life in this or that way, internal functioning might be delineating or destroying a functional mechanism. The inwardly directed mechanism is capable of deviating from its course and taking adjustment measures in order to lie in another bed as it were; it can repair damaged or blighted fragments of its equipage in a way that keeps its functional operations intact; it may thus be brought to deviate from its course and enter onto a slightly shifted one. Thus the adverse may be taken in and changed external conditions participated in without life's being at any point — within a specific range of virtual transformations — destroyed because of a rigid operational schema.

This self-repairing virtuality is essential to life's ontopoietic continuity amid changing conditions, transforming and yet maintaining life through generating different types.

With the question of "type" we enter into *the third defining feature of the ontopoietic unfolding of life — the generative continuity from an individual to its offspring.* The replication initiative from within the living individual does in itself constitute a radical characteristic of life. Though this initiative inevitably draws on forces in the environment, it is not haphazard. It does not lead to the production of just any type of living beingness. On the contrary, here is the pivotal rationale of the logos of life: the living being replicates its own type. Herein lies a new more complete and specific conception of "sameness" and "alterity"; "sameness" is sameness of type.[5]

The further, more elementary characteristics of life are all to be understood in terms of the inward-outward functional complex of the individualizing process and its soil, the fertile ground on which the process draws and which it accordingly fashions; all are modes of this interchange, which prompts further operations and development.

When we attempt to differentiate "that which is life" from what is not life, we ask ultimately about the main constructive streaks that carry the self-individualizing progress onwards in their interplay. There are five streaks of the vehicle of life that manifests its presence, its distinctive presence in its kindred but nonliving environment.

We have just distinguished two concrete operative complexes in which numerous specialized operative systems of a contingent nature function in tandem. In addition, there are three preestablished setups of organization and functioning that are intrinsically encoded and are followed by the concrete embodied systems of operation through the sedimentations of life.

These three crucial operative complexes are 1) photosynthesis; 2) so-called metabolism, the distinctive double-faced inward/outward exchange and processing of material for life that orders forces between the self-individualizing life and its circumambient circuits and from which unfolding beingness grows, the engine of the individual life in life's "macrocosm"; and 3) the intrinsically unfolding functional system of self-reproduction, procreation, by which the individual produces out of its own resources and capacities another living individual. Thus is the continuity of life seen to, for the expanding of the macrocosm of life.

Metabolic processing transforms what is appropriate into life's resources. This operative system is composed of a network of operational segments, each of which performs a special function through a specially developed operational knot — organ — with all of them being tightly imbricated in an orchestrated sequence of operations. There are three main lines in this transformational process: the first being the direct intake of the circumambient atmosphere, of air, with the crucial absorption of oxygen through the entire respiratory and circulatory systems. The second is the absorption of light, and of heat and cold. The third is nutrition obtained through so-called metabolic processes.

The metabolism proceeds by its own rhythms, which while established within the organism are nevertheless dependent in their fluctuations not only on the concrete specificity of the functioning organs but on the conditions of a favorable "supply" from without as well. In this, the rhythm of metabolic functioning is orchestrated with those of the second main operative system of the individualization of life, the reproductive system.

This system of ingesting, processing, and transforming, of absorption and rejection runs through all the circuits of life's complexity. It is a pattern followed not only in the vital — natural — circuit of life's self-individualization, but also in the psychic circuit, in which we not only fulgurate with feelings, emotions, etc. welling up from within, but take in, work through, and absorb or reject, sometimes according to our "needs," "preferences," and sometimes depending on our "weaknesses," or "strengths" when it comes to fending off "invading" elements. We thus "nourish" ourselves on other people's actions, words, gestures, etc., from the mass media, from books, from the general "intellectual climate," and work through that material, either inventively refashioning it and throwing it out into humanity's common pool of thought and art or rejecting it indifferently, defensively, or inimically. This exchange is the vehicle of human personal relations. But, first of all it is the means of our psychic "sustenance," for a human being cannot live without communication and personal growth.

This feeding on elements from "without" and processing it through our own ideas, feelings, etc. in a practical or creative way is the nerve of societal and cultural life as well.

At each of these circuits the inward/outward exchange, the transformational absorption/rejection apparatus projects its own way of scanning/spacing through its rhythms. At the vital/animal level of complexity the lines of sustenance and growth proceed with direct relevance to the cyclic movements of the cosmic solar system, "synchronizing" as it were their own rhythms with it; in the psychic/personal circuit, the parallel transformational apparatus synchronizes with the rhythms of others; in the societal/cultural circuit it synchronizes individual processes with societal/cultural interaction cycles and rhythms.

Although it is specifically in the digestive system that nourishment occurs, that is, the absorption of newly refashioned synergetic material by the relevant "pores," there is parallel nurturing/nourishing interplay with the environment in all our systems: the air we breath, the cold/warmth we feel, the humidity and aridity we experience, the light and dark are of crucial significance, especially for plant life, simpler living beings, etc.

Despite the fact that life seems to be suspended upon external conditions, since the specific modalities the life system are implicated in the environment and its sustenance hangs on it, this is not so in the same way in which inorganic formations depend on it. On the contrary, the nurturing/nourishment of individualizing life is measured up by an inward agency. The type and degree of intake, transformation, and absorption is measured up by the needs of the living being's inward unfolding of its cyclic phases, its specifically developed "organs," its acquired habits and tastes. A further striking feature characterizing life is that the processes of nurturance find a response throughout the entire system of functioning proportionate to the role they play for each other and the whole.

The measure of nurturing/nourishment is thus the most intrinsic measuring system of life. It is played out in the innermost core of the individualizing process and stretches into the proximate spheres of ingrownness with other living beings within the unity-of-life.

The substances processed in digestion aliment life's individualizing progress. In its three phases: ingestion and decomposition in ways relevant to the end of nourishment, transformation into needed substances, and absorption of appropriate amounts of them, the alimenting system of the individualizing functioning accomplishes a crucial role with respect to the unfolding, sustaining and growth of the constructive progress. We have in mind here what is called the system of respiratory regulation, which maintains the intake of air indispensable for the functioning of life; then there is the photosynthetic system, which is equally crucial for the nourishment of plants and some one-celled animals; and then there is the most complex system called metabolism, which sustains and carries on the main exchange of forces of the inner with the outer field of life upon which the individualizing process is suspended. The digestive system becomes more and more complex with the advance in complexity of animal individualization.

The digestive processes intrinsic to the individualizing beingness were prepared for by a series of operational circuits that began already with the first limning of an interior sphere of beingness and its autonomous taking off on its own. In fact, the material to be processed into "nourishment" proper has first to be "recognized" as relevant at large for the specific kind of living beingness. The grass and other plants of a meadow have to be recognized by the rabbit, goat, sheep by the "surface" objective manifestations caught by the organs of perception: smell, sight, touch. Taste is even more discriminating. It mediates directly with the singular individual.

And taste prompts ingestion, mastication, and the swallowing that brings food into the visceral core of the transformational/digestive processes. The circuit is completed by the absorption of synthesized fuel for life's synergies and growth.

3. *Vital Relevancies*

The "recognizing" integral to the absorption/rejection of fodder is already a bridge between the sphere of ingrownness and the spread of the unity of life. In the perspective of the wondrous unfolding of the logos, this is an undisrupted continuity harmonizing the qualitative and quantitative ever more tightly. Finally focal processors appear that we call "organs" (the stomach, the liver, etc. with their subsidiary digestive system; the brain, with its nervous system extending throughout the body; the heart, with its blood conducting circulatory system; the respiratory system; etc.). Strikingly enough, all these seemingly singular systems come together and are indispensable to each other. This logoic life-sustaining apparatus that is uniquely each individual's at one and the same time delimits the living being within itself and makes it partake in external life resources; it involves the individual in processing those resources into material appropriate for the individualization process.

We witness here the intrinsic law of the logos manifesting itself in a game of opposites: the meaningfulness of one horn of an opposition depends on the other, but in their opposition they go into one another, they do so in an infinite gradation like light and darkness, moist and dry, strength and weakness, etc. This holds for the direct play of opposite qualities. But it seems as if there is an analogous law of "substantial" transformability from one chemical ingredient into another.[6] Contrasts are yoked by the transitions between them, but contrasts have networks of family resemblances, so the passing over is most complex.

No wonder that from the beginning of Western thought the question emerged: What are the further untransformable, primary "substances" from which all springs forth. From the heart of the logos there springs forth the differentiation that sunders as well as successively unifies in harmony.

Yet what is most striking in this harmonious interplay of the differentiation of the logos though life is that at the level of the organs of a living being the working of each is intimately interconnected and all are subsumed under the highest authority, the brain with its nervous system that reaches into every segment of operations giving and receiving mute signals, signals that then are relayed to the functioning of the respiratory and circulatory systems, to all the very specific systems of organs carrying life's processes.

TYING POINT ONE

THE MANIFESTATION OF LIFE THROUGH THE NATURE-LIFE COMPLEX AND ITS RADIUS

TYING POINT ONE

NATURE-LIFE

I. THE MANIFESTATION OF THE LOGOS OF LIFE THROUGH THE COMPLEX OF NATURE — ITS CORE SIGNIFICANCE FOR PHILOSOPHY

After a period of considerable neglect the question of nature has reappeared forcefully as a focus of scientific as well as everyday reflection on life. This is because the culture of contemporary society has been very much informed by technological advance and for that very reason society now finds itself at a breaking point. Under the guise of discussing the environment, natural resources, ecological questions, even interplanetary investigations, we are at bottom asking this question: What is nature?

Now, we could say that all great issues that philosophical doubt raises and philosophical marveling at the world of life, the universe, nature, and human beingness provokes have already been raised and that in these so-called postmodern times we really only corroborate them. But if we grant that philosophical inquiry always responds to the life situation of a historical period, of a given culture and society, we cannot maintain such a view. In our times such râdically new types of sociocultural situations have arisen as provoke novel doubts and induce novel kinds of marveling that shake hitherto valid explanations down to the very questions that led to their formulation. Philosophy confronts new points of reflection in perspectives hitherto not encountered or simply not clearly seen as being relevant to human concerns.

New existential-practical problems emerge, and the human mind is tested by the challenge of formulating them and seeking conclusions. The philosophical treatment of them could, of course, limit itself provisorily to the clarification of situations and issues. Yet in such attempts we deal only with the surface of the situation and do not gain any clear and distinct foothold in the subject matter. Without such a foothold, we cannot participate in the quest for an adequate formulation of the novel situations in which humankind finds itself in its existence in the world-of-life, the biosphere, the cosmos.

Recent philosophy has given up on the vocation of providing such a foothold. With its radical "dismantling" of the logos — of foundation, truth, absolutes, ideas, the ideal — it slides over the surface of things, just given the relativism to which such a renunciation leads. Rejecting critical forms of rationality, contemporary philosophies seem to rely exclusively on "common sense," but in fact, they go radically against it. I attempt to establish in my metaphysics of the manifestation of the logos through the edifice of life that there would be no world, no life, no human beingness, and no possibility of them without the universally valid relevant systems of the logos to which the processes of the world of life refer. Without some meting out of the forces that unfailingly provide the structural devices and without the intergenerative synergies of the logos of life, there would be no life. Furthermore, without some universally valid schemata for the logoic junctions that are maintained in our equipage for participation in the world of life, we living beings could not make the distinctions that are indispensable for our basic survival and life-enacting operations. This is the evidence of common sense. To ignore it is to launch out upon the clouds.

Nevertheless, it has to be conceded that the classical formulations of issues, and also of the nature of 'foundations,' of 'necessity,' of reason itself, seem to have to a great degree been outpaced by the radical developments that, after having matured for centuries, have come into full light in recent decades; the very philosophical decadence of the various forms of relativism quite appropriately testifies to this. And yet the great intuitions of the ancients, of Plato, Plotinus, seem to coincide with those that I will develop.

A novel starting is required by the acuteness of the contemporary situation, and, in my view, the most promising one is that offered by a new inquiry into the ontopoieisis of life itself through the access that we gain to it via the creative sources of the specifically human differentiation of the logos of life at the primogenital point of the Human Condition as a station in the evolutionary radiation of the logos.

This new inquiry into the systems of relevance operative in the ontopoiesis of life with their infallible selectivity (in life's broad advance, if not in every instance), universal transformability, generative and intergenerative resources and their laws, tension between excess and moderation with norms for apportioning, systems of relevance that serve reason, rationality, logos, the logos of life, while providing a new conception of what has traditionally been called 'foundational' philosophy, has to meet the test of being able to accommodate the perspectives, discoveries, intuitions that have arisen with the novel existential situation of human beingness within the network of life, nature, earth, cosmic forces that throws new light on reason.

In what could be attributed to the excessive prodigality of the logos of life in its creative human differentiation, it appears that human inventive ingeniousness in bringing on an enormous advance in the deciphering of the rules of cosmic and natural forces, allowing their manipulation, has broken the timeless pattern of surrender to nature and gone beyond the equipoise established through millennia of life between nature and human beings and between the gifts of nature and their use by living beings.

This equipoise having been shattered, the vitally significant belief in universally lasting and unchangeable factors of life, in the unshakable evidence of life conditions on earth — in the enduring and indubitable certitudes of natural endowment — is replaced by the prodigies of the inventive logos, which plays with the rules of life itself. With the loss of the natural faith of the human being in its world of life and in its natural position among living beings, there appear new hopes and expectations given the stupefying inventiveness of science and technology, which gives the impression of an endless advance in the perfecting of the natural endowment of living beings through the transforming of the conditions of life. But the marvelous spontaneous unfolding of the creative logos that has allowed the human being to enter into the mysteries of nature, of life, and of the cosmos has uprooted the vital experience of the stability of life. Today's thinker is struck by the urgency of the need to understand the vital facts, which suddenly after so many millennia of human civilization have become enigmatic and lost their foundational role so that human beings are now at a loss to envisage their own existence and future on the natural soil of the earth, another planet, or who knows where?

Struck by our own inventive powers in relation to natural conditions, we have lost the experience of our proper place and role, and the moral motor of praxis is pushing us to the wall creating an existential urge to take the situation in hand. I am talking, of course, first and foremost of all the issues that have emerged with the stupendous unfolding of genetics, ecology, nuclear science, computer technology, etc. Again, it is up to philosophy to undertake the task of a new investigation into nature.

But in what sense is it a new task?

1. *The Need to Formulate Anew the Issue of "Nature"*

In as much as individuals have lost their compass in the play of the natural and inventive forces of the unfolding logos of life, the central issue of theoretical and practical science, as well as of society and culture is precisely the question: What is nature? Other questions follow in short order. What are the natural forces that human genius intrudes upon and refashions according to its own intentions? What is the balance to be struck between the prerogatives of nature and those of human existence? Lastly, what is it that human beings have to confront in their existential exigencies now that they have become fully aware of their own inventive/creative virtualities?

Even that contemporary philosophy that has not succumbed to skepticism and relativism does not seem prepared to meet this challenge. We are still heirs of the modern hiatus between nature and culture in philosophical topography.

It is Husserl's last work, *The Crisis of Culture,* that has sharpened the phenomenological stand on all these matters. Husserl — along with the great majority of phenomenologists of all persuasions and scholarly disciplines — brought his vast inquiry into the human powers, human rationality, human cultural/scientific/technological unfolding to the culminating concept of the human lifeworld. When all is said and done with respect to the human universe of conscious constitution of the objective world, and this in the various stages of human unfolding along scientific and cultural lines, Husserl sees the great difficulty into which the present world has got in its obvious estrangement from what is called "nature" and with the loss of a firm point of reference for human individual existence given the "artificiality" with which throughout centuries of scientific-cultural development we have imbued, covered, and perverted our natural, ingenuous relationship with the world of our life.

We have to return to the original stage from which the world and our relationship to it took off. From the artificial, unduly rationalized world of culture, we have to retrieve the supposedly underlying world of life.

Husserl's and his successors' search after this primordial lifeworld, the peeling back of the sedimentations of rational cultural foundations — hence, hermeneutics as an outgrowth of phenomenology — really aims at reaching the border between human constitutive activity, the specifically human takeoff, and rudimentary, elementary, prehuman Nature.

The search for this border is, however, upon careful scrutiny, dubious if not altogether misdirected. What would the human lifeworld be before the human faculties that transformed it went into action? Could there be a world before the individualizing process of life? In phenomenological discussion of human historicity and of the questions of the lifeworld, especially of the differentiation of the lifeworlds according to geographic conditions, the development of civilizations, and individual-personal development, one would expect that we might ultimately ascertain a primary, primordial givenness in exterior reality that would bring everything ultimately together. However, it seems contradictory to expect to find that everything goes back to any primary world or lifeworld had in common. The very idea of a human world means already a humanly constituted environmental self-enclosure of the living individual. It is precisely through the different types of such self-enclosure that human civilizations diverge. Thus we may get very close to the border between seemingly independent nature and the human being's constitutive activity, but we cannot really reach it. We cannot through the notion of the world or lifeworld establish the generative, grounding mutual relationship between what we call external nature and man's internal nature.

We need another key to open the gate of the seeming rational wall between them.

2. *A New Formulation of the Concept of Nature-Life, One Open to the Cosmos and to Culture*

It is with the same key by which the human genius opened the Pandora's box of natural forces and their operational rules resulting in the inventive outburst that then upset the equilibrium of vital/existential forces that we can open the gate between the seemingly separate realms of nature and culture. This key is the creative/ inventive act of the human being. By taking the investigation of the creative/inventive virtualities of the human being as the starting point of philosophy, rather than the cognitive act, we enter into the heart of the ontopoiesis of life in which both nature and culture are situated.

The human being as inventor/creator emerges from within the system of vital forces and concurrently disrupts and distances himself from it, destroying his essential ties with it. This system of vital forces 1) carries the self-individualization of life, 2) accounts for the unity-of-everything-there-is-alive, and 3) points to a system of existential relevancies to the cosmic logos with its laws, sets down rules, and prompts synergetic resources toward a functional establishment for founding and carrying on the evolutionary progress of life in all its variety. Among the forms of life, the human being appears to have attained the most advanced complexity as the free, that is, creative differentiator of the logos of life.

There seems to be nothing new in recognizing life as the crucial significance of nature. Aristotle, Leibniz, and Hegel in succession conceived of nature in a way that quasi-identified the essence of nature with life. Both Aristotle and Hegel situated the concept of "nature as life" at the center of their philosophical systems. Both insisted that nature means "all living beings." In this there is nothing new concerning the concept of nature. Certainly an urgent question today is that of the totality of living beings. However, the actual urgent situation of humankind brings out new perspectives from which this totality of life can and has to be envisaged.

In fact, the emphasis now falls not on the totality of life, which Hegel could still consider as a totality closed in on itself, but, first on the vital modalities that allow the differentiation of living beings from each other in their becoming and then on the modalities of interchange necessary for maintaining them all within a vital/existential network. This amounts to saying that the emphasis falls on the individuation of living beings in their existential progress and on their co-existential interchange in their advance.

Scientific inquiry now focuses on questions concerning the ontopoietic relevance of the spontaneities and forms of living beings to a set of three correlative systems: a) the system of rules, schemas, etc. of the generative and intergenerative forces and of the seminal resources from which the individualization of life may emerge (the "biosphere"); b) the system of cosmic laws, and resources upon which the planet is suspended, since the planetary situation cannot be omitted from the extended conditions of life, for life stands and falls with it; c) in the opposite direction (that of the unfolding individualization of life, with its generative and evolutionary rules, the rules of "coming to be," of entering into vital connections, and of passing away through the loosening of vital ties in order to leave space for other living beings to come), there are the rules, forms, patterns, and generative forces of the logos of life itself, which once it takes off, establishes the status quo of life with a system of provisions for its unfolding, for the constructive advance of its types, for its continuing progress.

In other words, in speaking of the "totality of life," we can no longer consider nature to be a "closed" system. We can only think of it as a web of concentration of the vital synergies and intellective forces of the logos of life that is suspended upon several circuits of forces, laws, rules — those of the set of systems just covered that must be correlated if life is to be — as well as upon several systems of relevancies.

Furthermore, the perspective that approaches nature through life does not even ask: What is nature? It is not a universal concept that is asked for. The clarification that the radicalization of human nature's involvement calls for is: What is nature when envisaged as the vital milieu of all living beingness?

Emphasis now falls on the interrelations, reciprocal exigencies, the interchanges that in the perpetual flux are made, loosened, and dissolved among living beings as well as among the modalities of forces and relevancies of the biosphere and the cosmos through which living beings differentiate their living forms tying further webs of existential interchange, webs indispensable for the generation, growth, flourishing, and fruition of each living individual.

The dependencies of generation and interchange might basically refer to the vital significance of life and then to cosmic laws. However, they also lead in the opposite direction with a certain loosening of the vital significance of life for the sake of acquiring several circuits of the specifically human significance of life. Indeed, nature in carrying the burden of life's unfolding enters into the specific dependencies of the intellective circuits introduced by the human condition into the universe of life. From the vital circuits oriented exclusively to survival, we pass by almost imperceptible gradations to the — while life sustaining — not vitally but intellectively (morally, aesthetically, socially) significant interaction among human beings. This is equally supportive of the vital progress of life and the "natural" unfolding of the living human being. We have just shown how crucially important this infinitely gradated and nuanced cooperation of these two lines, so differently significant and yet most intimately bound together in their tasks, is for the unfolding and sustaining of life in our denunciation of the breaking up of this cooperation in contemporary culture and the radical menace to life itself that that constitutes.

The task of philosophy with respect to the critical situation of humankind today, although it has come to light only recently, brings us genuine enlightenment about the cosmos, bios, and the human being — an enlightenment that constitutes a crucial break from the tentative searching of the philosophy of the past. To pick up the challenge presented by the sciences as well as by the vital concerns of humankind and to formulate a novel conception of nature-life along the lines of the above-outlined ontopoiesis of life is to indicate philosophy's new parameters.

II. THE WOMB OF LIFE: NATURE-LIFE AS THE SEMINAL/GENERATIVE GROUNDWORK OF LIFE'S VITAL SIGNIFICANCE

The lines that design the routes of the logos of life as it reaches out and branches off into innumerable avenues circumscribe its central vitally significant zone. This zone has its core in seminal and generative synergies that, once activated, surge and prompt the vital/organic significance of life to unfurl. In prompting individualizing life on its course, these primary forces acquired organic/vital significance, first, by providing seminal resources, and, second, by providing nurturing resources also. They, thus, constitute a "womb" for nature, the nurturing soil from which the seminal virtualities emerge in individualizing (form projecting) constructive outlines, and from which they draw (and continue to draw during the entire course of their growth) background stimuli, spontaneities, energies. The functional system of individualizing life in its constructive progress returns to this reservoir of life forces to retrieve propensities needed to redress or compensate for capacities twisted, weakened, or lost in the process of growth.

However, how should we understand our metaphor of the "womb of life"? In an analogy to the "chora" of the story that Timaeus narrates in Plato's dialogue, "womb" suggests a "receptacle" that is the locus within which elements for pending generative processes gather, confront each other, and join to initiate the generative process. But in contrast to Plato, our thesis is that this generative outburst initiates time as well as space. This amounts to saying that the womb of life as a receptacle is neither in time nor in space; it cannot be a site, a locus. Yet it is certainly the moment — an atemporal one — of an "encounter" of "unattached" elements, "free" forces that are endowed with both the seminal load of primeval generative, material and an impetus to crystalize that material in one or another way according to the available conditions for reaching a measure of equipoise in an encounter and "click." This click is not an event, for it is outside time.

The spermatic logoi are first in line and then concomitant subsidiary logoi. The logoi occupy no space nor are they in any temporal network.

They are neither entities nor beings, neither living nor nonliving. Anticipating our analysis of the origin of space and time, let us state that "they are."

This is the "absolute prius" at the threshold of temporality, which will lead toward the setting up, as if in a "preter-time," by energizing its cumulating encountering moves, marking steps in what I would call "the primal generative matrix" of life. Essentially, the generative process, its incipient phase, is a play of logoic forces endowed with seminal virtualities and seeking to coalesce constructively among themselves according to the law of attraction/repulsion that governs the logoic dynamism; among other things these virtualities seek to acquire an equipoise between the motion that the forces in play initiate and that then carries their play and the stasis attained in the constructive fulfillment of an "encounter" in which tasks become defined and accomplished. Thus, although motion is the conductor of the incipient and ever subsequent moment of the generative progress, this motion is itself determined by its directional (constructive) vector, by the seminal "material" that is the logoic forces, and by the constructive task virtually referring to the appropriate logoic schema. Thus the motion as well as stasis within the generative processes and within all ontopoietic processes assumes in its conducing function innumerable modalities expressing the innumerable nuances and kinds of sense that life projects in its exuberant abundance, all spun in this preter-time of life by the surging confrontations of the generative logoi driven by their inherent impetus in search of the "fitting" generative "response" among themselves.

But what about the generative "material" of the womb of life? I submit that these are the spermatic "substantive" logoi proceeding in line together with the first logoi. The first logoi bring in the generative synergies, dynamisms, qualitative content, and the driving force to bring all into the constructive process; the second logoi mediate between and among the generative elements and the atmospheric conditions to mark the meeting points in their attunement. The generative encounters occur in a "preter-space."

What appears to us, on our ontopoietic analysis, as the metaphor of the womb of life is a generative phase, the setting down of incipient elements, and the crystalization of the crucibles within which the self-individualizing processes will begin on their course.

It is the phase in which the selective "material" still waiting-in-readiness is energized by the very coalescence that prepares the first stirring (as Plotinus says about the radical beginning). To a large degree this energizing of the waiting-in-readiness "material," that is, of selective so far inactive but virtual logoic synergies, takes care of the preparatory degree of activation that will at an appropriate point of completion constitute a matrix of generation. This matrix preparing for the self-individualizing process bursting forth thus consists of the selective seminal, logoi as well as of concomitant logoi appropriately called upon to mediate between the seminal and constructive proficiencies and the conditional elements calling for adaptation or attunement in order that the novel self-individualizing outburst may be received within the already available system of life. The completion of this material is crowned by the triggering of the "first stirring" out of its own intrinsic virtualities. Within our perspective, which focuses on the self-individualization of *life* as the absolute criterion of life, becoming and beingness, this matrix of generation is not considered (in contrast to the ancients) to be the site of the all-becoming, of the entire spread of the macrocosm.

The criterion of self-individualization indicates not only the further features of this matrix but also different zones of life's generative-concomitant conditions and spread. First of all there comes what in science is called the "biosphere," its "nurse," to use Plato's term for the womb of life; second, there comes the "web of life"; and third, the vital relevant system of the cosmic logoi. We differentiate them according to their more or less proximate involvements with the generative matrix. It is unavoidable that at this point we mention similarities and contrasts with the conception of the generative matrix proposed by Timaeus in Plato's dialogue. Plato's "chora" matrix is, as it is for Plotinus after him, a "site" in which takes place the origination of the entire spread of becoming, beginning with cosmic formation and descending toward the world of life. It culminates in what Timaeus considers the "world soul" that then spreads out the whole of becoming. It is this living world soul that is the active principle that brings about the first stirring.[1]

Should we consider our generative matrix to metaphorically be a "world soul" of life, it would stretch only so far as stretches the direct self-individualizing ontopoietic process. It would not include the cosmos, which in our perspective is essential to the self-individualization of life as a zone of relevancies for the fashioning of the originative molds but does not participate directly in the generative process; it participates in it only by the intermediary of the concomitant logoi.

The reason that Timaeus evokes for assuming the living world soul to be the principle of life and becoming is that, being the matrix of all becoming, the "chora" as the shaper of formless material must be an amazing "all-embracing" intelligence.[2]

However, on our analysis, the forming and molding that occurs within the generative matrix of the womb of life is endowed with intelligence as ordering proficiencies within the material being molded. There seems to be no need to assume a special agency of intelligence to effect this; the spermatic and concomitant logoi of which this material consists carry within themselves their constructive devices. One has to recognize the logoic, that is, intelligent self-sufficiency in the articulating, ordering, measuring, molding procedures.

At this point, however, a crucial complex of problems emerges given the contrast between the concrete unraveling of the hidden generative moves that the logos itself performs and the speculative cosmogony of the ancients. For Timaeus the generative matrix of the *chora,* the receptacle and the "nurse" that enters with this special intelligence, is a special site, a special sphere of reality situated at the center of the net of becoming, on the one side of which lies a model, unchangeable and intelligible, and on the other side of which lies the concrete, visible and changeable reality generated thereafter.

The *chora* is thus a special site in which the generation, the modeling takes place, a middle term between the model and the copy. It is seen then as being formless, plastic, and as such capable of receiving the imprints of the molds of the models That is to say that the *chora* as the generative matrix is a special type of reality between the timeless and unchangeable and the contingent

In contrast, since we attribute the molding function of the multiplicity of concrete beings to the logoic nature of its self-realization in life, to the logoic elements carrying within themselves virtual modeling according to their coalescing opportunities, the question of the model assumes for us a secondary position. We see the generative matrix as a self-patterning of the logoi falling together into a constructive coalescence of forces, and as a consequence, the womb of life, this matrix or patterning network, cannot be either a separate reality in itself, nor can it be it be seen as a reality or a sphere of the logos as such in isolation from the generative processes themselves. As a matter of fact, it cannot be considered to be either eternal — as Plato had to see it — or temporal. And the generative coalescence of forces, logoic virtualities, is not a temporal phase because only with the initial stirring of life into a self-individualizing process does temporality set in. We could grant it then a pretertemporal status insofar as it has precedent in engendering the temporalizing event of the generative self-individualizing course. It is the phase within which the temporal is being prepared. That amounts to the fact that the temporalizing stretching forward also stretches "backward" postulating an in-itself pretemporal conditioning circumstance.

We have to return to the problem of the model again. We proposed that the modeling of changeable beings comes from the entire logoic system of crystalization in the self-individualizing process. As a matter of fact, this modeling is in concrete terms nothing other than the crystalization of logoic virtualities within a common constructive pattern. It is a spontaneous activation of the inward constructive proficiencies of the logoic material; however, this material delineates a design that then conducts the unfolding and growth. We may wonder what the status is of this ontopoietic design, or as I have álso called it this entelechial design, in performing the indicative, directive function within the concrete work of the womb of life.

Is the ontopoietic or entelechial design "pregiven" to the generative process? Or does it emerge with the coalescing of synergies within the generative matrix? It seems the latter, that the logos of life is a *postulating* device.

On the basis of the generative evidences that indicate in a postulational way a prime position for an ontopoietic directional, vectorial factor, we may infer priority for the ontopoietic design. As a matter of fact, we have recognized the seminal logoi as synergies bringing into a constructive play of forces content/quality as well as the logoic impetus to crystalize their virtualities. Each logoic force brings with itself its own contribution to the constructivity.

Yet the ontopoietic design that they all follow in tandem in attunement and interplay does not use up its constructive vector prerogatives with any single click of the self-individualizing process that initiates it. As will be emphasized below, the self-individualizing process is directed by its inner self-fashioning devices. But this design itself is multiplied in an infinite number of cases; it is a representative of a "type," or "species." As I have indicated elsewhere, within the evolutionary progress of life as such, this design holds the regularity, relative stability of the individual process, and as such would appear to be static, immovable, unchangeable, even though it is not. First of all, it itself participates through its tentacles of synergies in the development of the living being. Second, it is being "transmitted" in its vectorial and regulative role to these innumerable instances of the type, but within its "transmitting" pro-generative phase, it undergoes the influences of the pro-generative factors. In this pro-generative inheritance, as it were in process and under the pressures of concomitant elements with respect to changing life circumstances, it undergoes a transformation as well. That is to say, not only does it take part in the evolutionary unfolding of the individual but it also undergoes transformational changes in the multiplication in the evolutionary progress of life.

III. THE SPONTANEITY OF GENERATION AND ITS CRUCIBLES: THE ONTOPOIETIC DESIGN

Coming back to the metaphor of the "womb of life," the receptacle and matrix of the generation of life, and before we enter into a further descriptive vein, we have to raise the question of the generative logoi again in another perspective. First of all, the status of the logoi needs further reflecting upon. In the Timaean perspective, the generative work of the *chora* necessitated a "model" and a "cause." From this highest point of the logoic spheres, as we will outline in the following pages of this work, it is life itself that times and spaces itself in its infinitely intricate spontaneously thrown designs of the routes traced by the self-individualizing course.

In ancient metaphysical speculation generation was set forth as involving an unchangeable, unmovable model, and a mover, the demiurge. The movable, changeable, generated world proceeds from these. The exemplar is assumed to live eternally.

Our perspective, taking the creative phenomenological insight as its lead, begins, not with an overarching metaphysical scheme, but precisely from the statement of life. It painstakingly follows the lead of life's arteries as those of the logos of life, leading to its workings and its conditions. On this path, instead of universal principles, we come to the point of recognizing the most singular spermatic logoi as the synergies simultaneously bringing content and quality into the game of existence as well as the impetus to crystalize it. In this view the question of the "model" of the generative work does not seem to appear. Each logoic force brings with itself its own constructive element.

Yet, the encounter/click within the generative matrix of the womb does not stop at the singular work it has carried out; this work is just a station meant to initiate, to ignite the surging of the life process to which all the singular works contribute: the self-individualizing progress of beingness.

At this point, the question of the Platonic "model" after which the generated work is fashioned comes up again in a different guise. As has and will be emphasized again, the self-individualizing process is not an unfolding proceeding according to the whims of singular preter-moves and developmental phases; on the contrary, its moves are directed, but not by being fashioned after an outside model, but from within. What we have here is an ontopoietic sequence, or entelechial sequence, enacted step by step by the singular moves. The ontopoietic sequence appears to surge into the arena of generation at the crucial point at which the singular pretertemporal operations trigger the surging of the dynamic project of individualizing life as it detaches itself from the womb of life, cutting the umbilical chord, as it were, to navigate upon the stormy ocean of life as compact, self-individualizing beings.

We come to this intriguing question later on. For the time being let us emphasize that it is the ontopoietic sequence that orients the generative moves surging within itself. Yet the surging of the ontopoietic sequence is not an event; to the contrary it is its appearance as an active force in the arena of the generative web of life that initiates motion, change, happening, event, occurrence, temporality. The progressive unfurling of its route is its enactment. This enactment is, however, a self-enactment, a realization of intrinsic virtualities in reality. It determines the individual as much as the "type."

But the issue of the recurring generative crucibles or forms is not solved. These generative crucibles reappear from time to time in transformed states. They must be accompanied by a range of the "possible." Beyond that range would occur the breaking of the type. What is the furthest limit for the logoic forces and constructive devices in the generative symbiosis of the incipient agents with their endowment for generative formation and their attunement with circumambient conditions? Beyond what limit of their fulgurating schema of constructive progress will the vital impetus of self-individualizing propulsion no longer find a proper response and have to abandon the task? This would amount to the disappearance of the type, or of the species.

IV. PLATONIC ESSENCES REWORKED — THE ULTIMATE ROLE OF THE LOGOS

In fact, if we follow the models for the constitutive structuring of givenness, essences, or *eidoi* intuitively given by the transcendental mind and extrapolate them into the logoic ontopoietic design as a regulative structural device for both the ontopoiesis of life and for the transcendental logos of the mind, then, to use the terminology of structural ontology, we could see this limit as being circumscribed by the "constants" of the essence, even while made open to change by its "variables."

The essences or *eidoi* seem to appear to be, like Plato's and Plotinus' generative models, atemporal and eternal and thus to carry with them the stamp of "necessity." And essential eidetic organization of concrete reality in its phenomenal manifestation as well as the mental construct of "essences" and *eidoi* are undeniable. However, I have attempted to bring to light throughout the entire unfolding of the logos of life — as well as of the universal logos — this essential core dominating the molding of the ontopoietic unfolding of the manifestations of living beingness in complexes of logoic dynamism with full logoic constructive endowment that command of necessity only the variation that finds within their prompting and dynamic impact an appropriate response, a basic "fit" with the available conditions. (See my monograph "The Moral Sense: A Discourse on the Phenomenological Foundation of the Social World and of Ethics," in Anna-Teresa Tymieniecka (ed.), *Foundations of Morality, Human Rights and the Human Sciences,* Analecta Husserliana 15 [Dordrecht: D. Reidel, 1983], pp. 3-78.) There is an "intelligence" in the logoic generative complex that effects an equipoise between the two contrary tendencies: that of a driving force toward shaping and forming and that of strict implementation of abiding logoic rules. It appears then that there is no "necessity" outside of the becoming, that the logos alone is the absolute factor of becoming — and of being — and its ultimate principle. It proposes and initiates, it performs its task, it prescribes the limits to its own dynamic strivings. Within its "intelligencer" lie all the possibles and all the actualities. All is projected, prompted, initiated, and ordered in innumerable negotiations with other logoic projects and elements, and carried out within the compass of the logoic powers.

The tension between freedom and necessity lies within its innermost nature: the absolute freedom of the impetus to prompt versus the regulations for ordering.

There are the tensions between the contraries that set the limits to the ordering as prompted by the impetus; they outline the possible frames for the self-individualizing design. Thus we cannot put forth "eternal" ideas or *eidoi* ahead of self-individualizing, or design ahead of the logos of life, as "models" to be imitated or mirrored. It is the logos of life that applies its own modeling — molding — rules as the measure of its dynamic spread. But is there a universal logos prior to its own dynamic unfolding?

We may, however, seek the ontopoietic design of the individualization that makes the process in its temporal unfolding participate as much as in the permanent logoic design of virtualities as in the virtualities' being in their concrete crystalization adumbrated, adjusted to the relevant conditions as a principle of "modeling" in another way, through all conditions. In this process the modalities of concrete individualization remain in the processes of change and transformation. While the core of the constructive design persists through all the variations of concretization, this latter may then be considered to be both the activating as well as constructive, modeling factor within the temporal life process. It may also be seen as being "outside" of it in its permanent endowment, which passes from one cycle of generation, growth, and decay to another, serving as both the activating force and the model.

V. THE ABSOLUTE AND THE INTRINSIC TRANSFORMATION OF DESIGN. THE QUESTION OF ETERNITY

To say so amounts to raising the question of just what the status of an ontopoietic sequence that pertains to the changeable as well as to the perduring is. Does it also perdure beyond time? If it did, we could then consider it eternal, that is, without beginning or end in time. However, most interestingly, as we have seen in our analysis of the ontopoietic design and the evolution of types that occurs within the advancing, cyclic unfurling of life, although they perdure in their core, ontopoietic designs undergo intrinsic transformations at the margins following a spiral line. It is the logos of life that is here in question. Its crucial factor, the ontopoietic sequence, is certainly not "absolute," but may we not grant it the status of "eternity" insofar as its entire life course in time is concerned? Is there, however, such an entire life course to be even considered? Although the intimate mechanism of life is cyclic, that is circular, yet — in contrast to the views of Plotinus — nothing ever returns the same; the seasons of life return, which means that the same cyclic pattern of becoming is implemented by the logos of life, but the definitive course of singular beings is altogether irreversible, its cycles moving ever onward. The vector of life's unfolding is the future, which is an open horizon. It is in this open, undetermined horizon that the nature of life in its temporality consists. This brings us to surmise, what we have already substantiated (see "The Womb of Life," present chapter, Sec. II, *supra*), that we can not with Plotinus speak of a "complete revolution of life or the world," of an entire "cycle" or *aion*, because should it have an end, it would not be measured by criteria of time, but by those of the logos, the primeval, universal logos. Indeed, we have come to the point of considering that sphere of the logos lying between the primeval logos and the logos of life.

Turning back from these speculative distances to our concrete reality of life, let us recall what a wondrous sight is the spontaneous sprouting of new plants from mother earth in early spring, at any possible vacant place! The dance of vegetation, the symphony of nature. Indeed, each plant stands on its own appearing as an actor upon the stage of the world of life to initiate its play with steps meticulously choreographed, following the sequence of a virtual script, drawing on it like an actor on his/her memorized directions. Yet although they stand on their own and get from within their very individual set of moves to perform, they are not insensitive and separated soloists, but bend to the moves of other players, who attune their own moves to theirs so as to participate together in the logoic performance, in the dance that the logos of life stages and enacts, initiates, and guides. The logos of life appears in the great spectacle backstage; there it works and manifests itself

VI. MOTILITY, BECOMING, EVENT, CHANGE, CONTINUITY

We have reached now the pivotal device of the universal logos in its constructive grand project, as well as the crucial device of life — motility.

Prima facie, we recognize life through the perception of motion. Yet this bearer of motion that we identify as "living" has, first, to exhibit motion that is in accord with a rule, a constructive design, or to some fit with the circumambient world, in short a logos. Moreover, in order for this motion to characterize its bearer as being alive, it has to proceed from its very own inward constitution and impulse. We extrapolate the movements that we observe onto a vague but coherent network of the inner or circumambient conditions of its bearer.

In the case of the generative matrix, the primary moves seek appropriate response toward a constructive encounter, and once this response is found, they come into their own as moves of confluence, absorption, etc., that is, as incomplete moves. Even these incomplete moves integrally refer to a logos, a project, on the one hand, and a constructive network with its rules, on the other. The project would then abort, and the network dissolve. A parallel but inverse situation may be seen in the disintegration moves of the transformational matrix. The disintegrative moves follow in reverse those of construction. They emerge from the inside of their bearers and are moves only until those bearers are still endowed with some, be it the simplest, coherence and its logos.

Thus movement in its various modalities is the intrinsic pivotal device of the logos of life, first in its individualizing, generative project, then in its entire project of cyclic life turnovers, of generation, growth, decay, and dissolution, a turnover guaranteeing renewal, distribution, and evolutionary advance.

The basic characteristic of primal motility is to make a passage between the inward and the outward. In this way it accounts simultaneously for spacing and timing but also for the coherence of the entire logoic grand project as well as for the peculiar discrete continuity of ontopoietic individualization.

I have here in mind the discrete continuity of the various realms and levels of the functioning in a fully developing individual in spite of the different modalities of sense that they imply — whether organic, vital, sharing-in-life, existential, or intellectual, of primary and secondary consciousness, of the motility of the body with its kinesthetic system — bringing them all together into a unity of "body and mind."

However, motility, motion, is also the thread that ties up our lifeworldly existence in the grand project — in its own dynamic turns and system of ordering.

It may be said that it is logoic distribution of types of motility in its innumerable types of creations that pervades the entire grand plan, that we living beings feel resounding within our existential-vital vibrations the reverberations of the cosmic moves of the stars, the planets, the nebulae, etc.

Not without reason does humanity attempt to scrutinize ties with the stars and planets in their presumed influence on the turns of our existence, our personal dispositions, our moods and destinies. Unwittingly, but most judiciously we feel that our daily life with its rhythms of night and day — activity and repose — participates in the great movements of the skies. They constitute basic, elementary principles of our earthly existence, of life, of our world of life. We may even feel like puppets dancing on the strings of the grand system of universal motion, whose telos and modes we ignore. They escape our loftiest grasp.

It is movement that ties all together. We will yet come to this point. (See "The Embodiment of the Logoic Life-Dynamics and the Phases of the Conversion of Sense," Part Two, Ch. 1, *infra.*)

For the time being let us turn to movement's two crucial modalities: life's timing and spacing. From that perspective we will analyze movement again.

VII. THE PRIMAL TIMING AND SPACING OF LIFE

Even with respect to the transformability of the ontopoietic design we find the tension of the logoic contraries. In one direction with the current of the logoic impetus of life and its advance, it tends toward transformational changes; in a contrary direction, that of maintaining the relative stability and the fundamental order of life, this impetus is arrested at a limit of its pattern of basic ordering, structuring, maintaining the blueprint of reality. Life's advance is progressively arrested in steps upon a spiral line. Indeed, it is in virtue of this core of the ontopoietic design, in virtue of this entelechial core resisting transformability, that the entire grand plan of life is devised. Otherwise, we may say, the entelechial design is the regulative principle of the entire constructive drive of the logos of life, though not an external model to be replicated.

Now, since we have entered into the heart of the issues concerning temporality and spatiality, we have to prepare groundwork for it in treating these two devices of the logos directly. The timing/spacing device of the logos may be considered its primeval device; however, it seems, as I will attempt to substantiate, to diversify according to logos' constructive spheres into the macrocosmic, stellar timing/spacing of becoming, and the specific timing/spacing axis of the microcosmic becoming of life. It is this latter that is our central concern. Along the timing/spacing axis of the generative matrix of the womb of life are situated all the issues of generation that we have been discussing so far. As mentioned above, it is with its life-promoting moves that the logos initiates the timing and spacing as inverse moves are carried out upon the timing/spacing axis. In this perspective timing and spacing are a concrete reality, the axis intrinsic to the most essential core of the matrix of generation. This amounts to saying that as life's vectorial and regulative axis timing and spacing emerge together with the first stirring of the self-individualizing process. But from its very advent we have to postulate conjecturally that this advent is not prepared by covert, invisible real moves in the full sense of empirical reality. In these preparatory pretermoves of the logoic material in their mutual discrimination of appropriate propensities and their selective coalescence of organic prevital forces, the matrix of generation performs its basic work. How can we in our perspective focusing on the individualizing process consider these moves otherwise than as participating vicariously in the ensuing spatiotemporality of becoming and still being invisible, hidden, unmanifested not belonging directly to the spatiotemporality of manifestation? We will call this borderline situation, pretertemporal.[3]

Yet should we not distinguish this pretertemporality and preterspatiality of quasi-moves and operations performed by spermatic logoi in the generative matrix from that of the persistent core of the entelechial design? In both phenomena there lies the subjacent device of impetus being checked, arrested, mitigated by the rule of equipoise. The primary, originary forces and energies of the logos that press onward and compete for inclusion within the constructive life network are stepwise mitigated by their own compatibility or incompatibility and the external life conditions. However, in this search for equipoise the persistent core of the ontopoietic virtual design is sustained by the grand constructive plan of the logos. This is a crucial device of the basic ordering of life. The invisible, hidden, premanifestation operations of the generative matrix have no organizing vector as precise as the central entelechial design. Their ordering, albeit proceeding from their own forces and propensities, occurs as a function of the entelechial design, is suspended upon it. In this sense the pretertemporality of the virtual existence of the entelechial design is of the primary sphere of the universal ordering of life, while that of the pre-initial, preparatory moves operations belongs to the postulational sphere of the manifestation of life.

VIII. THE LOGOS IN THE COSMIC SPHERE: THE GRAND INTELLECTIVE DRIVE

So far we have been focusing exclusively on the generative matrix of life. But what about the generation of the macrocosm of the myriads of galaxies, stars, nebulae, etc. of which our planet is but a tiny speck?

What about the cosmic system per se whose fixed "east to west" motion we witness along with the motion of the wandering stars and the streaks of comets and the fall of meteors; and what of all the movement in the cosmos as a whole system that the tools of astronomy have revealed, the expansion of the universe to infinities that even our mind cannot follow? It seems that we experience and measure the motion of the stellar system as much as its spacing in relation to our life systems, finding in the motions and placings of the cosmic system that includes our planet earth explanations for the nature of our own life's becoming.

In exploring the varying environments of different galaxies, stars, planets, science prompts us to surmise that different planets might possess "times" and "spaces" different from the time and space we know here. Time and space there might display quite different features. In our perspective, where there is no life in one or other form, there would be no timing and spacing proper, that is, no timing and spacing as we experience it on our ground and in our situation within the cosmic orbit. So if we should still assume there to be "cosmic space" and "cosmic time," they would belong to the intentional extension of and be the absolute prius to our human view of the spheres of life. Now in the perspective of contemporary science and astronomical technology, it is in fact evident that the timing and spacing of the stellar expanse is available to our human experience as being part of its context insofar as it lies within the circumference of the laws of gravity, is warmed by the radiant heat of the sun, is swept by the solar wind, is subject to the impacts of comets and meteors, and, indeed, the earth's own gravitational field, which holds our atmosphere. But what is beyond our earthly experience is altogether alien to our rules for moving in "space"; the sidereal distances have no proportion to our time-space axis with its ontopoietic regulations.

The ancients used the image of a dome to picture the heavens. Today they tantalize our imagination, which falls short of envisaging its extent. We witness our intrinsic ties with our life-system with its sun, moon, and stars in motion. We realize that there are intrinsic ties with the expanse of space and our life system. It is with these ties in mind and even owing to them that we see the starry firmament seemingly motionless and yet regularly on the move. And this very own sphere with which we are tied up is itself an integral part of this incommensurable expanse with multitudes of galaxies, nebulae, each having stars with systems of circling, rotating planets having moons, with systems disintegrating as well as being "born." Each is a simile, a version of our own system with its spacing and timing axis of life. However, these are far off similes. Each carries its own "measure" of spacing/timing, of disintegration and coalescence. Plato, like the ancients in general, saw an intelligent factor in the regularities of the heavens, the entire system of which he pictured as constituting a dome. Plato, and Plotinus after him, identified that intelligent factor with the soul. The movements of the heavens were interpreted as signs of life. However, in our perspective, which takes self-individualization to be the criterion of life, we deny that the heavenly bodies are living or that they possess individuality. No doubt that there appear to be self-enclosed centers of energies; we can identify some insofar as they correspond to those forces that enter into our earthly orbit. Yet we may not reach beyond this.

There is an overall grand system of forces, laws, regularities displaying intelligence that holds it together and accounts for its varied measures of timing/spacing, its generation and disintegration. And certainly we may expand and extrapolate from our conception of life's generation, surmising that out there too there is the question of a generative matrix, womb for coming to be and passing away again into the transformational reservoir of the logos. Nevertheless, it seems that the entire sphere of outer space could not be an integral part of the generative life matrix as Timaeus saw it. In contrast, the generative matrix of the cosmos appears to belong to the constructive grand project of the logos of life. We recognize it as a special logoic sphere with its own laws and regulations, as a gigantic network of constructive forces within which our life system appears just as a grain of sand in a desert does to us. However, this tiny grain of sand appears to us who are involved in this context to vibrate with infinite dynamisms that are so wondrously articulated and so vigorously move onward leaving us always a step behind, as the very nucleus of this gigantic network of the buoyant forces and energies of the constructive entanglements of becoming.

The cosmic sphere in its furthest extension from this planet and the solar system appears to be an indispensable playground for the universal logos in its specific undertaking as the logos of life. We may say metaphorically after Plotinus that the logos in enacting its very own creative impulse is the *choragus,* the master of this cosmic dance that it plays on the stage of the cosmic spheres as well as the "backstage" that is the womb of life. It sets up the scenery and directs the greatest of plays: the life play stretching from coming to be, through the evolving rhythms of the cycles and stages of life, through the struggles for existence, to the weakening of spontaneities and the exit from the boards to the "backstage," the return to the transformational web.

IX. THE DISINTEGRATING TRANSFORMATIONAL MATRIX

The generative process we are attempting to describe is as continuous/discrete in its progress as life itself. It is enough to leave aside the abysmal question of the beginning of life, of the earth, of the cosmos, and just consider what we are in: the onward flux of generations of living individuals, an epic enacted through the cyclic nature of individual life, the seasons, stellar movement, etc. Seen in its entire spread, whose beginning and end remain unthematizable, we could generalize by saying that the process of life at large is that of transformation. Nothing remains the same. Everything at a given instant is being transformed, broken down, and processed into the elementary material of logoic forces.

They remain as a "reservoir" of material forces, substantial energies that the generative processes of the womb of life draw on. It may be seen as a subsidiary and yet indispensable factor of generation, going before it at least one step. The reservoir of transformation is both pretertemporality and real life temporality. The reservoir of transformational material is not just a passive "deposit"; on the contrary, as has been pointed out numerous times, each logoic element is endowed by both the stimulus for activation and a set of propensities allowing a constructive encounter. That is to say, this reservoir also constitutes a type of matrix in which the transformational processes that will occur in the time of the individual's unfolding is held in the waiting position to be activated, as a comet grows a tail as it approaches the sun.

These two matrices, the generative and the transformative remain in incessant interplay. The first draws on the second for supplies as much as for vital generative stimuli; the second, receiving its task from this interplay takes up its work when the life course reaches extinction. It is no wonder that philosophers pondering this wheel of transformation and rebirth were tempted to see it as an "eternal return." It is to be hoped that our concrete analysis clarifies this issue. Nothing returns the same as the great wheel of change turns.

Both of the matrices share the timing/spacing axis with its margins. Both of the matrices share the spontaneities of seminal and concomitant logoic material, substantial, dynamic shaping, on the one hand, and the constructive axis of time/space, on the other hand. Yet their relation to the temporality of becoming is different. As I have pointed out, the generative matrix is pretertemporal. The transformational matrix, in contrast, dealing with disintegrating material that has impaired motility moves along a weakening life's temporality till its total eclipse by the post-time period. This is how the two matrices appear to be interlaced on the logoic canvas of life, between coming to be and passing away.

At their intersection lies a logoic "territory," pretertemporal and preterspatial, that reaches deep down into the generative womb. This "territory," which might be considered the reservoir of the logoic forces, seminal and concomitant, as well as of materials all ready and waiting for the appropriate conditions in which their tendencies may unfold, for the implementation of their spontaneities, for being situated in networks leading to their crystalization. The matrix of disintegration and transformation shares with that of generation the crucial spacing/timing logoic axis of constructivism; upon this they meet.

In this fashion the great plan of life finds a means of accommodating the impetus of the logos of life in the equilibrium that meets the advance of the becoming. The two matrices present two contrary flows by which is put into motion the "wheel" of becoming that proposes ever new beingnesses; what emerges through the one is progressively evacuated from the scene of manifestation in reality by the other. Thus an equilibrium is being perpetually established between the ever further driving forces of the logos of life and its actual crystalization on the stage of reality. Most significantly this equilibrium remains proportionate to the elementary groundwork on which the stage of life is erected: proportionate to mother earth, which has its very own logoic ordering and set proportions among the proficiencies of its elementary logoic groundwork for the constructive enterprises of the logos of life.

The sphere of the interplay between these two matrices guarantees an equilibrium, curbing the excessive drives, compensating for the losses. In short, it takes care of the "economy" of resources. In fact this equilibrium has as its measure the proportions of the earth itself and the system of life on it. Hence although some species disappear in the evolutionary course of life, though civilizations disintegrate and exit the stage, though human beings wage incessant wars against each other and sometimes wipe out entire societies or groups, these are replaced by new incoming species, civilizations, societies, and so life continues. It is the law of the logos of life, in step with its nature, to advance while retracing its steps, to advance in constructing while destroying in proportion to its means.

Nothing is lost. All is retrieved in an encounter that clicks and makes a step toward the further deployment of its new, already transformed material. Thus, it is in this preter-time of the logos of life within the circuit of our earth as a planet of our sun, a planet having its very own atmosphere and atmospheric conditions, what we may call this system's own "station." It stands within these atmospheric conditions, where are had the basic "elements" that account for the deployment of life: earth, water, air, light, and heat and other prerequisite cosmic-atmospheric conditions — what is called in scientific terms a 'biosphere.' It is in virtue of all this that the generative pretertemporal quasi-moves may emerge and the generative matrix of the womb of life be established.

The biosphere works as the nurturing nurse for the womb of life in its throwing out and setting up the web it makes from the elements and other cosmic factors attuned to generation. We have anticipated its role and prerogatives within the life context. Let us summarize them.

1. *The Vitally Relevant Web — the Biosphere*

Life's forms emerge on the planet earth from the germinal soil-nature called the biosphere, which comprises not only the concrete "soil" that serves as a constant reservoir of life-convertible resources spread over the surface of the planet, but the entire schema of specific elementary conditions that allow planet earth to become the site of germination and sustenance. In order for life's generative forces, synergies, virtual seminal factors, the primary endowment of nature at the so-called prelife level, to be brought into generative interplay so that the entelechial program of life be instituted, a specific set of conditions has to be maintained: temperature, moisture, and for most forms of life, light.

2. *The Zone of Life. The Existential Relevance-Systems of Nature-Life: Reaching Out to the Vast Cosmos and to the Cultivation of the Human Spirit*

As has been emphasized all along, this seemingly compact complex of nature-life is not closed in on itself. On the contrary, it implies two major systems of relevancies extending in two opposite directions and upon which it is suspended. Its circuits are only partly involved with these systems, and yet it is basically dependent on them: the cosmos, on the one hand, and the human spirit, on the other. Indeed, the forms of life that individualization unfolds point to the system of cosmic relevancies in one direction. And the symbiotic networks of vital interchange reveal at their summit the human condition, the creative act by which is instituted the novel life-significance of the spirit, which is nestled in life like a butterfly in a chrysalis; in the network of circuits, it links all the living circuits together.

It is through a cosmic network that elementary conditions — warmth/cold, moisture/dryness, light/darkness, etc. — are apportioned and an appropriate balance established, such that allows life to take off. The atmosphere is a network of forces and synergies reflecting cosmic laws with the forces in play being of proportions suitable for and vitally relevant to nature's germinal nurturing conditions.

Just as the germinating and nurturing systems point to a system of vital relevance in the cosmic realm, so in other ways does the individualizing constructivity. This we will look at later.

The deployment of seminal and generative synergies implies their going hand in hand with appropriate circumambient conditions. It is obvious that these quintessential germinal factors of life are not self-contained or self-sufficient. On the contrary, they rely in the crystalization of their virtualities on subsidiary factors; through the exigencies indispensable for their "animation" or "activation," they indicate dependency on networks of vitally relevant factors extending beyond the core of the generative milieu. This web of interdependencies projects an expanded zone of life. There are the climatic factors that are indispensable for life's germination and nurturing. There need only be maintained the propitious mix of the contraries of warmth and cold, moisture and aridity, light and darkness called for by some measure of the laws of life itself.

3. *The Universal Life System*

The unfolding of the entelechial principle's design through the self-individualization of life is not only the central but also the centralizing factor of all organically activated energies, and of the environmental networks upon which the germination, sustenance, and growth of living beings are suspended as well, for life, as astonishingly adaptable as it is in its constructive becoming, is by no means self-enclosed.

To the contrary, in all the processes that take part in the core of life's nature and in the networks of differentiating exchanges between living beings, the living individual is oriented toward the life system as such. It is upon the universal life system that the individualization of life is intrinsically suspended. The cyclic phases of growth — which encompass the essential phases of the individualization that invariably accompanies it — manifest in their various typical processes the system and laws of life. The living being is born and grows. If it reaches the apex of the fruition of new life, it must then decline and die without exception.

4. *The Web of Life: the Unity-of-Everything-There-Is-Alive*

I will echo and paraphrase Heraclitus, "One is All." We may apprehend this as meaning that the life-schema of nature seems to manifest the absolute necessity that the living individual be inserted in a web of life, that it be what it is owing to its insertion in the "All" of life. Once again, the very individualization of life is a differentiation of an autonomous, distinctive life route that occurs within and without a web of other living beings and their existential ties. It is within these ties that differentiation projects and within those into which it inserts itself that the symbiotic exchange of organic-vital and vital-intellective elements among the beings caught in individualizing becoming proceeds. This symbiotic exchange in its various circuits — first, those of vital energies and virtualities; then, those of affective gregarious links; then, those of psychic-empathic circuits; and lastly, those of empathic-intellective circuits according to the various levels of constructive complexity of the living individual — insinuates the becoming that individualizes in ways so intimate that to disrupt them would be to tear the very "flesh" of beingness, would be to slash the innermost arteries through which life's juices flow and be the cause of extinction.

The symbiotic exchange among living individuals reaches the apex of its vital significance in "procreation," in which exchange all of the symbiotic lines meet. Then, the "incubation" of a new, germinating and form-unfolding life in the womb — in another individual being — is again a web of symbiotic/sympathetic linkage in every circuit of life.

Symbiotic linkage and interactive exchange throw up the web of life — as it were, a nervous system for life among other things — and make the unity-of-everything-there-is-alive, without which no living being would come to exist. In this sense, "One is All." Given this state of affairs, we may see also the validation of Leibniz's intuition that each monad reflects all the universe.

Thus, as it begins in the germinating womb of life and radiates into the divers circuits of vital relevancies, the extension of nature-life brings to light the intrinsic, essential unity-of-everything-there-is-alive. Since this unity is suspended upon the individualizing differentiation of singular living beings, it allows for this individualization. As these beings integrate themselves into the common pool of energies, forces, virtualities, synergies, in the germinating soil of the womb of life, individualizing beingness inserts itself into the circuits of intergenerative pulsations, constructive processes, and common symbiotic vitalities. Hence, we experience nature as a seemingly compact sphere of life, without gaps and disruptions.

X. THE VITAL RELEVANT SYSTEM OF ARCHITECTURAL LAWS OF THE IMMEASURABLE COSMOS

The insight that we have gained into life reveals to us life in the making, constructive making, at that. This means that the progress of life is not a formless and haphazard flux of energies. On the contrary, it is the differentiation of the logos of life itself. If we speak of "differentiation," we imply differentiation of forms. Briefly, the existential fusing and interchange of primary energies, the synergies of the entelechial principle, the very impetus of the germinal synergies of life, that is, its innermost workings in the womb of nature, aim at the constitution and the following progressive transformation of some form, an extension in which life will manifest itself.

Form, in its contrast /opposition to the formless flux, is the principle to which the entelechial contructivism of life is absolutely committed.

Form, as the principle of relative repose and perdurance amidst incessant transformation, as end or object of that transformation, is the vehicle of the constructivism of life. Furthermore, it is life's manifestation as much as it is the accomplishment of its endeavors. The ontopoiesis of life aims, in fact, at an exfoliation of forms in nature.

Forms of life — forms in nature — manifest fundamental characteristics; they are of their nature suspended between the postulates of the system of life as such, on the one hand, and an individualizing ontopoietic design, on the other. The constructivism of individualization, the means for introducing an order of ever advancing becoming that promotes the ever expanding spontaneities of the Logos, is the key factor of life, for it satisfies the postulates of both.

This constructivism, which essentially refers to the formal in nature's expression, is founded on a special type of form: extension. Extension is the "volume" and "size" of the concrete beingness that meets the requirements of life. These requirements comprise in the first place, motion/change/transformability/kinesis, that is, change in quality, quantity, substantiality, that is, awakening, generation, growth, corruption, decay, renewal.

Thus, at its origin, life is on principle already suspended upon a system of vital relevance in which the laws of nature govern the cosmic forces and synergies in such a way that extension is the proper and indispensable means for generating and implementing the designs of the Logos through life.

It is "volumed" being that, having received the orienting laws of the cosmos (laws such as that of gravitation by which earth is situated in the solar system), implements them by fitting them to the postulates of the system of life and to the entelechial principle of origin, growth, and transmutation, assuming a fundamental role in nature.

As nature-life sets out on its constructive route, that is, as kinesis with its spacing/timing gets underway, volumed being with its three-dimensional form underlies specific formally grounded kinesis; thus is nature-life suspended upon nature's system of relevance to the cosmic laws that govern the planet.

In conclusion, even if nature-life be, as the classical thinkers Aristotle and Hegel believed, self-generating and even if it contain in itself its final cause, it is certainly not closed in on itself. Both its existential principles and its progress are suspended upon several relevant systems of linkage and interlinkage that reach all the way to the laws of the cosmos.

XI. NATURE'S RELEVANCE TO THE SYSTEMS OF HUMAN CULTURE AND THE CONVERSE. THE SYMBIOTIC/AFFECTIVE/EMPATHIC LINKAGES THAT GRADUALLY LEAD TO THE HUMAN DIFFERENTIATION AND CULTIVATION OF THE LOGOS OF LIFE

It is to the other "extreme" that we will now pass, going all the way to the strikingly different sphere of the specifically human significance of life, of the human inventive/creative differentiation of the logos of life and of its cultivation (human culture). This sphere was gradually fashioned. This is the crucial point concerning the adequate formulation of the notion of 'nature-life.' If we approach it as situated in the context of life, as the manifestation there of the logos as it unfolds progressively and differentiates into a series of types advancing in complexity, we reach progressively — and without an artificially introduced hiatus — the apex of the advance: the inventive/creative differentiation of the logos of life through the strikingly novel apparatus of the human mind.

Indeed, there is no existential hiatus or disruption in the functional lines of the operation of life between the vitally significant circuits of nature-life and the inventive/creative significance of human culture (science, technology, art, etc.).

The drastic differentiation of the logos of life into two different modes of life-significance, based on different principles of valuation and modi operandi does not sunder the grand design of its unfolding. First, this drastic differentiation does not occur abruptly; it is mediated by a gradual differentiation of the modalities of interlinkage. Second, at each phase, the linkages of life's differentiation are carried by the vital circuits that preceded them; it is the same with the furthest inventive/creative development of the human differentiation and cultivation of the logos of life. Third, the linkage systems of one phase conversely bear the promise of a phase still more advanced in complexity, so much so that the specifically human circuits are carried by the vitally significant organic/sentient/sympathic interlinkages that have immediate relevance to the specifically human intellective interlinkages.

The vast network of vital linkages and relevancies comes together in the crucial web of the unity-of-everything-there-is-alive. Manifesting the vibrant workings of life pulsing through all its circuits, this web comprises all types and concrete individuals into an infinitely flexible, movable, changeable, transformable whole that is necessarily tied together for success or failure, for flourishing or destruction.

The enigmatic situation of the human condition within this web — a situation that has given rise to unwarranted views on the autonomy and existential independence of nature and human culture, and to an equally unwarranted existential reduction of all that is original in the human spirit (culture) to vital nature — consists in its existential position within the web: the human condition acts as a sieve through which all the synergies of the vital significance of life flow.

Human beingness, which crystalizes the Human Condition, is existentially dependent even as it projects a new "intellective" significance of life, one autonomous and free of entelechial constraints, one through which new avenues of life are opened and the inner workings of the organic and cosmic synergies in their logoic ordering are cognized and employed to promote those avenues.

This entering into the workings of the logos of life itself may promote or hinder life, advance or destroy it. It is now to some degree left to human genius to superintend the logos of life as it discerns, evaluates, and decides.

PART TWO

EMBODIMENT AND THE TRANSFORMATION OF SENSE

CHAPTER ONE

THE EMBODIMENT OF THE LOGOIC LIFE DYNAMICS AND THE PHASES OF THE CONVERSION OF SENSE

I. THE EMBODIMENT OF THE LOGOIC LIFE DYNAMICS: THE FIRST PHASE

1. *Embodiment As the Core Manifestation of the Logos of Life. Expressing the Primal Constructive Devices of the Logos*

What else could what we call "body" be if not the seeming repose in self that arrests our attention, the "something" that persistently stays "there" and confronts us, a phenomenon that does not vanish or recede but resists our pressures and invites our reaction to it albeit on its own terms?

And what else could this opaque and "substantial" "phenomenon" be but an "outward" manifestation within the context of the world of the "inwardly" operating self-poietic individuation of emerging life?

It is precisely in embodiment, a crystalized "appearance" that at its culmination attains an equipoise, that the strategies of the logos establish their crucial devices, the cornerstones of life.

In the passages from the "hidden" to the "open," from the universal (cosmic) to "concrete" individualizing living beings, lies a secret device: prescriptions for the logoic constructivism — for its relentless striving — which converts into "restrictions" on this very constructive drive, restriction that ultimately aims at introducing measure into its progress, measure that calls a "halt" in its swing for the harmonizing of the "one" with the "other." This basic distinction elicits another primal device of the constructive strategy of the logos, that of the tension between structural opposites: differentiation/unification, complexity/simplicity, singularity/multiplicity, "the one and the many."

The "inward/outward" duality is itself a most complex phenomenon. It is more ramified and extended in its constructive powers than are the "inner workings of life" that sustain it.

As the inner workings of the individualizing life form a self-enclosed functional system that then spreads into the outward conundrum of the play of forces that impact on life — the universal life schema — the functions of this system expand in outward oriented moves that by participating in the environs take in material that provides energy for its own aims when processed, with rejection of useless elements following. What is ingested contributes to the "shape" then crystalized in the living being's outward "appearance," which reflects not only interior dispositions but also "reaction" to external provocations and the impact of circumambient energies (e.g., heat, cold, etc.), according to its own reactivity. The peculiar manifestation of "life reactivity" (the *Erregbarkeit* of John Brown and Schelling) will provide protection from potentially destructive direct impacts by throwing up a "surface shield" keeping external forces from reaching the functional inner system that the living being embodies.

With these preliminary remarks we pass to the expressive manifestation of the logoic constructive drive through embodiment.

(a) The body, as the expressive manifestation of this inner functioning, is first of all a set of instruments by means of which the self-individualizing process simultaneously extends life's constructive spread and establishes a vital platform for the sake of a full-fledged vital operation. It is through these instruments that there is established an extended participation in and cooperation with all that is "other" to life, with the emerging or already established circumambient world of life. By means of bodily shaping the logoic life strivings devise and form further means of the constructive advance. Through corporeal motility the tentacles of the individualizing life stretch far beyond the lines that form the "environmental" conundrum; they enlarge or shrink it, or move towards another vital circuit following what the interplay between the inward functional system and its ambiance calls for. Embodiment serves the aim of stretching out while projecting circuits of interplay that invigorate the initial spontaneities in multiple exchanges, that by meeting challenges gather intergenerative energies for the further unfolding and variation of self-individualizing life.

(b) Embodiment serves the logoic drive by carrying it precisely "onward," that is, by framing irreversible steps, one upon another. It expresses and manifests the logoic by reaching out in a twofold fashion: its spacing grounds the move "onward" within and among bodies on a "surface" or in multiple dimensions. Simultaneously, its innermost "timing" brings forth moves already at work within the hidden self-individualizing operations; these works time the unfolding in embodied shape in the intraworldly operations, that is, the progressive stepwise primal establishment of the "before," "simultaneous," "after" set order of succession in constructive procedures and phases. A fundamental irreversibility marks this progress and lays the groundwork for the motility of embodiment, which thus becomes its carrier. It is then through bodily "instruments" and on the ground of corporeal reactivity/affectivity and the motility that expresses it that processes and events are joined as they mutually impact each other. In this most ingenious strategic device, the scanning of moves is synchronized between interior bodily functioning and the external bodily motility that expands the living being and shapes it into an organically cogent vital unit. This compact and self-reposing unit deploys its ties with the surrounding life system and synchronizes its corporeal motility with it to form a common world of life.

Let us, however, emphasize that the most prodigious phenomenon of the logos of life is precisely this passage from the hidden intrinsic schema of operations that subtend it to its phenomenal "appearing" in the full-fledged individual expansion. In this most complex construction all of the functional lines of the operational system find "expression" in the emergence of the individualized body's rigorously instrumental organs. In this dimension lies the "monadic unity" — to use Leibniz's expression — of the living being.

(c) The "embodiment" of life's constructive/constitutive processes crystalizes/manifests them in a striking way: the embodied life comes forth in a self-enclosed, "substantial," "opaque" fashion.

In fact, by its very emergence the living individual circumscribes and "occupies" a measured place for itself among other common beings, delimiting itself to its proper "size." In other words, the individual "appears" with "volume" and shape. It is through the primal instance of volume and shape that there occurs a "spacing" of life's territory. Hence each and every beingness, or thinghood, is crystalized into a primal form, is "volumed," and this occurs according to the relevant cosmic laws of the universe and our planet earth.

At this point we witness the primordial twist of the logoic constructive strategies mentioned above in the convertibility of its devices; the crux of these is the rule that "prescription means restriction."

Prescription for the embodiment of the logoic striving toward a formal shape means restriction of its exercise of powers in a measured "halt" to the logoic drive with the achievement of an equipoise harmonizing the individual proper with the other, singularity with multiplicity, within a common logoic network that emerges. This crystalization of the individualizing processes with relevance to the laws of the cosmos (e.g., gravitation in a specific sphere) takes striking form in the singular, primeval "volumed being." It sets the entire multifaceted framework for the logoic spread of life even as it sets limitations that circumscribe it. This limitation is self-imposed for it derives from the volume and size of the embodied singularity that constitutes the mainstream of life's domain.

At the primal phase of the logoic establishment of life, the inward prescriptions that follow from the volume and size of the living being are the major strategies of the logos for the implementation of a life system.

Having volume brings the functional system of the singular being into self-enclosure, substantiality, and opaqueness as to its inner workings, as over against its outward "appearance," brings about that appearance itself and its safeguarding surface resistance to anything that is outside of the organic frame. From this set of determinants of voluminosity flow the main and crucial lines of life.

As a matter of fact, and as we have already announced in a preliminary fashion, owing to the implementation of our planet's cosmic laws, this volume assuming crystalization of the logos may already at its original instant unfold its basic devices for life's constructive spread in the scanning of life's moves.

The body's virtual mobility as a distinctive self-relying unit allows it to assume a "dynamic position" among all that is "other." Through the extremities of its members it enters into the play of forces of the outward sphere of life and extends its mass in various ways, transmitting dynamically the telic directions of the self-individualizing effort in the efficacious moments of "actions."

The strategy of the logos in accomplishing life's propagation does indeed install through this volume assuming crystalization of the inner functional system of the individual singular beingness a major vehicle for its inward/outward expansion, extending the inner workings that otherwise focus on their segment of life's operations toward the "external" play of forces thus creating a circumferential sphere of the multiplicity of forms and forces that environs it Furthermore, by extending its own functional tightly organized operational system through its bodily members and beyond its bodily extension in its interplay with the surrounding forces, the logos gains a firm foothold, sets itself up as a station within the dynamic flux, wringing out of it the energies to set a firm course. Thus does the logoic striving attain the existential status of equipoise, its innermost quest.

Thus embodied and in concourse with the embodiment of its surroundings on the plane of existence-in-becoming, the ontopoiesis of life and beingness advances in all its phases.

Having a platform upon the whirl of life's forces, the embodying crystalization of life's functional schema follows its generative progress and begins already its first embryonic phase. Furthermore, the unfolding of the inward/outward oriented functioning of life proceeds along with a progressive unfolding of organs that shelter as it were differentiated functional systems (the brain, seat of the nervous system; the heart, center of the blood circulation system; the stomach, workhorse of the digestive system, etc.) and the limbs that serve as tentacles into the external sector of the world of life.

Inherent to this is a constant dynamic tension between strivings and equipoise, at the heart of which is the living being's reactivity. This reactivity/responsiveness itself becomes a further constructive device of the logos of life, becoming a unique factor in the unfolding of life. There is synthesized an attentiveness that synchronizes all the dynamic states of living beingness, permeates them all, holds all "in sight" and distributes at all times the "signal" to the proper limb, operational segment, organ, etc. their appropriate reaction. Thereby the entire embodied living unit assumes the role of being its own "agent." It serves in the first place to survey the state of the individualizing progress and safeguard the equipoise of its advance. It is an outstanding feat of the logos to assume through its individualizing apparatus a measure of dynamic unity-in-progress that takes into account the special dimension of the equipoise between the interior mobility of the body and its exteriorization in the kinesthesis of its movements and its surface senses. It is the organs of the sensorium in particular that serve the expanded reactivity characterizing life as such. All the activity of the limbs and organs of the body and its entire realm of "sensibilities," that is, specialized reactions/responses to inward/outward provocations, gather as it were, in the awareness of such.

The living being is always attentively engaged in a pursuit to maintain the outward equipoise of its position within the play of the forces that life confronts and conjoins to its project as well as an inward equipoise among its inner dynamisms.

As the complexity of living beings grows at various levels and their powers become differentiated, and as reactivity expands into psychic circuits of functioning and becomes self-awareness, it is the living individual that is the fulcrum of the process. "Processing reactivity" becomes vigilance.

II. VOLUMINOSITY, THE ONTIC "FORM" OF BEINGNESS

Maintaining the traditional notion of Being as an intellectually abstract and imaginatively concrete ultimate factor unifying under the same umbrella all entities, we cannot do justice to the main type of entity, which is living beingness. Under the umbrella of Being, we can see it only as a static segment of an overall universal system.

To this, the "fact" of existence, to use Ortega y Gasset's expression, is radically opposed. We have gone far above Ortega with the creative intuition, leading us through the labyrinth of life's development; we must first assess the living beingness as a geyser of synergies streaming, forces being harnessed into a constructive design while they unfold.

In this perspective, in order to understand what it means to become a "beingness," we have to begin with the argument already made here that the living being, and here we will focus on its most advanced complex type, the human species, reflects in its constructive schema universal fundamental life conditions. In fact in its "voluminosity" — which appears to be the most striking stroke of genius that the logos of life has projected in its manifestation, and which is congenital with the "substantiality" of the living "form" in the triad of manifestation — the living beingness installs itself into a living space that it has already unfolded from the beginning of its generation insofar as its organic, psychic, and spiritual functions although inward bound, are "outbound." They function, draw, interact, exercise themselves in an active, movable, changeable fashion that traces a spatiotemporal spread. Although in last analysis this spread is contained within a neutral/abstract spatiotemporal sphere — the sphere of the lifeworld — this spread itself constitutes a specific spatiotemporal construct. This is a domain for life, as it were, a self-enclosed domain, as if self-encapsulated in a perpetual inward/outward transforming interaction. The living being seemingly reposes in its own "substantial" enclosure as it encloses itself in its own self-made shell even as it remains in uninterrupted interaction with the outside. In this perpetual inwardly and outwardly condition of transformation it maintains its identity through its voluminosity.

And the striking phenomenon of volume assuming embodiment is that it in each and every process of transformation it expresses the vital awareness of the unfolding agent, of its very self. Animals, from the simplest to the human being, are constantly aware. For humans this awareness is the core of self-consciousness. Self-awareness is unjustifiably often considered to be a matter of consciousness, but all living beings, self-conscious or not, are aware in the sense that they are constantly in search of optimal proportions of forces and conditions, are constantly adjusting their "position" to achieve equipoise with other living beings, objects, and forces about them. This positioning goes with the "voluminosity" of a living beingness as such. Whether it is sheep grazing or people shopping, living beings have to be positioned within a spatiotemporal region that has its own positionality. Hence the living being itself is ontopoietically predesigned for interaction with forces, obstacles, energies, nutrients. These all become knitted into a living being's vital circuits, be they nervous or mental.

On the psychic level, I may thereby revive my vital/psychic network by retrieving sensations and feelings that immerse me as a living being in my primogenital soil and that have been in the shadows of my attention until now.

We all know, of course, that we can take this reverse course owing to the strength of spirit we have acquired in societal or spiritual spheres. Part of the value of this exercise comes from our having gained this strength at the expense of some blunting of our sensory acuity and impairment of our bodily strengths.

Recapitulating, we may conclude that the entire ontopoietic itinerary of life's unfolding is the constant effort of living beings to establish adequate equipoise in their situations, first with respect to maintaining the innermost proportionality in the internal functional life system carrying central and basic tasks. These inward bound functions must be maintained in an appropriately measured balance among organs, with growth and procreation being provided for. Meanwhile, external positioning must be established to optimally obtain the benefits offered by the working of cosmic laws, the potentialities of the flesh in its substantiality, and the resources of the environment.

III. SURFACE AND VISCERAL EMBODIMENT, AND NOURISHMENT

1. *The Inner Regulatory System*

There is, however, another twist of the logoic genius to be brought out, one equally crucial for the self-individualization of life's advancing in formative fullness and simultaneously throwing tentacles aimed at rooting it within the play of forces that is the world of life. The inward/outward orientation of the living being's functioning assumes a most striking role in the exchange of materials between it and the constructive schema of the outer realm of life. This exchange of materials is most significant for the origination, genesis, and further deployment of the ontopoietic design. Called in medical science "metabolism," it, first of all, means that after life's ontopoietic initiation and genesis, "nourishment" is needed. Material for growth is indispensable. Thus, the individualized beingness becomes a processor of substances ingested into its inner functional organs, assimilating those that it needs and sending superfluous material back into the outer realm of life. This nutritional system of individualizing life emphasizes even more the ingrownness of each singular life in its primal foundations, considered apart from its integration with world dynamisms.

In this connection we will distinguish between the surface and visceral dimensions of the living body. While the viscera consist of the body's main substance processing organs, they are in communication with the circuits of reactive sensory posts on the body's "surface."

Metabolism is the motor of the individual unfolding and growth. It is the instrument by which the beingness in progress maintains itself upon the flood of forces. It is the means by which the impetus of the initial moment of life's inception is progressively tamed as the living beingness throws out tentacles into the surrounding conundrum of forces and so establishes itself as a relatively stable station therein.

While metabolism aims at performing these three functions — growth, sustenance in existence, and the establishment of a distinctively and organically operative force/beingness upon the flood — the further function of self-reproduction is *sui generis*. And yet the reproductive system's functioning is most intimately dependent on the metabolic system. It is a unique phenomenon of life because it is purposively oriented, that is, its unfolding is meant to lead to another, absolutely new living individual within the world of life. The growth of the procreative functional system goes in tandem with the line of growth of the individual. Its development and health depends on the work of the metabolism and the schema of growth phases. Its fulfillment depends on inward and outward conditions. And yet, it is a system that surges autonomously.

Here we have to pinpoint the schema of logoic equipoise that is the direct processor and qualitative and substantive transformer of energies, the digestive system. In the processes of the digestive system specifically predetermined assortments of mixed material is processed and transformed; "useful," that is, life sustaining material acquires in the process the required qualities for its absorption by the body for nourishment. Irrelevant material is then eliminated as waste. Thus does the logos split the body up functionally while harmonizing the functions in a novel unity.

In the entire progress of the differentiation of the logos in cosmic constructivism we witness a "dissociative harmonizing," a coming asunder and coming together. The birth of a star presents a most striking manifestation of this. However, there seems to be no intrinsic regulative mechanism accounting for it. There is just a random coming together of relevant elements, with their fusing and transformational harmonizing resulting in constructive effects. And only "external" conditions seem to be operative in their harmonizing unification, no inside, processing regulatory system being in evidence.

Life as such manifests itself above all by just such an inward regulatory system station that represents the logoic differentiation/harmonizing impetus at its most ingenious point. Indeed, the inner processing and transformation of substances/qualities/quantities into nourishment and elimination of waste marks a beingness as "living," manifests life over against non-life. In and through this exchange of matter or substances embodied being pulsates and throws roots into the "ground" from which it draws its nourishment. This is more than simple spatial positioning on the part of an individualizing being, but a growing along with corresponding life conditions.

2. *The Nutrients of Life*

The profusion and excess of the logoic spread after the advent of Life is owing to the immense endowment of nutrients — of synergetic material with qualities that can be used by the individual as it unfolds its entelechial pattern — that is spread before living beings.

Projecting itself along two lines of concurrent relevancies, the "seek-and-encounter" game/play of the logos proceeds in all the phases of its constructive progress. With the advent of life and its intrinsic entelechial principles, a strictly organized core is engaged in this play. For the intrinsic propensities of the synergies involved to unfold, a strictly preordained range of relevant elements must be available for their activation and promotion. But the most striking and notable moment of this concurrence of relevancies in both the inner and outer lines is the inner agency's unfolding of equipment meant to process in specific stereotypic ways the relevant materials available in the environment. The inner workings of individualized life are centralized above and beyond the stereotypic segments of chemical and organic exchange that establish and sustain this very equipment that we call the organism. The striking novelty with the advent of the organism — in seeming contrast to the other modes of formation in the universe, stars, planets, etc. — is first, the consistency of the inner workings through which the unfolding of the individualizing logos is sustained and, second, its being centered in the inward agency that springs forth with the animus that characterizes the opening of life within the individual's ingrownness in the unity of living.

This consistency is much more than the result of a concurrence of relevant materials; although it did evolve on the basis of that, it coalesced into a tightly knit operational schema implementing an intrinsic operational consistency; moreover, this operational consistency projects an expanse beginning at the frontier contours of the individualizing being emerging from the unity of life as an autonomous entity and stretching through the spheres of interlocking operational segments, each with its own constructive propensities. All the inward-and-outward-bound functions here must maintain a vertiginous fluency of performance one passing into the other such that no distance separates their acts.[1]

The organism's throwing out of hooks for the establishment of a "position" within the realm of life does not exhaust the repertoire of "spacing." As a matter of fact, we see here a particular existential modulation of the expansion of the logos of life. Establishing individual life in its activity and repose, its cycle of mental awareness and sleep, its closer and further circuits of vital interest, its reaching beyond the frame of the familiar in its surroundings, the living being through its spacing establishes its very own place: rest, shelter, refuge, fortress, etc. We will examine this in some detail in Part Four, in a section dealing with the generation of the human sphere of life.

We have been thus far focusing our attention on the spatial positioning of the embodied self. Equally important is the scanning of life or the temporal positioning of the volume related performances of life. It is not only the "spatial" positioning of our operations and actions that have to be taken into account, though undue distances do hinder appropriate operations or make them altogether impossible. We cannot distance our spoon from our mouth without disrupting feeding or making it impossible. But the same holds for the temporal scanning that is equally tied to the voluminosity of life's embodiment. Any "slowing down" of our movements or sharp disruption of their natural sequential flow makes us as living beings "out of step" with the current of life; it leads to our vacating our "place" and role and relocating or to the postponement of the fulfillment of the initial plan. As a final effect, it weakens the living being's grip on the arteries of life or altogether suspends functioning so that either other living beings come to its support or it is replaced by others in the carrying out of its vitally significant responsibility.

Concurrently, the self-awareness of the individual — the sum total of its existential life status — weakens in cohesion and intensity, and the spread and vigor of the spontaneities of this self-centered dominion diminish so that it seems to be just hanging on to life.

We are intimately related to the energy and matter in the universe of which we are a part. Newly developed instruments are allowing us to examine the origins and evolution of the universe as a matrix from which life evolved. Ambitious programs are being developed to search via spectroscopy for biotic or prebiotic conditions in extrasolar planets.

We are constantly exchanging atoms with the total reservoir of atoms in the universe. Each year 98% of the atoms in our bodies are renewed. Each time we breath we take in billions and billions of atoms recycled by the rest of breathing organisms during the past few weeks. Nothing in my genes was present a year ago. It is all new, regenerated from the available energy and matter in the universe. My skin is renewed each month and my liver each six weeks. In brief, human beings are among the most recycled beings in the universe. Life has made a relatively late appearance considering the total age of the universe and there are three intriguing questions that it poses in terms of the evolution of the universe itself: 1) In the evolving physical universe was it inevitable that life come to be? Was it by chance? Can it be understood?; 2) Is life unique to our planet?; 3) Is life at the level of intelligence and self-reflection an important factor in the future evolution of the universe?

IV. SOIL — THE LIVING SPACE

Cosmic laws and the concurrent strategy of the logos determine that a nutritional reservoir is necessary for the individuation of life. This means that life is grounded in the soil. Soil here is to be understood in several senses. First, there is the existential fact that the soil is necessary for life. It is quite improbable to envisage autonomous living beings in outer space. Even on earth birds might take flight for extended periods, but they obtain their sustenance ultimately from the ground in which the trees they roost in are rooted. The vegetation at the base of the food chain springs from a firm spatial groundwork. And then the space-time existence of living beings establishes its existence on the groundwork of the soil, the *fertile* soil, where life is hatched and the system of bios is nurtured, albeit built on the cosmic entity of the earth. And then living beings renew their originary bond with the soil when, after surging and unfolding, they return to the soil upon their extinction.

(a) The soil, as "living space," is naturally polysignificant, for the organic, vital, psychic, gregarious, and for human beings, intellectual and spiritual exercises of life all unfold, flourish, and perish on the soil.

These are simultaneous and congruent functions and it is only in their being abstracted from their concrete performance by scientists and philosophers that they may be separated from the connatural soil. Naturally we humans live these conditions in our bodily operations although we objectify them as being distinct from our bodily framework.

(b) The soil, as the dwelling place on earth shared by all living beings, determines our existential volume and provides synergetic vital mineral/chemical forces that flow in and out of the various individualizing channels whereby living beings seek and capture them.

Hence the figurative/constitutive schema of the specific senses, and what is more, on the human level the resulting subjective constitution of objectivity (objectivity being significant only for a subject), all logoic operative structures being honed to follow cosmic laws and attuned to the workings of the biosphere with its specific life nutrient factors.

Let us, however, not forget that the pivot of this shaping of life lies in the differentiation of the logos from within into significant operative sequences. In discrete continuity an entelechial principle incorporating the primogenital blueprint for life passes over from non-life through into the life of the spirit through a plethora of gradations, qualitative intensities, propensities, etc.

It appears manifest that it is intrinsic to the impetus of the logos to pass into opposites so that through gradations in quality and intensity, first, differentiations spontaneously occur and, second, a continuous harmonizing is worked, one grounded also in quantity and in its apportionment. Of course, with the advent of life's successive stages it is also grounded in temporality. But the ultimate succession that quantity is grounded in would not be temporal but would inhere in the primordial logoic laws of differentiation. Their basic categories appear to be: *quality, quantity, number, direction.*

(c) So much for the inward bound functional circuits. There is an equilibrium of the entire functional schema of the ontopoietic apparatus to be constantly checked and renewed with respect to the constant changes in the given external lived space-time situation. Each living beingness is encircled by its conundrum of forces and nutrients and in its exchanges with it extends roots into it. Then at the other pole there is for human beings the indispensable equipoise to be negotiated between the functioning of the individual and the ever new demands put on the unfolding individualization of life by evolutionary transformations within the societal interactive network, transformations taking place owing to natural cataclysms, wars, social unrest, cultural shifts in the historical evolution of nations, societies, social groups.

(d) A strategy of repair to maintain the equipoise of living beings both within and without is a specificity of life. The life promoting functional operational circuits have to adjust in order to accomplish the balance indispensable for the continuation of the progress of living beingness at all the levels of complexity, culminating in the requirements for specifically human being.

Each living being, however, has numerous contexts of operation with specific referential systems of their own, each of which it must accommodate in order to keep afloat. These subnetworks of circuits have to be constantly developing an inward flexibility/responsiveness/reactivity as well as "reparative," "compensatory," and "complementary" inward bound mechanisms informed by the senses "located" on the external surface.

These mechanisms are indispensable to the living being's ability to operate in varying vital/social contexts and varying conditions. The inward bound capacity to replace a sclerosed or dysfunctional organ with another or to repair injured or deteriorated parts is of paramount importance in maintaining the development of living beings in identical continuity. This capacity is one of the main features distinguishing life from non-life. To this "repair" flexibility of life belongs also its ability to compensate for a non-operating or missing organ by appropriately developing greater acuity or capacity in other organs. Organ transplantation is the ultimate instance of repair.

All these lines of revision, reevaluation, repair by which the equipoise of the operational schema is maintained are subservient to the major ontopoietic principles: 1) the constructive principle promulgating the originary spontaneities of the self-individualization of life; 2) the maintenance of the continuity of the growth and cyclic advance of self-individualizing life; 3) the spontaneous tendency of the logoic manifestation-in-progress to differentiate ever more complex forms in responding to the demands of the entire spread of functional virtualities instead of regressing; 4) the imperative of the entire inbound/outbound life promoting apparatus to constantly and directly follow the transformations in the individualizing beingness itself, its appetites, tastes, longings, aims, aspirations, forces toward an ever to be renewed equipoise-in-process. Although listed last, this imperative should be treated as the axis of all the transformations that follow from the adjustments that have to be made in order that the indispensable equipoise that constructively tames life's initial impetus as well as the current of existence be maintained.

V. THE GENERATIVE STRATEGY

The remarkable strategy of continuity and discontinuity (discreteness) in the logoic generating strategy needs pointing out. As the new process of self-individualization in life broke up into various processes given the living being's metabolic dependence on its environs, there was maintained throughout in the intrinsic core of beingness-in-process a continuing streak of life-growth-procreation progress. By it the the individual's time-fashioned endowment includes the self-unfolding — entelechially programmed — system/process of generation. That system is essentially informed by a set of tendencies comprising various striking or weaker features of the progenitors who "conceived" in a common procreative effort the initial project in its essential outlines and launched it empirically onward. In other terms, there are empirically manifest individual "resemblances" not only within types but also in the particular lines' existential endowments. The concrete empirical transmission of these existential individual endowments from one individual or pair to offspring makes a "chain of generations" that transmits functional features in a continuity that reaches over individual variations to show in "families" and races through the discontinuous empirical individualizations of life.

Through this inward/outward move, this time the process of inner individualization absorbs incoming environmental forces into its endowment, which is then maintained in the procreation of offspring, showing some flexibility with respect to the generative strategy of the Logos of life as it spreads its constructive rays further and further in a self-projecting ordering of the expansion of life. Simultaneously the territory of living beings is enriched by its ties to innumerable self-individualizations-in-progress, which generate fluctuations in their mutual interchange.

Here in life's system of generation we see yet another resolution of the opposites of necessity and contingency. It seems to be intrinsically "programmed," that is, to intrinsically belong to the entire system of growth itself, following its rhythm of phases as well as its most intimate synchronizing of its moves and operational segments. It follows its own specific line of development and growth apart from the growth phases of the individual carrying it, "taking off" on its own and oriented toward its very own fulfillment. Thus in this progress of association/dissociation the metabolic process plays a crucial role as resource provider/ nourisher in the origination of new life. And yet the initiation of this new life does not belong in any way to the metabolic process; it has a unique significance with respect to the process of growth as well. It transcends all in its positioning ontologically the development and self-limited framework of the process of growth and its phases, its own tending toward extinction, its own life course. The newly initiated life course transcends its bearer and slowly, progressively makes itself not only autonomous but also more and more self-sufficient in its operations and its own metabolic system; it grows and unfolds within its bearer till all of a sudden it brusquely enters the current of life at large on having its umbilical cord severed and then having to actively take in sustenance for its growth.

In the higher forms of life, the bearer of this new individual then dismantles one support system and turns to other means of succoring the newborn (e.g., breastfeading with mammals). And there is initiated in higher forms of life a new vitally significant interaction between parent and offspring.

What we are describing here is universally valid, transcending the concrete forms it takes — cell division, procreation by means of eggs, viviparous reproduction, etc.

We can here point out two ways in which life dissolves the intellectual distinction between contingency and necessity. Procreation is an empirical fact composed of a series of processes beginning with the empirical event of the "inseminating" logoic, strategy that leads to the generative progress. However, even considering that the specific features of the unfolding individual are already stored as sedimentations of the previously accomplished individual growth processes within the seminal logos itself, there is still the dramatic strategy of the transmission of these individually and singularly elaborated constructive devices from individuals to their offspring forming the long lines of "family," "clan," "nation," "race," "type," and ultimately a species. Thus the strategy of the generative transmission of specific features from one individual to another is not mere contingency. It is not "ideal" either, because it depends on a contingent process. We would say that procreation lies in between as the empirical progress of Life and the indispensable device of individualization, which yet overarches each and every individual's progress.

Within the generative conundrum embracing the union of the generative agents, their intrinsically performed transmutational exchange of elements charged with individualized dynamis and endowed with resources, we see a specific scanning and spacing outlined by the logos of life, that they may initiate and then foster the progressive growth of a new life in accord with an unfolding functional schema. There is firstly the indispensable "opportune" moment in which, as Aristotle says, "the bodies are suitable for procreation" (*Nicomachian Ethics,* Γ 1, 1110a 13-14). There is then the kairic moment for the appropriate union of generative resources that launches the initial process of individualization. Hence again, numerous functions flow together in their respective roles of inward/outward differentiation and growth, which brings in the forward scanning of moves into steps, phases of constructive progress; with individualization the new functional apparatus progressively loosens its intimate metabolic ties with its bearer, and with the strengthening of the new network, the continual scanning of the advance culminates in another kairic fulfillment with the delivery of a new individual onto the great stage of life at the opportune moment. In the cycle of generation we find, indeed, all the characteristics of the kairic: opportune time, occasion, right measure, and foremostly, the characteristic of "fulfillment" that our analyses have brought to prominence.

CHAPTER TWO

THE GATHERING OF THE DYNAMIC LOGOIC THREADS

I. THE GATHERING OF THREADS IN THE LIVING AGENT

In evolutionary progress, the unfolding of the functional apparatus that serves ultimately the creative imagination has proceeded in step with a developmental increase in vitality of proficient organs, members, etc. Over time this has meant an extraordinary loosening of the ties of the entelechial design of the embodying functions of life at the experiential level where "phantasmic" functions are unfolded.

It belongs to the logoic strategy in the inner functioning of self-individualization to proceed by centering and dispersing, integrating and diffusing, binding and dissolving, with this rhythm creating new forces and synergies, the inward movement of self-individualizing progress, the inner agent that is the exponent of the progress of life.

II. THE AGENT AND THE SELF

Virtually launched in the initial moment of the the self-enacted, self-oriented functioning of individualizing life with its inward/outward orientation, inner direction accompanies the deployment of ever new functional modes in the progress of growth, strengthening the virtual ties between the various functional streams as the constructive distributor of roles and simultaneous conductor of the overall organic orchestration. This direction intensifies functional links and provides all systems with the focus of a common telic orientation. This coordinated orchestration first endows the living beingness-in-flux with a distinctive autonomy in delineating its life course and sustaining its "identity" amid the flux surrounding it. It leads also to a synthesized reactivity: under any particular provocation from without, addressed to any single part or organ, the entirety of the living being is provoked and put on alert. Reactivity is not "localized." All of the organism's territorial-individual expansion in functions coincide in this one point of synthesized reaction; this coinciding of the entire functional system in one intensified point is what we call the "self."

Let us emphasize that there cannot be living agency without a "self," that is, without this vital referential condensation of all the functional lines within one point of experiential unity in reaction/action. The self is by no means territorially or functionally in a "central" position; it is not a center of a functional pattern. On the contrary, it has no territorial position within the functional system because it is within it as a logoic knot of its vital functioning.

It is not important what status we attribute to the so-called "I" or "self" in any abstract distilled ontological framing of the living individual because what matters is the strategic twist of the logos by means of which a lived balance is always being sought, first, among the seemingly dispersed functional segments of the living individual-in-process and, second, between the segments of "inwardness" and the external flux of processes in which the functional apparatus of the individual works its way. Thus, what matters for our inquiry is investigating how this strategic situation of a crucial balance is achieved.[1]

This agency is not only the depository into which flows the logoic channel of the impetus dynamics for all the individual's functions but is also and simultaneously the distributor of the moves of action/reaction. Life's self-individualizing progress is promoted by its agent — as an orchestrated cooperation among all functional modalities and processes. In this agency is crystalized the entelechial principle deployed in the sequence of individual growth. The agent is also the movable center of alert attention, receptivity/reaction, and direction. It is the promoter of the orchestration of forces and function, the initiator of effort and movement, the coordinator of sequences of moves that we call the "self." In philosophical speculative ratiocinations we may dismantle reality and experience theoretically as much as we please. We may also invent devices to put it all together again. We may argue for or against hosts of abstract "centers," and just as abstractly for "decentering," to our fancy's content, but the firm reality of life cannot be affected by the vagaries of human imagination. The facts remain intact: without an agent and a self-gathering, combining, and thus centering of all the functions of the individualizing life, the philosopher would not exist.

The role of the agency, the agent-nature of the unfolding individual and of the self, comes strikingly to light with the the individual-in-progress' specific capacity for "self-repair." Indeed, one of the five essential specificities of life is self-repair from within. If a functional organ is damaged, in many cases it is repaired by the functional system. Of course, there is a balance determining the degree of damage that can be repaired.

In determining its moves the individual always has to weigh the degree of adjustment or readjustment that would be required of the "rest" of its functioning, to recalibrate the overall balance of its equipoise within the limits imposed by its volume. But also the equipoise that the flesh postulates has to be maintained with respect to the experimental circuits; sensory functioning has to be maintained with respect to the varying, spatially motivated contacts with other beings; physical functioning has to be maintained with respect to our visceral preexperience of our bodily/physiological functioning, on the one hand, and the coordination of our inward moves and mental prompting with their physical enactment, on the other; and our psychological functioning has to be maintained with respect to achieving sufficient balance for the adequate appreciation of our ever changing situation.

In our perpetual motility, whose instances and segments are distanced from each other only by intervals, the agent maintains constant attentiveness distributing its attention to various of the body's limbs, organs, and senses, which are all gathered in awareness of the bodily self. The agency constantly at work to secure equipoise in its essentially precarious ontic spatiotemporal situation is so carefully, so tightly melded with our living functional schema from our entry into the world of life, or to put it another way, from our assuming a volume possessing position within the same, that we become aware of its entire scope only through reflection.

This effort at equipoise has in that voluminosity its reason and point of orientation within the spatiotemporal field of life; it can be mentioned only in coordination with the efforts of other living beings in their schema of mobility — this in accordance with the same basic life principles that voluminosity entails.

We may say that it is through a specific function reserved to the body in the buildup of the whole of living beingness that the agent or self establishes in the interactive world its presence. To say that highlights the pluridimensional reactivity/responsiveness located within the bodily "dimensions." And here we arrive at the particular means of this reactivity/affectivity or reception/expressiveness in which is prepared the various gradations of the circuits of what I shall call bodily experience.

Primal affectivity, the rudimentary "experience" of the agent or self at any level of complexity in life registers a provocation and formulates a response as a provocation to and response of the whole body. However, in the body's pluridimensional functioning, some organs, like the digestive system, are situated in the hidden core of the body, while others, like the preparatory organs of that system that perform the operations of ingesting, biting, chewing, liquifying, are accessible. Thus we may acknowledge the crucial life strategy of installing at the body wall the organs of direct sensation: sight, hearing, touch, smell, taste, in which inheres the foundational reactivity/sensitivity constituting bodily experience. The self, the agent acquires awareness of them as an awareness of itself as their performer, but this in an unprecedented contraposition with "what" is being seen, heard, touched, smelled, tasted, etc. — that is, in a contraposing thraldom to the "outside" world of objectified, distinguishable entities.

The body as such does not just embody the self-performing systems of organs; owing to its springing forth, a panoply of circuits of relevancies crystalizes in the overall cosmic life strategy of reason. Given the body's voluminosity, its situatedness, "positioning" is the exponent, "porte-parole" of experiencing operational reaction/response. This being possessing voluminosity envelops its differentiated organs in a protective "wall."

It is through this "surface," our skin, etc., that provocations from without are "received," and it takes an interval for their impact to reach the proper organ (see Socrates' delayed registering of the pain brought on by the hemlock), and when a response is formed it again operates through the bodily envelope.

However, and here an important point is to be made, contrary to Descartes' view[2] this interplay of bodily systems is not "mechanical" and the organism in which they operate is not a machine. As has been emphasized throughout this discussion of the autopoiesis of life, its workings are initiated spontaneously and the initial impetus of forces is carried on as well as continually replenished by the processes of an organism's unfolding. While a machine stands in need of an external propelling force to begin operating, it is of the essence of the ontopoiesis of life that it is self-initiated and propelled by its own energies, that is, by energies generated by its own operations. In its performance, the living being "experiences" itself as a body. The embodied organism is through and through self-enacting, and the embodied agency as the self that coordinates it is self-enacting as well.

Thus to its inner becoming aware of itself through the very performance of its processes, we have to add the expressive bodily experience that is founded in the body's surface organs and that in a connatural way we ignore within the natural attitude toward life, unaware of the knack with which it is operated, unaware actually of the reception/response/expression nexus between the volume possessing bodily established beingness and surrounding objects and forces, the interacting connectedness that we call a world.

It is precisely this bodily experience that is foundation to the objectified expression of the sensing reactions, that is, the foundation for the erection in a further circuit of unfolding of an entire operative system in which the inner operations of the ontopoiesis sustaining the operations at the body wall constitute the volume manifesting expressions that the bodily agent performs in the surrounding world. We will come later to a fuller discussion of this constitutive expressiveness that yields the culminating accomplishment of the logoic strategy in the natural circuit of life — the full-fledged constitutive experience of human beingness.[3]

As we mentioned above, the agent responds to the outside provocations received by the bodily senses by constituting his own expressions in the categories of a volume possessing stationary construct. Husserl distinguishes the "body as experienced," which we have analyzed above, and the body as "objectified." We will come to this while discussing the specifically human type of beingness.

III. THE STRUGGLE FOR EQUIPOISE IN MY EMBODIED SELF

Let us state at this point that experience, whether at the high level of complexity reached in the human condition or at any level of life, is in the first place "bodily experience" and that all else about experience is rooted in and derives from bodily experience. We have been discussing so far the "inner working" of life's unfolding, and have stressed the point that indispensable to its progress is "agency" and "selfhood," "centeredness" and "autonomy." We wish to emphasize here that all these organizational and regulative devices of the logos are essentially cofunctional with embodiment: they go together with the operations that make them "embodied." That is to say, they manifest their performative proficiencies not in a limited operational circuit but as "bodily functions" within a "bodily orchestration" of "bodily functional systems" or "bodily organs." The agent and the self, are those of the "body." The four bodily systems — the nervous, the circulatory, and the alimentary systems distinguished by Descartes, plus the reproductive system, which Descartes considered to belong to the nervous system — as well as the organs that they represent, are the foundation of the body as a locus, a frame, a vehicle by which life runs its course, a wholeness manifesting the autonomous ontic beingness in its existential interplay with the surrounding forces of life.

In contrast to inorganic physical agglomerations, volume possessing bodies whose inner directionality appears to be nor more than a bundle of mechanical circuits, the embodied living organism cannot be considered a machine. It directs itself.[4]

We have so far been talking about the "surface" reactivity and action of the body. Let us now compare and contrast that with the functioning of the body's viscera.

The body's visceral functioning is "automated" to a considerable degree. It is experienced in the rhythm of the heartbeat, in the pulsing circulation of blood, in the "lightness" or "heaviness" of one's head in the morning, in digestive discomforts, satiation, etc., etc. But it does not possess expressiveness, its "voice" or "graph"; it remains mute, leaving to surface organs most of our sensing and reacting, given our need to experiment with our environment to meet our basic needs.

And yet in my bodily grounded self-awareness, I rely essentially on the smooth operation of these mute inner processes. When I receive from my so perfectly coordinated organic/vital conundrum of functions a signal of alarm indicating that an interruption of its smooth operation has occurred, that is when I feel bodily discomfort, pain, kinesthetic incapacity (e.g., inability to rise from a chair), dizziness, etc. My self-direction then loses its grip on life, becoming uncoordinated.

Special training adjusting mental and spiritual circuits and physical effort is then needed to restore the regularity of the inward processes that support one's existential state and so maintain a level of equipoise in self-awareness robust enough for one to ride upon the flood of life's forces and not sink into isolation, alienation and disappear from participation in life's affairs.

We directly experience it only when it breaks, but this mute running of the visceral circuits of the body constantly works to maintain the organic/vital/sentient/psychic/intellectual equipoise of my embodied self.

CHAPTER THREE

THE EMBODIMENT OF THE LOGOS IN THE SECOND PHASE: TRANSFORMATION OF SENSE

I. THE TRANSFORMATORY STATION OF SENSE

We have thus far been studying the logoic constructive elan as it is embodied within the forms and patterning of the play of forces uniquely specific to the individualizing living beings inserted within the great logoic schema of the cosmos. We have so far touched on their fundamental relevancies in the grand context of the cosmic forces of the logos, and on the harmonizing correspondences between the organic/vital play of forces — those involved in growth, in particular — and the cycle of the seasons following on the cyclic motions of the heavenly bodies. The existential necessity of life's creatures to combine periods of activity with those of rest harmonizes with the diurnal and annual cycles of the heavens.

Now we have to turn to the second phase of embodiment, that in which the logos of life makes far reaching moves to convert these primary process of generation and growth by endowing them with new significance. Living beings, in contrast to the heavenly bodies that follow their prescribed course without being actively involved in their conduct or path, not only act to promote their course, but, what is most significant, to control it and direct it. No logoic move or device is ever conceived in "isolation." On the contrary, the stars remain in place even if moving; they do not "fall" and collapse inadvertently, owing to the action of the laws of attraction/repulsion given their particular weights and inner composition. Even so, the logoic strategy for the embodiment of life carries with itself an essential existential communication/interconnection device. So in this new phase of life, of liberation from numbly following of the rules of nature and entry into a self-directive stage, what is really in question is the conversion of the organic/vital sense of life into a vital/communicative/self-directing sense of life.

We will briefly sketch to this effect the entire sweep of the evolution of life's stepwise progression from the least autonomous species of animals to the creative human being possessing a high level of autonomy.

We will bring out the two most significant stages of this progress. The first is the phase of self-revelation. That phase, however, is just a foreshadowing of life's establishment of the great transformational station, that is, its unfolding of a full psyche crowned by the mind and the soul. This station embraces the foreshadowing phase, is itself complex, and comprises a vital soul, a bodily/organic brain, and the thinking mind.

The intermotivation of these three factors is of the greatest interest and has throughout history created numerous difficulties in thought, especially the so-called mind-body controversy.

We will bypass controversial issues that do not occur within our perspective and center on the main problem of the conversion of sense.

II. THE FOUR STAGES OF THE ACCOMPLISHMENT OF LIFE'S SPACING/SCANNING — WHICH INTRODUCE VALUATION AND MEASURE, WITH FULFILLMENT BEING THE CRITERION

Life's unfolding and expanding of the logos advances through most minute operational moves that form in their concatenate efficaciousness segments punctuated by the accomplishment of an aim, a purpose, a telos. It is indeed through the constructive orientation of the entire project of life that this gigantic play may unroll the initial virtualities of forces and energies. Telic constructivism appears to be the engine of the drive of the logos to expand, the key to the reign of reason. In fact, and what is most intimately relevant to the spacing/scanning organizing principles, its proportioning and measuring draw from the evaluative criteria that the telic schema of life's individualization installs. With reference to the stage of accomplishment that the telic schema of life's constructivism entails, a scale for evaluation is established. Life's estimating and directing values spring forth.

The scanning of life's unfolding with all its minute movements and their arrangement in larger segments, joining with each other, fusing into each other, prompting each other's transformation, intergenerating, etc., falls into three complexes-in-becoming. The incipient complex-in-becoming released with the surging of the spark of life or the animus is followed by the span of growth in the self-individualizing course of life, and this latter leads into the span of maturation/fulfillment/fruition.

The incipient-complex-in-becoming itself is made up of three intimately related phases of becoming: the igniting of the spark of life, incubation, and positioning for individuation in the circumambient world. When incubation is accomplished, the living being breaks through the eggshell or away from the womb, separates itself as an independent complex of cells, and is outfitted with all the synergetic forces, *in actu* or in readiness, needed to begin its life course. How marvelous it is. It already has not only the endowment of forces and spontaneities needed for navigating its course, not only the mechanisms of spacing/scanning with which to project ahead its operational system, its measuring and weighing, and not only the individualizing design that is there in readiness, but also a providential arrangement by which the individual through serving itself serves also the continuity of life as the universal manifestation of the logos through its proliferation and multiplication, its trimming and spacing continually throughout the intrinsic cycle of individualizing unfolding.

In fact, we barely exit the heteronomous phase of incubation and enter into autonomous existence and we are already in a phase of growth expanding our virtualities in an interchange with the circumambient world of life and the organism with the development of various faculties unrolls in a way proportionate to its full potential and in a way that is equally proportionate to the the telos of its growth cycle: generation, fruition, accomplishment.

The telic accomplishment finds its apex and central primogenital significance in the unfolding of fecundity as life's innermost sense of continuation and renewal in ever repeated cycles: life's ascertainment of its absolute dominant set of values.

On reaching the phase of fulfillment, we have reached the zenith of the manifestation of the Logos in the self-individualizing progress of life. Slowly the phase of maturation and fruition yields to the phase of the diminishment of life's processes in the individual, the shrinkage of their modalities and energies, followed by the disintegrative processes of self-individuation, the extinction of the spark of life, the closing in of life's premises, and then passage.... Yet this most complex stage of life follows from the instauration of a unique type of intergenerative apparatus, the differentiation of the sexes, differentiation into a pair of opposites, male and female. They attract and repel each other in turn, remaining indispensable to each other. The "reason" for this separation and juxtaposition is so that by surpassing it couples may unite in the creation of a new individual life. The phase of fecundity is not only the period in which nature shows its capacity to generate a new individual life; it is in the first place a carefully prepared outburst of an urge to do so, a galvanizing by this urge of all of the strings of the individual's inward-outward life propulsion. It is only in the simplest animal life that generation takes the form of self-contained parthogenesis. In all higher forms of life, generation is not the simply mechanical reproduction of a type, but involves a period in which the individual reaches into the vast reservoir of possible partners mixing among individuals of the same species — and at the fringe on occasion roaming to an individual of another species. An intergenerative range encompasses vaster and vaster circles leading to innovative mixing, fusing, which generates novel features, nuances, forces, routes of life.

Here is a reaching out with all the lines of the individual's interiority, with all its sensibilia, with all the tentacles that relate it to other living beings, into the reservoir of moving forces and energies, links, and antennae, into the living process of the life course, in search of the most appropriate answer to the appeal to join in the intergenerative event, the event crowning the constructive progress of individualizing life, that which gives it particular value, the flagrant manifestation of life's prodigious dynamism — into the Logos of Life's wondrous ways, its schemes and fulfillment. Fecundity is certainly the greatest prodigy of the logos and fruition-fulfillment is the best of life's accomplishments, its higher criterion of valuation; it continues, promotes, and gives meaning to life from within. The scanning of life finds its exuberant climax in this fruition. In the passage that follows, that of diminishment, shrinkage, disintegration, the way is prepared for the extinction of the animus.

The disintegration of the unfolding individual's articulations, the vanishing of the dynamism of forces and energies sees the withdrawal of the logos. However, when the spark of life, the animus, dies at the end of the life cycle, it will burst into light animating another beingness that already stands ready. But the nature of this changes with complex human life, for here a spirit lives. We may then conclude that in its initially endowed ways of unfolding individual life is most precisely proportioned to the unity-of-everything-there-is-alive, and is in its cyclic evaluating phase relevant to the universal economy of the entire life system through which the logos operates. First, there is control in the discrete continuity that is the uninterrupted progress of life as a streaming advance of individuals that renew the codes of life and cede their space to new living beings that they themselves prepare; second, the valuation that the logos of life projects through its telic progress culminates in the fulfillment of the higher criterion of fecundity/fruition that brings about new lives to replenish the empty space.

There is, however, a great question to be raised. Since it was clear from our outline of the four stages of the individualizing unfolding that there is no "absolute" beginning of the incipient complex of life's becoming/incubation and that its readiness is a continuation of a generative progress that itself has its beginning in the synergies of the generating individuals that come together in procreation, whose species characteristics and even singular features are transmitted, should we then conclude that withal there is no entirely complete beginning? After procreation in the realms of bios and zoe, extinction, the loosening of functional ties, frees the synergetic elements, dispersing them, but only after their having been transformed by organic processes. They perdure conserved and lay in readiness to enter into a new individualizing work. Should we not then wonder with Leibniz whether all is not animated, or at least in readiness, and whether animal death is in the perspective of the flourishing economy of the Logos something more than the completion of one cycle and concurrent conservation of life's resources in another?

Should we, however, go with Leibniz's conclusion that, in view of the successive but never clear-cut, never entirely complete, beginning and endings of life in its unfolding, there is no radical interruption in the process of life, neither birth nor death?

CHAPTER FOUR

VOLUMINOSITY CRYSTALIZING THE VITAL DIMENSION OF BEINGNESS

(a) The functional role of the visceral "inbound" body seems to be tacit one hidden in the shade. But it actually overshadows all the sensory action on the body wall in a special way that we will come to later. Its "securely" positioned physical-organic circuits perform in unison their appointed roles in a mute, seemingly subservient but in reality basic fashion. This yields the platform to the individual's sensorial activities and experimental contacts with its environment, leaving to them the gathering and transmission of data crucial for interactive symbiosis with other individuals "out there" and for negotiating forces "out there" that have to be taken seriously into account as the bodily agent formulates its directives.

This mute visceral conundrum of functional circuits in their semiautomatic regularity, regularity that challenges the chaotic play of life's outward forces and harnesses them, albeit indirectly, is indispensable to my natural spontaneous equipoise and self-awareness, for my keeping myself afloat and wringing for myself a specific role amid the flux of life. The mute regularity and equipoise of these self-adjusting physico-psychic processes guarantees at every instant my "position" in the living space-time continuum of life and allows my projection of an entire gamut of soil relevant existential meanings as well.

If this regularity be jeopardized, my entire rootedness in the soil is put into question and has to be revised and reworked if there is to be a new start.[1]

(b) My bodily self-consciousness is the culminating point of our integrated existence, integrated because we tacitly encompass all functions. I am the bodily expanse that encompasses me, and just as much I am my psyche and mind, I stretch in my self-awareness through all the organs in place on the body wall as well as through the body's internal organs, of which I feel "in command" even though I do not maintain a vivid and constant awareness of their presence.

(c) In this awareness of my dominion, exclusively mine, spread within and along the arteries of living, pulsating spacetime, and in my incontestable, identifying with a firm grip with my tentacles in the arteries of the struggles going on in the current of life, I become truly myself, the autonomous monad with which Leibniz and after him Husserl identified the living individual. Yet as we see, the living individual in process cannot be conceived as an altogether independent monadic entity; this entity is viscerally bound to its circumambient conditions and through them, to the unity-of-everything-there-is-alive.

The conundrum of the circuits of my visceral functioning extends into the prominent and significant archipelago of senses operating in the head, senses so smoothly orchestrated that they almost automatically perform their conjoined task without drawing attention to themselves so long as they are not hindered or disrupted. Each enters into the common task of the entire performance. It is owing to internal direction that each sensory organ performs its task in harmony with the others. This inward directedness incorporates itself into my voluminous body and projects as well as carries my voluminous beingness.

(d) At the end of our focusing on embodiment let us assemble some principles of logos that have emerged. They culminate in the role of motion. We have pursued some of the basic dynamic phases of embodiment as the foundational operation of the logos as it aims at the establishment of life. It has been clear throughout that this process is identical with the generation and unfolding of the self-individualizing process. That process consists, first, in a series of phases; all of them are phases of motion. Second, these phases aim in their discrete continuity at the performance of various constructive tasks, at the implementation of a universal applicable model, that of living beingness as such (although this implementation always occurs in concrete processes and the model is just our abstraction from them). There is, in fact, not a "living being" as such; it is always endowed with a set of properties, being situated within parameters specific to it. The point that I want here to stress, however, is that there is a pervasive continuity in all of the phases in this construction, no matter how divergent they may appear to be in their prerogatives and limitations. The conductor of continuity is, on the one hand, the constructive entelechial design. On the other hand, in the concrete performance we find motility as such to be the dynamic thread running through the whole operation. As I have pointed out earlier (in Tying Point One, Section II, "The Womb of Life" and *passim*), motion is not only the first experiential individuation of life but also its innermost conductor. We can say that motion is the primal, primogenital, and pivotal device of the logos of life as it projects the great plan of life. It is pivotal not only for the essential lines of the plan of life but also for life's conditions within the grand context of the cosmos. Furthermore, as the pivotal device of the logos as such, motion with its intrinsic axis of spacing and timing appears to be the essential apparatus, the primal factor of the crucial logoic device of impetus and equipoise. It is motion that sustains the entire dynamic of becoming and is its manifest bulwark.

In what follows the point on which I will concentrate is that of the continuity between the different functions and faculties operating within a living being. This continuity accompanies innumerable transformations, going through and conducting even transformations of sense.

Whether it be in the generative phase of the womb of life or the phase of the full unfolding of the virtual endowment or that of decay and disintegration, all these phases are carried on, as I have emphasized before and do again, in motion, in moves of absorption, diffusion, confluence, radiation, association, etc.

The motion of this project ramifies bridging the dynamisms of constructive forces. There is the vast realm that the functional — if we may so call it — motion traverses from the organic to the vitally significant operations of life. And even more significant is the realm of functioning that extends from operations significant for survival to those that distill the responsiveness of the living individual into an "agency." This is most complex territory, one requiring innovation to cover while modulating the innumerable sense bringing moves, transforming their aims and directions, attuning them to new functions and roles. The most dynamic task of embodiment for the process to project is the move that stretches from the responsive agency of the living being to the full psyche of higher animals with its realm of consciousness, and finally to the already unfolding Dionysian logos of sharing-in-life with its "many delights," and to fully human, reflective consciousness, which introduces the conscious intellect and its Apollonian distancing of the emerging spirit from matter.

The inner directionality of inorganic "bodies" appears to be merely a bundle of physical mechanical circuits. The living body cannot be considered a machine; it directs itself.[2]

(e) As said, our visceral functioning is "automatic" and to a considerable degree mute. We are aware of the rhythmic heartbeat and our pulse, of the lightness or heaviness of my head on rising, various digestive discomforts, etc. But by and large our internal organs are not expressive. It is the sensory organs of the body wall that are at center stage.

And yet in my bodily grounded self-awareness, I rely fundamentally on these mute, smooth running, internal systems. When out of my internal conundrum of functions comes a signal of alarm indicating some malfunction, such as pain, discomfort, odd sensation, dizziness, or incapacity, then I lose some of my self-direction, my grip on the ways of life loosens, my participation in the web of life becomes uncoordinated.

It then takes special training and adjustment of my mental and spiritual circuits and much physical effort to reestablish equipoise amid the flood of life forces and participate again in life's affairs or even just to keep from sinking into isolation, alienation, and despair. Although for the most part we are vividly aware of our visceral circuits only when they break down, their mute operations constantly sustain the organic/vital/sentient/psychic/intellectual equipoise of my embodied self.

But it is in fact in kinesthetic movement that all the pieces of our self-awareness fall together. This prethematic awareness of ourselves as "a center of spontaneous ability to move," in the words of Landgrebe, impresses us even more deeply than our internal circuits of functioning, so much so that it is correlate to the "agency" of the living being. Kinesthetic movement is key in transmitting from the sentient body the foundational moves leading to the emergence of transcendental consciousness, that is, to the emergence of full intellectual perception. This movement has complex origins in the body's organs and leads on to the individual's fully grasping the "objective," that is, that which is positioned outside the dominion of the individual's emerging subjectivity. This complex synthesis transmitting all that is objectively presented the person does so insofar as the person can receive them determined by them. Perception is qualified by this complex of operational moves.

It is with the emergence of the synthesizing self-awareness of the "agency" that we indeed have the means for a differentiation within the individualizing living being of an absolutely "subjective" dominion that seems to stand drastically apart from the circumambient, "objective" sphere of external forces. Yet as I have been indicating, this ownmost subjective sphere appears as such precisely as it engages the play of external forces. Agency is in the kinesthetic interval between the "inner," "subjective" self-control and dominion and the "external" play of forces within which it is bound to deploy its moves. The kinesthetic network agency is spread through the whole of the individual's functioning. It builds up an entire functional station within the equipage of the individualizing being. This agency becomes amplified with the increase in complexity following on evolutionary advance and becomes an entire apparatus in the conversion of sense, which it already initiated with the first kinesthetic movements. These movements can spread through an ever more complex assortment of instruments performing the role of "objectifying" the world and bringing together the inward with the outward in a conversion of sense. The hidden, mute operations of bios yield a dynamic stream of motility that is ever more attuned to the task at hand, progressively passing to a clearer and clearer presentational performance in which the inward and the outward come to stand in a unique tension for the living being with each sense organ and extremity stretching into the external realm. What is brought forth into the light in presentation is as much the individual's capacities as the "image" of the other.

This network of transformational, converting moves in the functioning of the living individual we call the "brain," which we consider an organic/physical factor of life. However, as I have attempted to show, while the brain is an essential factor of embodied functioning, it cannot be identified with the organism — seen as an essentially physical dimension of life. It is also an essential factor bringing about a new communicative sense of sharing-in-life; mute self-enclosed organic material functioning is liberated from blind bondage, brought out into open, communicative interconnectedness with other living beings.

This marvelous, most complex, and adept organ works through all the living being's functioning as its coordinating, energizing, directing center and is receptive to happenings within as well as without. The brain networks all spheres of the inner, mute functioning of bios and prepares through its own qualitative progress the emergence of consciousness. The brain is so extraordinary because it is an entire station for the conversion of the significance of life.

It acquires, indeed, a special turn with the development of the conscious powers that it founds and yet endows them with a quasi-autonomous role with respect to the processing of the experiential moves that we, on the one hand, consider to specify the living, animated being's "soul" and, on the other hand, consider to specify "mind" in its presentifying of "objective" reality. We may venture to presume that it is the interval between the experiential level of presentation and that of thought, imagination, and reflection that is the passage from the station of the brain to the further step of the drastically novel station of the creative condition of the human being. We will turn to this subsequently in its proper place.

CHAPTER FIVE

THE DIFFERENTIATION OF THE LOGOS IN CONSTITUTIVE AND INTELLIGIBLE EXPRESSION

The next passage in the ingeniousness of the logos follows. However, at this point let us point out that the living beingness in considering its selfhood by the very performance of its bodily processes identifies itself with them. But "visceral" processes are in themselves "mute." They go largely unperceived by the sense organs on the body wall. They are perceived when they are disrupted, and then they have a "voice" uniquely their own, for example, nausea, pain, dizziness, etc. In our "constitutive" attitude we "objectify" our bodily voluminosity; we constitute and hence experience in perception, our head, trunk, breasts, limbs, hands, feet, etc. as specific extended, shaped, substantial forms. And we identify with them. Together they are "our body," the body through which we perform and which we direct as we lift, kick, kiss, push, squeeze, wrench, smooth, approach, move away, sniff, taste, etc., which are our very own performances and activities. The bodily self lives in the entire set of one's functional circuits, visceral or sensorial, but expresses itself in motor performances within the interactive, object-filled, and "populated" world.

However, the full-fledged life of the self, in which the bodily self performs a foundational role, is progressively achieved through life's unfolding phases of complexity and with the specifically human creative mind and the intelligence that it brings.

At life's primal phase, however, the originating type of beingness is restricted by its basic life functions.

I. THE VOICE ISSUING FROM BODILY SELF-AWARENESS: A SIGNAL, A SIGN, A CALL

The equipoise of the inbound/outbound cooperation is the key to the process of life. This balance seems to assign priority to the outwardly oriented functions of experience that serve as lookouts and scouts for the advancing army. The core of the body then appears to assume the second place with respect to the embodied self. But with the further development of life, this is completely reversed. Deprived of "voice" in the process of deliberation, in decisions about life's course, meant only to decipher obediently the pregiven entelechial code, the body core then comes into its own.

This reverse occurs when the living being temporarily dims the lights of sensorial experience and withdraws from its environs, loosens its vectorial outbound interest and attention, retires to its own bodily sphere.

With the drastic move of uttering a sound, making a gesture, etc., the individual functioning of life, as it were, completely activated its powers and hence the organically/vitally significant network of the logos triggered a novel expanding device for life's progress.

The uttered sound issues from a self-enclosed "sufficiency" of functioning and makes a bridge of communication, in a vital/existential sense, with other living beings in order to lend and obtain mutual support in life. At its very outset it proceeds from a strictly individual agency to connect that agency with others, from what now maintains itself as a subjective agent amid other parties in communication, the "objective" side.

In this interval three logoic streaks of utmost significance for life's progress delineate themselves. *First,* there surges a logoic connectedness in the agent's attentiveness, observation, and synthesizing of the vitally sensed situation within the web of life that prompts the issuance of a "call." This prompting throws a logoic connectedness over the common net of intrawordly existence anchoring the individual and his very own functioning within it, allowing him to lift himself above the body's mute, hidden operations, processes, moves, organs, etc. into a synthesizing, all-comprising, novel, "intersubjective" logoic sphere of sharing-in-life. This lifting establishes a novel significance of life; life goes beyond self-enclosed agency to a communicative intersubjective network. *Second,* the conversion of agency into a germinal transcendental subjectivity sees the unfolding of a host of communicative-intentional circuits that carry with them not only communicative ties but equally a variety of presentational activities that establish a vitally evaluated but "objective" intercommunicative world of life. *Third,* these communicative and presentational streaks of the logos coincide in the ordering of the living individual in its world of life.

This order is the basis for the transcendental ordering of the so-called "constitutive" functioning of transcendental consciousness, that is, its presentational — "manifesting" — schemas, for the logos' "expressive" system of language, and for logic, which combines them.

Transcendental consciousness has forces of conversion brewing in its generative womb.

II. THE SENTIENT LOGOS

(a) If with Aristotle we may identify what we call the "animus" of life with the breath of the soul (*De anima* IV, 4; 416b 11-12) and then with Leibniz attribute degrees of animation — in our parlance, of consciousness — to living beings in a gradation extending from the simplest organism to the most complex, the human being, we may distinguish as manifestations of life at least appetite, sensation, locomotion, and thinking. Already, beginning with appetite, the purely organic, automatic moves of the logos are left behind in an advance toward the operation of a "sentient" logos of experience attuned to the qualitative requirements of the constructive project. Motion goes ultimately in the direction of quality.

We have in our perspective to differentiate matters further. The sentient attains its highest levels of complexity. There is a sharply felt advance along this line from feeling sunshine to feeling emotions. The latter distance themselves from the prehuman feeling proper to animals by involving the unfolding of a rudimentary psychic inwardness, a subjective awareness of oneself, with even moral valuation taken as a point of reference. This line of transformation is carried on by innumerable moves, operations, and processes stemming from the organic and transmuted into novel, original sense. It culminates in its most multidimensional task, the establishment of the full phase of human transcendental consciousness and the mind. As I have described it in my earlier work, the emergence within the evolutionary advance of life of the Human Creative Condition called for the setting up of an entire transformational, processional station.

The logos' dynamic transformational carrier is the system of movement that stems from the physical, organic base and sets up in the midst of it a uniquely complex organic apparatus for pushing organic forces toward higher levels of logoic manifestation.

(b) The organic/transcendental brain and the transcendental/reflective mind make up one dynamic system processing the entire line of experience. On account of its striking accomplishments, we naturally single out the fully experiencing human consciousness with its emotive, cognitive, reflective powers as a complex that Aristotle calls "thought" and juxtapose it with the vitally significant circuits in order to project human subjects as central in a transcendental universe of our very own, one, however, that we share with all living beings. There, we single it out in contrast to its foundations as if it could stand upon its own marvelous accomplishments.

The mind-body problem, however, overlooks the foundational situation of the transcendental mind and all its prerogatives, the thought and reflection unique to its manifestation of beingness.

Predominantly, these foundations are the organic processes of the brain. The brain's moves established themselves as a system through the dynamic constructivism of the logos, which places it as a specific station, intermediary between the organic physical organism and the vitality bound psyche. Freeing itself from rigid bondage to matter, the station of the self-individualizing brain plays a pivotal role in the transformational dynamism of self-individualizing life at this level.

III. THE ORGANIC BRAIN AS THE VITAL STATION OF THE CONVERSION OF SENSE

In calling the "brain" the conversion station of the sense of life, an "organism," that is, an integral instrument, we emphasize its twofold directedness. On the one hand, the organ of the brain occupies a central position within the functioning of the living being. It sets out from its incipient moment as the motor of the articulation of internalized moves and advances toward the self-directedness of the individualizing being. It goes through all the phases of this individuation in its self-directing modalities. But doing so, it assumes the direction of the establishment of the sense of life and it carries on its modulations and conversions throughout the entire vital as well as psychic course. Its second direction transgresses the organic significance of empirical, physical, organic operations. In this direction, through innumerable instruments that pertain to its organic/physical bedrock, it unfolds an entire system of modulation and conversion of the significance of life.

As such, the brain is a transitional station between bios and psyche, and the psyche and the mind. But it also serves as the foundational instrument of the radically novel station in life's developmental evolution that is the creative human condition.

In its transformational power and operation, the brain is the conductor of the logoic forces of nature into the creative forge of human turmoil. In this task the movement of the brain transmits the yield taken in, transforming and molding it in accord with an ontopoietic entelechy, and thus transmuted that yield generates new movements (processes, operations, etc.). All that is taken in passes on to the station of the creative mind with its apparatus for bestowing transcendental significance, and there it contributes to the completion of the constitutive work of the intelligible manifestation of the world of life. The great secret of the misunderstanding called the Cartesian "mind-body problem" lies in ignorance of the uninterrupted yet discrete continuity of this dynamic schema in which each phase engenders and carries the next. In short, the motion that is the carrier of the primeval device of the tension between impetus and equipoise exercises its power connaturally in the quality, form, and direction that determines life's direction.

1. *Life's Station for the Conversion of Sense: The Brain, the Embodiment of the Mind and Sense*

Already with the synthesizing-synchronizing coming together of the entire spread of the sensitivity and attentiveness of the living being, we have made an essential step in converting sentient moves by liberating them from their unidirectionality and particular attunement to the task at hand toward the all-embracing move of attentiveness combining them all and compassing them all in one attentive glance. This passage from one-to-one connectedness of psychic moves that were qualitatively constrained to just one vector and form of fulfillment into an all-comprising unifying move means already an essential qualitative transformation of sense. The single focus senses of a lower order are converted to a higher order of sense involving the identity of the individualizing beingness in which they are transmuted. These new senses, when set over against the circumstances and influences of other beings round about, have a unique place within the individual, a place that stands out given the individual's very own being and tendency to maintain it.

This sense of selfhood is the great step in the advance toward the establishment of an inward agency, a psyche. And along its lines the great phases in the conversion of sense, of the logos itself, of life occur. The first is the psychic or soul phase upon which we have just encroached; the second is the great transformational network that reaches into the deepest embodying mechanisms and is centered in what we call the brain but that rises above the brain's organic-vital sense bondage. As a matter of fact this agency of the conversion of sense is operative already at the kinesthetic level of the living body. Every motion within an entire network is already organized as a directed whole; kinesis expresses the animus of a living being; all kinesis within the network partakes of an orchestration coming from the living being as a directing agency. As such each and every motion bears a threefold character: it partakes of the network's program and displays its characteristics; it is endowed with a direction entailed by its specific function; it expresses the vitality and spontaneity of its agent.

The moves by which we perform our existential tasks are embodied in that not only are they performed through our members but they belong to the workings of the great station of conversion that is the organic center of the brain and specifically refer to it for their articulation among themselves and for their intelligibility. But it is worth emphasizing that from the start the kinesthetic movement is, so to speak, ambivalent: to exercise our bodily limbs requires participation in a non-corporeal system of reference, in the psyche as well as in the brain. The latter can be analyzed in physical/organic terms; however, the sensory, sentient, kinesthetic *experience* is not organic.

IV. VOLUMINOSITY AND FLESH IN THE SELF-EXPERIENCE OF THE AGENT

(a) Retrieving my "forgotten" innermost sphere means withdrawing from world bound concerns to the shadows of the inbound visceral body. With the unfolding of more and more complex and flexible specifically human types of functioning up to imaginative and speculative significant operations, the emphasis in self-awareness is oftentimes "transposed" from circuits close to the soil through the organic/vital/psychic functions popularly called "my body" (which should be considered after our analysis rather as "myself as bodily being") to less soil bound psychic aesthetic, intellectual, and moral circuits with their great flexibility and consciously intense concatenations. These circuits naturally claim, then, an autonomy from soil bound operations, owing to their distance from them. They revel in their self-reflective force, in the vigor of their seeming independence, which owing to its flexibility/freedom does appear to be superior to soil bound functions. In this realm the human creative faculties project inventive devices that transform even our soil bound relations; instead of following blindly the natural beaten track, the person finds new avenues in living spacetime. Indeed, human beings, drunk on their creative powers, tend to expand their self-aware dominion far beyond the reach of their limbs; they tend to see no bound to that reach, as if their arms were wings. And yet this vision remains just an experiential fling. In actual fact, human beings do not and cannot avoid oscillating in their living selfhood between the poles of their existential spectrum, now putting the one out of sight, now the other. Human beings constantly oscillate in their existential performances — at every instant, with every step of scanning — between the efficiency of their mute soil referring functions and the vocalized inventive/reflective and spiritualized experience established by their presentational and re-presentational circuits: in the manifestation of givenness in the conscious constitutive processes, the re-presentation of life's constitutive effects in cognitive forms, and lastly the creative role of the elemental, that is, the aesthetically, morally, and religiously informed passions.

But if while feeling ourselves to be "elevated" and even powerful in this state of our complete selfhood we are suddenly struck by a severe malfunctioning of our organic/vital circuits, our elevated spirits deflate like pricked balloons. And if the malfunctioning be prolonged and impossible for us to contain, we may be forced to cede to others our place behind the wheel, so to speak, for our mental powers have lost their center of gravity.

(b) Here we reach the crucial point of our inquiry. It is, in fact, in the pervading fusion of mental/conscious performances and in inward bound organic/vital functioning, in the linkage of the operations of the will and aesthesis, that lies the harmonizing of all the circuits making up the territory of the embodied beingness. Neurology as a science sees this orchestration as being performed by the brain and the nervous system. But in order to move our legs in walking, our hands in lifting, our eyes in scanning, we need not only a system of distributive links but conscious propulsions, the impetus of will.

If such a willful initiative be taken and not spontaneously effected, owing to some disruption of our kinesthetic system, we feel helpless and thrown out of the game! When our incapacity persists and we become discouraged in our helplessness and feel alienated from our surroundings, we may turn toward a reflective, speculative, mental source of power for help.

On the other hand, when our psychic/spiritual endeavors in the societal realm are rejected, if we feel humiliated, abandoned by our close friends and associates, we may break off communication, relax the intentionality that keeps our sensorium on the alert, fold our arms, curl our legs, close our eyes, withdraw into the inbound domain of our body, its warmth, its silence broken only by our heartbeat now heard loud and clear, the solace of our entire beingness lying in this enclave all our own, in our most intimately felt identity with ourselves. We retrieve ourselves from the realm of our dispersal, where we met contempt, rejection, humiliation, defeat in some contest or where unfriendly forces have made our conscious and spiritual sphere shrink, weakening our self-confidence. We fall back on the physico-psychic domain of our visceral bodily base. That base, which is otherwise dependent in its performance on our mental circuits, nevertheless reveals itself to our feeling as our root foundation in living spacetime. We rediscover our deepest bodily self within the flesh that pulsates with imperturbable fidelity no matter what we may suffer psychologically or physically, no matter what our losses in terms of power, respect, credibility, reliance, no matter how intimately personal the disappointment suffered in the social scramble, no matter how "burned" or "scorched," we feel ourselves to be and what upheavals we are going through. On our retreating to our visceral bodily self, all these become ephemeral and recede from our existential orbit for a while as we abandon ourselves to the comfort of our flesh.

(c) Many forms of so-called narcissism, such as aesthetic or erotic narcissism are natural intentional moves properly directed toward others as objects/subjects of desire returning in truncated form to the subject of origin, who is then oriented toward his very own resources.[1]

(d) It is also in the tacit intimacy of our flesh as a receptorium of our inner self that the natural sexual drive seeks the completion of the amatory human encounter. Indeed, this most elevated spiritual rapport between a man and a woman in love remains incomplete unless they share the innermost intimacy of the flesh within the entire network of the body's functioning; this completion is itself on its way in a "carnal gaze," a caress, a kiss.

(e) All those of our circuits that achieve distance from the soil, such as our moral, spiritual, esoteric circuits, are infinitely variable, weather vane dependent. In order to take on significant shape they have to be subjected to reflective searching out and intellective interpretation. We may mistakenly take a violent outburst of love for hatred or fear when it is mixed together in the turmoil of our psychic life with everything else. While our feelings need to be sorted out, our shadowy, our mute visceral functioning remains regular, working all along in an essentially reliable fashion, substantially manifested in our voluminous solidity — within which our self-awareness solidifies. Our volume possessing embodiment is thus our anchor within the flux of the world's life forces, the anchor for our Promethean quest, an anchor to which we may turn as long as we live.

(f) While our thought, our joy and anger, our friendly or destructive attitudes remain hidden in the psychic sphere, that sphere may be altogether ignored in our intraworldly exchange. In order for it to be acknowledged, revealed, it has to manifest itself in the intersubjective language of action. We may disengage from our societal lived space instantaneously, but our very own vital space, our organic place within living spacetime is relatively stationary and is secured by its very own nature: by its direct relatedness first to the life system and second through its voluminosity to the soil in which all living beings are rooted, which is life's reference point, measure, compass.

When because of dysfunction or injury to our vital/organic inbound system, the embodied living being loses its foothold and reference system in the soil, then organic sentient vital functioning dissolves and no efforts on the part of the conscious outbound functional apparatus may restore it because this apparatus too dissolves concurrently, not to be revived.

(g) Our deep bodily spheres faithfully carry on in silence the tasks that maintain us in existence during our flourishing and expansion into societal spheres while being all the while harmoniously orchestrated with all other circuits, so harmoniously that we do not even have an opportunity to pause and wonder about the interval, if any, between ourselves as embodied and "our body," since this latter appears to be a set of instruments subserviently incorporated into our complex spiritual beingness. Still, when the seemingly all-encompassing sphere of my outbound circuits seems to have burst like a balloon or sparkling soap bubble, I may withdraw to those deep bodily circuits as a bastion, stable and secure, redirecting my attention to my innermost visceral sensations, to my pulsing blood, my sure step, the smoothness of my skin. In the full solidity of my "substantial" experience of my voluminous dwelling, I may relax from the intensity of the spiritual, intellectual, moral, societal ties woven by my outbound operative systems.

TYING POINT TWO

ANTICIPATING THE MANIFESTATION OF THE LOGOS OF LIFE

CHAPTER ONE

METAPHYSICS OF MANIFESTATION

Logos in the Individualization of Life, Sociability, and Culture

I. INTRODUCTION: REASON AND MANIFESTATION

We may approach the question of reason from three different points of view. First, we may follow the Greeks and make a distinction between certain and dubitable cognition, between episteme and doxa; or, second, we may adopt the critique of reason initiated by Kant and focus on the powers and operations of the human mind assumed to be the source of reason. In this study a third approach is proposed, one that is a counterpart of my previous critique of reason. In contrast to the manner in which the two approaches mentioned first divide rationality into two types (episteme and doxa, on the one hand, and "pure reason" and the "rationality of Nature," on the other hand), I have in my critique of reason taken the human creative act — and not the cognitive act — as the Archimedean point from which reason and rationality may be differentiated into several types and innumerable strands that sustain the cosmos, life, and the orbit of specifically human life circumscribed by the radius of the human creative genius. Having already investigated the origin of Logos, we will now raise our sights to *the vision of the entire spread of the Logos in its manifestation.*

I have already elsewhere ascertained *the priority of vision over "proving" by methodical argumentation in the actual work of a scholar, scientist, philosopher.* The proving of tiny fragments of cognition is meaningless if they are not placed within an all-embracing vision.

Here I want to bring out another crucial point further supporting this view, namely, the fact that meaningfulness-rationality takes place, occurs, is "opportune," congenitally implies *manifestation.* Forces, energies, virtualities, potentialities, proficiencies, etc. become "meaningful" as such and come as such to be "activated," defining themselves in their nature, only when they assume rational "forms" in their manifestation. The formless, the mute, the soundless, the invisible, the transparent, the weightless are in themselves meaningless; these concepts acquire meaningfulness only in contrast to their opposites and with reference to their generic quality. The "hidden" acquires meaningfulness only with reference to the "obvious." It is only in the gigantic manifestation occurring, according to our myth, in the sudden explosion of the cosmos into being, that its generic forces acquired/revealed themselves in their forms.

The "hidden" and the "obvious" in their opposition lay the basis for the enigmas of life and of the human being. The "concealed" has meaning only with respect to the "manifest." The play between the two, which is one of the threads in the dramatic play of human reason, lies at the very heart of the Logos of Life.

In philosophy, we are so captivated by conjecture about the hidden and the obvious that we get absorbed in it, digging our way like a mole into ever deeper corridors. We embroil ourselves in labyrinths, in the obscure, and we forget about getting back to the light and the great vision of the manifest.

It is time we turn to the great vision of all that is manifested. After we have seen how the human creative act spreads its constructive rays — like the sun at its zenith — gathering the relevant synergies from the virtualities of the Human Condition, on the one hand, and from the realm of bios, on the other, as well as constructive relevancies from the laws and rules of the cosmos, then this vision of the universal play of forces takes precedence over the two classical approaches to rationality/reason mentioned at the beginning.

It takes precedence over the narrow focus on human cognition that dwells on certitudes and fallacies and restricts the horizon to the world of one living species and has but limited grasp of comparable things among ants and eagles; this horizon bounded by the faculties of the human mind introduces inflexible and illegitimate borderlines between man and bios, nature, cosmos. *Our approach, however, supercedes the classical approaches in that by it we gain access to the hitherto ignored or little considered unity-of-everything-there-is-alive.* Carried by the rays of the human creative act we reach beyond the limits of the human cognitive grasp and enter directly into the channels of our human participation in the existence/subsistence progress of life and of the universe. No doubt, cognition makes preparatory forays into the inner workings of the universal forces. It fails, however to reveal our Human Condition as a major station in life's advance. The overall play of bios and cosmic forces and laws in our innermost workings appears in the manifestation of the Human Condition, which is incomparably fuller and more of the essence than whatever our cognitive grasp with its abstract forms and approximating conjectures may yield.

Within the full panorama extending before and within us opened to view by our approach, we see clearly the fallacy of the classical philosophical approaches. We realize that *a sharp division, even sharp distinction, between the rationalities of the cosmos, life, nature, and those of human creative genius cannot be made*; further, we see that cognitive rationalities are just a specific modality of human creative genius, and that in turn *the human being in all his functioning is himself a manifestation of the forces, powers, constructive rules, and proficiencies of bios, life, and the cosmos.*

II. VINDICATING A METAPHYSICS OF MANIFESTATION OVER CLASSICAL EPISTEMOLOGY

1. *Cognition on Trial*

Before we enter into the exfoliation of the vision of Life, we will probe further into the manifestation of the Logos of Life in its underpinnings as we sustain the contrast between cognition and the creative experience of man.

To begin with, let us state that while in cognition we are caught in the intellective rationalities constituting the specifically human universe of life, society, and culture, we, in contrast, through the human creative act that gathers all the proficiencies of life, discover innumerable threads of rationalities sustaining the entire life system that lies within the schema of cosmic forces and the carrying on of life's progress; these rationalities manifest themselves through the genesis, growth, and advance of life. Cognition, having its fulcrum in the human mind, meets with the manifestation of life, especially with life's individualizing genesis; cognition and the individualization of bios constitute, thus, just two different poles around which the Logos of Life organizes rational paths of life. The creative act of the human being is the meeting point of the two paths. Through investigation of the workings of life as they appear in the creative act of the human being, we may discover the shortcomings of cognition. We may also see life's hitherto shadowed features in a novel, revelatory light.

The very question of the conformity of our cognition of things, beings, events, processes, etc., of all from our surrounding world, to their assumed "objectivity," that is, to their status independent of the cognitive processes — the crucial question around which all the epistemological questions turn — appears in this perspective to be itself inadequately formulated. Already Edmund Husserl, nurturing some hints at the inventive nature of the mind, rejected radically any identification of cognition so understood with the "constitution" in which, according to him, the objectivities are formed and devised; the referential dependence of the cognitive processes (understood as constitutive) on any assumed "referent" lying outside of the cognitive process is thereby disclaimed. "Objectivity" as such is shown by Husserl to be precisely the effect of constitution. The assumption of a direct correlation of the cognitive process with an "object" or "thing" lying beyond the "object of cognition" that would represent it and mark the end of the fulfillment of this process seems to be an effect of an undue and hasty narrowing of the focus that cognition commands, and of a forgetting — or ignoring of — the fact that cognitive consciousness is, first of all, just an outgrowth of the multiple functions of human consciousness, even if it be the major one.

2. *Cognition among the Life Functions*

When we envisage cognitive consciousness within the entire field of individualizing life opened to us by the creative act, and, consequently, within the evaluative progress of types, we follow consciousness from its emergence as it takes on progressively varied degrees of complexity. On the rebound, we discover the dependencies and existential interconnections of the cognitive processes with the rationalities of the life process. From the simplest phases of life, from reactivity to sensitivity, to psyche, consciousness surges as a functional complex prompted by the inner workings of self-individualizing life as an organ necessary for processing life itself — an organ stemming from nature, belonging to nature, remaining in its service. Consciousness has reached its highest level in full-fledged human experience and self-awareness. In this perspective we see that the clear, distinctive intentional intelligence proper to human consciousness is the crowning point of the evolutionary progress of individualizing life, but also that *at all phases of life's complexity there is essentially present a network of intelligence at work.* Following the creative tentacles of the self-individualizing progress of life, it seems that the simplest forms of life show already a shadow of intelligent activity, that is, of *intelligence understood as a sensing reactivity to inner or outer stimuli, as a modality of reason that sets itself apart from the rationalities governing cosmic and prelife functional spheres.*

Here we have reached the culminating point of our inquiry. The human intelligence proper to intentional constitution is inventive and creative in establishing spheres of life with rules, laws, and structures having a flexibility unprecedented in the evolving course of life. But despite the operative input of the main engine of the Human Condition, that is, of Imaginatio Creatrix and the creative act that seems to soar above all the regulations and constraints of bios, loosening or breaking its ties, and in spite of the fact that the distinctly human intellectual qualities set the human being free from the confining provinces of instinct, again, allowing him to exercise freedom through his power of invention and to leave restrictions behind as his imagination takes wing, yet — and here lies the insight to which this discussion leads — cognition/constitution does not introduce rationality per se into the orbit of life. It constitutes merely a new phase of intelligence within the spread of life.

3. *The Integrating Vision*

This type of intellection so extraordinarily different from the rationalities of the cosmos, bios, nature as well as from the processes of the simpler life functions cannot, in the last account, be set apart from them. What cognition puts asunder, what epistemology, believing in narrowing the focus and splitting the hair in ten to ensure artificial certitude, separates, the creative vision, by drawing on the very source of all in the differentiated powers of the Logos, encompasses in a discrete unity. We have to recognize, with Pascal, "the grandeur of the human being." And yet as autonomous as human creative powers are in the full gamut of their "vocal," "graphic," "motile," "emotive," "sentient" metamorphoses and in the infinitely varying voices of their expression (in marked contrast to the mute and stereotypic rationalities of the cosmos), and as revelatory of the universe in and around us as are the thrusts of conscious intelligence (while the rationales of the forces and powers of bios are bound by narrow tasks and repetitive cycles), it cannot be overlooked that the wondrous faculties of the human being even as they inventively and creatively transform and expand Life's orbit, and even as they have the power to open an entirely novel sphere with respect to the simple realm of bios, are themselves the manifestation of life, nature, and the cosmos, being the most striking affirmation of their power and driving forces.

What we want to argue here is that in the novel perspective that here gives a radical anthropologico-metaphysical treatment to the nature and role of the creative function of the human being, cognition is put on trial and a radical change of perspective occurs. Instead of choosing cognition as the entry to philosophical investigation, we have opted for the creative act of man, opening then a diametrically opposite door. Through this door we have entered into the workings of life itself. Cognition proves to be just one of the moments — a crowning one, to be sure — of the functioning of life.

By this discussion we have been moving toward the discovery of the manifestation of the Logos of Life in its accomplishments — that is, toward a metaphysical modality of inquiry into reason and rationality — and contrasting it with one of the factors of that manifestation, the cognitive-epistemological modality. This vision, which places all of the perspectives on life within their assumed common ground, does not imply, nor does our argument warrant, any hasty adoption of naturalistic, vitalist, or pantheist views. In the final analysis, which lies far ahead still, life itself might turn out to be beyond these categories, necessitating and calling for a special formulation.

4. *The "Hidden" Vortex of Manifestation: Virtuality/Actualization; the Outward and the Inward*

Can we in any way describe the significance of "manifestation" other than in contrast to its opposite the "hidden"? In the circularity of their meaningfulness, crossing from hidden to manifest is first of all "bringing out of hiding," but this implies that what is thus coming out of obscurity into the light has already been present in its fullness. When we talk about the manifestation of the Logos through life, we mean indeed, the works of Life seen in full light. But our very revelation of life is grounded in the notion of the progressive genesis of life, that is, in the conjecture of a passage from the prelife realm to a breaking forth in the crystalization of life.

Furthermore, the crystalization of life is partly synonymous with what Husserl called "presencing" (*Vergegenwärtigung*). Crystalization means formation and implies simultaneous or successive steps — actually both — resulting in motion, and bringing about temporality. Here seems to lie the very knot of factors by which manifestation may be grasped — or approximated — in its nature. "Presencing" means the first, the clear, diaphanous appearance, apparition of intentional objects within the vast horizon, internal as well as external, of the human intentional universe;

this "apparition" is understood as an end result of the entire intentional process. Nevertheless, in our perspective we have to envisage presencing as also occurring apart from the constitutive/cognitive processes, apart from subjectivity. Presencing, which in the subjective sphere means "presentification" to . . . "appearing to us," means the progress of life's crystalization in the sphere of life's genesis; there is a passage from the "hidden" but virtual to the appearance that brings about spacing and scanning as life takes shape within a phase of constructive spreading and expands the dimension of space. The manifestation of Life consists precisely in this passage from the virtual to the actualization self-presencing its constructive generative scanning.

However, to echo what I said earlier in my cosmology (*Why Is There Something Rather Than Nothing* [Assen: Royal Van Gorcum, 1967]), in the passage from the virtual to a crystalization in the cosmos and in the realm of bios, progress is inner-outer oriented. And yet, the "inner" workings of the cosmic forces remain concealed; hidden, they explode into light only when attaining formative effect on a spacing/scanning axis.

Herein lies a distinction between the manifestation of life and that of the cosmic rationalities. The inner-outer orientation of the life processes leads in self-individualization-in-existence to a double dimensionality that opens and unfolds: the "outer," the spacing/scanning of life's progress, the world of movement, formation, duration, and succession, of the spacing/scanning common to all cosmic rationalities; and the "inner" manifestation proper that concurrently occurs within the living individual and makes up the other side of the coin, namely, the psychic forms establishing human intelligence in the specifically human universe of existence. This will be discussed later on in more detail. As for now, let us emphasize that it is the inner/outer direction of all life or prelife processes that offers the axis for manifestation, just as much as it is spacing/scanning expansion that accounts for its forms, shapes, "appearance," and presentification.

To conclude and set the stage for manifestation itself, let us summarize. As we follow the unfolding individualization of life, we find the Logos of Life expanding in innumerable modalities setting off from the universal schema from which life surges and then projecting progressively articulations for life's unfolding from its smallest and simplest forms through more complex forms to its highest complexity, that attained in the human creative phase.

When, proceeding in the contrary direction, we gaze out from the peak of the expansion of the Logos, that is, from the human creative phase, from us in whom the Logos of Life is differentiated into infinite strands through rationalities that are autonomous with respect to bios, we discover the human mind's consciousness to be the center of its work, as it processes all rationalities in all directions so that human cognitive intelligence crowns the work of creation.

Raising our sights to the metaphysical plane, we are able at last to put the epistemological plane in its proper position and to place the question of reason in its proper perspective, for manifestation covers the entire spread of the Logos as it extends now in one direction from its central point, which is the origin of life, and ramifies in its unfolding and extends too in the other direction, toward the cosmic logos, going from the origin of life backwards to prelife conditions indicative of the constitutive network of the realm of life and reaching the cosmic laws upon which life's forms and play of energies are suspended. All life is suspended on the cosmic parameters. Cosmic Logos/Logos of Life/the logos of specifically *human* intelligence, the Promethean Logos, all of these projects of reason and rationality radiate in life's manifestation.

To treat the question of ratio in its due magnitude, we turn to sketch the provinces of reason in the metaphysical vision of its manifestation.

CHAPTER TWO

SPONTANEITY, CONSTRUCTIVE DYNAMISM, AND CIPHERING IN THE HUMAN CONDITION

I. FROM THE BECOMING OF NEW LIFE TO THE HUMAN CONDITION

In the perspective of life, the emergence of new living beings and their self-individualization poses new ontological issues and gives new essential ontopoietic insights. First of all, we may think about the emergence of a new being as the emergence of a set of individualizing germinal structures, or rather virtualities with constructive propensities and propulsions. Yet when we turn to the progress of the evolutionary process, we witness a specific ontopoietic pattern, such or other, that maintains continuity between the carriers, a carrying system of this continuity itself — evolutionary continuity within intrinsically individualized carriers.

It is owing to these carriers of continuity that a species continues from generation to generation. And strange to believe, the strategy of the logos of life resides in establishing these footholds of genetic continuity of the type within the intrinsic endowment of the single self-individualizing being, in the self-individualizing process, even in the heterogeneous ontopoietic phase establishing the individual's existential autonomy. This culminates in the continuity's being concretely performed, accomplished within the innermost functional network of existence.

In this functional polyvalent organization — first and foremost, carrying on the self-individualizing course, second, promoting the individual being's unfolding and growth and progressive diminishment and extinction, and, third, carrying a distinctive, functional schema for the performance of the emergence of the novel living self-individualizing being

— is the key strategy of the expanding logos of life. In the established rules and laws of the inward-outward oriented processes of growth and vital existence, there enters a selective and deliberative function that gives space for flexibility and this at the very heart of the proliferation of logoic forms. But this flexibility, which in the animal kingdom is circumscribed by the operation of natural selection, expands with the human individual. Not only is there a far-reaching selective variety in the basic vital circuit of complementarity — it remains the law that the human type operates procreatively within its own type — but there is a vertiginous measure of novel features, qualities, forms, taste, etc. in the circuits of selection for the procreative project.

And it is not only that the establishing of the complementary, dual, cocreative unit has a vastly varied circuit of selection, but in the intergenerative process that combines, adjusts, transforms the received propensities, the radiating vibrations of the logos of life find their crucially significant workshop. This is manifest in the innumerable shades of color, the endless nuances in features, etc. in the entire spectrum of humanity's races and ethnic groups at any historical phase.

While remaining within its own species, the human individual manifests the proliferating logos of life within the space of freedom: freedom in selecting the specific nature, character, bodily appearance, temperament, inclinations, predilections, etc. of one's partner, a selection that occurs in part on the basis of spontaneous attraction and in part on that of attunement of character and dispositions — on both emotional response and rational appreciation, then.

There is not enough wonderment over seeing this station of life's transformatory work, this nodal point in the continuation of its effects, installed within the functional crystalization of the Human Condition. It is from the creative orchestration of human functioning, which brings in the flexibility of vital moves and above all the space of imaginative freedom, that the logos of life accomplishes its innermost constructive aim: to expand in imaginative rays through the very vitally significant ontopoietic design that crystalizes the basic differentiation of beings.

This frees the human being from a fixed dependency on one type, establishing through the procreative complementariness functioning within the creative orchestration of the human condition a sunburst of specifics — of shades, shapes, features, dispositions, abilities, etc.

Let us now emphasize the ingeniosity of the ontopoietic design in its establishing the dynamically progressing reality of life and its relevancies. The secret of life lies undoubtedly in its "renovating" course, that is, in the succession of the phases of growth, diminishment, extinction accompanied by the emergence of succeeding new life. The death of living beings has become the innermost condition of life. All is grounded within the ontopoietic design of self-individualizing beings, all the focus of life's unfolding seems to lie in this individualizing progress and its cyclic unfolding. This cyclic unfolding is precisely the foothold of the secret of life: emergence-extinction.

Without this cyclic law of unfolding, life would become extinct in its first steps.

II. THE HUMAN CONDITION

In its unfolding the human condition inherits from the system of life some basic ontic constituents:

1. those characteristics and virtualities that pertain to the cyclic design of the constructive unfolding of the living being as such;

2. those that inscribe the individualizing being as an entity, ontically autonomous and reliant on its intrinsic agency-in-progress within the constructive progress of the species;

3. what is most specifically the dominant ontic prerequisite of life: procreation; the human individual is fundamentally outfitted for procreation to carry on life's evolutionary advance; and

4. the body's intrinsic mechanism for replacing its matter, which leads to decline, corruption, extinction.

The essential expression of all this is the basic dualness of the human individual. The cycle of individual life is paired with the procreative virtuality of the human individual. This latter, however, involves the complementary endowments of a pair of individuals. From the perspective of life in its active flux onwards, the life progress, life preservation, the powers of life, life as such follows an ontopoietic law of cyclic progress. Life sees individuals carrying it onward in complementary pairs.

III. SPONTANEITY, THE CONSTRUCTIVE DYNAMISM OF LIFE, AND CIPHERING

Every serious philosophical endeavor consists, first, in taking up anew, that is, *ab ovo,* the great task of the philosophical reconstruction of the universe within which the philosopher as a human being is enmeshed, that is, of grasping within an intelligible discourse unfolded by the human mind's expressive means and rules the progress of the constructivism of the logos of life that is in the mind, in the Human Condition itself. The human mind then — which is differentiated and identifies by its own means the concrete among the significant factors around it and probes what remains hidden to the mind's analytic powers, this mind seeks to discover links among the signs it discerns, conjectures about those that remain hidden, and attempts to discover frameworks within which signs, concrete and hidden, have specific functions. In the process and progress of life itself the primeval logos ciphers the signs that the human mind seeks to decipher. The signs, the significance of the universe as well as of the cosmic establishment-in-process and of life's flowing advance is, on the one hand, that of the differentiating logos; on the other, it is being projected, ciphered by this progress itself, as Romantic poets such as Novalis, and philosophers as well, have clearly intuited.

The impetus of the logos from its incipient bursting forth in initial spontaneities is harnessed by its being ciphered through the operations, moves, pulls, fusions of the universal constructivism into opposites, intensities, measures, formations, seminal and generative diversifications that culminate in the strategies of equipoise of life's origination and unfolding. Thus, it is the manifestation of life in constructive progress that is the very factor of sense, the ciphering of the logos.

IV. THE HUMAN SCRIPT: THE HUMAN BEING INSCRIBING HIMSELF WITHIN THE SCRIPT OF LIFE AND NATURE

Insofar as we envisage the specifically human unfolding as occurring through the person's individualizing himself amid the social, cultural, and personal lifeworld within which he develops his system of meaningfulness, this is no longer an anonymous deciphering of the life system in its generic conditions, laws, and aims. The existence in which a human being differentiates his concerns, aims, desires, and ways of satisfying them in his life progress and in relation to other human beings unfolds a most complex schema of conscious deliberation, choice, decision in which he experiences himself as the deciding agent. Selecting and determining his life course, he differentiates himself from other beings with their individual radius of pursuits by bringing them all to the central point of his involvement and consideration, where before seemingly infinite possibilities, by making his selections, he establishes his seemingly very own life meaningfulness. And yet this life meaningfulness is nothing but a variation on the universal theme. If we look from a certain detached distance at the life of the great majority of men, or, for that matter, at the life meaningfulness the "typical human being" carries onward, we might be tempted to see all the acts of the human agent as being circumscribed in their reach and as being as repetitive and forseeable as those of birds or of ants, who also live in "colonies," within a social environment. Their occupations keep human beings moving in a circle in their lifeworld; within each circle acts are repeated infinitely, acts such as anyone could perform just as well. This is the accepted specific situation that like a piece of canvas, or a piece of land, offers well-defined possibilities for embroidery and no others — an accepted well-defined frame or script that assigns the individual a role to play on entering into the lifeworld. This is a condition indispensable for our existence's unfolding, which is, on the one hand, the condition for the individual in existence (the part of each being just a tiny fraction of the universal lifeworld script to be played) and, on the other hand, a total commitment to it and resignation to this one and only loom at which the individual maintains life's equipoise and is "happy."

The question that remains, however, is that of whether man can definitively close his eyes to all the possibilities that actually exist. Can he blind himself to them in the conviction that the part he is playing is the only accomplishment of a significant existence possible? Would another really not be open to him? Would he necessarily have to suffer should he discover and see other possibilities and feel no capacity to actualize them? Would he have in such circumstances to reject his own, limited, significance of existence as futile and himself as unfortunate? We will come to these questions later on. For the time being it seems opportune to note that this universal significance of the individual human script of existence despite being the fruit of his toil and decision, toil in which his entire agency operates, is still not uniquely his own: it is the role of life common to all men, and so is so to speak "neutral." As André Gide says, this is the condition of every man, existence on the level of a "neutral territory," which lives in every man. Gide gives in the personage of Palude an example of the "universal and necessary" human existence. He is himself inspired by Vergil's Tityre and points out that Vergil uses the expression "Tityre recubans," that is "a man lying down," which is meant to express this territory at which the human agent is plowing the field of his existence and which is in every man.

In fact, no matter what we do, what our profession is — whether just a common laborer toiling to barely satisfy our vital needs, to wring out of our life conditions the bare essentials (which, however, does not bar unfolding a higher meaningfulness in moral terms!) or the occupant of some seemingly exalted station in the lifeworld, and whether or not we can rise above the survival significance of life to unfold our artistic imagination and have complete freedom to choose our life route and meaningfulness free from contingent constraints — we enact of necessity, however we go about it, the universal script in some or other typical manner.

The great question that emerges is this: Is human existence limited to that script? Is the human individual bound to close his eyes, bow to the ground, and submit unconditionally to its demands?

V. THE UNIVERSAL COORDINATES OF THE SCRIPT FOR THE INDIVIDUAL HUMAN BEING'S LIFE COURSE

This "neutral territory" of the human individual's life script is laid down (projected) in its great lines by the principles for life significance established within the lifeworld, which decide the lifeworld's entire progress.

First, there is the life script and significant life course of the most general distinctive social group in which the individual plays a significant part, that is, a nation, a tribe. A nation is a human grouping rooted in common vital interests and above these a significant system of life orienting ideals, devices, and convictions uniting individuals. The nation gives, so to speak, the most general lines for the individual's significant inscription within his lifeworld and a choice of roles to play within it.

Inversely, the human being's life course cannot develop other than within the significant life course of his nation. Within his own radius he partakes of the line of its development; in his own unfolding, he is subject to the confines imposed on his nation's course. His meaningful unfolding may be twisted, uprooted, annihilated, but also enhanced, promoted, and brought to its full fruition by the development, confines, distortions, hindrances of the life of his nation. That is to say, he is in his own individual unfolding participating in the "common fate" of his nation. Should he like to shake off that fate, he has to dissociate himself from the nation. But his social and vital existence is so deeply rooted in his "native soil" that should he dissociate himself from the fate of his nation, he would never be able to grow such roots elswhere. At the utmost, he may link his roots with the roots of beings having originary existential connections with him (e.g., those of his children) or he may throw a bridge above them in the form of higher ideals by which he may link up with another nation's existential script.

Above and below that is the individual as a significant element of an originary human community, one who shares in its common dependence on Nature's cosmic conditions. He is subject to violence from the elements and the blind forces of the inorganic cosmos, over which he has no jurisdiction — earthquakes and other cataclysms — and he is helplessly thrown into the common fate of his native community.

But to his universal coordinates as an element of the social lifeworld belong also the coordinates of the destiny of his social group. He either thrives with his once initiated life script if it falls in line with the destiny of his social group or with the situation of his profession, or he is thrown off the saddle and has to find a completely different line along which he may plot the meaningfulness of his existential route.

In actively resolving his existential situation, he has to cipher its dynamic moves with reference to their constructive virtualities and his own dispositions and endowments. His existential route designs his individual script.

Thus the primogenital sense — prius in genesis — in which the edifice of the universe and life manifests the logos is the sense of the work itself: the workings of its constructing, through which the logos differentiates itself in being ciphered. Hence, the human mind's de-ciphering of life deposited signs cannot be entirely adequately accomplished, for the signs deposited by the very operations of construction cannot be grasped by the mind's representational functioning, the mind itself being the last touch in the series of life's constructive unfolding. The cognitive/constitutive apparatus by its very operation adds its own intellectual significance to that which has been deposited by the workings of the unfolding bios and embodied psyche (if we may artificially, that is, by the specific operations of this mind, provisorily draw some clear distinction between them). Indeed, the human mind as de-cipherer, reader of signs, is in the first place the factor — the last factor — in the line of the sign-making and depositing itself. What allows the human mind to deal with, to manipulate and discover the message of the signs that life deposits is precisely mind's, the human being's participation in this process. The philosopher as a living being belongs as an integral segment of this process and carries within himself all the segments of these sedimentations.

This truth, although, intuitively implemented by philosophers dealing with natural processes — Aristotle, Leibniz, Hegel, Bergson, and later Whitehead — has been expressly stated as a postulate for philosophizing only by Peirce. And yet Peirce did not get at the crucial aspect of this insight, namely, that it is the creative functioning of the human mind that should be followed — and not its cognitive functioning as traditionally assumed — in the inquiry into ciphering, and that should be consulted in the proper approach to deciphering. It is the task of creative phenomenology to introduce this work.

The access to ciphering is barred to our cognitive/constitutive functioning. *The light of the intellectual significance given objects within the frame of the lifeworld is so dazzling that the human manifestation of life's script overwhelms the evidence of the other manifestations of life's script, covering up and blocking access to all preceding phases and ciphering sedimentations. Hence the creative function of the human mind, that which itself continues the ciphering progress, provides the only path toward authentic recovery of the original ciphers.* All human cognitive and artistic endeavor seeks to discover the original ciphers of existence. The sciences, the arts, technology, and philosophy alike aim at discovering the hidden springs of a sector of reality. Cosmology, natural history, the arts, and above all philosophy aim at interpreting these ciphers using the intellectual tools of human discourse. The creative phenomenology being expounded here directly engages the access to the original ciphering of the universe of life: its path is the itinerary of the creative workings of the human mind, which prolong the itinerary of unfolding bios and its cosmic substrate.

Hence the unique encounter between the creative phenomenologist and the natural scientist on the construction site, as it were. Hence the extraordinary cross-pollination of most significant insights and intuitions.

However, we must be attentive not to identify the insights and intuitions of each. There are several points at which the endeavors of the philosophizing phenomenologist and the natural scientist both recover phenomenological insights and incorporate them into his/her theoretical framework with its specific focus. Therefore, even if the intuitive insights coincide, their interpretations of signs diverge in their focus and in their emphasis etc.

PART THREE

MANIFESTATION AND DIFFERENTIATION

CHAPTER ONE

THE SURGING MANIFESTATION OF LIFE

I. THE "PHENOMENAL" SPHERE

We have so far been pursuing the unfolding of life in its "inner workings" and this within the context outlined by the strategic points of relevance: with nature-life (the womb of life) being the center arrayed around which is the prelife zone of cosmic relevancies and the unity-of-everything-there-is-alive. However, in the final chapter of Part Two above, we progressively reached via embodiment the central complex of the brain as the promoter of the transmutation of sense, and the living agent as the core of all the functions of the living individual "at one glance" and "uttering a voice." With that another sphere emerged into the light — the sphere in which the inner workings of self-individualization become "manifest," a novel existential sphere. We had encroached upon a new sphere of multifarious sense, that of the manifestation of life proper. With the opening of this sphere, the context of principal life relevancies expands. Role distribution undergoes a transformation as well. In the previous sphere of self-individualizing life in which the inner workings threw tentacles out into its support system, the emphasis fell on the singular individualizing origination. But in the next phase of life, in which the inner workings of the already projected individual-in-formation prompt it to insert itself by its own spacing and timing into a conundrum of other singular beings and occupy a "space" among them, the "phenomenal" sphere of the manifestation of the logos of life emerges. In this sphere the individualized being "appears" among others within the context of manifesting life. In this sphere we become aware of the differentiation of kinds.

The novelty of this "appearance" consists in two main points. First of all, although the individual is always in progress it appears fully delineated in a pattern of unfolding and endowed with a dynamic battery of forces, energies, synergies promoting it. Furthermore, it is not a haphazard conglomeration of forces — on the contrary it always emerges as one of its "kind"; it shares with other singular beings similar (or identical) vital characteristics requiring similar (or identical) vital conditions. In other words, we recognize them not merely as singular beings but — and foremostly — by the general features that they share, that is, in their "type." The term here is "re-cognize." With the embodying strategy of the brain, we have already entered into the functional spheres of the higher living beings culminating in those of the human being. That entails for those higher beings on the one side, the "presentation" — in the sense of manifestation — of life and, on the other side, a corresponding "recognizing" function of this presentation. This is the basic operation within the entire gamut of intellective intercourse, of the logoic strategies promoting ever higher and more complex networks of life.

The simple presentation/recognizing — one implies the other — strategy we will call "phenomenal." Indeed, the world and its beings are to us humans — and to a lesser degree to all other living beings — given in direct, immediate networks of presentation, in "phenomena."

We will to begin with a quick overlook of the phenomenal presentations of manifested life, which are prior to all philosophical interrogation of them.

The first and foremost feature of the phenomenon of manifestation is the differentiation of life.

II. THE DIFFERENTIATION OF LIFE: PLANTS AND ANIMALS

First of all, there is to be seen an overarching strategy to organize as well as to order the constructive inner/outer mechanical device of the individualization of life, of its interlacing dependencies.

As a matter of fact, putting emphasis upon the individual, we cannot overlook that the individual could not surge forth unless there were appropriate prelife conditions and that its complexity unfolds in accord with those conditions, which their appearance and development then changes. The overarching ontopoietic device of generation attracts our attention to the fact that types of living beings generate in conjunction.

The metabolic mechanism of life is not limited to the sustaining of one individual; on the contrary, each individual throws hooks that fall into the constructive progress of other living beings. Supplying the requirements for survival does not just satisfy the needs of individuals, but promotes further external exchanges. The individual then unfolds within a network of life to which its own progress contributes.

However, there is also a more basic life strategy, one that leads to the establishment of coexistential networks. This is the very differentiation of types and their various complex specifications. Looking at the vital interdependencies of types that ensues, we may see an extraordinary strategy of the logos of life to establish equipoise with respect to its various schemes.

The first differentiation of living beings into plants and animals offers the most striking example. It has been usual to point out that the most fundamental difference lies in mobility: the animal is endowed with mobility, whereas the plant is fixed to the ground. Hence, as Aristotle pointed out, there is a relationship between mobility and perception: both contribute to expanded flexibility in selection, choice, satisfaction, understanding.

However, we cannot deny the plant all mobility; it moves in the process of growing or shrinking, it is moved by the wind or by other agents, in numberless minuscule moves it may daily follow the sun. What it does not possess in comparison with an animal is the capacity to "propel itself," to move of its own impulse from the place in which it is grounded to another. It may be transplanted but not of its own impulse, initiative, and direction.

Yet, it should be considered that plants are compensated for this shortcoming with respect to animals: the self-sustaining mechanism — metabolism — of the generality of plants draws simultaneously on two external "spheres": the soil with its water and nutrients and the air with its energy supplying sunlight.

The plant's fixedness in the soil and the element transforming system that it establishes in the soil crosses the border — what border there is — between the vital and the nonvital, vegetable and mineral, etc.; indeed, it is through its unique position with respect to the "ground" that the plant plays a twofold role with respect to the universal strategy of the logos of life:

(1) First, it is the immediate transferer of energies from the nonvital to the vital. In a way it "animates" the ground, endowing it with the significance of being "fertile" soil. And in its transformational function at its simplest level it returns to the soil what it cannot use in a direct exchange. Lastly, the plant returns everything to the soil in its putrefaction so that the whole individualizing progress of life may be renewed. It "recycles" life's resources.

Thus, the plant plays the role of the mediator of life resources being recycled over and over.

(2) However, there is also another even more fundamental role that the plant uniquely plays, namely, that of offering the ground for the ordering of life's individualizing strategies as well as for the entire projecting of life's interfunctional networks. Its fixedness in the ground is such a basic condition of spacing itself. In order that life may advance in a progressively more flexible exchange between the individualizing agents as well as the factors that provide resources for growth and sustenance, there must be a relatively fixed groundwork that provides reference. We may then orient ourselves — all living beings do — by the spaces determined through the growth of trees, of crops, of water plants, of grass or of other innumerable species of plants. They constitute the primary source of nourishment for animals and humans who perpetually recycle their substance.

Indeed, the rootedness of plants is the "watershed" event through which the mobility of life was distilled from the inert matter that serves it.

(3) But there is still the question of the different types of mobility among complex living beings that has to be considered.

As said before, mobility emerges with the process basic to life; it emerges together with the scanning of time and the spacing of extension. This mobility, specific to constructive growth, that is, of the scanning that involves of necessity kairos and chronos is specifically characterized by its surging forth of itself and maintaining the inwardness of beingness-in-progress. This mobility characterizes life as such and brings about the radical distinction between life and nonlife processes: flexibility within the progress of growth.

In contrast, the mobility of the animal is of a self-propelling type. The animal establishes a system of inward directed satisfaction of life necessities. That goes hand in hand with a capacity to initiate a move that will space extension in accord with a directive given by the agency within. Thus in contrast to the plant, which is limited as to its interiorly oriented motion for the satisfaction of its needs as dictated by its structure and metabolism, the animal can seek to satisfy its needs by roaming about on the ground that is home to the fixed plants, and through its crucial capacity to propel itself extensively it may unfold its inward agency into other functions as it measures its choices and measures as well itself and what it needs. It also on the rebound enjoys the satisfaction of its needs to a degree not expected in a plant. The same holds for other vital functions. In the continuation of the individuation of life, a plant grows a "blossom" at the time of its maturation — a specific outgrowth that allows for the spontaneous production of seeds for further promotion of its kind. It can also allow for fertilization involving appropriate seminal elements from another plant or tree. Here the fertilization has to be effected through an intermediary in motion: either through an insect, a bird, or the wind. Thus the plant, although in its own unfolding and life-course it is fixed in one environmental position, may in its generative functioning not only spread its germinal capital "around about" by itself but may also participate in the interfunctional network of life through the activity of other beingnesses, namely, birds, insects, frogs, human beings. By such mediation, that is, through partners whose moves it cannot control, a plant may spread its seminal logos.

III. THE SOUL OF THE PLANT

The plant may be fixed to the ground and as such in contrast with animals — especially those of a higher order than corals, algae, etc. — be deprived of the possibility of spacing its existential functioning by a further unfolding of organs for observing, comparing, seeking, and ultimately choosing, that is, of the capacity to discriminate with reference to a self-directed agency.

However, the plant must possess such a self-operating agency because its reactions to external provocations are very much like those that we call sensing. It may open the petals of its buds to the sunshine and close them at nightfall. It takes in a measured amount of water to offer sufficient resistance to the buffets of the wind. It is "aware" of these provocations and is affected by them. In John Brown's and Schelling's terms, the plant manifests "irritability," a sign of life. But beyond that, this irritability as affectivity assumes specific directions, as observed above, or otherwise assumes specific functions in the expressiveness of the plant.

This expressiveness touches the various life functions and in overall effect indicates how these life functions reach in their vital interactive dependencies either to the existential maintenance of the plant or, in adverse conditions, to its regeneration, or to its death if the vital/organic functional sphere is incapable of repairing grave damage. Hence we may consider this directional agency the "sentient soul" of the plant.[1]

CHAPTER TWO

THE STRATEGIES OF DIFFERENTIATION AND HARMONY IN THE SELF-INDIVIDUALIZING LIFE PROCESS

I. SETTING THE ISSUE: FROM TRANSCENDENTAL DIFFERENTIATION TO THE DIFFERENTIATION OF LIFE

As recurrently pointed out, the great issue that philosophy shares with all branches of knowledge and practice is that of the "differentiation and unity" of everything there is. In Antiquity budding scientific research shared with philosophical reflection its main points of departure, ways of approach, and key points of orientation within the universe of human being. Then with Kant and later with Husserl philosophy took its very own course, namely, one that treated the issue of the differentiation and unity of all there is with reference to human modes of constituting reality, that is, with reference to consciousness. Husserl, as we know, ultimately conceived of consciousness in relation to the lifeworld, that is, to the ways in which conscious being spreads its existential tentacles through the contexture of the world.

The differentiation of things, of beings is — as I have shown on the basis of a reading of *Erfahrung und Urteil*[1] — brought back to its center within his schema, that is, back to apperception, and we may certainly pursue it at that analytic level. However, as I have tried to show, his genealogy of logic is not limited to transcendental consciousness but touches against the borderline between transcendental consciousness and physiological consciousness (which he then sought to explore further). In an analysis of Husserl's genealogy of logic, I have attempted to show how the last instance of "dynamis" that he invokes, that of transcendental consciousness — which is par excellence constitutive — opens upon the all-embracing and fundamental "ontopoietic schema of the self-individualization of life."[2]

Indeed in my phenomenology/philosophy of life I have moved to a deeper, more generative level of the differentiation of life into distinctive structures, functional systems, modes of interrelation and interaction, etc. We will come to these things later on. What is important to bring out at this point is that the ontopoietic self-individualization of life is grounded in what the natural sciences call the "biotic system" and even more deeply in prebiotic conditions, the material available to the processes of life with its primary pulsations and potentialities.

While my conception of the ontopoiesis of life renews the classical and modern views, it stresses that we cannot fail to acknowledge the prebiotic grounding of life and has as its reference the most recent scientific thinking.

When we ask the question of the differentiation and unity of all, it is obvious, that first we have to consider four approaches to reality in its becoming: 1) direct intuition and description, 2) scientific investigation, 3) transcendental constitution, and 4) traditional ontology. Whatever the approach(es) adopted, we must still frame the question of differentiation and unity in the basic terms of life's manifestation so that in our interaction with entities within reality we distinguish between and among "phenomenally" (not "phenomenologically"), that is, in a spontaneous and "naive" way, and in terms of efficiency and of relevance to our own existential enterprises and those of others that we discover to be entangled within the same web. We have also to grasp the innumerable elements and aspects that enter into the vast sphere of our existential reach and beyond, the distinctive, although interrelated, forms of beings and things, of rocks, plants, animals, and human beings like ourselves. This phenomenal manifestation[3] is more than the outcome of a generative synthesis of the four approaches just spoken of. It embodies their work in a lived, experienced "spectacle" within the human experience; it endows it with the inward/outward dimension — the "within ourselves" replayed without and the without replayed within.

As Husserl considered the lifeworld to be the ground of evidence, the measure and proof of reality, I propose that phenomenal manifestation serves this function and is the instrument for testing reality. All previous approaches have had to draw their evidence from reality *as manifested* and obtain from it "confirmation" of their adequacy, legitimacy, validity.

And yet in order to understand the phenomenally present world, thinkers throughout the history of philosophy have also proposed ontologico-metaphysical approaches and solutions to the problem of differentiation and unity. We do have to consider these solutions along with those other approaches even though they might not coincide. There might be cause to search for connective links between all of these approaches. Even if no connective links are to be found among the lines of thought projected by the human intellect, nevertheless there are many reasons for pursuing our quest for reality and life along all these channels hoping that they ultimately will converge in one gigantic blueprint of the real.

The great reason for being so open to a plurality of approaches is that the human involvement in life and reality that is always our starting point draws on numerous capacities and faculties of the human being as a living creature and each of these opens a special perspective within which this being can expand. As different as these perspectives might be, they all fall ultimately together to coincide in a being, self-identical and distinct. Here lies the great issue that we are raising again: what does this selfhood, this identity, this distinctiveness amount to with respect to the whole of the reality it stands against even as it participates as an entity in it? Are the concepts reverently handed down to us still sufficiently univocal to help us handle the question of differentiation and unity? Can an Aristotelian type of ontology with venerable notions such as that of the "individual" still be considered beyond question within the pluridimensional inquiry that we propose? Should we seek patterns, criteria, principles applicable univocally to all the modalities of reality and life? Should we seek to universalize them when we find that they have particular applications, forfeiting thereby fidelity to a reality that might be differentiated generically at varying levels, each making its own claim? For Pico della Mirandola there is indeed a sameness in unity. He avers that first of all there is the "unity" of each thing and being that makes it "one in itself," consisting of itself and "consistent with itself" and that then there is that unity "through which each creature is united with the other and all parts of the world comprise a single world." In virtue of what, however, can the linking of that sameness and identity that distinguishes a thing or being and that by which it is related to all other things and beings be accomplished? Instead of getting at our inquiry with ready-made tools of concepts and notions, let us, in contrast, seek them while we progress.

But first of all as we attempt to disentangle this sameness/otherness and ask just what this divergence as well as relatedness consists of, we will have to ponder numerous questions of a basic nature.

The crucial one seems to be the nature of the relation between the single living being distinct in itself and the community/society or "population" of beings in which it is existentially/vitally enmeshed — and that in a twofold way, enmeshed in interaction and primordially existentially/vitally enmeshed. In the latter enmeshment the living being's traditionally attributed "autonomy" is at stake. Does any living being exist in itself in such a way that it supports itself in all its necessities and unfolds all its functions without being dependent on circumambient conditions, or even more without interactions with others that necessarily support it in its own progress? Does this support infringe upon the distinct entity's autonomy, is it existentially dependent?

The classical ontological differentiations: autonomy and heteronomy, dependence (structural or existential) and independence, essence and existence, etc. are put into question, if not made obsolete, in the light of the new trends in viewing reality, life, Nature-life, inorganic nature, the cosmos.... That is to say, in order to investigate anew the differentiation and unity of all there is, we cannot use such traditional notions without testing at the same time their adequation to the real state of affairs.

Then there is the question of the "method" or "methods" we will have to employ in this task if all of the above-mentioned approaches are to be given a voice and heard.

In this study I will focus on only some of the concepts that help us enter into the conundrum of relations within the web of life in all its spheres and I will give voice to a type of evidence that gives us reality in all its cross sections. But we have to begin by bringing up one of them at the start. The concept to be held up for consideration will be one that is at present the object of great uncertainties, the notion of individuality.

The "method" or evidence that I will identify as being uniquely serviceable to philosophy in its present uncertainties — as it has been in all great philosophy of the past — is that of "direct intuition." This evidence has fallen into disrepute in recent times, which favor hermeneutic approaches that seek in historical processes the meaning of reality. But to avoid the infinite regress of the historical sedimentation of meaning, we need to retreat to the pristine source of direct intuition, which is available to the analysis of reality at various levels.

Our disentanglement of the conundrum of questions concerning the differentiation and unity of the real is made possible by recent scientific investigations/illuminations that are of primary significance for our reorientation within the world and life given that the phenomenal manifestation of life no longer satisfies our scientifically inquisitive mind. Not only is the question of differentiation and unity of paramount significance in the clarification of our status as living beings as well as human beings within the web of life as it extends through Nature, spirit, and community, but it has also become a burning issue for the sciences, especially those concerned with the evolution of life, which are seeking footholds and guideposts for their "taxonomy."

How do we either cognitively or practically obtain access to all-that-there-is? Although we do not subscribe to the approach of classical Husserlian eidetic intuition, we will yet adhere — in contrast to most of contemporary philosophy with some phenomenological flavor — to the Husserlian faith that "direct evidence" genuinely gives us reality in its various guises. That intuition has the capacity to obtain this direct evidence in the various analyses by which it penetrates reality goes without saying. Here alone is access to nature-life, access that none of the circumventions of hermeneutic or semiotic procedures can ever attain. We have come to this conclusion through analysis of the human creative process, which enters into the very individualized meanders of life. We find confirmation of this on quite different grounds in Bergson, who saw in intuition the pristine instrument of life in contrast to the objectifying artifacts of the intellect that but skim the surface of reality.

The real is not given to us and does not assert itself through either the sclerosed representations of the mind (eidoi, essences, concepts) or their meaningfulness for it forever escapes the conclusive stage of the mind's interpretation. The exaltation of the spirit does not give it to us either, for it surrounds us all in full flesh and blood and is not to be thought away, to be disseminated into thin air. The real impinges on us, is within us, and absorbs our faculties.

Although it seems that the first phenomenological investigation of Nature made in this spirit was pursued by Hedwig Conrad-Martius, an investigation to which we will return, she did not dissociate or even differentiate sharply "eidetic" intuition from the all-penetrating *life intuition* that she was by and large practicing. This latter involves far more than the exercise of the essential structures of the intellect. In establishing the phenomenal manifestation of life, the transcendental constitutive system of the human being builds on the vitally significant rationalities of living beingness, thus reaching/skimming the frontiers of its own transcendental exercise; nevertheless that system does not have the means to thematize these frontiers nor to estimate or appreciate the gap in significance that lies between its own apparatus and the schemas of the play of forces shaping the vital existence of living beings, their interdependencies, their relevancies.

Life intuition, which I have isolated from the creative process peculiar to the human being, shows itself adept at penetrating and giving due place to all other methods and approaches to the real, at appreciating all the peculiarities of their findings whether connected or seemingly disconnected. In short, we will follow along the spontaneous run of life's intuition in pursuit of the meanders and relevancies of all-there-is-alive, seeking ultimately the differentiating as well as uniting articulations of life's origination and becoming at the cross section of various perspectives.

Although we will attempt to detect and specify the principles and models of inquiry in their pristine forms within the differentiating progress itself, we have first to discuss some basic terms of differentiation as such.

II. PRELIMINARY SORTING OUT: DISTINCTIONS TO BE MADE IN GRASPING THE INDIVIDUAL AND THE TYPE

1. *Singulars and Types*

Do we distinguish in a primordial phenomenal way first the type, "horseness," for example, or the singular horse, "lilyhood" or the lily? This question seems at first to concern the distinction between the universal and the particular. However, upon a fresh look there is more to the question than there at first seems to be. First of all, a type and a singular being of that type are existentially grounded in a reciprocal determination so that their meanings stand inseparable.

Although to distinguish in terms of categories between the singular and the type is a way in which at the higher level of constitutive human activity we may organize the phenomenal manifestations of reality, nevertheless at the basic levels of this constitution the discernment of features, aspects, elements of the singular instance goes hand in hand with and depends on the phenomenal manifestation of the type: this singular horse, this singular flower cannot be considered to be a unique kind. Likewise the type of horse or flower cannot be conceived other than on the basis of the singular horse or flower.

This is also the great issue with which the natural sciences are now wrestling: What is the nature of the "singular" elements, functions, particles that make up the substructure, the grounding of the order of life? Maybe there are no singular beings but only "families" of beings, only interrelations, interactions, intergeneration, etc. With the great progress of natural science, we have indeed to review how these distinctive entities, "forms," etc. emerge from and are integrated with the web of life's order in a development somewhat analogous to the phenomenal precritical level of perception and the differentiations it makes as it moves from a singular object to its existential grouping, to its existential "population," as well as from the population within which it acquires its specific, distinctive features back to the singular entities that make up the population as such.

But if we seek the distinctness of the singular, we have to go to its type, no matter what their interrelationship. We do not advance without finding some measure of distinctiveness in the individual or without drilling below the phenomenal to its grounding in the order of life within and below the vital circumference, an order that is partly fluid, escaping any grasp, and partly recognizable in its recurrence.

2. *Structures and the Life Process*

The differentiation, the singularizing of life does not depend on or directly refer to the laws of logic, nor, as Aristotle sees it, to a prior metaphysical stratum. Its correlative partner is the striving of life itself. Life is understood by us to be at the center and also to be the ultimate point of reference (for metaphysics). *If the ontogenetic nature of life is asked after, it is a "constructive functioning."* Its modality is multiple, consisting of innumerable types of "operations." Yet each functional operation points to a closer or further relation with an overall partly fluctuating, partly perduring schema. This schema is intrinsic to the entire functional progress of a living being, maintaining its continuity, its fluctuating sense of direction. In fact, if Monod's concept of the "teleonomic,"[4] that is, the conjoint direction of the functioning of the individual in all its intrinsic processes toward one individualizing aim, is thought through to the end, it calls for just such a constructive schema intrinsically unfolding through the operations of living beings themselves. I have called this schema the "ontopoietic design."[5] A series of functional thrusts are made out from the center to the periphery, and after gathering forces from their peripheral orbit these impulses reach back to the center thereby establishing an inside/outside oriented functional axis that runs through the entire functional system of the living being singularizing, differentiating its progress in an individualizing process and maintaining the purpose in that becoming.

The question here is that of life's elementary ordering and of the basis and means by which we may conceive, experience, and reorganize consciously the rushing life progress so manifold in its variety within the entire spread of the world that is in front and within us and in which we are willy-nilly caught. That is to say, it is an issue as much about the progress of life itself as it is of the experience constitutive of reality for the human being (as Husserl treats it), as it is of logic's struggles with it as it works from the base of Aristotle's metaphysically rooted conception of the individual.

We have tried to outline how this self-individualizing process proceeds, but we have still to ask: What does it consist of? Does it aim at — and accomplish — an "absolute" singularization? Does it ultimately consist in "cutting out" entities that are then discontinuous and alien and aimed at goals separate and distinct?

As I have voiced it before elsewhere, this life individualizing progress consists in processing forces, energies, synergies with their germinal affinities, floating, in wait, around a seminal profile, which falls together like a dynamic jigsaw puzzle having at any moment in its emerging unfolding a reservoir of support to draw from, the support of the already accomplished phases with their inherited and acquired proficiencies. Thus a living being draws on already established reservoirs of forces with their propensities and tendencies to fall into dynamic modalities. There are constructive devices hit upon, on the one hand, and an "inherited" reservoir of life nurturing resources lying in wait to be drawn on, on the other, the foundation of life wrung out from the crude stuff of prebiotic and biotic existence by the work of generations of living beings.

In the middle, lies the intersection of a living being's seminal promptings and ongoing dealings with a circumambient sphere of life already set in its forms and resources. Here the seminal nucleus of the self-individualizing beingness seeks to insert itself with its very own functional program into the anonymous play of forces proceeding from already existing beings and things in becoming.

3. *An Overview of the Scene*

To prepare the way for our argument we will now give an overview of the entire situation out of which it emerges. In looking at scientific inquiries into the intercourse among individualizing factors and their ways of singularizing themselves within groups, i.e., "populations" of beings of the same kind, into the interplay of their functional needs and the intimate network of interdependencies to be found at several existential levels, we cannot fail to acknowledge the contribution/participation in the constructive progress of the individualization/singularization of the innumerable functions that develop the distinct organs that together make up a living organism. But each of these organs in turn consists of innumerable "families" of those prime individuals called cells, each of them having a singular endowment, functional orientation and performing a role of its own within the entire functional schema — the metabolic system, for example, which secures growth and development, or the reproductive-genetic system, which assures the perpetuation of the species.

And yet, although the actual functioning of existential progress may in its unity express the cooperative work of an innumerable number of "families" of those primal individuals that are cells, functional sequences, singular organs, etc., we still cannot deny that the whole is functional and constructive, self-oriented, "autonomous" in its organization, carrying within itself dynamic resources as well as an ontopoietic design to be fulfilled jointly by all the otherwise independently endowed coworkers coming together for the fulfillment of a common aim.

We simply must grant to this fulfillment oriented functional wholeness a special type of "individuality." Its unique character stands in contrast to the "hidden" works of nature performed on its behalf. There emerges a complex living being, an individual assuming a phenomenal appearance that its primal constituents lack and that its subsequent functional segments or organs have only with respect to the whole. Indeed, in its phenomenal autonomy, the whole living individual stands out from the amalgamation of functional families, stands out amid the mesh of the population of its own kind, amid whatever grouping, and punctuates the factors that work as its conveyor by its discernment and distinctive presence, so that it is to be recognized as a definitive center of forces within an otherwise polarized multiplicity. We must ask whether individuality as a modality of life should not be assumed to be spread through numerous degrees of constructive efficiency, numerous spheres of accomplishment, through innumerable primal and segmental entities and to be differentiated accordingly. Indeed individuality appears to instigate and preside within more or less complex life processors, definitive life agents up to the complete living beingness, a whole individuality phenomenally manifesting itself within the world of life.

With this we return to the question, "What does the individuality represented in the various spheres, degrees of self-promoting life stand for?" Differentiated obviously in its manifestation, it maintains its constitutive factors as acquired within life's self-individualizing, ontopoietic progress. What does the individualizing progress of life seek? By what means does it proceed? Should we, with Bergson, assume that the living individual is caught within a dynamic development and is merely a relatively stable foothold of life, and that life consists in the movement that it carries, the living individual being merely a place of passage?

To conclude our preliminary queries let us emphasize that what is at stake here is the concept of singularity, or that of the distinctive moments that establish amid the otherwise undifferentiated steps of life a relative stability for the organization of life within the discrete continuity of the primeval flux. We are here engaged in a dialogue with the science that proposes as such moments of constructiveness, of relative perdurance, first of all atoms — devoid of qualitative features but endowed with what are thus far considered to be the "ultimate" constructive factors. In its inquiry into this constructive progress, science persists in referring to the classical notion of the "individual" to differentiate the constructive processes, their progress amid propitious or hindering conditions and accomplishment. It sees the individual as representing life-in-progress, already there, that is, as capable of performing, or as consisting of its performance of the functions of life on its very own so that it carries the life process and maintains it. The individual so conceived promotes its own existence and as such is the center of an exchange of forces, energies, dynamisms, and constructive entanglements. The question here for the sciences is that of life and world progress, of the footholds of this process and their elementary ordering.

III. THE SINGULAR AND THE MULTIPLE

1. *The Individual and Individualizing*

The following inquiry will intermittently return to the crossing of the four approaches to reality named at the outset of this study. Standing before the phenomenal manifestation of life with and within its world, we move by making a "map" of what surrounds us. And first we come into contact with, lean on, and find our pathways and orientation schemas by distinguishing what we traditionally call "individuals." In their coherence, resilience, and self-propulsion as they exercise their ability to act, they engage our action, become for us centers toward which our action is directed. As the center of the exercise of its own intrinsic force, the individual is the cornerstone of varied processes; even as development proceeds from the individual's intrinsic dynamic organization, which serves as its launch pad, that organization is the stronghold of the relative stability of life.

With this phenomenal description we have laid down our groundwork. This is not to adopt the metaphysical notion of 'materia prima' by which as a substratum the individual would by acquiring a form, become singular, be differentiated as an ontic abstract skeleton of an "individual"; nor is it to assume the postmodern view that absolutizes "individuals" as reposing entirely in themselves. Our views are diametrically opposed to dispensing with the individual in favor of a theory that knows only classes. Our ground is life, from within which the notion of the individual presents itself.

I assigned a crucial role to the notion of the individual as early as my cosmological inquiry[6] into the problems of constitution and have done so ever since throughout the entire unfolding of my phenomenology of life. The individual is simply the singularizing principle that creates diversity in the progress of life, and this differentiation causes life to advance, with individuals serving as guideposts and footholds for ordering as they shape and invigorate life in ever new exchanges of forces and dynamic interaction.

We will here focus on life's *individualization,* rather than the "nature" of the individual, but first we have to discuss the notion of the individual, which has received particular attention in the forge of Western culture through all its shifts. As a matter of fact the concept of the individual has become quite a controversial philosophical notion in the discussions of scientists, especially of those who seek to study the ordering of life and evolution. It is my purpose in this study to subject the notion of the "individual" as the primal singularizing factor of life, to new scrutiny.

The individual certainly does stand out in the phenomenal realm for being stable, resilient, resistant to pressures or intrusions as well as for having its own promptings from within, for the individual follows its own intrinsic rules and its own agenda. However, already at this phenomenal level, when we consider living beings, they appear as unities but by no means as uniform. On the contrary, as Goethe (whose work on the metamorphosis of plants Hegel saw as the start of a rational approach to the nature of plants) already observed, although a living thing may appear to us to be an individual if we do not delve below the phenomenal level, each living thing is existentially multiple, that is, subsists in "an association of living self-sufficient beings, which though alike in idea or plan, can in their manifestations be identical, or similar, unlike or dissimilar."[7] Taking the plant as an example, Goethe emphasizes that although plants or trees appear to us as individuals, they nevertheless consist of numerous individual parts — roots, trunk, leaves, etc. Each has its own "nature," organization, role in making the unity that is the plant. Thus, we cannot conceive of the unity of the living individual as being indissociability. We have to grant to it a "wholeness" in which each part is associated by a common project in which it performs an assigned role and function.

There is then the question to be raised of the distinction between the "whole" complex, multiple individual and the many primal individuals, a question that demands the further investigation of the very notion of the individual.

So much for such a simple creature as a plant. What about the wholeness of animals, the organs of which build up the whole organism, with each organ being again made up of a great number of functional parts having their own particular roles, all enmeshed with each other and with the participation of each being indispensable for the accomplishment of the organ's function? No wonder that a great scientist like Virchow calls the individual "a commonwealth of necessary constituents."[8] "Only in their togetherness, union and community do they guarantee the total expression of individuality." We will come to ask what makes this whole seamless in its collaborative task. For the time being let us pass from the level of phenomenal observation to that of experimental observation, to the level of natural science, which in our times has passed from mere hypothesis to certainties having their own evidence and which are to be acknowledged by philosophy and by phenomenology in particular.

2. *Individuality and the Unfolding of Life*

Our quick sketch of the individual has taken us far from the abstract ontological notion of it inherited from the metaphysical tradition and still to be found in some contemporary phenomenological analyses. This notion, having lost its metaphysical substratum in matter as well as its grounding within a framework of objective eidetic structures (as in Ingarden), seems to hang in the air and be at the mercy of direct phenomenal observation.

So it may seem. But actually we have brought the individual directly to its ultimate ground: life itself. When we acknowledge biological science and take the cell to be the structural as well as functional conveyor of life, we recognize "below" phenomenal appearance to our senses and our existential involvements that these smallest life processing individuals are building blocks of life, though invisible to the naked eye, imperceptible to touch, and able to be thematized only in relation to life itself. Rudolf Virchow, the founder of cell pathology, whose work initiated a transformation of medicine as the science maintaining life, emphatically insisted already at the end of the nineteenth century, "All life is bound to cells, and the cell is not only the vessel of life but the living part itself."[9]

Life is present in everything of cellular origin, and the presence of cells indicates life.

The important point, however, is that the cell possesses all the basic features of what we so far call an individual and that it is in these features that resides the elementary "engine" of life. With the great progress of biology in our times we can no longer say with Virchow that the cell is the ultimate building block of life beneath which there is only change not life. Nor may we draw a line between the spheres of the "organic" and the "inorganic." This borderline is blurred because we acknowledge today numerous "biotic," that is, life carrying, elements in the composition of the cell as well as prebiotic but life oriented atomic elements. These distinctions refer all — as does the cell itself — to the genesis of life and to its processing capacities.

So whilst we may not distinguish cells as the first building blocks of life as Virchow assumed, they do seem to be the first individuals, and their appearance is the watershed within the self-individualizing progress of life, the structuring line leading to the completion of a plan of individualizing development. Cells have an existential individualizing project within themselves — the first structuring factors of bios — through which the advent of the biotic sphere is prepared as the primal forces and innumerable pulsations and moves of the prebiotic realm are carried on a subterranean course until life emerges at the phenomenal level of the world. Attributing the essential role to the cell's work as such, we must still with present-day science consider the outstanding contributions of prebiotic elements in directing the self-individualizing process *in toto*. We might discover that we have more than singularity and multiplicity to discuss. The role of these elements with their apparent autonomy and determining power forces us to pause in our consideration of individuality.

We will be able, in contrast, to grasp the main features of the individual in its concreteness as well as distinguish *several modalities of individuality depending on the evolutionary progress of life.* Let us now consider one more puzzle of individuality, one brought to light in contemporary biological studies.

3. *The Individual and the Complex of Life*

Are the growth determining biotic and prebiotic factors autonomous individuals? Assuming at the start the individual to be the basic unit of consistent life functioning, we have still to consider the individuation of this unit wholeness in the perspective of life's unfolding. We are so far tacitly attributing to it some type of existential autonomy. A lucid study by the botanist Canullo, however, prompts us to think differently.[10] Envisaging the life process of a simple individual plant — one already an individual having at its existential service innumerable primal individuals and their specialized constructive segments — within three perspectives, the architectonic, ontogenetic, and functional, we find, first of all, that in each perspective there appears to be active in the plant a set of "suborganismic" elements (e.g., modular units). In the architectonic formation of the plant's growth these determine the key "strategic" limits of the species, thus acting themselves as autonomous individuals. In the ontogenetic perspective we discover again that from the two genetic factors at work (genetic and vegetative), from which morphological individualization follows, a single genotype may be fragmenting itself into innumerable varying forms and renewing itself while carrying on its work in interaction with circumambient forces, and this indefinitely. Both of these factors seem to exhibit an individualizing autonomy beyond that of regular cells or primal cells. For Canullo there is yet another "functional unit" of special efficiency to be considered. Identifying the plant with its phenomenal manifestation, its "above ground" construction, its shoot, we see that this shoot may possess a specific functional efficiency in relation to the spread of the plant in vital space, inter and infra specific relations with other plants. The role of the entire population of a species may be, in fact, decided in the given community of a living plant.

With these observations showing how some intrinsic factors of the individual may play a dominant role, going beyond the singularity of one plant in determining, directing, conducting its growth and characteristic features, we must acknowledge that even though these factors are not individuals according to our provisional conception of an individual and are not autonomous or existentially independent, they exhibit architectonic as well as functional and genetic growth of their own accord. If not individual beings, are they not to be considered individual factors of growth? But a further issue yet emerges from these descriptions.

4. *Individuality/Singularity*

In these new views the concept of the autonomy of the individual has become greatly differentiated, and yet insofar as these factors of growth and reproduction are understood to be intrinsic to the whole that is the plant, we may consider its autonomy to be maintained. However, in the case of the "functional units" intrinsic to the plant, which undertake various generative as well as interrelational roles "on their own" and may even determine through infraspecies relationships the path that the entire population will take and the place it will assume within its community of plants, the boundaries of the individual that is the whole become questionable and views on its autonomy veer between the "substantial," phenomenal perspective wherein autonomy consists mainly in selfhood, in self-promotion and self-functioning, and a perspective wherein there is autonomy also in specific elements of the whole plant, which elements have each some specific and uniquely significant power, propensity, role, capacity, etc. that makes them stand out amid the plant's natural endowment, they being, in effect, within but not restricted to its singular confines.

This passing beyond the boundaries of the existential realm of the singular plant is particularly accentuated in the case of the type of functional unit discussed above. That leads directly to the question of the individual's existential place within its "population," its place within the plant community (and maybe further out into the living kingdom), a question prompted by bringing out the coexistential modi of plants within their specific concrete territories and culminating in a move to the borderline of the phenomenal and the scientifically observable where a crucial challenge to the individuality and autonomy of the whole living unit of a plant is being raised. Meanwhile we gather from this some clarifying hints for the investigation of other types of living beings. We are now alert to the possibility of an existential relationship between the individual/autonomous being and the complex of life in which it participates, on which it draws, and to the expansion and maintenance of which it contributes.

As a matter of fact, following the scientific presentation of Canullo, we will acknowledge a "social behavior" in plants. They may be mute, but they communicate by special signals as well as receptive organs. (Some plants, potatoes, tomatoes, etc. use "volatile molecules" by which they in tandem activate defensive genes while responding to immediate danger or mending damage.[11] But there is, first of all, the physical integration of individual plants that occurs in their occupying space that other intermediary organisms (e.g., mushrooms) contribute to in the existential continuity of life-in-space.

And here comes the most provocative insight into the relationship between the individuality/autonomy of living beings and the complexity of life's existential modes, that is, life's emergence and self-maintenance in dense existentially entwined and enmeshed complexes. Indeed, consideration of the all-embracing dense networks of imperceptible but strongly efficient mediating vital elements that prompt and promote the growth and existence of the trees and bushes that constitute a forest prompts the question of what constitutes the individual whole. Is it the individual trees, bushes, and plants or the forest itself as a complex of life? It is in the forest as a whole that trees find their proper grounding and in turn participate in a functioning system of vegetation, contributing to and maintaining it, and it is not the case, as we would be tempted to assume, that singular trees individually make up the forest.[12]

The existential growth factors beneath the ground and the enmeshment of functions and roles that we see in the forest point, first and above all, to the crucial significance of the greater life complex for the individualization of all the quasi-autonomous, self-organizing centers — trees, etc. — in which the life forces are processed, for their growth, maturation, and proliferation.

Second, it is obvious that we cannot operate with an "essential" concept of the individual or of existential autonomy but must see that individuals diversify with respect to the various types of living beings about them; types "individualize themselves" as they establish niches beside each other in a common evolutionary wave that all are caught up in.

The final conclusion to be drawn from this multilevel investigation is that instead of assuming an "absolute" ontic universal and unchangeable individual to be key to the differentiation of the types of living beings, we should now be alert to the distinctive aspects of individuality that point to the individualizing process of life. On this basis we should at least intuit how the essential nucleus of individuality — if we still use that ontological term when speaking of the ontopoietic process — shapes, enriches, and expands itself in type producing features along the evolutionary line of life. We will begin by adopting the phenomenal perspective, which for us human beings always remains a referential schema of reality.

In the phenomenal perspective we may join Husserl, who affirms that the evidence of the world that surges from the lifeworld is the foundation of all experience and draws from that the conclusion that it is participation in the world that determines types of consciousness. Hence, there are to be distinguished various degrees of consciousness proper to animals, one type of which is that of the human being. In this diversity of consciousness along the evolutionary ladder of types, man's would be "complete" consciousness. But Husserl does not identify consciousness with life; rather he sees its emergence and presence in a "ray of awakening," "attention." In the final development of his thought he introduced animal instinct as well as sensory pulsations into the orbit of consciousness.

IV. THE DIFFERENTIATION OF TYPES IN THE ONTOPOIESIS OF LIFE

1. *The Radical Novum*

Our preceding inquiry makes it clear that the differentiation of types of beings is a matter of the self-constitution, self-individualization of life as much as a matter of singular beings themselves. Husserl proposed that types of living beings may be distinguished and compared with reference to the manner of their participation in the lifeworld. In the framework of the phenomenology of life we have expanded on, we arrive at the level at which this lifeworld emerges together with beings and things in the ultimate life process. That means that we must seek the differentiation of living beings not in their participation in the world — which in itself would undoubtedly be quite instructive and has to be taken into primary account — but in their ontopoietic processes and in their self-individualizing as much as in their integrating or gathering reach. We may learn as much about the ontopoietic process from inquiry into this differentiation as we could about the latter by inquiring into the former. We cannot assume much a priori without vitiating our approach to reality. But such a full-fledged inquiry lies beyond the scope of this essay. We leave it for future investigators to pursue. My concern here is merely to examine the differentiation of types in the concrete terms of the four perspectives of reality outlined at the outset and this with particular focus on the line of self-individualization.

From our point of view we must ask, "What about the plant?" Conrad-Martius, the first phenomenologist, along with Max Scheler, to devote considerable attention to the natural sciences and the question of life, treats this question with great subtlety in her work *Der Seele der Pflanze* ("The Soul of the Plant").[13] Precisely on this point she contrasts the functioning of the plant and the functioning of the animal. The plant has a soul: it receives signals of light, humidity, pressure and possesses an intrinsic apparatus for responding to them. The plant assimilates some forces as propitiously bettering its existential condition. It responds to the attacks of predators, it bends before the wind, it may even repair some degree of injury it suffers, regenerating a damaged part, a capacity that seems crucial among the prerogatives of living beings. Thus, a plant manifests its being a sentient being, a self-promoting and controlling unity no matter how many organs, parts, and cells it is made up of; all of them perform their assigned function, assigned by the aim of the plant's individualizing course within its life community of plants and the greater milieu.

Lastly the plant propagates itself either by the seed that it produces and offers to birds, insects, and the wind for appropriate dissemination, or by budding, in the manner of strawberries, which then establish new independent plants. The plant adapts to the climate, soil, humidity, and to other plants around it. In short the plant manifests itself as a living individual. And although a strict borderline between it and the lowest, simplest forms of animals (corals, etc.) cannot be drawn, this does not negate its individuality as a type.

Although it may have a sentient soul, the plant does not possess one of the main benefits of consciousness, one that animals, even the smallest of them, do possess, namely, memory. It responds instantaneously to the relevant signals, but it does not grasp the appropriateness of its reaction, it cannot learn. Its action/reaction schema is strictly confined to direct practical instances in which remarkable promptitude fulfills a point of a plan, a plan that is strictly identical with vegetative growth and unfolding, with the *hic et nunc,* with the aim and end at hand. How could it be otherwise? For the simple reason that a plant cannot move from its place, how could the plant entertain alternative choices in its reaction other than those that arise when direct contact must be directly met. To be precise, a plant is not immobile since it reacts to light, the wind, birds, insects, etc. with the gentle motions of its limbs, but it cannot change location. It is existentially fixed to the soil. As Aristotle said, mobility belongs to life, but fixedness to a spot of ground is the limiting existential condition of the plant in contrast to animals, which are not attached to any one place unless a human being so attaches them for need or pleasure. The mobility of the animal includes motility and with that comes the devising of an entire inward apparatus, one that varies from species to species in degree of efficiency. Here is the principle that introduces into life the freedom that Bergson considered to be life's own.

Before we delve further into the differentiation of plants and animals, let us first get an overview of the individual in the perspective of life's generation.

a. *The Novum in the Web of Life*

Scientists seem to unanimously believe that the replication mechanisms of the living being, be it the simplest or the most complex, is a life promoting system quite distinct from those systems promoting the growth and unfolding of the living being's singular existence. Each system goes its own way. An essential coordination of the growth directing functions and those of replication is nevertheless indispensable for the adequate performance of the latter. Be that as it may, there is in generation, specifically in continuity of type, a serious threat to the individualizing, singularizing autonomy of the living being

It is indisputable that a perpetuation of type is to be found at work in the genetic dimension of life. Each living being emerging into existence bears within itself a functional model of the replication of its kind to be put into action as its unfolding occurs with the cooperation of the growth functions that support this aim (or does not occur, if those functions are hindered owing to poor nutrition, etc.). Thus, the replicative system definitely carries within itself an "aim," a "task," to be accomplished; this aim amounts to the conservation over time of the species as a constructive line of life. And although existential conditions leave room for almost imperceptible and yet significant transformations in the long gene sequences bearing hereditary traits, it would seem that there can be no "radical" or "essential" deviations from the preestablished model in life's individualization or any "leap" in propagation.

At this point, however, an observation is in order. If this were actually so, we would see a uniform or at least, relatively uniform progression of life moving, within fixed perimeters and leading not only toward monotonous uniformity but toward the exhaustion of available energies and the extinction of species. Of course, things have gone differently.

It seems that the genetic model assuring inheritance in reproduction is not mechanically repetitive in its workings. On the contrary, instead of a monotonous repetition (doubling) of the original, each individual, although itself inalterable as such, bears the possibilities of another spiraling line of development, such as I describe further on (see Tying Point Three, "The Great Plan of Life"). One might presume that given such progressive transformability in the genetic-hereditary program of the species, we would have the continuous appearance of transformed kinds. And yet this is not so, for cogent reasons. As a plant reproduces/replicates itself a specific set of features is maintained despite variations in them.

Life is still a radically new reality, as we can see with particular sharpness when we consider the now classical striking juxtaposition of instances of life and crystals. Crystals also reproduce themselves through means of their own, following a model *ad infinitum.* However, they do not manifest a line of growth and unfolding in action/reaction to circumambient provocations. They do not manifest the feeling and sensibility that we see in plants. This sensibility is in line with a life promoting and conserving functioning and manifests the light of the "awakening" that we see in the simplest life forms. This sensitivity that characterizes life as such is an awakening to the inward and the outward. This inward sensitivity stands in striking contrast to the inertia of the crystal and allows attunement to both present and possible conditions and provocations for the purpose of pursuing a line of constructive development, the steps of which lie waiting within its endowment.

This manifestation of the movement of life is a radical novum with respect to the prebiotic sphere. If we assumé with science the necessity of an extremely complex conundrum of prebiotic and biotic forces and preparatory schemes for the emergence of a living individual in the constructed instance of a cell having primal sensitivity and constructive motility, we are presented with such a radically new normality within nonliving nature that philosophy and science have pondered for centuries just how and why this entity emerged. If the singularizing of life is genetically predelineated, what about the original autonomy of the individual? And what about that of the novum?

Should we grasp the evolutionary line as a self-individualizing process both differentiated in and differentiating types and at work in specific and singular beings, we cannot fail to acknowledge at each significant break from the preceding model the emergence of a radical novum. *This radical novum might be destined or prepared for by the spiraling genetic program at work in the preceding type, yet this does not dispel the significance of the radical leap that occurs in the progressing line, each step of which is characterized by a radical novum.* Paleontology and the biology of evolution attempt to reconstruct with precision the links supposedly missing in the progression in complexity resulting from life's self-individualization. Some of the differences in individuals accompanying reproduction seem to be but variations on a theme, while others are truly radical, if not as radical as the leap from senseless matter to life. Then we are faced with the appearance of a new order. Crying out to be discussed is the leap from the self-individualization manifested in plant life and that manifested in animal life.

b. *The Passage from the Vegetative to the Animal: Subjective Consciousness and Freedom*

A sharp demarcation line between strictly botanical and strictly biological investigation is difficult to draw since we know of aquatic creatures and corals that possess characteristics of both plants and animals. There are besides plants with life promoting strategies similar to those of animals, e.g., carnivorous plants.

Still, there is in the animal individualization of life the unique specific of "subjectivity" (or consciousness) and "freedom." The radical novum of freedom, the ability to range through time and space selecting food, shelter, companions, mates follows from the new apparatus of consciousness. This consists first in a capacity for representation, however rudimentary, and then in "self-consciousness," which allows discernment, remembering, and choice.... In short, the animal individualizes life as a conscious being, while the plant individualizes life as a self-centered agency having an intrinsic coalescing organization. Through feeling and sensing, in reaction, animals have been able to diversify and evolve into much more complex forms of life as conscious subjects having capacities going beyond the action/reaction sensitivity of the vegetal "soul." Here are individual complexes of life of a higher order, such that manifest life phenomenally. While the cells constituting a plant do not "appear" in global form, the animal stands out as a full "subject," that is, individualizes life in a self-enclosed, self-reliant selfhood. This selfhood means not only relatively stable, resilient action and reaction within the environment but, more particularly, an ability to process events in their temporal sequence and record them in sequences of memory that can be retrieved as guideposts for action. This ability allows apprehension, comparison, deliberation and, combined with an ability to move freely through space, strategic searching for the satisfaction of vital needs.

As for animal reproduction, whether it is meiotic, ovular, or viviparous, there exists inscribed within the reproductive system of the individual on this higher level a spiral trajectory promoting the conservation of the species but encouraging transmutations of form when needed.

This apparatus makes the individual an "agent," the "subject" of the life involvement we call consciousness. What a hiatus exists between the individual "selfhood" of a plant as it enacts its own life course and replicates itself and the individualizing "subject" self-enacting its life course within a circumambient situation that it instinctively or representationally recognizes. The difference consists foremost in a dynamic locus of relative freedom that is the basis for the animal's deliberative and discriminative functioning in its life promoting and conserving activities. The sensitivity of the plant is directed at one specific point, all of its beingness is "sensitized" to light, humidity, and its coexistence with the circumambient world, its selfhood being then defined by all these factors. The selfhood of the animal, however, proceeds from a special novel factor, that of the "ray of attention" that not only moves among all the sensations the subject experiences in its circumambient situation, but also brings them back to the "center" as if synthesizing and estimating their global significance. It incorporates sharp, direct insight-reactions into an estimate of the situation at hand. "Instinct" is the name we give to this unique direct reaction to life situations, to this instrument of life itself, an indicator of just where the individualizing being stands, alerting it to vital crises, to the need to react, to the path to take, etc.

There is great controversy as to the significance of animal instinct for cognition, recognition, and reasoning.[14] Even if we have to deny that the most complex of animals have a complete representational apparatus like our own operating with ideas and concepts, there remains vivacious debate among cognitive scientists as to what the representational capacity of the animal consists in.[15] For example, in the case of the bat it has been observed that as it hunts in total darkness it hits upon its prey by practiced and then perfect recognition of its location, recognition not achieved through using representations. There is a great need for calibrating the gradation in the conscious proficiencies of the various animal species. Nevertheless, the smallest insect is endowed with what I call a "vigilant ray" that keeps it attentive to the benefits and perils of its environment. All animals, then, are individualized in a primordial mode of life, which is none other than that of an "awakened" subject. However, as already stated, there is a vast gradation within the animal manner of life's self-individualization according to the distribution of conscious proficiencies.

At the actual peak seem to be human beings, whose representational and conceptualizing communication seems to set them apart from the rest of life. The human creative condition, to which we will turn at the end of our probing, especially evidences this.

c. *Individualization within the Mesh of Interlocking Existential Ties and Life Communion*

Thus far I have briefly reviewed the main individualizing differentiations of life in a fourfold perspective. This merely indicates the forms that the self-individualizing process of life takes. It has become clear to us that self-individualization is not suspended in an ontic vacuum. But should we not raise the question of the dependency of each living being's individualization on the type within which it constitutes itself in the web of life and its genetic, universal patterning?

We have already indicated some of individualization's basic and decisive features: selfhood inwardly articulated and expressing itself at every stage of life's complexity in striving for orientation toward self-preservation; inwardly promoted self-repair; the conservation of the singular life and the species; and the innermost striving to promote life as such as evidenced in the mechanism for spiral regeneration at work in all types of reproduction or effected by other inwardly conditioned means. All these features are exemplified in the individualizing enactments of life's design by singular living beings. Strange as it may seem, it is within the singular individual acting on its own and struggling with the limited means nature has endowed it with to preserve the spark of life by unfolding a life promoting and conserving apparatus that serves the entire system of life. The individual is at the service of life as such, first, through its integrative functioning within a life complex and, secondly and more importantly, in its response to an inner preestablished drive to go beyond self-interest in promoting the continuity of the progress of life.

The above-mentioned features of the individualizing process put individualization and the notion of the individual in unique light. Might we actually say that the individual consists in an "in itself" but not a "for itself" and so minimize the aspect of the individual's autonomy (its reposing in itself but drawing on other factors for its existence)? Does life in essence consist in the performance of the singular life promoting task even while not being confined to that? While completing its own life design, its singular path, is it as it were entrusted with the simultaneous propagation of the entire line of life, of not only its own species through replication, but of the circumambient community of all plants and animals as well? It has actually come to seem that the very individualization of life means its opposite — the formation of complexes, a social existence, communing.

At this point let us return to our previous biological discussion of the relationship between the individual and the life complex to which it belongs. The example of the forest, where were there no forest, no complex life community, then no individual trees could flourish, is a striking example in favor of my argument here.[16] Science has given us several layers of evidence telling us that there is a point in asking, "What comes first, the singular or the whole?"[17]

d. *The Human Creative Condition: Culture, an Overflow or Prolongation of Nature, or a Novum*

Following indiscriminately this line of thought, we might be tempted, as several prominent scientists are, to consider the fruition of human self-individualization in a "complete" subject, or as Husserl would have it "complete consciousness," to be a natural prolongation of animality in an evolutionary advance. Indeed, given the relative freedom of the animal as a conscious subject, we may assume that there is a passageway between the animal and the human individualizing schemes of life.

Considering the findings of paleontology and their interpretation in theories of the evolution of life, which emphasize a long chain of mutations and transformations contributing to a spiral progression of animal types, a progression so long and so intricate that when we come to the "first" human beings and hence speculate and make inferences about the variations, metamorphoses, and mutations that must have occurred on the way to the "complete" consciousness of the modern human species, we may well be tempted, as are many scientists, to consider the life individualization of the human being to be the fruit of a natural evolutionary process — as a work of nature-life that is entirely understandable in evolutionary terms.

And yet, as in the case of the passage from plant life to animal life, we find a radical novum, the emergence of which cannot be explained by the laws of nature known thus far, namely, a relatively free subject of individualization. We find at the point at which we may ascertain the emergence of the hominid a radical novum with respect to the animal. This is a novum no doubt emerging from nature-life and yet characterized in its specificity by its having powers and abilities to shape its own life course, something not only unprecedented in the spiraling line of life's generation, but also, while certainly anticipated in its emergence, constituting a leap over its preconditions, it being unaccountable for them. It is as if life proceeds by fits and bounds and yet always proceeds into a more advanced field for its expansion.

By this novum I mean, of course, the emergence of the unique human condition within the web of life, a condition signaled by the surging of our creative/inventive virtualities.

Seeking life's relatively stable footholds in its individualizing course, we have proceeded from the blind turmoil of the game of prebiotic forces that find constructive engagement in the plant, to the novum of functional wholeness that is the soul of the plant as it processes those elementary forces and on to the novum of the self-deliberating, self-directing, seeking and selecting agency of life that is the animal subject. We advance again as we follow the progress of self-individualization to the radical novum of the human creative/inventive mind.

We might infinitely scrutinize the whole evolutionary scale, arriving at the highest, most complex animal closest to the human being, and we will find that animality consists ultimately in an intimate closeness to the processes of life. The animal's representational and communicative abilities are applied solely in its close and most intimate engagement with the play of life's forces, within the dense world of living creatures encircling it, in brief, in a communion with material existence that is mute but inspires total enjoyment.

It is the mark of humanity that it has distanced itself from the world in its density and lost this communion with it. The conceptual sphere irremediably constitutes our world. With the human being there bursts forth an inventive/creative profusion of representations detached from existence.

Animals exist in the range of life's spread and depth, while human beings soar above it and float on the waves of their own mental sphere. Sensing this, humans domesticate animals in refined conditions for the sake of recovering some affective attunement and coexistence and actually believe that they then somewhat bridge the gap between the fullness of fleshly animal experience and the volatile sphere of human existence abstracted from fleshly existence.

In fact, it would seem that owing to the human mind, life has become bifurcated into two separate realms: the natural life of plants and animals with its own embedded rules and laws, and the life of the human world with its own freedom — absolute freedom — to imagine, to think, to project, to discover the rules of the game of nature, and to fashion a "world," the human lifeworld, around us.

This bifurcation is actually a simple illusion of the mind itself: there is no way to disentangle the self-individualizing existence of the human being even with all its projects and freely pursued aims, a search unique in the realm of nature, from the natural ground within which the human animal is embedded.

Still, the uniquely human sphere of life, human culture with its social types and behavioral styles, is man invented, our unique dominion.[18] Would it not be too hasty then for us to conclude that the human animal in his cultural expansion is just continuing the spread of the individualization of life? Should we really consider culture to be a simple outgrowth of nature-life and in its difference in kind from nature to be just an excess of nature? Or do the clues given by the self-individualizing principles of life compel us to consider the individualization accomplished by the human creative mind to be a novel phase in evolutionary progress? Is the human being, a radical novum that stands out from the relatively autonomous unfolding of the rest of life, a sphere of life entirely novel with its own rules, laws, principles, its own generative powers for promoting and conserving life, still but a phase of life most intimately embedded in the entire spiraling process of life's differentiation and transformation?

V. SPECULATIVE REJOINDER

I believe I have brought my argument to its concluding point. Although I conducted it "in the concrete," that is, avoiding the usual abstraction of ontology and ratiocination in presenting and conducting it, nevertheless it is deeply steeped in the traditionally established concepts within which philosophical discourse is usually caught. Thus it will be in order to engage in an ontological speculative discussion on some striking points. At the risk of repeating some of that which has gone before, underpinning some of the striking issues in which the concepts used are embedded will throw a brighter light on the preceding analyses as well as help to explain their progress.

It is the conception of the "individual" that comes first and is central to our considerations. Only a living being can be seen as an individual on account of its selfsameness and autonomy, and this in contrast with "objects" that are heterogenous because in their coming to be and maintaining themselves in a quasi-existence they draw on other entities, that is, on autonomous individuals. Thus intentional objects (societies, institutions, works of art and of thought, etc.) cannot be considered individuals; in their very "essence," their "selfhood," they do not manifest a "selfhood" or an intrinsic mechanism peculiar to living beings. Nor may we consider "ideas" to be individuals; "floating in the air," they seem to draw their enigmatic status from their position in between factual reality and the transcendental system of the human mind.

Let us then enter into more specific features of the individual.

(1) What imposes itself first on our attention is the notion of "autonomy." We will be discussing this notion throughout our outline. To begin with, let us state that the autonomy of the individual "living being" consists in its self-enacted development from conception to extinction.

(2) This development proceeds inseparably with the individualizing being's very own functional system, which sustains it.

(3) In virtue of its intrinsic material and functional endowment — one giving it "volume," "embodiment," "flesh" spreading it "in space" and the other, a line of successive phases and steps of operations that stretch the unfolding life "in time" — the individualizing existence takes its "shape." Simultaneously it establishes its uniquely own domain circumscribed by the reach of its vital powers. Stretching, it acquires also a distinctiveness from the circumambient sphere. Lastly, in so doing, it acquires "inwardness" vs. "outwardness" and establishes its "selfsameness."

(4) Relying essentially on its very own intrinsic elements, the living individual does not share any of them with any other being or thing, and yet its functional system is basically oriented toward the vital exchange of life's materials, substances as well as energies and forces, as nourishment that sustains this very functioning and its progress onwards. Let us emphasize again first, that the very own territory that the living individual thus occupies it owes to its intrinsic matter's remaining within tight constructive functional ties. Although each being is self-determined and circumscribes its own existential realm, so that we may with Leibniz

consider it "self-enclosed," this is only by the above-mentioned — and other — standards, so that working its own way into the world it forms a fortress-under-construction. But it is essentially opened to the world by the above-mentioned means. Indeed, it is building itself up, establishing itself within the web of life, while pursuing its very own intrinsically predelineated life course precisely by reaching out and opening itself incessantly through its functional apparatus to this web, throwing hooks outside to bring it in, and after processing its material to its advantage, eliminating what it cannot use.

Second, we have to bring out a further point concerning precisely this "constructive" unfolding, as I have unravelled it before.

(5) We have above — pursuing our inquiry in four lines: functional, structural, generative, phenomenal — attempted *to show that that which is being constituted is in virtue of what it becomes flesh and body, an energy palpitating beingness.* And this beingness is by no means an abstraction, a universal, an eidos that would "concretize" itself in identical singularities. However, concerning this being and the intimate workings of its individuality a few items have to be identified precisely in contrast with the abstract, eidetic descriptions of ontology.

The question — and the difference between us and the ontology of, for instance, Roman Ingarden — is that of this constructivism of unfolding, which I have brought forth as the original and essential point of a phenomenology that envisages beings and the real from the perspective of life. In this respect two moments of the constructive becoming of the living individual have to be elucidated. First, it has to be emphasized that in this ontopoietic unfolding the functional consistency of the unfolding stays in a most intimate cooperation with the "material" endowment, that is, with the energy, force, matter, substance, etc. that flow in and out, beginning in an initial nucleus and continuing throughout the entire unfolding progress. These two: the functional operative system and the material on which it operates with their peculiar propensities for constructive operations appear to "imply" each other in the performance of their constructive project.

This synthetic cooperative harmony cannot be a haphazard deployment of life. On the contrary, when we consider the specificity of each living being, its unique self-identity, which already attracted the wonder of Leibniz, we cannot fail to see that the consistency of this common effort is such that, although the situations in reality may call for adjustment and accommodations, there will not be any essential deviation from the course: and this course appears predelineated within the unfolding of the individualizing progress itself. Here we cannot help but share the Aristotelian intuition of the vital forces that carry with them constructive propensities lying-in-wait to be unfolded — in one word, the entelechial element directing from within the entire progress. This constructive line that is being followed we have appropriately called the "ontopoietic design," and its embodiment in a constructive force we have called an "ontopoietic nucleus."[19]

It has to be kept in mind that this constructive line, its consistency, etc., are moments of a pulsating process of energies and forces.

(6) The ontopoietic nucleus pertains in a unique way to the selfhood (as well as to the autonomy) of the living being. This unique concrete selfhood of the individual consists precisely in the central functional system, on the one hand, and its embodiment in the corresponding functional energies with their propensities, the "matter" that they produce and reproduce, on the other. These material elements embody their own inclinations that together are central to the entire individualizing progress. That is to say, all the operations of the life process refer to its synthetic guidelines. It is in this way that the unique selfhood, self-identity, of the individual comes to be and maintains itself. It is owing to this functionalized matter that the individual carries an "existential" weight in manifesting itself as a relatively stable being-in-process.

(7) But the individual is also the bearer of life's awakening as well as of the "light" that surges as a glimpse giving awareness, the glimpse of self-awareness that brings together all the vital spread of the living being and makes it stand out as itself, as the selfsame and unique.

(8) It is a functional synthesis with living matter that carries the individual. This is of paramount importance for our differentiation of living types according to the degree of self-awareness in their selfhood, which varies with the progress in complexity.

(9) With our emphasis on matter we seem to agree with Ingarden's

structuring of the individual, but by interpreting the role of material elements within the individualizing process, we place ourselves on the other side of the river than that on which stands Ingarden's conception of the individual.

In our perspective favoring the life process over the static sclerosed eidetic structures of objects, we have recognized that these "materials" that give the living being its embodiment, the flesh and blood that integrate it within the community of life, and that are the means of its life manifestation, realize themselves only in an instantaneous reality. They stand out only insofar as they are the fruits of a constantly advancing functional growth and an unfolding of the ontopoietic process. That means that we cannot with Ingarden attribute to them a crucial role in the "constitutive nature" of the individual. These materials come, perform their roles, and go. The living being's constitutive nature, if we insist on using the term, would in our perspective be found within the complex of endowment, functioning system, a material spread, with the emphasis falling on its ontopoietic unfolding of an entelechial design. Indeed, in this incessant change, exchange, transformation, nourishment and elimination process in which matter perdures, it is the ontopoietic design that is enacted through the modalities of the functioning-in-progress and on being embodied in matter is actualized that represents the quintessential specificity of the being-in-progress being constructed. This specificity in a marvelous way represents simultaneously the singularity and the type.

To conclude our survey of issues concerning the individual as a living being, we must return to its most striking propensity (faculty), that which makes it stand out radically from any other type of entity, namely, the faculty of replicating itself via its own intrinsic means and according to its own devices. In this replicative accomplishment it follows its type, its generative hereditary features, and singularizes them in its own uniquely unrepeatable style. Notwithstanding the variations in the phases preceding the initiation of the replicative process, it is still initiated as such from within the individual. It ignites its own different flame, but one that carries with itself its own fuel — to be, of course, constantly replenished — as well as its very own constructive propensities, a constructive mechanism to unfold them, and a direction to proceed upon.

Lastly, but importantly, the individual carries with itself the blueprint of a type — the same as that of its generative individual factors — to unfold and embody with each step of its progress. We witness here a most fascinating situation: there seems to be a "generative split" between the constructive tendencies primed to constitute a singular individual in a "repetitive" and yet "selfsame" autonomous being and also to bring into this constructive constitution a set of features, propensities, elements that propagate through the individual the individual's type and thus throw out hooks for the continuity of the species, a hereditary tendency.

In spite of this apparent split into two lines of accomplishment within the generative progression, these two tendencies complement each other and, more, sustain each other. In the tight knot of the initial generative endowment and subsequent unfolding, they are inseparable.

Here we have reached a point at which a theoretical, ontopoietic discussion of the generative phase of the ontopoiesis of life focusing on the most essential complementariness of the individualizing/typifying processes within one and the same generative progression would be most revealing about the nature of life.

CHAPTER THREE

ONTOPOIETIC DIVERSITY AND THE UNITY OF APPERCEPTION

I. THE ISSUE OF DIVERSITY AND UNITY IN THE TRANSCENDENTAL PERSPECTIVE

1. *Proto-ontological Diversity and the Unity of Apperception*[1]

Throughout history the greatest question philosophers have wondered over has been that of the immense diversity and simultaneous unity of everything within and around us. Treatments of the question abound in Antiquity, the Middle Ages, and modern times. Today's philosophers — thinkers concerned foremost, if not obsessed with, our innermost powers of intelligence to form an all-embracing unitary vision and maintain it as the fulcrum of our existence — have sought the sources of this universal unity within those very intellectual powers of the human being. Hence, the unity of apperception has come to be unravelled from within the specifically human mind.

The question of the unity of apperception stems from marveling at the cohesion of our human existence and experience. There is our experience of the cogent world within which we dwell, a world in which the closer and the more distant all "hold together" without "holes" or gaps jeopardizing their coordinated cohesion. There is the cohesion of our life praxis, which keeps us alive only if our needs and performances are coordinated. We pass to the unifying coordination of the inward and outward orientations of our beingness, our bodily-psychic-mental functioning in which all our organs work in tandem with the networks of vital, organic processes and forces that lie outside the entelechial core of our singular being, and also to our involvement in the life processes of other individuals and with the gregarious/societal processes of their lifeworlds.

Let it be made clear at the outset of this inquiry that we will mean provisionally by the "unity of apperception" just this coherent global presence of the totality of things, of the beings (in their intertwinings, dynamic processes, etc.) in which all our experiences are immersed and of which we are only partly clearly and partly dimly and partly just marginally, even unwittingly, aware, given the totality's vast and deep dimensions. In agreement with Husserl, we will uphold each experience in its immersion in the coherence of the world horizons comprising all possible givenness. That is to say, we will consider the unity of apperception as awareness of reality that is part and parcel of the totality of experience, in its cohesion, in its harmony. This is our working hypothesis, one embracing the totality of the issue in a pretheoretical way. We will discuss this hypothesis within the perspective of Husserl's transcendental philosophy juxtaposing that with our own philosophical insights.

Since Kant, who attributed the unity-in-diversity of the objective world, of the human universe, to the constructive powers of transcendental consciousness, experience has been the key to the issue. Consequently, the diversity of things and their unity come into focus in the unity of apperception, which is understood to differentiate as well as hold together because that is the way in which the system of consciousness — transcendental consciousness — organizes and establishes it.

What makes this notion central to the present inquiry? The notion of "apperception," which came into prominence in Husserl's investigations into the workings of consciousness in its constituting the objectivity of the lifeworld, the conscious subject and its insertion into objective time and the world, and in Husserl's discovery of other subjects as well, is to be traced to Leibniz, to his distinguishing between the "petites perceptions" we are not aware of but that found our conscious perception of an objective state of affairs and those perceptions that bring everything to the level of consciousness in an organized way. The distinction is drawn between "insensibles," preconscious multiplicity, and conscious apprehension of objective givens. This distinction, albeit with a difference, is maintained by Kant in his concept of the 'manifold' that precedes the activity of experience — which then makes it fully conscious — as well as by Husserl. He, after having adopted the term 'apperception' in his early period, would attenuate its sense later. However, the distinction between preconscious multiplicity and the formal organization of it by the conscious operations that it prepares the way for is one he continued to make throughout his development. For Kant and Husserl both, it is conscious sense bestowing (*Sinngebung*) that through formal constitution introduces sense, meaningfulness, rationality to the preconscious multiplicity that proceeds from within and without the conscious subject. Sense giving is the work of the conscious subject, reason, itself. This is how the transcendental perspective answers the metaphysical question of the diversity of being and its unity, by means of answering the question of the origin of rationality and its laws, of logic.

2. *Husserl's Treatment of the Unity of Apperception*

In line with Kant, Edmund Husserl refers this unity-in-diversity to the nature and order of experience, understood principally as cognitive experience. That cognitive experience comes first, before any concerns of the praxis of life, accounts for the formation of the objectified universe. However, *Husserl probes much further than Kant into the analytic outcome of the processes of objectification/constitution of reality.* On the one hand, he a) *differentiates the preconstitutive manifold into various pregiven spheres* with a borderline between passive pregivenness and active ego-centered constitution and b) at the same time sees on both sides of this borderline a reference system of paramount importance for constitutive formation — the world. *The world assumes the role of the groundwork of experience* as well as of its several constitutive horizons.

In other words, although Husserl reaches far deeper into the concept of the manifold than Kant when he distinguishes between the many sedimentations of the preconstitutive, "passive" synthesis of the manifold preparatory to the active intervention of the ego's objectifying constitutive work proper, *the unity of apperception is still reserved to conscious-*

ness as a system without the work of which no cognition/objectification or volition, appreciation, etc. would be accomplished, with the work of consciousness culminating in and tending in an indispensable way toward the activity of the ego as its constructively regulative center.

Without the ego actively entering the scene of the preliminary passive conscious synthesizing of the world, no identification of objects, no unity, no objective world would be accomplished.

Husserl gives a sweeping treatment of this latter issue in his genealogy of logic, which he does not confine to the functioning of judgement but conceives as stretching throughout the spread of experience.

It is then most appropriate that we will refer chiefly to his later thought as transmitted in *Erfahrung und Urteil* ("Experience and Judgement" — hereafter cited as *EU*),[2] in which he draws on most of his previous work in establishing this genealogy.

In this Husserlian perspective the concept of the unity of apperception pertains to the great metaphysical issue of the One and the Many as it is pursued along four critical lines of questioning: 1) the question of the emergence of objectivity and the role of consciousness in it; 2) the question of the source of rationality (*Sinngebung*), both specifically human rationality and that of Nature; 3) questions about animal subjects and the gradations in their measure of consciousness up to "full consciousness" in humans, questions raised by the immersion of the subject in the world; and 4) the question of the role of the world with respect to the constitution of objectivity and, ultimately, to rationality and universal order.

It is at the cross section of these questions that we situate *our inquiry*, carrying it to the all-embracing field of the phenomenon of life. *Its upshot will be to bring forth the ontopoiesis of life as the primal and generic factor of constitution while placing the lifeworld together with its correlate, transcendental consciousness, in a secondary genetic position.* This amounts to proposing a Copernican revolution in phenomenology.

We will maintain that Husserl's analyses, a work of genius here submitted to new scrutiny, contain valid intuitions that within the vastness of his field of inquiry *give hints of interpretations other than those he drew* within his transcendental perspective. We will then progressively draw out those pregnant insights that pertain to our search for the key to the diversity and the unity of all. We will emphasize these steps in our itinerary: 1) the extension from ego-centered transcendental consciousness to its dynamic conditions; 2) immersion of the subject in the totality of the world/nature, thus recovering animal subjectivity and the extension of differentiated consciousness through the span of life and its evolution; 3) focusing attention away from the ego and onto life/agency; and 4) *bringing out the self-individualizing principles of life as the primary coordinating factor of genetic constructivism and its primary individualization of beingness as well as their crucial role in the ontopoietic unfolding in which the individualizing coordination of all beingness introduces order throughout.*

II. A BREAKTHROUGH FROM THE LIFEWORLD TO LIFE IN HUSSERL'S *ERFAHRUNG UND URTEIL*

At the outset of his genealogy of logic, Husserl emphasizes strongly that there is no strict distinction between experience and judgement, two functions of the mind that are usually treated separately as different tasks performed by seemingly distinct sets of faculties: sensory and intellectual. But both actually stand in the service of experience: the experiential process (e.g., that of sense perception) cannot be accomplished without the intervention at a certain point of the function of judgement and this latter could not take off other than from the ground of the sensorium. (One cannot help but think of Kant's dictum: Experience without concepts is blind, concepts without experience are empty.) It is not to be thought that the faculties of judgement enter on the scene as those of sense perception depart. There is a "passage" in which they are intertwined. We will concentrate our analysis on the space of that passage.

For the present, let us ponder a statement Husserl made in embarking on his genealogy of logic, one that gives an overview of its entire spread. Husserl characterized the spread and scope of experience, the central idea in his theory of constitution, by saying, "Wir können auch sagen: vor jeder Erkenntnisbewegung liegt schon der Gegenstand der Erkenntnis als Dynamis, daß, zur Entelechie werden soll."[3]

Here is delineated the entire itinerary from the experiential genesis to judgement's contribution with an indication of its source at the one extreme and the telos of its accomplishment at the other. The itinerary simultaneously takes in source experience and the genesis of order.

1. *The Dynamis that Opens the Way out of the Constitutive Circle toward the Energies, Directions of the Ontopoiesis of Life*

Indeed, in the Husserlian orbit of thought, the full-fledged experience (originary and foundational sensory perception) spreads its varied processional "synthesis" of the pregiven (here we see already the corroboration of the Kantian "manifold") between these two parameters: the dynamis that is experience's origin, and entelechy, an already constituted "object" that represents its telos or fulfillment. In between these points lies the aforementioned intermediary space into which we will enter later. Let us begin, however, by raising the question of the "dynamis" that initiates the process.

In the Husserlian perspective, this dynamis cannot belong to the intentional system of constitution itself. It lies at its origin, it is its initiator, lying beyond its own mechanisms for propulsion. Husserl would further consider the groundwork for the constitution that he sees in the lifeworld. However, this initial force is not active at random; on the contrary, it is gathered within an individual subject and only as such may it instigate the performance of a cognitive/constitutive process. This means that the dynamis is in a primordial sense the force of the agent performing the cognitive process. It might well be that the agent is prompted by occasions in the circumambient world, just as it is correct to assume that the agent is embedded in a network of forces, activities, processes belonging to an area larger than its own in the lived world and that it is only in view of this participation that prospective objects are "pregiven," but this dynamis has to be processed within the individual subject in a way appropriately geared to the subject's own functional schema, which differentiates it from all the other centers of force in its surroundings.

It is only when processed through the individual subject's specific functional system that this dynamis may acquire its direct orientation in initiating the perceptual process. This dynamis is not itself conscious. Otherwise it could not initiate a conscious process as such. It lies then at the borderline between the intentional system that it is meant to put to work and the pregiven, even preexperiential realm. This indicates that this dynamis is in its emergence itself motivated by the functional individual system of its carrier, the system that I have called the ontopoietic, self-individualizing delineation of the life course of each subsistence.

We may further conjecture that it is within this living ontopoietic process that this dynamis itself originates as a distinct jet of forces and thus not in a neutral form but in the form of a singular, specialized "synergy," a cogent bundle of forces informed by virtual sense. It is only as such that it may distinguish the various experiences that it will initiate and so direct the experience's focusing on such or another object in perception. This explains how the initial dynamis is neither intellective intentionality nor blind. The background of the lifeworld remains valid as the field of play of the ontopoietic game, but it begins to appear that only this game itself could lead to the channeling of forces down the appropriate avenues of individualized and uniquely meaningful experience.

2. *The Self-Given Evidence Leading Us away from the World and toward Self-Individualizing Ingrownness in the Ontopoietic System of Life*

a. *Partaking in the World — Husserl's Measuring Stick for Determining Animal Species' Degrees of Consciousness; Differentiating among Species and Adopting an Evolutionary Approach*

Opening the inquiry into the genealogy of logic, Husserl set forth the crucial and decisive role played in experience by the originary self-givenness of its objects, evidence.[4] It is on account of this self-givenness of sensory experience that he attributes to it a founding role for all types of experience. He seeks to establish that the self-evidence of the objects of experience follows from the generic, pervasive, and persistent ingrownness of the experiencing subject in the world. He emphasizes that the world is therefore the groundwork of *all* experience,[5] of all ordering and that logic as the exponent of rationality is the logic of that world.[6]

We will focus on Husserl's substantiation of the thesis that the world is the ground for experience from three perspectives. First, we will ponder the fact that self-evidence is the result of individualizing experience. Second, we will show that the differentiation of types of consciousness proceeds from their partaking in the lifeworld. Third, we will appreciate the twofold nature of evidence, the fact that the evidence of the objective telos and that of the modality of experience lead us via the genesis of the experiential subject through experiential constitution to recognize an opening toward the evolutionary genesis of types of conscious subjects.

b. *Objects That Offer Themselves in Self-evidence Draw Substantially on the Lifeworld, but Are Ultimately the Fruits of the Genetic Activity of Individual Subjects*

Husserl emphasizes, as noted above, that the definition of experience is principally the "evidence of individual objects" (*EU,* p. 21). This evidence is of primary importance for the entire field of cognition and serves as one of the main channels of the continuous passage between experience and judgement.

However, it is not the case that any evidence would be sufficient for judgement. Judgement too has to be "evident." For this reason there is need that the objects be, as it were, the content of judgement in their self-givenness (*EU,* p. 11). In fact, statements in which we express our cognitions, our experiences, are already fruits of the judging activity; in regard to these statements the focus of our attention falls on the objective content that they present; they are fruits of both cognition and acts of judgement. In this way the judgement that the statement represents is "treten auf als prätendierte Erkenntnisse" (*EU,* p. 7) and the knowledge of whether a judgement is a real experience/cognition or just a "pretended one" is there before any logical analysis. Hence, our primordial judgements are judgements "with individual substrates, judgements about the individual, judgements of experience (*Erfahrungsurteile*)." They go back to the prepredicative givenness of individual objects. Inversely, in this sense, the theory of prepredicative experience (cognition) is simultaneously the first phase of the theory of judgement.[7]

In short, Husserl's point of departure in his pursuit of the logical structuring pervading the entire constitutive network is that all knowledge/cognition as well as judgment is based on the evidence. And while an evident criterion of evidence is seen by him in the self-givenness of the objects of cognition to the cognizing mind, as well as in the possibility of tracing the statements of judgmental activity to this immediate evidence, self-givenness, which is the entire bulwark of cognition/ constitution at all of its levels, goes back to the experience of individual objects. The definition of experience is principally "evidence of individual objects."[8]

Together with this we see that the givenness of individual objects of experience occurs on the ground of the lifeworld. Each experience occurs against the backdrop of the world of objects pregiven (*vorgegeben*) in a "simple certainty" (*Gewißheit*) "als uns vor der Erkenntnis schon geltende, und das in verschiede der Weise. Da sind sie als schlicht vorgegebene Ansatz und Anreiz für die Erkenntnisbestätigung, in der sie Form und Rechtscharakter erhalten, zum durchgehenden Kern von Erkenntnisleistungen werder, deren Ziel heißt 'wahrhaft seiender Gegenstand, wie er in Wahrheit ist'" and then further "Weltbewußtsein ist im Modus der Glaubensgewißheit."[9]

All praxis of life, like the theoretical praxis on a high intentional level, presupposes this belief in the world, in its being there as naturally assumed, unquestioned. *Each experience, whatever its mode, carries with it this belief in the world. And yet this very belief in the world is the fruit of singular experiences in which the pregiven* (vorgegebene) *objects of the world were established in the process of the ontopoietic unfolding of life.* (Husserl himself conceives of the world in terms of sedimentations of historically established experiences.)

Ultimately the lifeworld as the circumambient world of singular constitution, a singular ontopoietic course and performance, as well as *the conception of the lifeworld at large goes back to these singular experiences of individual beingnesses.* Belief in the world has to refer ultimately to the performing subjects, to their performances and system. Let us now follow Husserl's path further along in the constitution of the intelligibility of this world.

It is beyond doubting that each singular individual initiates its own constitutive process against the backdrop of the lifeworld into which it is launched, and its unfolding constitutive progress inserts itself into the already established constitutive forms of the world just as the individual's conscious/constitutive apparatus inherits the conscious mechanisms already developed in the previous stages of humanity and of prior living subjects. Yet, *this world, although first in givenness, is by no means first in the "order of things."* We have now a clearer awareness of the place that our world occupies in the cosmos; *it is a "world" because our planet as a cosmic entity allows for life. And it is life that determines a world.*

c. *The Evidencing Experience Carried On by Life Performing Ontopoietic Functions*

As mentioned above, it is not only the objective content of experience — its object — that in bearing absolute self-evidence is the basis of reality; evidence of experience is just as much a matter of the *modality of experiencing.* The evidencing experience as we may call it focuses on an objective moment: a state of affairs, a thing, a living being, etc. For example, I see in front of me a particular tree as being there and my experience extends into the garden in which it then lingers and then to the fields beyond, to the woods at their edge and on to the far-off horizon of the world of which all this is a part. This tree, a pain in my leg that I cannot think away, etc. carries a belief in the factual reality of the whole state of affairs. But beyond — or perhaps parallel — to these concatenations of objects stretching from object to object ahead of each experience *ad infinitum* is the evidencing experience that accompanies, better essentially accompanies, carries along the "positing" of the objectifying act of consciousness and that is itself embedded in the entire sensing/feeling experiential subjective system of functions, functions that carry the empirical experiential constitutive-objectifying progress. This progress offers an objective moment or state of affairs in question, distilling it from the sharply acute moments of empirical attention. However, as I have emphasized, all this is carried by the empirical processes that belong to the entire presentational objectifying system of apperception. This latter is thus embedded throughout its performance in the entire experiential system of the living subject.

In short, and this coincides with Husserl's analysis, although it loosens the sharp intentional edge of his interpretation, evidencing experience (originary experience to boot) draws throughout its performance on the entire system of the experiencing subject. In this way it is most intimately interwoven — tributary to and grounded in — the subjective lifeworld organization processes and enmeshed in its processional fabric, in its dynamic, sense processing synergies, which are present in each experience. The evidencing experience in its synergized performance draws on these energies, sensing items not just as led by the random conditioning of the particular circumambient world situation in which it is enmeshed, nor just by the universal lifeworld horizons to which it constantly refers to check its bearings, nor just by the system of intentional consciousness, but as led by the very fountain of processes and functions promoting the living being's subjective discriminations.

First, it draws on the empirical system of sensing/feeling, on organically grounded processes that in their individualizing, life prompting functioning perform the self-differentiating acts of the self-individualizing life course of each being-in-progress. As such, each evidencing experience is ingrained in the system within which the universal life spontaneities seek and meet appropriate life forces, structural devices, etc. and delineate their routes accordingly, bringing these together as ground laying concordances. Simultaneously, the individualizing concordances that are species/genus constructive devices have to be arranged there. But foremost, the differentiating processes have to seek and establish accord among organic/vital exigencies allowing for a) "diversity concordance" and b) "diversity fullness" among the indispensable particularizing intertwinings of life's synergies and generative forces in their indispensable selection for an individualizing life. They carry each other in tandem in ever more expanding circles within as well as outside the individualizing beingness toward the unity-of-everything-there-is-alive, the great and ever changing network of life's advance, into which enter all of life's synergies, forces, constructive devices, through the final concordance of each self-individualizing existence.

There is no doubt that the lifeworld as the constructive median of life's advance belongs with its substantial and interactive fabric, its structural network, its intergenerative reservoir ever in transformation, its palpitating "flesh" to the generative and referential system of life's constructive groundwork. It is life's bulwark. However, with the shifting of the emphasis from the objective reference of the evidencing experience to the "subjective" performing system of experiencing itself, which carries the positing of evidencing experience, we delve well below this median sphere, piercing it, as it were, and going below toward the life system as the dynamic source and ever spontaneous spring of everything.

Our further investigation of Husserl's vision of the relationship between the lifeworld and individual consciousness/subject will allow us to advance its specifications. But before we enter into that investigation, let us summarize. It is beyond doubt that each singular individual initiates its own constitutive process against the backdrop of the lifeworld in which it emerges and that this unfolding constitutive progress inserts itself into the already established constitutive forms of the world inasmuch as the individual's conscious/constitutive apparatus inherits the mechanisms of consciousness already developed in the previous stages of humanity, in previous living subjects. Thus this world, although first in the order of givenness, is by no means first in the "order of things." In itself the world is the offspring of experience as the living subject's means of self-differentiation amid other life processes. We have now a clearer awareness of the place that our world occupies in the cosmos.

It is a "world" because our planet as a cosmic entity allows for life. And it is life in its spontaneity and interconnectedness that defines a world. Thus, ultimately, the lifeworld as the circumambient world of each singular constitution, of a singular ontopoietic course and performance, and, therefore, the very conception of the lifeworld, goes back to the singular experiences of individual beingnesses. Belief in the world has to refer ultimately to the self-evidence and system of those performances.

3. *Recapitulation*

The differentiation of types of consciousness with respect to the lifeworld leads to the extension of consciousness into the animal sphere of life as well as to their evolutionary advance. Conceiving of the experiencing subject as being immersed in the world and as drawing its existence from it, Husserl simultaneously immerses consciousness in the multiple world/nature complexes of sense. Indeed, it is partaking of the world that defines, and determines, in Husserl's view, consciousness as such. Consciousness is conceived as being essentially correlated with the world, and inversely the world is conceived as being essentially correlated with consciousness. Hence, insofar as all living beings are intertwined with their circumambient world and take it in, they are to be conceived of as being conscious. Husserl emphasizes strongly that whatever is real exists within the sphere of the world and that in this way to "exist" means always to "exist in."[10] Each experience, each apperception carries with itself not only its own differentiated object but together with it the sphere of sense in which it is embedded as it partakes of the world. By the same stroke there is in Husserl a naturalization of human consciousness (*Naturalisierung des Geistes*), and with that all of the works of the spirit — all *Kulturgüter* — are plunged into nature, with human and animal beings being put in the same sphere of sense. Animals are equally "subjects" (*tierieschen Subjekte*).[11]

This view offers us two crucial insights supporting and specifying our above-enunciated ontopoietic breakthrough.

First, grasping that the conscious subjects that emerge within the natural sphere of life vary in their manner of apperception with respect to the degree, mode, and complexity of their involvement in the lifeworld leads to distinguishing types of living individuals with the human being then being seen as being endowed with what Husserl calls "full consciousness."

Second, since, as we have attempted to establish above, the lifeworld itself is the result of the self-individualizing progress of the living subject, progress uniquely its own, and since, as Husserl himself establishes in his genetic phenomenology, the modalities of apperception as well as those of the conscious functioning of the experiencing subject are forged in the genesis of its experiences, the door is open to the evolutionary progress of not only each experiencing subject in itself, but most significantly, since we "inherit" prior experiential forms of partaking in the world from other subjects, to the evolutionary progress of subjective forms as well.

With respect to these two openings, first the opening to the "reality" into which all that is, to use Husserl's expression, 'worldly' is plunged (and what other reality could it be for us, given the spheres of sense preceding apperceptive intentionality that we have already revealed access to?) with the dynamis breaking through to the ontopoiesis of life, to life's constructive becoming, and then the opening to the same spheres of sense with the raising of the question of the evolutionary genesis of types of living subje*cts, we are bound to postulate intuitively a principle of life, of its self-individualizing poiesis.* Indeed, *our analysis leads us to intuitively conjecture an ontopoietic principle that functions as an axis for the coordination of the preconscious, vitally significant life carrying operations that, although they remain at the circuits of the preintellectual, carry on the mute order of nature that will then nourish and launch the intentional system. This principle may be called equally* a) *the point of reference* for the distribution of sense initiating operations at the horizontal level of the life carrying processes that operate the self-individualizing differentiation, b) *the entelechially charged indicator of direction* for the genetic construction of individual beingness, and c) last but not least, *the measuring rod for the constructive attainment of types of complexity,* that is, of the sense bestowing levels (levels of conscious functioning in the experiencing subject).

In short, from our analysis and new interpretation of Husserl's genealogy of logic, we find, on going beyond Husserl's ground, an emerging *factor of self-individualization, a concrete metaphysico/ontological factor of life* as such, making it self-centered, inwardly/outwardly oriented, and entelechially pregnant. This ontopoietic factor of a principle is indispensable for any accounting for the origin, genesis, differentiation, and dynamic unity of the individual life course within a stir of competing forces. We will now pass on to a discussion of this "dynamic unity," which is apperception within Husserlian bounds, but goes beyond those bounds.

III. THE TRANSMISSION OF SENSE: FROM DYNAMISM TO ENTELECHY — THE UNITY OF APPERCEPTION AS THE KEY TO THE ONTOPOIESIS OF LIFE

With the above discussion focusing on the differentiating line of the apperceptive process — and in which there occurred a shift away from the strictly cognitive constitutive approach to one encompassing the vaster life individualizing complex of processes — we already broached the question of the unity of those processes, that is, the question of the unity of apperception as such.

In this inquiry it cannot be forgotten that for Husserl the strictly cognitive telos assumes a primacy when it comes to apperception — even if he considers volition, desire, feeling, aesthetic appreciation, evaluation, etc. as belonging to the cognitive attitude — over against pragmatic sources of sense as well as those of vital unfolding. When we now pass to the analysis in which Husserl establishes the genesis of objective sense and the objectifying unification of its processes, the emphasis falls on the object to be constituted in virtue of which these processes organize themselves. We will now follow Husserl's line of thought, extracting from the analysis in which he forces the plurivocal influx of sense into a narrow channel *certain conjectural intuitions that lead us away from the narrow objectifying path toward a full-fledged restoration of that plurivocal diversity to its adequate role in instituting the grandeur that is the manifestation of life.*

1. *The Diversity of the Manifold and Its Progressive Concordance in the Unity of Apperception*

The issue of the unity of apperception within the Husserlian framework is, as in Kant, the issue of how the manifold preceding experience may be brought together into a coherent, objective form. To put it another way, the question is that of how the incontestable diversity — the discrepancy of the diverse — is brought into the unity of apperception. These very formulations of the issue prompt the question of the role of human consciousness in harmonizing the diversity. However, mining an analytic vein above that of Kant's work, Husserl distinguishes clearly between the pregiven to human experience and what is given to the human consciousness that gives the pregiven form, making it pass from the "hidden" to the "apparent," to the *given.* Although both thinkers work at the pregiven, non-apparent level, with what precedes the work of consciousness, that is, with the diversity of the manifold that the conscious activity of the mind then takes in and brings to formal significant unity, Husserl goes beyond Kant in treating the pregiven manifold as being already in explicit relation to consciousness. We would call to mind here the differentiation of the various stages in the constitution of the sense of the object in a series of "passive syntheses" at the pregiven level, syntheses that advance to what Husserl (and Merleau-Ponty) calls the "transition syntheses" (*Übergangssynthesen*) of "passive" consciousness and then encounter "active" ego-centered consciousness. It is upon this transitional phase of constitution that we will focus our attention in what follows.

Let us recall that the genetic progress of objective constitution is divided into two phases, as it were, two phases, however, that pass over into each other without sharp division: the series of passive syntheses of the manifold of pregiven experience and that experience's encounter with the constitutive activity of ego-centered consciousness. In the progression, the judgmental activity of ego-centered consciousness gives to the content of that experience objective form, objective sense, an "entelechy." There is a selective unifying process leading from the elementary manifold proposed to constitutive consciousness, one that is initiated by a dynamis that is lost in oblivion at one end of the process and lost in the judgement of the constituted object, the entelechy, at the other end.

There are several questions that this Husserlian conception of experience provokes. Among them are some that are still vigorously debated and are most relevant to our argument: "What is the elementary 'manifold' conceived to lie before constitutive-passive consciousness to be worked on?" and "What type of unity is accomplished in the final constitutive synthesis yielding the object?"

The latter question pertains to the *mode* in which the unity of the diverse is accomplished in the last phase of apperception: "Does this unifying role belong to a preparatory stage such as passive synthesis, or is it the constructive inference of the active ego-centered consciousness?" (See the discussions of this question by Ogawa and Tugendhat.)[12] Although it is of interest for our present discussion to consider first the question of the nature of this unity, we will leave it for our concluding recapitulation of the results obtained by our argument and *focus here on the question, so decisive in Husserl's thinking, of the origin, role, and diversification of attention in the constitutive progression.* This discussion will give us crucial insight into the origin and transmission of sense in the entire constitutive process.

2. *The Series of Passive Syntheses of Constitutive Consciousness Carrying the Life Significant Lines of Sense Preceding the Judging Activity of Ego-Centered Consciousness*

As we know, preceding the constitution of the objective content of experience proper, there lies a series of steps in which the manifold proposed for experience undergoes in a "passive" way, that is, without any direct formation by consciousness, a sifting and amplification by lifeworld influences as well as by the previous experiences of the subject, as well as a sort of accommodation to the dispositions of the subject itself. Thus are formed sedimentations of sense that then pass into the next synthetic step, wherein all undergoes a similar sifting. In this way the organization of the originary manifold being is prepared gradually to take the final entelechial form that its encounter with active consciousness bestows. Husserl calls these syntheses 'Deckungssynthesen,' which lie in readiness contributing to the definitive objectifying synthesis by, as it were, announcing (*angebenden*) the arrival of the individualized object. Our perception of this table is already in the forefront, its appearance pregiven; this is to be expected on the basis of the sedimentations of the "covering synthesis" (*Deckungssynthese*). The question that arises first is that of "Upon what ground does the sedimenting/distilling of sense in the covering synthesis occur?" Husserl sees the run of the syntheses of the pregiven as being chiefly oriented by cognitive interest; for him the anticipated sense is oriented directly to the objective form, the entelechial form. At this point precisely the transmission of sense comes into question. When we reflect in an unprejudiced way (one not oriented by a transcendental vision), our anticipations of the experience to come based on the pregiven syntheses of the object to come are not directed only to objectivity as such. To the contrary, my pregiven approach to this table comprises foremost my usual (habitual) utilitarian evaluation of this table; I seek it to put my books on and to write at as well as for those of its features that are suited to my body's functioning, to which I may easily adapt — and that are therefore indispensable for its *lived sense,* which brings my body and the table into commerce. This *living sense* is that of *vital life significance,* into which a specifically human sense will pass that diversifies in multiple modalities, of which this is just a sample. To generalize from this example, let us state that first of all through the syntheses of the pregiven life significant lines of sense, the foundations are laid for the unity of apperception to be accomplished at a later stage.

3. *The Ontopoietic System of Concordance Running through the Continuing Series of "Passive Syntheses" Transmitting the Plurivocality of Sense*

If we consider the entire run in which progressive distillations and then the formation of the objective intelligible manifestation of an object in apperceptive unity is accomplished, it is the continuity that is most to be marveled at. Husserl brings us from the Ur-synthesis of the first fusion of sensory material according to "sensory fields" up to the already lurking objective discrimination of the object by linking this continuity first to the dynamic prompting of the initiated process itself and then to the Ur-form of consciousness itself, pinpointing first the spontaneous differentiation of the material to be synthesized according to its intrinsic homogeneity and then focusing on the progress made in qualifying the sedimentations objectively according to their teleological tendency with respect to the constitutive laws of consciousness.

I propose in contrast — and this on the basis of the previously described opening toward the plurivocality of sense that breaks with the one-sided Husserlian focus on the "objective sense" — that there is a threefold dimensionality of the objective coincidence among the qualitative elements that progressively constitutes the objective manifestation, a threefold dimensionality that stretches like an accordion. Indeed, the hyletic material experiential element of which Husserl himself speaks and in which the Ur-experience phase of sensory perception is grounded is to be identified with processes carrying life's differentiating poiesis; that is the first vital/sensing meaningfulness embedded in life-oriented functioning. Here is the circuit of functions within which subjective experiences flow and the appearance of innumerable aspects occurs giving the hyletic experience of fusion the groundwork of experiential continuity as well. To this continuity belong at a more advanced level the bodily concordances we described above. Fusion and synthetic "grasp" are its work. But the principles of the objective constructivism, which govern the succession of the phases of passive synthesis and their appropriate coincidence of singular elements, themselves indicate a threefold system of reference.

(1) The emergence of singular elements in the form of a series of impressions (of color, tactile qualities, and other sensory qualities) obviously already functions according to the life promoting prerogatives that we usually attribute to the bodily/physiologically grounded (motivated) sense organs and belongs to the system of functions that carry the self-individualizing poiesis within the unity-of-everything-there-is-alive — all while seeking formal and material adaptation to the conditions of life set by earth's objective conditions, as set by cosmic laws.

(2) It would be precisely this self-individualizing concordance, a necessity/requirement for the life process, that prompts the functional individualizing system in its unfolding and in its perfecting, in its dealing vitally with all the circumambient factors with increasingly sharp lucidity. This is of crucial importance for the evidencing experience, as discussed above, as well as for the cascade of vitally significant existential processes. "Firstness" lies then, not with the world, but with life's conditions for unfolding, life, in turn, devising in its progress the very nucleus as well as the entirety of lifeworldliness. Only in its turn does the lifeworld become the ground for life's poiesis.

(3) There remains still the third point of reference for this continuity in the transmission of sense in the progressive unity in diversity of apperception. Principles that recall the two faces of a coin — the individualization of life/the lifeworld, cosmos/life — emerge as subtending in some essential way the differentiation as well as the unity of life. This will have to be treated elsewhere.

IV. THE "VIGILANT RAY" OF ATTENTION AS FOUND IN ALL LIFE FORMS — A NEW CONCEPTION OF THE UNITY OF APPERCEPTION

We have now to return to the phase of passage (*Übergangssynthese*) in the apperceptive process as expanded on by Husserl, but from a different point of view. We will focus now on the moment at which supposedly undifferentiated clusters call for the turning of our attention to some of their strongly stimulating features and the actual turning of our attention to them. No doubt this affecting tendency of the preconstitutive fields points, calls, stimulates our mind toward further involvement. However, Husserl interprets these stimuli as being correlated by and directed toward the ego. He uses expressions such as, "Das Aufdrängende kommt dem Ich näher oder ferner" (*EU*, p. 81). But there seems to be no direct step from this stimulus to the ego as the center of constitutive acts and the ego as the central agent of the mind. Instead, following the line of ascension that the initial dynamis takes through the spheres of pregiven synthesis, it seems more adequate to interpret this affecting force — a distilled form of the initial dynamis — as a tendency directed *not* toward a central agent, *but toward a further step of fulfillment, that is, the next step of the constructive fulfillment of the ontopoietic life design.* It is only too clear that *moving from the primal differentiation of animal recognition and orientation in the lifeworld, this further step of fulfillment would be intellectual cognition, a complete objectification of givenness* in clearly identified. individualized, and essentially explicated form. However, do we need to assume the final involvement of an agent that incorporates the entire constitutive system in order for these stimuli to be received?

At this point Husserl's analysis again comes in handy even though we are calling for an interpretation entirely different from his own. The question just asked lies at the heart of two major contemporary challenges to transcendental phenomenology: both the empirico-physicalistic as well as the deconstructive approaches dispense with the centralizing agent of the ego altogether. Can they account on their own for the unity of apperception? How in the empirico-physicalistic view, which locates all conscious and mental processes directly in the performance of the nervous system and the parts of the brain, can perception and judgement occur? May all the constitutive functions be reduced to physical processes? Even recognizing their extensive role, the apperceptive unity of the functioning system that unfolds the ontopoietic progress of living individuals integrated into the world of life that they constitute is something we cannot dispense with.

Indeed, the "affect" of stimuli on our mind is such that taken in its entirety it impinges on the mind to the degree that there is a "tendency to give way" (*zur Hingabe*), to respond to it, all leading to "compliance" with it, which draws our attention (Churchill trans., p. 79) and seems to be the obvious link between levels of functioning, between the preobjective and ego consciousness. But this the "last" passive synthesis "affects" or stimulates ego-consciousness and "activates" the ego in its objectifying mode. What is here at stake is the origin and function of attention in the unity of apperception.

1. *The Passivity vs. the Activity of Consciousness*

Going back to the pregiven, let us assess the essence of the prepredicative efficiency (or as translator James S. Churchill puts it, "performance") that Husserl sees as being correlated with "receptivity," which is attributed by him directly to the ego. By the "ego" he means an absolute transcendental pole to which all intentional acts directly refer and from which they stream. The ego as the source and center of the intentional system of constitution assumes in Husserl's thought a crucial, decisive function throughout. From her flows the univocal rationalization of the constitutive system as much at the realm of incipient perception as at the higher mental realms where objects are constructed. Passive "pregivenness" has its correlate in the "active" directedness of the ego, its interest and receptivity, which is to be found in basic sense perception as much as in evaluation, willing, pleasing, etc.

Thus, we are faced first with the "field" of passive givenness and its associative structures at even its lowest stratum, where there is not chaos but already the products of the performance and synthesis that is time consciousness. These syntheses are the lasting (*dauernd*) content of the unity of identification that is being operated at the higher strata of constitutive performances. Time plays an essential role in bringing the data into a synthetic form of simultaneity, succession, etc., as a universal order, the coexistence of "all immanent givenness" (Churchill trans., p. 75).

However, these pregiven syntheses of givenness are not "simply passive." On the contrary, they have an "affecting force" in their various modalities. We distinguish several spheres of synthesized data that sediment the phases of constitutive falling together and that through their interlocking carry an "affecting force." They step forth from the background of passive pregivenness as they are moved by an "affecting tendency" toward the higher level of constitutive differentiation and synthesis. This higher level would be that of a progressive identification of an object.

We recognize in this analysis the intermediary stages between the unadumbrated presencing to our attention of the loosely cogent but already differentiated into spheres constructive becoming of individualizing life and the further transformations of givenness. It seems also that beneath the primal and lowest synthesis that we may understand, those performed by the singular sense organs — if we focus on sense perception — as functions of the body's life promoting instruments with their kinesthetic schemes, there abides the mutual existential intertwining and interactive intertwining making up the life network of the world (Churchill trans., p. 72). There is no doubt that the neurological sciences have located the various functions of experience in the brain system. And it seems accepted and well established knowledge that the extended empirical/sentient experiential differentiation is located in different loci of the physical extension of the brain, that no central consciousness — consciousness as a centralizing performer of these different experiential schema — is to be acknowledged at this point in the game.

Does this allow us, however, to dispense with a *tertium quid* between the synthesis of sentient pregiven experience and fully conscious, intellectual constitution? This reduction of the entirety of the work that has been attributed to consciousness to the physical organization of the nervous system seems as unwarranted as the telescoping of the steps between the syntheses accomplished by the senses and the constitutive work of the intellect that is assumed by Husserl. We will now offer a new analysis setting this issue on a new basis.

2. *The Priority of the Activity of the Ego As Opposed to the "Vigilant Ray" of Attention*

The passage from dynamis to entelechy calls for a further distinction. Here, at this point, however, Husserl's analysis again reaches a level of indeterminacy, a point at which various interpretations are possible. As a matter of fact, it is at the point of the differentiation between the affecting tendency, the "stepping out" of some items of the passively pregiven in their exercising a stronger or weaker stimulus and the reception of that stimulus, the awakening attentiveness to and interest in that stimulus, that Husserl introduces the ego. The intrusion of the datum into the emerging beam of attention is according to him an obtrusion upon the ego, for here is the latter's receptivity (*Hingabe*) to the datum, its attention and interest. The ego is here assumed to be the pole toward which the stimulus tends; with the ego being thus activated it is assumed to initiate an active course, bringing in, in its performance, an entire apparatus for collecting appropriate data and holding in focus as temporal succession unrolls ever new aspects of the object and then constituting the essence of the object in static form.

This "holding in focus" as a passive side of the active unrolling of "apprehension" is the concrete enduring of a perceptual process (e.g., a sound) and belongs in its structure directly to the activity of the ego, according to Husserl, as an "activity inserted in a series" (*EU*, p. 108). It is from the ego that "its invariable self-coincidence," its continuity stems. However, at this point Husserl makes a strikingly significant distinction between the "active ray springing up continuously and a fixed passive regularity." Although this analysis pertains chiefly to the temporal line of the perceptual performance, yet at this crucial phase of the "simple prehension" preceding the identification process of the constitutive work of consciousness in its own right, *there seems to be no need to speak of the awakening of attention and interest, for the material impinges itself upon receptive faculties of the living agent that seem already to be a distinctive system, nor ts there any need to speak of the agency proceeding toward the pregiven affecting material as the ego's making a jump, for in fact there is merely a differentiating progression in life promoting junctions springing forth with the advance of life's complexity.* We acknowledge the emergence of the "vigilant active ray," which at this point surges from the *organizing agency* and pervades its higher functioning like a "roving eye" attentive to each and every move — according to the degree of the force of the impingement — of the sentient apparatus with its dispersed organs whose functions are now brought together on the surface of the brain according to how things fall together. This falling together follows types of relevance: there is vital ordering according to the aims of the ontopoietic process and there is the ordering led by the structural laws of objectivity that produce a static entelechial form of a being in the mind's eye. *The activity/passivity aspects of the active ray, the "roving eye" makes the crucial step of the discretely continuous passage between the vital but mute order of nature-life in its ontogenesis, on the one hand, and the specifically intelligible order of its objective manifestation* as accomplished by conscious operations, on the other. The roving eye, the vigilant ray incessantly moving among all that makes itself available to its attention and draws its interest is a limit factor that expresses life itself, but simultaneously it initiates the person's higher order functioning and accomplishments.

This active ray vigilantly roving from one moment of attraction to another or taking in all of them together in distinguishing an object is at the service of the organic-vital-psychic needs of life. It is, in fact, the crucial life factor of self-individualizing animality. Its performance does differ according to the degree of complexity of the ontopoietic pattern of the individual, but it is the key factor in life's progress.

In it, nature extends into the dual mode and brings together passivity and activity, and brings together too into coalescence the varied life significant givenness of the circumambient world that is already being processed by the sentient functioning of the living agent. It completes the living agent's work of interactive exchange and coexistence, sharing and drawing on the available circumambient conditions while reworking them with its own capacities.

V. THE "VIGILANT RAY" AS THE SYNTHESIZED FUNCTION OF THE "LIFE AGENCY"

Here is the place to ask: in what relation does this active vigilant ray stand to the "living agency" of a self-individualizing life as it responds to striking stimuli among otherwise mute functional data, elements of sense perception before they are acknowledged in any constructive schema, e.g., colors, sounds, smells, textures, pain, inner body sensations, etc. and as it surveys them and pieces them noting prospective elements for the construction of wholes, a process in which individualization occurs? And what is the role of the ray with respect to the conscious constitution that then takes over in order to construct from the surveyed and acknowledged elements an entirety of appearance, a whole static objectivity?

First of all, the vigilant ray should be identified neither with a simple function of life or the all-pervading life "animus" nor with the conscious intentional system of constitution. With respect to the vitally meaningful life agency, the roving eye as an active surveying and discriminatingly retaining factor is an element pertaining to the animus but one that brings in a unique type of efficient performance: it surveys all that comes forth and, bringing it into a focus that it itself exercises, retains it until it passes to the next scene whereupon it is already detached from direct absorption in physiological processes — even though it is rooted and carried along by them — and can function on its own.

One cannot overestimate the significance of the roving eye for the unity of apperception — a unity that integrates all the spheres of sense — which emerges as the order of life is established by the circuits and series of its articulations. The roving eye is triggered by the organic synthesis' having advanced in complexity; it emerges as the first factor of vital significance from the organic coordination of functional performances. To begin with it brings the elements of this complex organic articulation of functions into focus, readying them for the rational linkage that will strengthen the lines of the central agency at a higher level of self-awareness and control owing to the flexibility that it introduces. The roving eye usefully promotes circuits of moves for sensory receptivity as well as protective action and emotional sharing. In this way, the roving eye is a factor of essential coherence for the various ontopoietic circuits contributing to the vitally significant comportment of the life agency, of its vitally significant spheres of apperception.

The eye's new original moments of performance, of ever renewed attention, of focusing and retaining in focus in an anticipation of the articulation of a new type of coherence, are geared as much to vitally significant physical mechanisms, to the strictly pragmatic requirements of the ontopoietic constructive progress, as to the lifting of the objects that catch its gaze to a level of significance higher than the vital/sentient level.

The roving eye may be directly rooted in and carried on by the physiological operations of the brain. It might well emerge from and retain functional links with the operations of the nervous system, operations occurring at various points in the body, *but it is not, unlike other physiological operations, absorbed by the operation itself, vanishing when the operation is accomplished.* The "vigilant eye" distances itself from its physiological bearer and deliberately detaches itself from any organic location; unbound, it surveys each of the elements on which it focuses its attentive "light" within a contemporaneity, within the vertical orderings of a series as the ray of attention singles elements out from a mute backdrop that remains unattended and hazy in its contours; going forward or backward it orders these elements and then conveys them toward an anticipated new "higher" coherence.

In what way "higher"? All physical functions that articulate the operational-constructive segments of the background order of nature-life in the ontopoietic design they crystalize, are, from the simplest to the most complex, fragmentary sequences of operations delimited by their singular task. As emphasized above, each of these functions is absorbed in itself and uses itself up in its performance of a repetitious task as it adjusts to the given transformational changes in the environment and continues, as if it were preordained by the imperative requirements of the ontopoietic laws of life, to "live" in its apportioned task. Its performance processes and means are exhaustible, and once they are exhausted, or once the constructive task is completed, it vanishes. Each of these functions thus keeps strictly to the area of its performance, its "locus." But the roving eye's attentive light, in contrast, frees itself from all strings binding it to the ground, without actually breaking them, and "flies" on its own in midair as it were. In its "activity" it is free from local constraints on it and roves in all "directions," surveying the entire orbit of the nearer circumambient sphere of life's coexistence, the "outer horizon," as Husserl calls it, of the self-individualizing beingness; it springs forth from the living agency's conjugated forces and energies without interruption or interval, incessantly, and on its own covers the areas of existence apprehended through all the senses. As a sense having unprecedented freedom from its organic ground, the roving eye brings an unprecedented flexibility to the self-governing life agency and an unprecedented promise of further spheres of existence to be opened.

However, the vigilant eye differentiates into two mute — not yet intelligibly manifested, that is — corroborating functions. One of these is in-grasp/holding while moving while spatially differentiating as well as temporalizing in succession. The other is the focus on the identity of these, thus generating and bringing forth constitutive series.

With this, we are moving toward those spheres that are none other than those of the manifestation of the mute and hidden spheres of life in the intellective rationality of cognition; the further differentiation of the constitutive operations into the cognitive degrees of experience culminates in a static manifestation of thinghood.

However, taking into account the evolutionary progress in the differentiating of types of living beings, of consciousness, *we have at this point to reach far deeper than Husserl does, namely, into the creative virtualities of the Human Condition,* in order to reformulate and exfoliate fully this crucial phase of the genealogy of logic, that of *the human logos.*

VI. CONCLUSION

(a) Concerning the nature of consciousness, it appears clear that, contrary to the tendency to dispense with it altogether, it is spread throughout the ontopoiesis of life and assumes different functional degrees according to the differentiation of the complexity of forms. If we continue with Husserl to consider it to be "transcendental," we have to differentiate its meaning according to its roles. Accordingly, the quest for its centralizing faculties, which remain undoubted, divides into a quest for at least three main centralizing factors: the living agency, the living subject (with a gradation of conscious faculties according to the gradation of psychic complexity among species), and full-fledged transcendental consciousness as a centralizing factor of objectification. If we assume its teleological direction, we have to couple that with an evolutionary spread.

(b) At all levels of life we see a ray of vigilant consciousness bringing into attention and bringing together all the functions of ontopoietic unfolding or being triggered by them. In the form of the "roving eye," it serves simultaneously as a receptive/responding recognition or overview of the lifeworld/individual situation or a self-identifying scouting factor to protect/promote life's unfolding. Giving up the favorite spatial imagery of philosophy, we have to deny consciousness territorial centrality. It is movable, going with the function that prompts its alertness and initiates the roving centralization of all functions. This freeing of the crucial ray of consciousness from any location responds to the present-day tendency to "decentralize" the subject, without, however, giving up the centralizing function of consciousness. To appreciate the roving eye makes understandable for instance the coordination and cohesion of sensory perceptions despite the fact that the sense organs' work may be located in the different parts of the brain. Simultaneously, appreciating the roving eye undercuts the arguments for dispensing with consciousness together with its central location. The role of consciousness remains that of giving sense — receiving information and coordinating responses.

(c) Concerning the role of the lifeworld, glimpsed at the juncture of functioning in the circumambient world, it is the ground of experience we have seen lurking throughout our analyses, peeking through the tightest of congenital ties within the analytic material of the pregiven manifold, going to the originary groundwork of the lifeworld as such prompting onward the spontaneities of the ontopoietic line of unfolding that each individual — a bearer of the world's constitution — works out through intergeneric, interactive, interrelational ties with other lines of development taking place. Individualized life spontaneity has a *force viva* (to use Leibniz's expression) as its motor and the very laws of the ontopoietic unfolding for its direction.

Granting to the lifeworld the role that it plays in the generation and actual unfolding of each self-individualizing entity as well as the essential role as a correlate to the ontopoietic forms and laws of this unfolding, we see lurking through all the entanglements that it has with the individualizing poiesis of life *the primal role of life itself.* It is beyond doubt that *the spontaneity of life, on the one hand, and its formal and material requirements, on the other, dictate and promote the nature of consciousness as well as set them in tandem with the basic conditions offered in the cosmos for correlating virtual life and a virtual world.*

(d) In this perspective it becomes obvious that the concept of the "unity of apperception" has undergone a considerable change with respect to the Kantian as well as the Husserlian conception of it. It is not simply the unity of experience viewed in the cognitive line as yielding us, the advanced species among living beings, a spectacle of the objective universe of life; in contrast, *the unity of apperception means here the unifying function that, while attaining its peak in the intellect's worldview, unites all the functional strands of the ontopoietic unfolding, not only the volitional, aesthetic, imaginative lines proper to the creative functioning of the human being, that Husserl acknowledges, but all the organic, vital, sensory lines that serve and implement the praxis of life* in a way that makes for cohesion among all of the innumerable moves and coherence between means and tasks as they serve in the grand phenomenon that is the self-individualizing ontopoiesis of life.

TYING POINT THREE

THE GREAT PLAN OF LIFE — ANTICIPATING THE TRIADIC LOGOS

CHAPTER ONE

THE ESOTERIC LOGOS

Human beings in previous phases of the evolutionary ascent — an ascent if measured by the yardstick of the most recently acquired creative/inventive proficiencies — attempted by rudimentary means to adopt to the conundrum of biophysical conditions by just refashioning segment by segment the available groundwork of nature and then submitting to natural conditions. At the present stage of the unfolding of their virtualities, they have become so powerful that Descartes' dream has come to pass and we are the "masters of nature." Our powers may now penetrate even the innermost human proficiency of generation and manipulate and refashion to a great degree the transmission of life's specifics according to selective decisions.

It appears, then, that the "natural" cycle of the expansion of the Logos — the Great Plan — has closed the circle in that the spontaneous manner of generating life can be eclipsed. Human creative genius is transforming the generic characteristics of animals — making cows that give more milk, hens that give more eggs, cattle with more beef — and it is quite possible that our skills will go beyond just "correcting" deficiencies in humans and come to elaborate different characteristics for humans at large.

With invention's entry into the scheme of inheritance that the Logos spontaneously projected for life, it appears that human inventive genius is taking over these strategies in an incomparable, free, and imaginative fashion. The great plan appears to have passed into the inventive and creatively open-ended prerogative of the human agent.

This succinct survey of the strategy of life's actual unfolding, which scheme appears to have been there in a virtual state already with the initial seminal impetus of the Logos indicating the presence of a "great plan" for the virtually predelineated crystalization of contingent life, dissolves the assumptions behind the traditionally distinguished pairs of opposites: determination and freedom, contingency and virtuality.

The existential obsession of humans with the great plan here finds its explanation: this is a primogenital passion stemming already from the primogenital singularizing of a certain, limited selection of germinal elements in the initial logoic "seeds," in which the entire virtual schema and even more the substantial unfolding of life's order as passed down through the generations is contained. The passion of a singularized, particularized existence is thrown into an incommensurable network that it does not encompass. The impetus of generating life then seeks a measure of equipoise. This is a yearning, a quest for an intellectual or imaginary whole in which the individual can find its ultimate place.

We have here the esoteric logos — esoteric because stemming from a logoic primogenital strategy that is not articulated or expanded along the constructive lines of life but is rather carried by the natural concern of the animal to get its bearings. A cat, a dog, a horse, a goat, or a sheep is disquieted at finding itself in a changed, unknown environment until it gets its bearings there. The creative spirit of the human being seeks his/her bearings via imaginative projection, relating it to the sense of existing in wider and wider circuits, until he/she seeks after the ultimate on other than "natural," wings. We seem to aim at situating our particular beingness within a wholeness that spreads further and further, an aspiration already virtually germinated in the primogenital existential singularization of life.

With the specific development of the human soul, the esoteric logos with its passion for the great plan imaginatively serves the intellectual crystalization of the soul's transnatural strivings. In its religious formulation, this quest after the ultimate takes the shape of a quest after participation in the great divine plan of creation and redemption.

Obviously, the great plan assumes a different significance than that of life's natural unfolding. Unfortunately, because the esoteric passion serves both, they are often brought together.

Thus do we find the great plan amid all contingency — life in its spontaneous bursting forth in a dynamic stream in the "first" (was there ever a “first”?) logoic seminal strategy, which recedes further and further as scientific/creative probing descends into origins. The scrutiny of the creative/inventive human mind yields merely the creative intuition indicating the seminal logos surging with its own new, unprecedented impetus of life at the strategic junction of the most elementary and simple forces, the earlier genetic trajectory being lost to the human analytic powers. Here is no prefiguration, direct creation of living forms, as Leibniz guessed, but a virtual deployment of forces present in the cosmic system so that when an appropriate favorable conundrum of forces occurs, they concurrently unfold, kindling the spark of life, which springing forth delineates such or other modalities and routes.

This intuition and desire, and tendency toward totalizing — which we may see in the idea of a “great plan” of creation as it was present in the mind of the ancient peoples of Sumeria, Egypt, Israel, and others — originates on the wings of the uniquely human and creative, esoteric logos. Anticipating here the distinction of the specific trends of the human creative esoteric logos, this is the proper place to briefly retrace the steps that led the human mind to the esoteric passion for the great plan. The reason for introducing it at this special junction of our exposition is that, having reached a culminating point in its unfolding, the human mind is today in its scientific exploration of the vast plane of life striving to grasp the “totality,” is testing its creative proficiencies for controlling cosmic forces and even refashioning the rules of life. In short, the human mind today, whether scientifically or just on a personal level, is seeking a new orientation within the scientifically transformed worldview — is more than ever tending toward the key that would open the enigma of existence. This is the esoteric passion for the great plan hiding behind all.

What lies behind this frenzy for the key to existence? Bringing the great plan of life into our exposition at this time is meant to tie the knot of the logoic forces so far revealed, as well as introduce us in an anticipatory way into the specifically human phase in which the esoteric passion has its source. It is a most striking phenomenon of life that man is now at the verge with his esoteric obsession with the Great Plan of Creation.

It appears that the generic transmission of the adaptive skills in interacting with, drawing on, and fashioning to advantage the forces of the environmental conundrum, devices that are detectable already in animals such as ants, bees that build their own worlds, birds, who fashion their nests, beavers who make housing, and other animals like mice, moles, squirrels, etc. who gather food to preserve themselves from the rigors of winter, is now advancing in inventive ways in circuits of expanded complexity. This proficiency took a major advance in apes, who invent rudimentary ways of using available material as "tools," and it culminates in the creative inventiveness of the human being. Humans, accumulating the generic sedimentations of deployed skills not only transmit them to further generations, but build and deploy machines that extend human skillfulness in meeting progressively more specified needs and in fashioning the circumambient conundrum, expanding human proficiencies beyond their natural limitations and actually refashioning the natural environment.

But even more striking, astounding, puzzling, enigmatic, and even frightening is that it seems as if the impetus of life acquires increasing intensity, dynamic force, and push onward. For example, although the division between day and night has remained astronomically the same for millions of years, the complex functional systems of the seemingly one and the same type of the human being is deploying greater skills and finding more opportunities for exercising them to human advantage and so tends to squeeze into the interval of activity — the "day" — ever more interactive endeavors, so that the rhythm of life's "everyday" business "accelerates" and the impetus of life drawing on the exchange of forces is stronger and stronger.

The history of philosophy may be seen as a constant drawing on or return to the ideas, arguments and insights of the predecessors. Just as classical metaphysics returns to Plato, Aristotle, Parmenides, *our metaphysics of life revisits the intuitions of Heraclitus: all is in flux, flux is the primal element.*

The originary insight of Heraclitus is that the gist of all things is flux, flux is the gist of all things.

Our argument is as follows: where classical philosophy has been entranced by seemingly static reality, static because manifested in objective forms before the human mind's functional system, a look under the abstract forms, structures, and principles universalized by the mind reveals the concrete workings of nature, a stream in which, as Heraclitus said, we may never step twice.

While the manifestation of life in universal forms appropriate to the mind is seemingly static, the workings of nature that carry it in its concrete becoming are in flux. The concrete is in flux.

While the history of Western philosophy has been fascinated by the universal and seemingly stable, static forms of life in its manifestation, attempting to find and capture their rationales, as rationales of these forms themselves, phenomenology of life, by seeing into the inner workings of life, returns to the concrete, and hence to the Heraclitean intuition that flux is the primal metaphysical element.

Looking at the Aristotelian conception of rationality, we see that it attempts to account for the structuring of this seemingly stable objective network of reality, while simultaneously groping to logically organize its operations into linguistic, universal networks. On the other hand, we see scientific inquiry, whether it be that of physics, astronomy, chemistry, or biology, attempting to enter into the workings of nature itself and frame universal laws of the concrete in flux.

The question emerges: is the concrete in flux to be radically opposed to the universal objectivity of its manifestation? Is the human being, who from his coming into the world until his last breath struggles to attain a measure of stability, a fixed foothold in what otherwise is constant change caught up in a complete illusion? Is this wonderful world just a passing dream, a phantasmagoria, a mirage upon moving sands? Or is movement, change, coming to be and passing away just a manner of appearing of the eternally present, never affected by anything universal Being? The sciences, in line with Leibniz, seek after the laws of the transformations and the motion, but they are far away from resolving this question.

They should first find the underlying common factor of the inner workings they look into, for the truth in the concomitant biological facts that substantiate it in the concrete, in short, for a metaphysical intuition. It remains to the creative imagination of the metaphysician to ask after and speculatively devise formulations of further questions. One such question is the quest for the prototypes of life's particular individuations and the particular prompting force that Leibniz saw in the essences of the monads stored in the Divine Mind ready to be pushed out into the existential life course if and when compatible conditions and operations allow. Another is the quest after the final reasons for the bursting forth of the logos in its gigantic game. Then there is the quest after the source of the seminal endowment of the logos. Several religious traditions, on the basis of that access to a different reality called "revelation," attribute the great game of life to the fiat of God, the Creator, and see the great plan of life as being doubled by the great plan of the redemption of this contingency, which plan has been often confused with the first one by religiously inspired metaphysicians.

Within the perspective of the phenomenology of life being developed here, our exploration of the great manifestation of the logos, we have to elaborate a legitimate ground for evidence before pursuing these speculative questions, and so we postpone them to a subsequent inquiry. However, this will be indicated already at the circuit of the manifestation of the Promethean logos in human creativity and its overcoming by the yearning of the spiritual soul.[1]

CHAPTER TWO

THE GREAT PLAN OF LIFE, THE ESOTERIC PASSION OF THE MIND

INTRODUCTION

Progress in philosophy is questionable. It is not only that we nourish ourselves on the theories and ideas of the thinkers who preceded us, that we then establish our own thought (or what coming out of our own reflection we consider to be "our own" thought) with reference to previously established theories, but when we find that our own ideas coincide with our predecessors' insights or intuitions, it is as if a stamp of authority, testimony that "things are this way," is conferred upon them.

Actually, each new thinker does contribute corroborations, nuances, new approaches to, or perspectives on, the real. Thus, I submit that instead of moving in a "circle," philosophy, like life, advances in a spiral movement. So it is with philosophy/phenomenology of life and of the Human Condition. Although, I declare that phenomenology of life as philosophy *par excellence* is opening, a new philosophical perspective, one new in Western thought, and is outlining an original system of thought that could have emerged only at the confluence of ideas, insights, results obtained by present day science, culture, and the entire history of philosophy, nonetheless, it is to the reflections of Heraclitus, which surged forth at the initial phase of Western thinking, that the crucial innovative intuitions laying the groundwork for the phenomenology/philosophy of life were first had and developed.

Heraclitus startled his contemporaries and has been tantalizing thinkers throughout the history of Western philosophy with his strikingly provocative statements going against prevailing philosophical attitudes and running in the face of common sense, that is, the "natural" attitude of the human being. Their enigmatic, puzzling nature intimates a mysterious depth often seen as an occult wisdom eluding the rational philosophical mind. Would this really be so? Furthermore, throughout history his insights into the nature of change have been swept aside by most and the gist of his thinking identified globally with his vision of the everlasting flux of all things: *panta rei.* Is this ready identification correct?

Is there not in Heraclitus another insight into the nature of things, which instead of leaving perpetual change free to reign wild in a haphazard way, counterbalances it with a search for order? It is actually this latter insight that animates Heraclitus' reflections throughout.

It is my intention to show here how Heraclitus' dispersed and cryptic statements betoken intuitions relevant to the basic questions of philosophy, inviting further corroboration and renewed questioning in depth. Numerous great thinkers in history nourished by Heraclitean insights have given them specific adumbration within their philosophical frameworks. It is enough to mention Hegel, Marx, Heidegger. However, it is extraordinary that a fully outlined phenomenology of life and its ontopoiesis has independently, out of an inspiration entirely its own, validated, reaffirmed, and carried further the foundational inquiries of Heraclitus as well as the main intuitions guiding his answers to them. It is as if the historical development of scientific research as well as of our Occidental culture, by which the phenomenology of life and of the Human Creative Condition was prompted, has reached a point at which Heraclitus' thinking comes fully into its own, flux and order being reconciled in our philosophy. In the phenomenology of life and its ontopoiesis, and of the Human Condition, it is the logos of life that accounts for the flux as well as for the order of the All in their indissociable nature. Its incipient focus falls on the innermost, primogenital nucleus of life's ontopoietic course and aspires to chart the inner network of that ontopoietic unfolding with its ties to all the factors that subtend it and so disclose the inner workings of life, the common genesis of both flux and order.

I. THE OUTLINE OF THE ARGUMENT

1. *The Intuitions Inherent in Heraclitus' Declarations and the Questions They Raise; The New Corroborating Answers Given to Them by Phenomenology of Life and of the Human Condition*

The cornerstones of philosophical concern of the Heraclitean vision of the All center on the nature, possibility, and distribution of order. First and foremost, there is the "great plan," the universal plan of all things that directs the universe, the "account" of all, or the logos.[1] Second, Heraclitus raises the question of how this plan could be discovered since the ordinary man lives without even being aware of it in his everyday practical cognitions.[2] Heraclitus considers the possibility of bringing the plan to light as the attainment of a special wisdom.[3] At the same time, in the Heraclitean view, which goes counter to classical as well as contemporary skepticism, this attainment is within human reach. In the philosophical expanse between these two poles lie Heraclitus' startling and enigmatic insights into the nature of reality, which in entering into the universal plan would be directed by it. Here we strike the crux of the matter: in the Heraclitean view the discovery of the plan calls for the *investigation of the real constitution of things.*[4]

However, how can things be distinguished, what may this "real constitution" reveal if, as Heraclitus insists in his vision of "panta rei," things are and all is:

I. a) in perpetual flux;[5] and b) this constitution itself lies at the cross-section of radical opposites,[6] all things remaining in a process of perpetual change.[7] How can we then differentiate them? How, furthermore, could the coherence of a plan, of the universal account (*tou logou,* brought out in Fragment 1) be conceived, since:

II. the differentiation of real things occurs through the play of "opposites" — be they opposite forces, qualities, relations; there is an intrinsic law of opposites — they remain in perpetual change.[8]

We are immersed in perpetual flux, then, but there must be — as Heraclitus' arguments indicate — an intrinsic law of the differentiation of things within it.

In fact, and here we come to the crucial point of his vision, which offsets (balances) the engulfing doom of flux, Heraclitus affirms insistently throughout that the differentiation of reality is measured out: a) there is a measure in all things in the universe and life;[9] b) the soul also is ordained by its very own measure;[10] c) and so is society.

In raising the question of measure as the indispensable factor of reality, Heraclitus challenges us with three significant and profound questions.

First, what is this measure that in its universality holds for all the sectors, nay, all types and singularities of the real? Second, how within this perpetual flux of change in which the very constitution of things unfolds and which is governed by the law of opposites can we find and ascertain their intrinsic measure? Third, assuming that, as Heraclitus believes, there is a special kind of knowledge that within the manifestation of reality common to all is available to only a few, what are the grounds of its possibility? How can we reach and legitimize it? How can we grasp and express it?

With this last question we reach the bone of contention. Indeed, Heraclitean views about the constitution of thing, of reality, of the world, the cosmos...run counter to our current rules of rationality and logic as they have been established in Western culture, and not without reason. Opposites as the constitutive factors of the real, should, it would seem, cancel each other out. In any case, unless substantiated by the intrinsic constitutions of things, they remain abstract notions. Heraclitus' insistence on the complementarity of their interrelation does not suffice to account for an ordering sequence upon which the Occidental type of logic, rationality, order as such relies: the laws of non-contradiction, the excluded middle, and identity, seen as guaranteeing the differentiation and identity of the real, and upon which the Occidental reconstruction of reality, of the manifestation of life, of human transactions, in brief, Occidental logic, relies. Hence, there follows the question: How could the human mind even, first, gain access to the core of being and becoming and so grasp the constitutive rules and norms, their separating and unifying links in any way but by dealing with and using the established rules of Occidental rationality and logic, and, second, express them in the rationale of a "plan," an "account of things," the Logos? Following the Heraclitean lead we seem to see even the possibility of philosophy understood in its full aspirations undermined and rendered helpless before skepticism.

However, in following Heraclitus' lead, are we irrevocably pitched against the laws of logic? It is up to the phenomenology of life to take up this basic challenge and to explain how we find access to the "hidden" springs of reality.[11] By what means may we follow the changing constitution of things with a mind shaped by traditional logic, and so grasp the plan of All?

Having reached the question of the seemingly radical opposition between rationality and flux, we will treat that question first.

2. *Flux versus Stasis: The Logic of Contradiction versus the Logic of Essences*

Is this opposition valid? Is the aporia that it opens at the heart of philosophy well founded? I answer in the negative. We do not need to introduce a new type of logic (e.g., the dialectic) in order to mediate this aporia. It remains precisely up to the phenomenology/ontopoietic metaphysics of life to transcend this aporia by establishing upon its groundwork the common essential core of the constructive logos. Let us begin by establishing a provisory equipoise between them, provisory because it is the intent of this study to lay down its foundations in depth.

Undoubtedly throughout the history of Occidental philosophy there runs a fascination with the seemingly stable, static forms of reality. In our common sense attitude, which is the leitmotif of our life conduct, we constantly struggle with change. Change, the fleetingness of everything within and around us, is our constant enemy.

We are perpetually concerned in following our intrinsic transformations that they conform to a "norm." We have to consider when planning action how it will adjust to the changing tides of the circumambient world; we witness our entire beingness growing and flourishing and then shrinking toward an irremediable extinction. In short, we are caught, together with all living beings in the world of which each of us is a member, in perpetual change. And yet we maintain our identity and steer our course in meandering moves. We tame the flood of change by projecting "objective" forms by means of which we project a network of interconnections among the moves of change; we articulate these according to the abstract patterns that we introduce into the flux. In brief, we differentiate and grasp instants of the flux of things in universal forms, structures, essences, and principles of our mind that are like honeycombs that we fill with our sentient and emotive stuff. We thus dominate the flux of change, at least partly but enough to project a relatively stable life network within which we carry on our life course. Our experience, our imagination, our ingeniousness fill the honeycomb to overflowing with marvelously expanding life dimensions; this is the manifestation of life that we project.[12]

It is this manifestation of life, the world around us, the heavens and the stars all moving in apparent stable permanency, this manifestation that is life's "glory," and to which we will return further on, that for obvious reasons constitutes the object of marveling, fascination, and unending inquiry for the Occidental thinker. And the Aristotelian logic that has fashioned our culture and science for centuries is rooted in and expresses rationality by means of the principles and modes of composition of the structures that manifest themselves by means of the work of the human mind.

Is the concrete-in-flux that reality remains in according to Heraclitus, with its logic of opposites, a logic of contradiction, to be seen as radically opposed to the logic of essences? Such an opposition is unwarranted. Reality is neither pure fleetingness, a passing dream, a mirage set upon moving sands, nor is it the depository of immovable, immutable, atemporal entities residing within an equally imperturbable network of interrelations, like the heaven of Platonic ideas. Here is an archetypal aporia. To elevate it is what I propose here. On the groundwork of the phenomenology of life and of the Human Condition, we will discover that the intrinsic means of the flux, of change, of advance are the selfsame means that bring about articulations, structure, and the entire apparatus of the manifestation of life (that is, of the logic of essences).

Sharing Heraclitus' insight into the real, that the gist of all is flux, that we cannot enter the same river twice, and even, as my unforgettable friend Jean Wahl used to emphasize, maybe not even once, we will improve upon Bergson in his belief that the flux of life is haphazard, that it moves by bounds and whims. I have proposed as *the principle of all principles, the constructivism of life.* Life is the Pandora's box hiding all the secrets of reality.

As Julián Marías stated, the recognition of life as the focal subject of philosophy was delayed till our century. He had in mind the voice of Ortega y Gasset. That voice, however, did not carry far enough and quickly faded. First of all, Ortega failed to find the key to this box and did not find the crucial links between philosophy and the sciences of life. It has remained for us to do so.

3. *Outline of the Main Tenets of the Phenomenology of Life and of the Human Condition. The Key Principles for Transcending the Aporia between the Fleetingness of Being and Its Essences*

I submit: *It is life that transcends our aporia and answers all the questions underlying the seeming opposition between the logic of essences and the logic of contradiction.* Heraclitus insistently affirmed that whereas common knowledge is "in the open" and shared by all, knowledge of true reality, "wisdom," is hidden. Yet we have access to it. His idea of the "uncovering of the hidden," of *aletheia,* has attracted particular attention in our times. It inspired the metaphysical speculation on the hidden that in Heidegger is identified with the notion of Being. Although Heraclitus' insight into the "hiddenness" of the roots of reality appears in the phenomenology of life as a very pertinent feature of double-faceted reality, it finds on that ground a nonspeculative, concrete crystalization. Not a hidden being, but a primogenital origination and genesis of beings as *onta* comes out of hiding, principles and vehicles of becoming, of living beings as such. What is revealed are the "inner workings" of life's unfolding, generating, progressing, which simultaneously bring in the differentiation of beings and their existential interrelations, temporality, space.

These inner workings of life do not reach the surface of life's "objective" manifestation but subtend it and carry on its progress. They offer the basis for its relative stability while remaining out of sight, hidden from the resplendent light of the accomplished life they carry.

We will then, in what follows, excavate this subterranean ground of life and all beingness and show how the inner workings of life bring in a constructive complex movement and order and spread tentacles toward the furthest limits of relevance giving meaning to the cosmos as well as to societal life, culture, and the life of the human spirit, in short, how they reveal the strategic coordinates of the great plan along which the Logos of life unfolds.

In the first place, *we will uncover the vehicle of the Logos of life — the self-individualizing process that proceeds from within, radiates without, and advances ahead.*

Second, we will follow the thread of life to the disclosure of the next primal means of life — its primeval web, which is woven over the perpetual flux and subtends its cyclic order, the cyclic order of unfolding life.

Third, we will see that while the phases of life return in cycles, their concrete individualized constitution-in-process is never the same, it having not a circular but a spiraling order.

Fourth, we will show that in the differentiation of living types life's innermost functioning progressively loosens the grip of repetitive entelechial direction. Hence, the Human Condition as a specific phase of the constructivism of life surges with its unique inner system of radiation and in evolving carries within itself all the strings of bios and yet is endowed with the virtualities — unique — of the vertiginous creative mind. Here the mystery of the possibility of wisdom is to be plumbed.

Fifth, we see that the spiral progress of the flux of life does not stop at a climactic point of accomplishment. Passing through its zenith, life projects a new spirit, that of the self-examining quest that dismantles the very significance of its accomplishments and launches the human being in search of the ultimate significance of all: toward the sacred, the divine.

It is along the lines of these five points that I will attempt to examine briefly the answers given to the Heraclitean intuitions by the scientific phenomenology of life, answers found in life itself.

II. THE WEB OF LIFE AND THE ORIGIN OF ORDER

1. *The Self-Individualization of Life: Ingathering and Outward Radiation*

From time immemorial philosophers have been marveling at the extraordinary manifestation of a harmonious, coherent universe within which individuals maintain relatively steady but always coherent courses of their own within life's turmoil. The great issue is this: How may a coherent universe of beings, things, processes, events in movement be realized?

Here is the issue of the phenomenology of life *par excellence*. Whoever says "life," says self-individualization-in-existence. When we attempt to differentiate between life and non-life, we distinguish as life a coherent, self-promoting, constructively advancing segment of development aiming at further progressive stages, over against an inert, static mass that undergoes changes only as provoked by outside forces and that projects no constructive designs. This means that life consists basically of a sequence of moves, operations that form a design that remains consistent within itself and is defined over against all that is external. No matter at what cost of trial and error, the initial spontaneity of life drives toward the surging of innumerable differentiating processes by gathering in the germinal seeds/propensities of energies, synergies, forces according to their propensities to fuse, mix, interact with each other and linking them together in self-individualizing complexes, such as owing to their intrinsic coherence of inner operations stand out against external forces and so form what we distinguish as individual living beings. I have elsewhere characterized the genesis of life and its unfolding as fundamentally and originally consisting of such a self-individualizing constructive progress. Let me point out here again that the vehicle of this progress consists of a *sui generis,* unique to the individualizing progress, "agency," which, to begin with, already carries the germs of the developing virtualities to be unfolded. But it does not contain them passively; to the contrary they are the equipment of its innermost propulsion to spring forth in a profusion of constructive moves. Finally, and this determines the orientation of those dynamic steps that follow, this "agency" contains an organizational schema, an "operational" schema that proceeds in a threefold movement that draws in forces, energies, matter from "without" for most specific processing and transformation within, all leading to the enrichment of forces and the generation of novel synergies, and that in a twofold outward radiation: the radiation of novel synergies for participation in life's traffic outside the agency's dominion, as well as the rejection of unusable material that in the outer realm of life contributes to the furtherance of the basic material processes. This core of self-propelling and transforming operations establishes the self-individualizing process of life. (Following Aristotle and Leibniz, I have called this process an "entelechial design" or "principle of life.") It as such performs crucial functions with respect to the establishment of life's expansion; doing so, it brings about the primogenital articulations of the workings of nature-life, that is, of rationality, order, significance.

To begin with, by its intrinsic self-oriented core, the self-individualizing process is a reservoir into which the wildly and idly floating forces and synergies pour and flow together. It gathers them into its own life-operational order and processes them into life's individualizing arteries. That is to say, this process employs them for life's propagating purposes. Then, the transformed and already life informed synergies and forces radiate outward with their acquired life significance, entering into and invigorating other arteries of life projected by other self-individualizing beings. Thus, the oriented operational nucleus of self-individualizing life is a processional station of life with a threefold orientation. First, life establishes a point at which forces are gathered, are transmuted into life significant energies; secondly, these energies are then radiated outward; thirdly, living beings in this way project the web of life — nature — a web that draws all self-individualizing beings into the existential interaction of the unity-of-everything-there-is-alive. Springing forth from the self-individualizing entelechial nucleus, the ontopoiesis of life proceeds. (We shall see what great significance the web of life continues to have in the stage of life that is the human creative condition.)

2. *The Ontopoietic Primal Ordering of the Inner Workings of Life*

We have outlined above the ontopoietic axis of life's arteries. However, having given an idea of the main tenet of the ordering of life, with the insight that life means propulsion, force, and order — the self-individualizing entelechial nucleus and its inner workings as well as outreach — it remains to be seen how it projects the cornerstones of the ordering of life. It is our contention that the inner workings of this primal factor of life lay down the crucial principles of differentiation and unity.

Entering into the concrete workings of the intrinsic ontopoietic nucleus, there are indeed to be distinguished two primal and concrete operational systems by means of which the above-outlined intrinsic nucleus of the self-individualizing differentiation and radiation of life proceeds. Although both of these systems consist basically in gathering in life material, transforming it appropriately, and radiating it outward again, yet they differ basically in the seminal virtualities that promote their initial differentiation. The first system is geared to establishing the primal rational articulations; it dissolves the seeming contradiction of opposites within the concrete progress of life. The other system, aiming at the promotion of the ontopoietic differentiation beyond the singular individualization of life, dissolves the seeming universal contradiction of necessity and contingency.

Let us first consider the singularization of the ontopoietic process. Its constructive initial instrument is the ontopoietic operation. It is identity and selfhood that are pursued within the multifaceted operations that establish distinctiveness. First there is to be distinguished here the ontopoietic moment of "operation." This "operation" is the performance of the basic function of gathering and processing materials, synergies, energies within the transformational complex of the nucleus. This is the poietic task of distinguishing between disjointed elementary particles of forces and then carrying out the first transformational move. Not only is the distinction here made between the "disjointed" and the "coincident," but simultaneously there is the introduction of the transformational "passage" between the one and the other, a passage created by an elementary, primal "move." The first pair of elementary opposites is thus dissolved within a change, a transformation, a move making them "looser." The operation as a primal category of the ontopoiesis of life performs the tasks of: 1) creating the above-mentioned passage between elements that it simultaneously differentiates as being disjointed; 2) releasing in activation some of their propensities, which, inert and dormant, are lying in wait as virtually present resources; 3) circumscribing closer and further circles of the individual's own dynamic proficiency even though as an operation it is per se restrictive and restricted as to access to the factors among which it works; and 4) waiting for the coincidence of appropriate elements allowing the living being to surge and to introduce its performance into a wider net of operations, that is, into a process.

Let us mark the point: the dynamic moment of the operation serves emerging and unfolding life by bringing in the basic articulations of order, rationality; it lays the ground for the significance of "becoming."

3. *The Game of Opposites in the Dynamic Unfolding and Construction of Forces*

As on the one hand, the basic operation lays the ground for what we would call life's "initial rational articulations," on the one hand, on the other, it serves to prepare the groundwork for the initiation of the self-individualizing progress that projects itself in a process that sends forth its branches like a young tree. We may see the dynamic complex of the process as the vehicle of the proto-ordering of life. Process is the constructive vehicle of order within the flux. A process, indeed, remains in flux while its phases differentiate from each other. Each phase performs a distinctive segment of operations drawing on the circumambient/ environmental field and carrying out its assignment with respect to its favorable, unfavorable, or fostering impact. This "efficient" assignment is "laying in wait" in the preceding phase, which, therefore, although not actually present, is and is not disjoined from its successor. On the contrary, its significance is adumbrated, amplified, efficiently reaffirmed with respect to the constructive progress going on, progress that with each phase delineates itself more clearly. Each actual phase, just now in performance, already anticipates the next, the one into which it passes.

If this flux differentiates into phases at all, it is first because of its constructive progress, and, second, because it is progress in a constructive outline that introduces the now, the before, and the after as successive stages, that is, introduces the succession of time with its sequence, its instant, and its measure. It is the delineating ontopoietic progress in its processional advance that introduces this constructive point of reference — constructive progress that meanders as a translacing continuity of the intrinsic design of life, which makes the river with its changing water the same. In the primogenital ordering of the logos of life there are no sharp edges between "what is" and "what is not"; the reality of the constructive becoming emerges within an artery of processes that is its "profile," its "design"; there is only the translacing continuity of the self-individualizing life whose essence is being-in-process. The human creative mind in its creative manifestation of life cuts into this incessant flux, and by establishing correspondences with its "logic of contradiction," it establishes objectified reality according to a logic of its own, the logic of structures and essences.

Multiple lines of processes with multiple arteries of operations carry the entire bulwark of the differentiation and articulation of life's primal poiesis. In the rhythm of taking, processing, absorbing, and radiating or rejecting the rational principles of "sameness" and "otherness," of "inwardness" and "outwardness," of "penury," "sufficiency," and "excess" are projected and installed. The fruits of a transformational function, they radiate outward as vital principles of significance.

The three concrete phases processing the substances that bring in the energy and material needed for the living progress to unfold are called "metabolism"; but, as we have pointed out, that which since Vico has been considered the focus of empirical nourishment is, however, not restricted in its significance to the organic circuits of the ontopoietic course. *Metabolism is merely a single instance of the empirical embodiment of the overarching strategy of the logos of life, which consists in drawing on all, transmuting all, and radiating onward.* The same transformational system of gathering, transmuting, radiating/rejecting runs through all the phases of the unfolding of life's ontopoiesis; it is its axis. As a matter of fact, it reaches, as we will see, its peak in the specifically human phase of the ontopoiesis of culture.

It should be emphasized that it is precisely in the self-individualizing principle of life's role as the axis of the entirety of life's expansion (that is, of its cosmic, inorganic/organic, vital, sentient, societal, and cultural significances) that lies its crucial role as the key to life as such. Philosophies that are restricted to one of these realms alone (e.g., Jonas' concentration on biology, Henry's on the inner self, Ortega's on the human realm) do not reach to the springs of the unity-of-everything-there-is-alive, that is, to life as such.

4. *Life in Its Ties to Non-Life*

There is certainly to be acknowledged a further reach of life's operations as well as processes throwing out cosmic articulations of order. They establish and maintain the cosmic order, on the one hand; on the other, they, from the perspective of life, prepare a profusion of synergetic material to nourish life at its advent and so coalesce in congenial networks forming a "nurturing zone" for life. They become life's relevancies. Their essential difference from the processes carrying life itself is that the latter consist in a most consistent, tightly organized processional nucleus that assumes the role of an operational agency reposing in itself and thus projecting an "innerness" juxtaposed to the "outwardness" from which nutrients flow in and into which rejected elements go. This operational consistency delimits the contours of "sameness" over against the "otherness" of what lies beyond, of "autonomy," which reposes in itself over against the "heteronomy" of what lies in the circumference of vital operations.

Indeed, there proceeds a differentiation of elementary significance with the advent of life that occurs in the bringing forth of opposites in forces, qualities, tensions, with an infinite gradation of intensity between them — useful and noxious, hot and cold, light and dark, moist and dry, strength and weakness, etc. Differentiated in the operations of the life process, they are, in fact, opposites in transition; they acquire their gradation of "opposition" in the play of vital forces that transforms substances (see Heraclitus, Fragments 11, 8, 51, and 126). With this transformability covering seemingly the entire field of Western thought, the question of what there was first in the universe, that is, of what is the irreducible primary substance, was pondered.

The game of life consisting in the play of trial and error, in the confluence and transformation of otherwise disparate elements unfolds the entire gamut of opposed tensions, and according to their intrinsic germinal propensities they are differentiated into an infinite gradation of qualitative or operational intensities, forcefulness, etc. — bringing about extreme points that oppose each other at the "opposite" ends of a common line. These opposites maintain each other's "meaningfulness," as the meaning of one horn depends on contrast with the other; they remain infinitely movable, in transition. This differentiation flows, indeed, from the heart of the logos of life initiated with the cosmos on the one end and culminating in the Human Condition at the other.

5. *The Surge of Scanning and Spacing at the Protogenesis of Life*

Indeed, in the nuclear protoexistential operations of life, those of coming together, distancing, processing, rejecting, transforming, generating, that is, in the "scanning" of life's genesis, growth, fruition, diminishment, extinction, "envelopment," and transformation again, etc., "scanning" and "spacing" (temporality and spatiality) unfold: succession, simultaneity, coming together, distancing, moving "here," going "there," processing material "in here," fishing for it "out there." Life projects proximities, distances, coincidences, etc. that are the "trading posts" of life's progress within the immeasurable network of the interchange of actual and future, past and possible.

Amid the tensions between the opposites in perpetual strife, the equilibria, that is, adequate proportions, measures, equivalencies are established; these equilibria hold all together and maintain living beings' intrinsically orchestrated autonomy, allowing them to float in a constructive progress upon the turmoil of conflicting forces in life's dynamics.

6. *Flux, the Constructivism of Life, Measure*

We have struck a sensitive chord: measure. We asked at the outset how can there be a measure within the Heraclitean flux? Leaving open the question of whether there is a supreme instance of measure, for which Heraclitus employs the metaphor of the Sun (the Sun being perhaps a god?), we have to observe that the inner workings of life are implementing a measure, that from the very start of the game of forces each operation carries in its performance a "limit," a "circumference" already lying-in-wait for the signal of budding life. This operation is itself embedded in a "complex" of forces that all together establish its "measure." Indeed, each element entering into fusion with another carries in itself elements for which there lies a measure according to which they may fuse and become transformed or reject each other. In short, to say "perform an operation" or "transform" — each dynamic bursting forth — means concurrently to evaluate.[13] A comparison, a discriminative "weighing against each other" and establishing the "right measure" of the dynamic bursting forth is concurrent with this valuation.

I would say that the constructive progress of life processing the flux and harnessing its forces means transformation, and that transformation means measure. The dynamic flux of life can be harnessed only because there are intrinsic "trading posts" for measure present in its forces, ready to negotiate.

III. THE GLORIOUS ASCENT, THE SPIRAL EVOLUTION OF TYPES, THE HUMAN CREATIVE CONDITION

1. *The Overarching Swing of the Self-Replication of Life: Dissolving the Opposition between Contingency and Necessity*

Life's differentiation proceeds from within. Amidst the crucial endowment of the entelechial nucleus of life's self-individualization, prime is the seminal capacity for singular self-individualizing progress, and second in importance is the capacity for the intergenerative replication of types. In fact self-individualizing progress carries within its system a replicating code of virtualities and a mechanism for unfolding; that is, it lies within the innermost virtualities of the entelechial nucleus to prompt generation within a new living individual and to bring it to an autonomous life course carried on outside the nucleic womb. The striking phenomenon characteristic of life alone is that the novel being replicates the type of its bearers (even when its origination is triggered by a confluence of appropriate seminal moments from two differentiated kinds of individuals of the same type). It is the replicating nature of this intrinsically orchestrated process that accounts for the spread of life in projecting ever novel nuclei for its expansion horizontally in concurrent "generations" as well as vertically in a continuous succession of generations, receding, vanishing, and incoming.

The replicating process rejuvenating life follows the life cycle. However, this cycle assumes a unique line of ascent. The recurrence of types does not guarantee their identity. This recurrence is not, as philosophers assume, circular, but spiral. Owing to variations in generative composition within and to transformational influences in conditions without, the type undergoes a long, winding transformational change. This continuity of the "same" type and yet the becoming of an "other" is continuity-in-transition. The sameness of the type loses the sharp edge of the necessity that it represents because the "same" is "in transformation"; and yet, it still guarantees a recurrence within the change.

The advance of types embodied in contingent, concrete beings is contingent in the concrete while it is perduring in the universal-in-transformation. Hence, we have the moving ahead with novel types emerging through long stages of spiraling transformations carrying a measure of change and advance.

In fact, it is by an essential transformation of the measuring modalities of life's progress that we may grasp the gist of the spiral ascent. To cut a long matter short, here is a loosening of the automatic apportioning of fused elements found in the operations of the simpler forms of living beings allowing the advance of a self-individualizing progress, which is what characterizes this progress most essentially. The automatic attunement of elements coming together for the release of appropriate "nourishment" that operates on the level of the least evolved forms of life using automatic closed criteria of "fitness" is, with the advance in complexity of life's inner mechanisms, open to several options in dealing with the matter available so that the selection of what is "fit" is left to discrimination, to choices made by an individualizing being that progressively establishes an ever more refined and increasingly flexible inner agency able to adjust the seminal measure applied in the selection. Through this process the freedom to project measuring criteria is attained. This attainment is accomplished with the glorious ascent of life toward freedom in the Human Condition.

2. *The Human Condition: The Flashpoint of Inventive Rationalities in the Creative Forge*

With the loosening of lockstep preordination in the mechanically adjusted confluence of elements within the entelechial nucleus of self-individualization and the life agency's concurrently taking on the task of seeking, supplying, and selecting elements for the transformational processes of life, there are to be distinguished numerous types of living beings ranked along the ladder from preordination to selection, from entelechial determination to creative freedom.

In fact, I have written too much about the creative condition of the human surge at the breaking point between determination and freedom in the living translacing continuity maintained by means of our creative virtualities to here dwell on the nature and composition of that august condition. Let me, however, for the sake of the argument running through this book, point again to the striking moment of the creative breakout of the specifically human rationalities by means of which the living being selectively projects its very own, creative self-individualizing life course. As the individual does so, he or she concurrently endows with significance, and enacts on his or her own initiative, each step of life's course. Here there is clear self-awareness in the life agency, the level of a self-directing and all-overseeing mind now having been reached, a mind that allows the human being to encompass the circumambient conditions to the point of forging singular routes, of proceeding like a spider on its own thread.

Within the creative forge in which the creative virtualities have surged as a novum in the spiral evolutionary course of types by establishing the phase of the Human Creative Condition, the inward/outward self-individualizing process occurs seemingly according to the same set of dynamic principles as always: ingathering, transformation, radiation. However, now there enter into the transformation inventive/creative virtualities that process in an unprecedented fashion the received material; they invent innumerable novel rationalities and radiate them without, projecting into the circles of circumambient conditions rays of novel significance: the specifically human significance of life.

In the Human Condition there is thus operated a unique encounter between the "old" and the emerging "new." At this transitional station of life thrown like a fisherman's net upon a stormy wave the mysterious potentialities of life are retrieved from the flux and through human creative genius are brought to a glorious achievement, the manifestation of intellectual life.

3. *The Creative Manifestation of Life. The Encounter of Both Modes of Rationality — Stability and Flux*

It is in virtue of the creative virtualities — aesthetic, moral, and intellectual — that the human being became "human," that is, having unfolded a specific type of nucleic life agency in the mind, they bring forth this glorious spectacle of life. This is life within a world extending in transparency in all directions as if encircling us, life as the center of its most fascinating intricate ramifications, the meaningful rays of which attract us, draw or repel, play on our sensibilities and desires, tantalize us when awake, and inhabit us while asleep. This seemingly unlimited realm of surprises awaiting at each turn of the road, and endless expanse of expectations, unformulated hopes, and wild dreams in which we are entangled with other human beings with whom we play ever renewed games, establish friendships with, or part from as enemies — all this so clearly "objectified" within the structures with which our mind frames all relations — bring everything forward and out of hiding into the light of reason.

Well, here it is that the human mind steering the individual human station set upon the flux has projected all means of a relative stability in the manifestation of the logos of life. We constantly seek to maintain that stability and the further progress in which the specifically human self-individualization-in-existence consists by, in fact, projecting ever new means for subjugating the flux upon which we float, by launching upon the waves ever new kinds of craft.

The recognition of the order of objective reality thus follows first — as was pointed out at the outset — from this urge to maintain oneself afloat, the structures of objectivity being the capital, principal correlate of this existential urge, necessity, device. There it is that the logic of essences finds its foothold.

We pursue our life facing and dealing with this glorious manifestation of life, and there are objective structures that uphold it that we have to deal with in forging our own life course.

Let us marvel! Is this not a glorious accomplishment of life to have begun with the nucleic dust, dust, dust and to have reached this infinitely varied and yet coherent world that radiates the spectra of colors and sounds and without dashing the harmony between them, that fulgurates with the ingenious achievements (societal, cultural, etc.) of human groupings, animal species, climatic zones, vegetation...and all this brought to full light by the objectifying creative genius of the human being?

Does this overwhelming achievement of the manifestation of life mean its radical distantiation from the underlying flux? And from the logic of contradiction? Here we may ask again whether "wisdom" is accessible to us. It is precisely at this point that having uncovered the breakout of the creative human condition in the evolution of types, and having focused on the creative act of man that explains how we can know, we enter into the spring of creative self-individualized life.

IV. THE SACRAL SPRING AND THE REVERSAL "BELOW," "ABOVE"; "NOWHERE," "EVERYWHERE"; "NOW," "ALWAYS"

1. *The Median Plane of the Human Soul within the Manifested Inward/Outward Operative Extension of the Human Life Agency*

The human being capable of the far-reaching awareness we discussed above is also capable of self-awareness, and consequently we distinguish functional territories within the human self-individualizing life agency. All living beings show themselves to be alive through a "spark of life" that runs throughout their frame, an "animus." Expanded in the creative human frame this "animated" region covers the entire functional territory of the self-individualizing creative agent. As anima, the soul carries as well the functions of vital organic significance as the creative mind; it serves them all. And yet, at the peak of this glorious creative manifestation of life in the specific creative endeavor of the human being, the creative propulsion of the logos of life loosens its hold on the forces of the soul that nourish and sustain it.

From within itself, its own resources, the soul, as I have shown elsewhere in great detail, releases a virtuality of its own, a virtuality that in expanding goes counter to and dissolves slowly the great swing of the creative logos: a virtuality seeking the "ultimate sense" that the logos of life falls short of in seeking

Here we meet the mysterious Heraclitean intuition of the unfathomable depth of the soul (Fragment 45). Indeed the quest for the ultimate proceeds again along the tension of opposites but does not have a measure common to either side. It proceeds "deeper and deeper" into the secret meanders of the soul and seeks "below" what Coleridge in "Kubla Khan" intimates to be an unfathomable depth of subterranean caverns inaccessible to the human being, through which "the sacred river runs." I will call this depth the "sacral spring."

Strange to behold, when the cycles of a singular life's individualizing course streaming from its entelechial nucleus and continuing in spiral fashion in the creative forge is completed, when the spiraling cycles of the evolving types of beings pass into one another in their striving onwards in the contingency intrinsic to transformation as such tend toward some or other finale, then the spontaneous onward swing of life turns from its zenith toward its nadir, from its ascent to its extinction, and is elevated yet again by the prompting of the sacral spring of the soul's very own unfathomable resources. In fact the specific complex of functions that phenomenology of life differentiates as the soul plays a role of its very own, meeting the Heraclitean vision of being without confines and augmenting on its own the human soul's lift-offs. Life's "conatus," as Spinoza and Unamuno after him would say, lifts up the self-individualizing device of emanation from within by shifting gears from struggling to survive to unfolding the soul's sacral destiny in a transnatural quest for another type of accomplishment in its cycles, not in the contingent now but in eternity. This sacral course spins its thread "above" life and "within" life. It moves with the all-pervading tension of the opposed tendencies passing from the spirit of life to the spirit of the divine.

2. *The Reversal of Life*

Given the tension provoked by this vision of an "other" life that is "within" and "without" the life of nature and culture, the human transnatural streak of the soul plays its innermost drama involving all. This tension of fleetingness and perdurance provokes an arduous struggle to salvage from the imperturbable pendulum of nature and the coming to be and passing away some "absolutely" valid moments. The soul forges its very own vocabulary of significance for life.

In a quest propelled by the sacral spring, myriads of new meanings alien to life's concrete and abstract meanings are arrived at: these consist of experiences with significance of hazy contour, feelings with vague content, foreboding as well as striving, that from our innermost depths lift up from a "below" not found on life's maps — a nowhere in the geography of the world — and carry the soul bringing up with it myriads of transformational intuitions, intimations that are "above" the frontiers of life's own most elevated creative meanings, that exist in a sphere also off the map of life. The sacral spring sends forth an entire new universe, which, although it appears to be the reverse of the one that we know and dwell in concretely, can in fact transform it in our experience of meaningfulness and in the enactment of life.

This new universe of meanings that are strangers to the vocabulary of the logos of life emerges amid this innermost tension between the logos of life and its nadir. The logos of life belongs to the older universe. The new vocabulary from an "other" world has crucial points of correspondence to that of the lifeworld. But faith exceeds belief as found in the world of vital interests, hope exceeds expectation as found in the world of vital interests, and charity exceeds love as found in the world of vital interests. And forgiveness utterly transposes hate in the world of vital interests, as does self-sacrifice, self-interest. Thus, this "otherness" breaks into the autonomy of the lifeworld revealing as it were a reverse side: at the limit of the tension between these opposing values we even speak of establishing in this world the "kingdom of the Divine."[14]

This "otherness" belongs to life, from which it appears to draw its natural dynamisms projecting its battle inside its arteries — and it belongs not to life, counteracting it and suspending its meaningfulness.

Where does the sacral quest come from? Where does it tend? No wonder that emanating from within the soul's intimate resources and enveloping nature-life with its meaningfulness, the soul can be seen as the source from which in stages there emanates the entire expanse, natural and sacral alike, that divinizes it all. Considering the fact that the soul is the foyer, the gathering and transforming forge, of all nature-life, there is a temptation to attribute the transnatural to the manifestation at the opposite horn, making of it an epiphenomenon of life. Yet being of life and not being of life, the sacral domain combines both horns moving upon the heart of this tension. It is, indeed, at the heart of the human being that the great game of salvaging life, the great drama of incarnation and redemption, is being played.

All seems suspended on this ultimate and crucial thread as it is pulled in opposite directions: "whereto," "wherefrom" remains open-ended. We here reach the edge of the final questioning of Heraclitus, who credits the Sun with "setting the final bonds" for life, doubling it, however, with the "primal god." Here is the frontier of the phenomenology of life proper and of the metaphysics of religion, a frontier at which we now stop.

V. THE GREAT PLAN

I hope to have in this brief outline at least evoked the great and glorious vision of life. The innumerable arteries of the fulguration, as Leibniz would say, of ever new self-individualizing beings, of their intermingling in dynamic unfolding, of their transformations in cyclic progress promote the continuity of life's swing...the myriads of significant articulations that the human creative mind brings into this great game of forces; all this proceeds against the tensions of seemingly opposed forces, energies being set on a collision course, and makes the possibility of a great plan for it all, a plan spoken of by Heraclitus, as remote as it seemed to be given his logic of contradiction. Here we have attempted to unravel the outlay of the dynamism of life precisely through its inner workings, which entail the logic of the concrete, the logic of dynamic becoming a logic at odds with the constitution of objectivity in the structures of the manifestation of life, with the manifestation of our human reality in which we pursue our own self-individualizing course according to the rules of the logic of this constituted objectivity, a logic of stasis and relative perdurance in contrast with that of change.

Yet I submit that we have now come quite close to envisaging the possibility of the great plan of All. As a matter of fact, phenomenology of life, having elaborated its entire field, is uniquely prepared to lay down the plan's cornerstones and describe its main outline.

1. *The Key: The Axis of Life*

First and foremost we have unraveled from the conundrum of life's dynamic interplay the axis that runs through all the phases of its unfolding: inward/outward oriented self-individualization and an intrinsically transforming scheme of operations equipped with its very own initial propulsion and an entelechial seminal unfolding design. a) This autopromoting, self-initiating progress of life carries an entelechial core in the empirical unfolding of nature-life: inorganic and organic, be it vegetal or animal. b) When it loosens its automatic ordering and the entelechial code opens up predetermination to the creative freedom of the human condition, self-individualization is then self-invented and self-enacted given the new significance of the conscious self. c) It is also from within and in a self-specifying, self-projecting swing that the soul projects her transnatural destiny in a direction that is the reverse of the empirical and creative courses both. The logic of contradiction is the axis of the manifestation of life as well as of life's reversal and elevation in a spiral. It is the axis of all the arteries of life's dynamisms as well as its great plan, the key to the investigation of science, culture, religion alike in their different conundra of significance.

2. *Philosophy, Metaphysics, Wisdom Made Possible Owing to the Creative Virtualities of Man*

We have shown how life itself defies the spurious opposition between the logic of opposites and the logic of essences in its inner workings, which ground the manifestation of life and offer pointers for structuring the spectacle. In virtue of what, however, do we arrive at the excavation and grasp of the inner workings of nature-life while dealing with the logic of this manifestation? Heraclitus attributed knowledge of it to an access to the real that is different from natural cognition, an access to the "inner constitution" of things. But how is it possible for the few who, he said, can attain this access to penetrate to the innermost arteries of life, which remain hidden, mute and do not reach the surface of the manifested world of life, cognitions of which are shared by all? This is a question that cannot be answered. With the constitutive powers with which we have established this marvelous edifice of the world of life, we slide above its surface. It is the creative act of the human being that offers the Archimedean point giving purchase in the quest for wisdom. The human being gathers within his or her creative force all the strings along which a human person operates on the level of nature-life; he ties them in a knot in order to reach further into life's own operational realm. With cognition/constitution geared to survival and vital praxis proper to hoi polloi, ordinary people, we slide upon the surface of the knot; we reach into it, however, and untie it for our own purposes, following the arteries of the creative act that not only draws directly on the inner workings of life but also refashions its resources. That is to say, with the creative act we discover the underlying cornerstones of the great plan and may proceed to investigate its articulations.

The human being, the creator, participates fully in all the circuits of the great game of life and has urgent reasons, more pressing at the present moment of history than ever, to strive to unravel and grasp its innermost sense. Phenomenology of life aspires merely to have offered the key to this search and to outline some of its main strategies.[15]

PART FOUR

THE EMERGENCE OF THE TRIADIC LOGOS: THE TURNING POINT

CHAPTER ONE

THE MANIFESTATION OF THE INTELLECTION IN THE UNIVERSE IN THE TRIADIC LOGOS: THE TURNING POINT

I. THE INTELLECTIVE OR TRANSCENDENTAL EXPERIENCE OPENING THE CONSTITUTIVE/COGNITIVE VISION OF THE HUMAN BEING WITHIN THE WORLD-OF-LIFE AND THE INNER-OUTER DISCOURSE OF THE SOUL

The emergence of inventive/creative virtualities with the advent of the crystalization of the human condition takes place within the realm of the sentient/animal soul. It encounters there an already organized functional system with its own meaningfulness, its own telic orientations and having its own guide in relevance to the entelechial code and its outline. The surging of the novel synergies galvanizes the already operative forces and stimulates in them random propensities thus far left untapped. It also disrupts, with the surging of novel forces, novel impulses, novel stirrings, the established rigid functional order followed up to that point in life's functional avenues and operational lines. The new impetus sprouts qualities of significance, shapes, sentient streamlets, and undefinable myriads of propulsions that do not fall into the available modes; these throw out hooks with the aim of attracting even as they repel by their strangeness. Emerging in contrast and opposition, they seek response; left to themselves they are a dispersed lot. In seeking to enter into a constructive cohesion, they have, first, to go through the sieve of the available vital significant system and have then to find ordering constructive intergenerative links, an ordering pattern, a harmonizing principle.

It seems as if it were the new tactic of the logos to give birth to a new type of generative factor: Triads of generative moments that both imply and oppose each other and acquire their respective meaningful proficiencies only in a divisive but complementary harmony. To put it briefly, an encounter has to be negotiated, a pact has to be reached between the fullness of the primogenital elemental/transcendental experience of the fully operative intellectual, aesthetic, and moral sensibilities, in which each side seeks a different outlet for its forces, and that transcendence's roots in the vitally significant functions of life.

Now, with the creative forge galvanized by the bursting forth of the inventive virtualities of the logos, the course of the constructive vision of life is radically transformed. With the igniting of the inventive factor that is not guided by the entelechial schema that is the foothold of order for the vital significance of life but instead by Imaginatio Creatrix, which brings in its own forms and rationalities, the vital continuity of the living beingness is upset and its schemas of meaningful intergenerative processes are disrupted. The vitally bound animus loses its hold as the creative/inventive factors prompt in its midst the bursting forth of an inner universe of the spirit/mind — a radically new type of inner agency of living beingness. Correspondingly, the vitally significant life apparatus of the organism is surpassed by growth that is inward in the direction of its own universe, on the one hand, and outward in the direction of action, on the other.

The inward agency of the living individual on expanding into logos processing mind/consciousness acquires, therefore, an unprecedented role in the evolutionary progress of life: it launches its own system of reasons, with its own generative and intergenerative rules, laws, and aims, and, more than all that, the human will is at the core of their prompting force. Through its volcanic eruption and spontaneity, the new significant synergies of inspiration promote the vibrant creative/ inventive push forward of the logos of life.

This new station of the inventive/creative differentiation of the logos of life emerges from the groundwork of the vital soul, using its synergies and forces, its system of operations, and yet its concrete processes do not follow its regulations and rational articulations and routes. They break radically with its system. They do not follow its aims or telic principles; still, they do not disavow the vital rights of these. Consequently, life on this level does not progress cyclically following an intrinsically grounded valuative system. To the contrary, it projects an entirely novel and most elaborated system of valuation and its own measures. However, this new system always finds relevance in the vital system.

In summary, the human type of beingness emerges as such from and is carried onward by the most intimately intergenerative, fused together and yet, at the extremes of their rationales, dramatically contrasting and distinctive systems of rationality — the vitally significant, on the one hand, and the inventive/creative humanly significant, on the other — which often distance themselves from each other (with one pulling in the direction of the unavoidable, indispensable chains of the rationalities of the biosphere and its animal order and the other soaring toward the open sky, toward the ever varying possibles of the creative mind) and yet come together, cohere, and harmonize, projecting sense-giving factors for the novel universe of the human spirit.

We may distinguish in this universe yet other modalities besides this tri-noetic/trialogical modality of the constitutive/cognitive logos that stems from the inner field of raw transcendental consciousness and accounts for the spacing/scanning edifice of the transcendental world that is unique in the sphere of life. We distinguish the Dionysian modality of the logos, which accounts for societal sharing-in-life, and the Promethean logos, which is nurtured by both of the other logoi and comes to the fore lifting them up, bringing forth the wealth of the inventive/cognitive riches that send the human being soaring into the freedom of the infinite.

This new turn in the unfolding of the logos of life can be seen as bringing about three absolutely novel significant lines of expansion: the intellective-noetic or dialogical-trialogical, the Dionysian, and the Promethean modes of rationality.

The first of these introduces two novel original instruments of intergenerative significance: first, the constitutive/cognitive intelligence, which ushers in intelligible communicability for the establishment of the specifically human inner/outer universe and, second, the judgmental apparatus that already resides in the first but that reacts and goes beyond its own office of valuating and decides on active human affairs. With the creative intelligence there emerges the intellective synergy of the will. However, the three modalities of the logos are only an abstractly distinguished set of sense endowing functions. In actual practice, although they represent three new sense giving faculties and stand in a certain opposition to each other, they do work together in a most intimate, harmonious, and necessary attracting-repelling-complementing intergeneration; no one of them is thinkable without the others. Therefore, the communicative intergeneration of meaningful functioning is in the first place to be seen as a constitutive/cognitive/judgmental activity of the human mind that raises life in its primitive circumstances to the highly structured presentificative modalities — which are sometimes a) representationally cognitive intellection, and sometimes b) the self-promoted (will-directed) inventive outwardly oriented act.

However, the very coming about of all this is conditioned by the sentient/emotive factor that comes to prominence with the specifically human unfolding, a mode of logos that I have called "Dionysian." But neither of these two modes could get off the ground without the inventive/creative act of the human being so decisive for this extraordinary phase of the logos of life. That act frees life from unyielding entelechial dependencies and inspires and lifts it toward imaginative flight and prompts it to bring out new forms, qualities, hints, etc. Thus, we call it the Promethean logos. With these modalities unfolded and harmonized the logos of life expands in "freedom" into a new universe of its own "invention," into the universe of the human spirit.

As we now trace the wondrous paths of the logos, we will briefly discuss first the origin and unfolding of the trinoetic/trialogical logos (that is, of constitutive cognition and knowledge) and then discuss the origin and unfolding of the Dionysian logos, that is, of the affective/sympathic sharing-in-life that culminates in human societal existence, and then we will finish by taking a brief glimpse at the nature of freedom and of the human spirit, that is, at the Promethean logos that is the fruit and crowning work of the two preceding logoi that are prompted by the human creative act and that flourish in the intergenerative circuits of human culture. The latter manifests the human spirit, which has been unfolding with the transcendental experience of the human all the way from the incipient instant of its origins. However, although with the onset of the human condition the logos sets out on such a different and autonomous course, and although there is no longer continuity in meaning, yet there is no disruption between vitally significant functioning and specifically human functioning. Not only do they encounter each other, not only does the vital have cosignificance in the inventive/creative domain, but the channeling of the logos into the human condition is by no means an abrupt, all of a sudden, unprepared for event. Although there was no highway to it, there was a stepwise progression toward it in the evolutionary advance of life along two of its lines: the triadic-dialogical and Dionysian logoi. With the appearance of the Promethean logos we get off the ground, but, again, not without some anticipation and stepwise preparation for it.

Following the passage from the vital to the intellective order in primogenital experience will lead us to see the surging of the novel logoic apparatus meant to install an intelligent agency for processing both the vital/entelechial forces and the intellective synergies and to create a platform for and then project the new order of the specifically human universe of life.

II. CONSTRUCTIVE TRANSITION ALONG THE LADDER OF EVOLUTIONARY ADVANCE

The Logos of Life Reveals Its Pregnant Rationalities: Voice, Graphic Sign, Word Intelligibility

As the most striking effect of its differentiation in the specifically human form of life, the logos of life manifests itself in its constitutive powers from within the functional expansion of each individualizing life. With the emergence of the act of intellection there surges the new universe of the human significance of life, which is a universe of resounding signs pregnant with significance, which no longer covertly lies within the hidden workings of nature but is right on the surface, "floating in the air." These are signs that stand for rational messages, that is, in contrast to the rationalities of the workings of nature, signs that remain strictly confined to their specific role, mute and nontransferable.

1. *The Synchronizations of the Scripted Network: Signals, Signs, and Languages*

As I have already explained, the meaning bestowing of the life process, that is, the life process unfolding itself, proceeds through different networks in which various levels of functioning are intertwined. With the increase of those levels the complexity of the pattern increases and with it the complexity of the meaningfulness.

It has already been pointed out that the actual individual's striving for life progress between the conditions of the given lifeworld situation and those of the Nature-Life system in which the real individual works out his route demands the flexibility of self-adjustment, of self-correction; those in turn presuppose a self-control at the smallest tracks with reference to the overall system of operations so that a retreat from an already taken step be possible and a new step readjusting to the situation be offered, discriminated, and taken. Such a control system operates owing to the "communication" effected between the different operational segments by various devices. The blockage of one of the routes by one segment that cannot unfold its schema of operations has a direct repercussion on the segment that was foreinstalled to follow it, and so on. But direct "communication" of a mechanical type would not of itself amount to the overall regulative system that seems indispensable if an alternative step is to be discriminated appropriately from among a ranges of possible ones. For one thing, direct communication left to itself would through its reverberations bring the entire functioning of the individual to a halt if it could not "in time" effectively get out of an impasse. "In time" means before an inappropriate command blocks the entire operational system. This entire system has, on the contrary, to devise alternate ways of "repairing" truncated and isolated operations. To begin with a functional network stalled at one particular point must receive through its antennae a signal that something has gone wrong.

2. *Signs and Sense*

In the unfolding of the self-controlling agency of the real individual his first trait is his conscious expansion through self-registered (doubly registered by experience) sensing, feeling, perception of his route, the capacity to follow it, differentiate its steps from the continuing unrolling, and, by detaching them from their embeddedness in an integrated complex network, single out the different articulations in their interconnectedness. The distinction of the various types of interconnectedness itself through which the movement that becomes progress is punctuated, e.g., the cause-effect-aim articulation that underlies life progress in its constructiveness, allows the agent a higher type of self-regulation. Indeed, owing to it, the agent becomes "interiorized" in a deeper way than that of the mere centralizing of all functional arteries. A certain detachment from the concrete pulp of life's progress serves to redouble it in experience as well as allow differentiation from its articulations, so that the agent may actively synchronize them at various levels in their shifting more and less repetitive configurations, may run back and forth through them distinguishing between the most repetitive segments of operations and those to whom they are subservient till finally their concluding ends are reached and the various lines of cause-effect-aims networks are brought together. That is to say, the agent may reproduce in a synchronized fashion, to the point of manipulating by his own intervention, the concrete unfolding of his route, the system of his life inscription and its progress.

His intervention in his own progress makes him a subject. Then the signals given by the system of his life operation become for him signs. When a signal alerts, it leaves "meaning" unexplicated, it does not contain the meaning of the articulation (in this case malfunctioning) at stake. When the signal appears to the subject in the form of a "sign," the truncation or disruption to which the signal gives alert can be sought out and apprehended. A sign has for the subject a sense to be explicated.

3. *Languages*

It is language that allows the highest and most complete synchronization of the unrolling of the life script. Existential becoming in its constructive-design-continuity then involves a fourfold synchronization: firstly, that of the structural conditions of life and unfolding nature; secondly, that of the segments of the operative functions and the order of the constructive progress; thirdly, that involved in the anticipating-completing nature of the interconnectedness between the process-like segments of operations; and, fourthly, that involved in the cause-effect-aim organizational orientation that underlies each operational segment and that runs through the entire life process in all its phases and that of its very nature consists in bringing about something that had not been there — the emergence of beingness in tandem with the individual being that is an absolutely discrete series of motions and changes.

The traditional potential/actual contrast fails to express this bringing about of what had not heretofore been. It does not have the metaphysical depth that the reality demands. No pair of principles like potential/actual can give to the novum a preexistence, because potential and actual are modalities that existence brings with itself.

At every point the segments of the grand network of the life script are discrete. Synchronized according to their own rules, the segments spontaneously unroll. Their unrolling is recorded by the script being constructed without discrimination, appreciation, criticism. Controlled and self-corrected, the course is recorded as it is.

Thus, it is up to the orchestration of the agent — orchestration and ever tighter self-regulation that is itself installed in the functional system by progress in complexity during the unfolding of the subject — to introduce existential self-control and differentiation. Owing to his capacity to take off from the concrete pulp of life while maintaining an existential foothold in it, the individual subject may reach the new level of synchronization at which life's progress is differentiated in its articulations, its meaningfulness, is brought forth and appreciated in its cause-effect-aim chain. While the concrete life script is being scrutinized with respect to the rationale of life conditions and its elements discriminated accordingly, infinite possible synchronizations in the unfolding of the script itself as the subject experiences it then appear. Although the same universal pattern is manifested at each juncture, infinite possibilities for variation open up.

Furthermore, only then does the individual agent become fully in control of his course. He may, owing to the different scanning of the previous stages of his unrolling and owing to the different appreciative synchronizations of his successive conditions, direct his course differently. The possibility of choice and selection opens at a level different than that of a mechanical reorientation.

At last, the self-unfolding subject in his capacity to synchronize grasps in various perspectives and in various configurations the rules, laws, proceedings operated by himself determines possibilities for his life script, and projecting them brings out its meaningfulness as relevant to his self-controlling concern. He brings out its sense for himself in the form of explicated signs. He unfolds his "subjective" script in the form of language for the sake of grasping and controlling his existence.

Language is the most complete synchronization of the life course, expressing the subject-agent attainment of control over one's life route with the option of changing course — of not only repairing it piecemeal while going in one direction but of changing direction if that be chosen.

Thus language expresses the subject's dominion over the life script. First, it shows his capacity to detach himself from the concrete pulp of existence and move toward grasping its rationale — in its meaningful articulations and their patterns. Secondly, it expresses the subject's capacity to appreciate the rules of actual existence, that is, the patterns of the universal script and conditions of life, universal meaningfulness, the state of affairs. Thirdly, language means the subject's emergence from complete adhesion to the concrete flow of becoming and absorption in his operations, even as they remain intact given his detachment. Fourthly, language expresses not only the subject's participation in his life course but also his existential condition, his being essentially intertwined in his progress with other subjects with whom he then shares the "objectivity" of the existential script, which is the script of all individual agents, that is, of all subjects profiled in their common progress.

Hence language is in itself the synchronization of all the elements instrumental in life self-interpretation: The rationale of life patterns, their meaningfulness, appears specifically in the subject's perceptive and cognitive experiences. These are guided in their operation by the subject's desires, wishes, tendencies, values, and will. Therefore, language is clad as it emerges; that is, emerging from the subject's vital commerce with other elements of the natural script and referring to signals mediated by his vital processes, language is embedded in sound and oriented toward "the other." But it should not be forgotten that its natural aim is to gain control over the course of the life script, over the origin and course and life of the "objective" world. Consequently language is the last stroke of being's unfolding, instaurating the system of objective communication. It explicates within its complex subjective form the hidden and otherwise reposing in itself meaningfulness of the life saga (not the meaning *for* but simply *of*).

However, if language is the vehicle of the real dominion of the human subject over his life script, allowing him to invent for his own sake alternative scripts, it still bears mentioning — what will be treated later on — that language is also tributary to the inventive and creative functions of the subject, which accounts for the expansion of his own meaningfulness beyond his life script.

III. PRIMOGENITAL EXPERIENCE AND THE EMERGENCE OF THE INTELLECTUAL SENSE: THE PASSAGE FROM THE VITAL SIGNIFICANCE OF LIFE TO THE HUMAN SIGNIFICANCE OF LIFE

1. *The Turmoil*

As I have already shown in the introduction to this chapter, the surging of the inventive/creative virtualities within the vitally significant framework of life's self-individualizing disrupts the schemas and lines of generative progress and creates a turmoil of dispersed, disorganized moments, fragments, elements, that are at odds with each other and yet attract each other at random, fusing spontaneously into haphazard mix-tures, a chaotic turmoil of manifold synergies racing wildly ahead seeking constructive outlets. Chaotic, yes, but living beings do not disintegrate this encounter. To the contrary, they jump the evolutionary gaps in constructive ways. The wildly racing forces seek new cohesion. But the question arises, how may this new kind of cohesion occur? At the level of human life this question is even more intriguing, for here there is a harmony to be reached among and between, on the one hand, the innumerable pulsations, forces, energetic and synergetic moves, and operational functions of the vital order in which all living beings are caught, and, on the other, the virtualities of the Human Condition with their own synergetic modes and propensities. This harmony between and among innumerable repelling and attracting forces is accomplished, however.

The terms of the final "pact" among innumerable partners may remain a mystery unfathomed by the human mind, but we cannot cease to ask: What new types of ordering perform the harmonizing task? What new transformations have to occur to accommodate the implacable calls for a transition as well as for appropriate outlets for the infinite variety of the propensities of the elements involved? Lastly, we cannot keep from asking: What may be the nodes and mechanisms of transformation that will perform the harmonizing function, laying the groundwork for a new logoic ordering? Finally, we must ask: How can this harnessing of the manifold reality of the turmoil and its transformation into a novel order keep faith with the logos of life and maintain the indispensable groundwork of the old vital order in the advance that sees an ever renewed radiation of myriads of novel intellective circuits?

The human intellect may remain helpless to explore the mystery of the ultimate pact between the organic and the sentient arteries of the logos, and then between the sentient and the intelligent, but we carry this pact within ourselves. We carry in our human/animal life the primal vital self-awareness that anchors us in living itself. Moreover, we already have gained — as explored in my previous writings[1] — a point of access to the "creative forge" wherein life's transformations occur. From there on we may bring to light the scene at which the game is being played, where the main protagonists in the harmonizing of the repelling and yet attracting elements of the initial turmoil appear.

2. *The Incipient Move Onward; the Passage to the Primogenital (Experiential) Sense*

The inventive/creative virtualities in the shock of the encounter with the manifold vitally *significant synergies,* fragments of articulation, impulses, prompting energies, etc., all of which repel and attract each other, are prompted from within their own synergetic resources into their specific act, that of projecting from within a set of functional moves. They first activate the entire complex of virtualities of the Human Condition, and, second, undertake an *organizing role*. In brief, there occurs an elemental (primogenital) sprouting of the new harmonizing/ordering apparatus: the conscious faculties of the human mind.[2]

In a vertiginously complex fashion, the logos of life in pressing relentlessly onward draws from within itself new differentiating and constructive synergies that harmonize the manifold of the initial vital/creative turmoil in a new key, projecting a new significance of life itself.

This novel move of constructive harmonizing encompasses the segments of vitally significant primal (animal) "sensing," feeling — be they of pleasure, pain, cold, warmth, darkness, light, etc. — in their own spacing/scanning framework (e.g., the primary "awareness" of presently felt pain not only as a "signal" of vital significance, of danger, etc., but also in its "location" in the "space of the body" as well as the joining of "my" pain with feeling "in my body") and lifts its elements from the opaqueness of just sensing and feeling to the "light," the light of intellection. It both arises out of and rises above the opaqueness of simple vitally significant signals of what promotes or hinders life's progress and brings forth *the sense of the act of sensing* in question by distinguishing in it the quality of its bodily spacing "in" and through bodily functioning, insofar as the differentiation between bodily spacing and its surroundings is made. Concurrently, and most importantly, this novel harmonizing agency in its incipient activity provokes the awareness — prereflective — that distinguishes "myself" as the agent of the sensing operations as it corroborates qualitatively the instance of sensing itself. Thereby a new sense surges from this novel circuit, one that is new beside the previously dominant vitally significant sense, which was proper to what had been animal sensing only.

From this first circuit of the now not vitally significant but elemental harmony, there unfolds now as the incipient step of the novel ordering system, and, indeed, of the entire system of actively constructive ways and means of the logos of life, a new operational agency, that of human reason. This assumes the multiple modalities and constructive means unfolded by the human consciousness that is concurrently forming itself. This latter unfolds around the novel central factor of an intellectual act that prods onward and retains the instrumental validity of earlier evolutionary developments while projecting a new one, the human mind.

3. *The Breaking-In of the Act, the Bearer of Intelligibility*

It is with this inventive/renewing step, the surging of the act, the bearer of intelligibility into the realm of life, that the logos of life makes its dramatic entrance. It brings with itself a novel and hitherto unprecedented modality of the intelligibility of reasons, a model of a novel apparatus and process for establishing intelligibility, and a new synergy for promoting its works.

Surging in the midst of the primogenital harmonizing of the manifold reality of the living being, the act, what we call the act of intellection, the act of apprehending the intelligibility of things, is the decisive move of the logos in its phase of radical transformation. It brings with itself, as if in limbo, the synergetic outline of the entire apparatus for the new ordering for establishing modalities and generating new synergies, synergies meant to proceed and unfold from the ground of the experiential manifold the mind, the conscious mind of the human being.

But before we enter into the significant elements of the operational proficiencies of the act of intellection, let us first see of what the overall essential significance of the act consists, of what "intellection" consists.

Indeed, the act that we have in mind, which surges in the midst of the turmoil and in the wake of its harmonization into the primogenital experiential sense, picks up this sense, which had been strictly singular, belonging uniquely to the inner experiential realm and nontransferable, and operates its passage into the constant form of an infinitely repeatable, universally accessible meaning representing experience — which thus emerges in all its dimensions — and its object, a passage into intelligible sense or meaning. The act of intellection, which differentiates itself into a variety of modes of which perception, sensory as well as reflective, appears to be foundational, is thus a factor of the intelligibility of reason, through the work of which there emerges for us, out of the hazy contours of the animal environment, human beings and a universe of life transparent in its meaningful structures. Here are objects of perception, representation, enjoyment, action, and will that are clearly communicable in intelligible, constant, and universally transferable meanings as well as the acts by which we constitute them. From the opaque animal life orientation directed by survival needs, we pass with the act of intellection and its work to the translucent orientation that is accomplished by discrimination and enlightened choice, to the intelligibility of life as such.

Translucent, because carried by acts that are self-aware, "conscious," this intelligible universe of life is grounded in experiential/meaningful givenness; it unfolds through three new and ever renewed logoic modalities that I designate the triadic-noetic logos, the Dionysian logos, and the Promethean logos.

4. *The Birth of the Human Spirit*

In the surging of the act of intellection with its generative proficiencies, which allow the primal experiencing of the animal logos to be transcended in primogenital (or perceptual) and then universally intelligible experience, we find the source of the human spirit, of the all-embracing principle of the novel harmony of life.

In its ever new waves of fulguration, the human spirit projects through its rays ever wider circuits in the ever new expansion of the highest manifestation of the creative, human logos. Before we come to exfoliate its significance, which culminates in the mode of the Promethean logos, we must first ponder the triadic mode, unfurling the central agency of human reason as initiated by the act of intellection.

IV. THE ACT OF INTELLECTION THAT LAYS DOWN A NOVEL PLATFORM FOR THE GROWTH, ADVANCE, AND PROLIFERATION OF LIFE

1. *The Logoic Triad and the Foundational Spacing/Timing of the Intellectual Modality of Life, of the Human Spirit*

The direct accomplishment of the act of intellection, which in philosophy has been called the "intentional" act, is the exfoliation of the intelligible but hitherto singular content of primogenital sense. Whether we emphasize the constitutive or the contemplative (in the Husserlian understanding) aspect of its workings, this exfoliation is essentially a differentiation of the constructive elements of primogenital significance and their appropriate reconstruction. This re-construction in the novel logoic modalities and forms is the fruit of the workings of the emerging apparatus of the human mind. The avenues of the human mind unfold from the incipient surging of the logoic synergy of the act of intellection and from its proficiencies. The apparatus of the mind is prompted to this act, but then it advances on its own.

The foundational work of the act of intellection consists specifically in a new splitting of the logos into a triad of factors that significantly contrast with/oppose each other and yet remain of essence complementary both in their existential performance and in their very meaningfulness; these are the subject, the act that springs from it, and the object at which the act aims. In fact, the intelligibility of the act as such is the fruit of this split. In the opposition between the so-called "subject" from whose innermost the act springs forth and to whom it returns in order to complete its task and the so-called "object" that the act distills and constitutes in its constant, clear, and distinct properties and finally grasps in a universal that is relevant to all possible subjects, "subject" and "object" are coupled, are cosignificant in their complementarity in the performance of the task.

The subject emerges as the actual source of acts, which spring forth from its "center" in the midst of the experiential turmoil and "tend" toward the primogenital sense to bring to it a universal modality of structure and establish it thereby in a "position." The "positioning" or positing of the object is an effect of the act's primogenital intellective spacing/scanning. It is grounded in evidence — positioned evidence — instilled by the act in its work. The act is brought forth by the subject for the sake of this grasp and this positioning. The act is meant to mediate between the subject and the objects of its world, the "subject" being the inner agency of a self-individualizing living being who, dissatisfied with mere groping among the outward conditions of its existential survival, seeks to penetrate their "reasons" for being, both those relevant for its existence and those that are simply proper to their nature. That is to say, the subject seeks to establish a "map" of its circumambient conditions and points of reference. Such a map with orientation vectors has to be grounded in relatively constant points; the positioning of the constant contents of primal experience — experience that relates the inward agency of a living beingness to the unity-of-everything-there-is-alive, that is, in the first place, to its circumambient existential conditions — is meant to establish objects as such constant points. The positioning of the contents of experience in objective forms performed by the act detaches these contents from subjective experience as such; in their universal form they stand on their own, facing the subject, as it were, in their presumed ontopoietic autonomy.

Thus, in coming out and separating itself from the subject, the act, while tending outward, beyond the subject's reach, sequentially differentiates, discriminates, and constitutes experience, distinguishing and separating it in turn from the subject and positing on its own merits a universally valid constant. And yet, this universal validity is meant for the subject; its positioning as a fixed point "on the map" is for the subject, the map itself being for the subject's sake. And so the act, which springs forth from the subject to serve its intentions, returns the object that it intended to itself, which object, having been separated from the experiencing subject, now goes back to him.

The movement of the act becomes the factor of "intelligibility," that is, of that unique mode of rationality that means the rationality of *something* as it is relevant *to somebody,* that rationality whose origin is to be found precisely in the split between the sharply contrasted and opposed "interior" aspect of the living being and its "external" world. Yet, just as with the other generative tactics of the logos that use diametrically opposed moments, subject and object are of their very essence complementary and connatural in their ontopoietic significance. There cannot be a meaningful "subject" without its "objective" ontopoietic counterpart, and vice versa. This essential correlation is the source of the intellective mode of reason. It is in the triple move of the act — the intentional act, if that terminology be insisted upon — that the intelligibility of the act itself, of the constant, essential (what Husserl called the "eidetic") form of the object, and lastly, the intellectual receptivity of the subject itself is established. The constant, essential objectivity thus distilled from the primogenital sense becomes intelligible because it, in contrast to the singular act of primal experience and the meaning to be detached from it, which reposes with its meaningfulness in itself, has acquired a universally transmittable form by which every human subject can receive it; it has become communicable, that is, receivable, not only by the one experiencing subject whose act positioned it, but by all possible subjects, by all human living beings endowed with intellect.

This objectivity brings the intelligible universe to experience and brings about our communicable cognition of it. Hence, this new mode of rationality in which the specifically human logoic apparatus surges forward in the triad of the human subject, the act of intellection, and the intelligible object deserves the name of the triadic-noetic logos.

The harmonizing/repelling tension between the three interdependent factors of the subject, the act of intellection, and the intelligible object is the most fruitful intergenerative arrangement in the logoic inventive/creative apparatus meant to bring about the full manifestation of reason. It lays down the platform from which the entire apparatus of the human mind — the processor of the inventive/creative logos — is launched.

V. THE ACT OF INTELLECTION

1. *The Formation of Concepts in the Act of Intellection*

Phenomenology is in its procedures committed to sticking as closely as possible to the "immediacy" of givenness. This givenness is not merely the manifested human universe but the inner workings of this manifestation. In the Husserlian perspective following the modern turn, these workings were identified with those of human consciousness in its constitutive perspective, its own workings and their ways. In our approach, we shift perspectives radically. We do not "invent" it, however. We take a strongly established route. The scientist, the artist, the researcher, as well as the engineer do not enter into the inner workings of the human universe through the objectifying/cognitive/constitutive channel of consciousness but take a central track that leads them to the workings of the Nature in which the conscious human being is embedded and of which consciousness is a product. Philosophers like Aristotle, Leibniz, Hegel, Boutroux, Bergson, who delved into the natural processes were close to the type of insights we are directly revealing: insights into the workings of nature. These workings are the groundwork, the carrier, and the promoter of the cognitive processes; these processes are embedded in the workings of nature, so that the cognitive instances carry in themselves sedimentations of the vital processes that contributed to their surging and now subtend them but remain mute and in their specificity do not reach the objectifying surface.

Thus, when we want to approach the way in which there reigns within the harmonizing work of the creative forge the turmoil of pregiven, preexperiential material and so have in front of us all the avenues into which the synthesizing act of intellection reaches — even if the most imposing trend be offered by the cognitive/presentational objective strand — we cannot fail to differentiate four perspectives from which this act is to be viewed and the four different lines along which the immediate insights received take form. They ultimately have to reach the constitutive level of the objectifying consciousness to surge in the communicative form of concepts, but they manifest themselves first and originally in their immediate adhesion to the workings of the entire human functioning in which they live and from which they give us sidewise self-evidence with respect to the sustained focusing of our attention.

1. First and foremost we have, as mentioned above, the imposing objectifying focus that is constitutive of the whole of reality in its presentational circuit. Its prestige lies in offering a relatively stable map of the ever changing and yet fixed in actuality games of life.

2. We may focus on the operational progress of the intrinsic workings of the individual life progress, such that we retrieve most immediately in these operations; we may enter into their rules and laws and select among them those that may extend our powers over the forces of life through human skill, craftsmanship, healing art, engineering "technique."

3. Penetrating the workings of nature, we may use the observable and immediately given as a springboard for our inventive imagination by asking questions, "discovering" answers through conjectural inference, and for our "postulational" imagination by differentiating among possible answers to questions and selecting the most appropriate "sufficient reason," appropriate to an observable state of affairs. Scientific discovery is achieved by this kind of intuitive insight into the primogenital workings of the forces within bios and cosmos.

4. Lastly — and here we reach the apex of the flexibility that the act of intellection manifests in contrast to the selectivity of the entelechial ordering of bios — there is the creative, free imaginative focus of the artist whose insights into the creative process run through the entire architectonic individualizing life system of the human being. The creative process frees and distances itself from the binding links of the vitally significant, while psychologically seeking to incorporate the creative vision with freely floating imaginings, tentatively working with them toward a conclusive constructive kairos.

Each of these immediate accesses to the inner workings of life in its various circuits — immediate because embodied and personified by the operations of the very observer — yields its own intuitive evidence. What evidence may we gain more surely standing for the authenticity of our insights than that which is the expression of the direct vital gist of the operation itself?

Can the evidence of essential intuition, which is an end effect of the constitutive process, give us an access to the complex springs of this process that the steps of the eidetic process actually cover over? And what plausible evidence testifying to deep reality and its reasons do we obtain from the sedimented meanings of the cultural/social evolution that the hermeneutic process mulls, which meanings feed upon each other in a circle never broken out of, a method that basically holds up mirrors?

What evidences are obtained through historical inquiry that compares, synthesizes, corroborates the concepts and insights of philosophies already established without probing the evidential legitimation of each? Moving within a circle of sclerosed, objectified concepts never gives us an access to the real.

VI. THE SENTIENT, DIONYSIAN LOGOS

1. *The Human Entrance into the Planning of Life*

By the work of the human animus, now become a spirit, new avenues of life, precisely those of societal sharing, are laid out by humans. The attention of the animus is raised from the constriction of strict commitment to the business at hand to emergent matters of human creation that lift the interests of life to immeasurable heights — the human experience of the freedom of the spirit as it sustains culture. Sociability and culture not only intermingle but also are in many ways connatural, coessential. There is, however, a nuanced distinction to be made, a different kind of emphasis to be placed on each of these, which we yet succinctly show. But we begin with the explosion of new avenues of life that manifest the logos in the sharing-of-life.

The crucial reason that brings about the radical step forward in the spread of the logos of life lies in the essential transformation of sense, which, now that the operational apparatus of the Human Condition is crystalized, receives new endowments: from the vital precoded sense of life we ascend to a specifically human significance of life. With this transformation of sense there surges three novel, utterly original kinds of reason/logos, those of sharing-in-life and their patterns of relevance.

2. *The Communicative and Dionysian Logoi of Sharing-in-Life*

The Stoics in their conceptual framework distinguished between the "unspoken" logos, *logos endiathetos,* and the "spoken" logos, *logos prophoricos.* We may refer the first to those previously discussed rationalities that articulate the vitally significant constructive unfolding of individualizing life. In fact, *logos endiathetos* was conceived by the Greeks to be "singular," constant, and "transcendent" to what we call "constitutive consciousness"; it also defined the rationalities residual in the entelechial code as well as those that are residual in the process-driven unfolding of life — which encroach upon each other in its continuity. In contrast, in the specifically human significance of life it is, as we showed in the first part of this study, the vehicle of the "dual" or "bipolar" logos that establishes the unifying links among living beings enabling the societal sharing-in-life and the embarking upon and continuation of inventive, creative activities, and this corresponds to what the Stoics called the "spoken" logos, *logos prophoricos,* which, as they saw it, combines thought and sound. However, we wish to emphasize rather the intrinsic meaningfulness of this logos, which may be "communicated" — being "communicable" — in utterance (written or spoken), which is to say that here logos has *intelligible* form.

But while the Greeks with their refined cast of mind already distinguished between the *logos prophoricos* and *logos endiathetos,* they did not seem to discern the uniquely sentient attunement of the ways and modes in which the animus in living beings binds and then puts asunder, fuses, prompts, diffuses, etc. the ties between and among individuals, convivial undertakings, enterprises, projects, cooperative work, which attunement plays a leading role in society.

Not only is it true that to a great degree the underlying layer of the societal logoic interaction is sentient/emotive/passional, but the impact of that layer often overrides, outweighs the force of the strictly intellectual stratum. Neither did the Greeks seem to distinguish the specific mode of the logos of human invention/creativity or that of sentient/societal linkage. Societal reflection, deliberation, and planning are not motivated by the elementary or subliminal passions of love and hate, but neither are they dominated entirely by intellectual reasoning and its concatenations in planning or decision making, not even in the fulfillment of societal obligations. Springing forth with the sympathetic symbiotic orientation of the animus, the *societally operative logos* maintains throughout its virtual congeniality with the sentient-emotive circuits, and it cannot be propitiously applied in human interaction otherwise than by attunement proportioned to the sensibilia of the partners in transaction.

Indeed, the societally operative logos does not consist of articulations in accord with the synergetic deployment of the vital code; it stands on its own rooted in the sympathetic symbiosis that ultimately refers to the sensibilia of the soul. Inherent to it also is a virtual call for passional attunement. Between the great human passions love and hatred there is an infinitely expandable spectrum of spontaneous inclinations, appetites, dislikes, inclinations and revulsions that enter spontaneously — unreflectively — into the logoic articulations.

To distinguish the specific passional, opaque, unprevenient nature of the societal logos, we call it Dionysian reason, in contrast to Apollonian clarity and luminous intellectual certitude.

3. *The Middle Ground — Basic Patterns of Relevance to the Cosmic Logos*

Although the specifically human style of life, societal life, is an outcome of a meaningfulness that, as it stems from the specifically human virtualities is autonomous with respect to the animal significance of life, this style maintains relevance to the style of the biosphere, and this on three major accounts. As in the primogenital unity-of-everything-there-is-alive where the proto-operations in life's unfolding are precoded with a core of individualizing synergies and virtualities, so in societal life there are now autonomous cooperative modes of coexistence, which we will call "sharing-in-life." Again, these will be prompted by the animus, although now in the specifically human phase of life the animus will reach its fuller, even fullest, expression.

Moreover, in discerning the organizational patterns of sharing-in-life, we will discover analogies not only with the patterns of the biosphere but beyond those with patterns of the cosmos that are relevant. In appreciating human societal life, which in its organization and structure manifests an almost radically novel spectrum of rationalities, we must begin by realizing that here again the great cosmic laws are manifested, being transposed to the way in which human beings settle down in organized groups with the intention to cooperatively conduct life together. The organization of physical settlements around a center, perhaps orientation towards a place of worship, toward higher forces or powers, is a manifestation of the gravitational powers of macrocosmic ordering and of the directional orientation of each living being as a center with circumambient conditions. We find this transposition of central orientation in societal organization around a chief, a shaman, a leader, a director, a teacher, a president, etc., though each of these manifests quite different rationalities.

However, when we consider aspects of the societal other than its vital and psycho/organic significances, we find that though these remain foundational, society sets off on its own and departs from the prescriptions of natural behavior.

Let us point out to begin with that the general tendency of the specific spectrum of rationalities with which the societal world sets out and unfolds essentially expresses the impetus of life to expand and unfold. In the extraordinary expansion of the human living being accomplished through the emergence of full consciousness and communicability, it is logical to expect that the human style of coexistence that would, first, be able to apply and accommodate these unique accomplishments and would, second, be apt to promote them and promote the further expansion of human life had to be prompted from within the very entelechial form of the human being. That is to say that this passage from the vital/gregarious to the moral/societal significance of life has to be acknowledged as a "natural" step in the evolution of life. And yet, although it manifests in this perspective life's push onward and its virtualities for transformation as well as the fertility of the Logos of Life, the innovation introduced by the surging of the human condition with its inventive/imaginative core and its apparatus of conscious mental faculties is with respect to the style of life hitherto flourishing radical; it establishes an autonomous agency of reason through which the logos of life becomes profligate in invention and differentiates into innumerable rationalities, rationalities that do not serve directly to prolong the lines of vital existence, directions the sense of which is not necessarily bound up with vital existence, directions that do not manifest cosmic or biocosmic laws and rationalities. They do promote life, but having their own significance, they stand out above life and in certain cases seem even to contradict life's interests.

4. *The Dionysian Logos and the Relevance of the Vital Unity-of-Everything-There-is-Alive*

There is a deep down logoic patterning in societal sharing-in-life. In the new register of sense that the manifestation of the logos of life bursts forth with, inventively projects, we witness the surging of an outgrowth of the vital ordering of life — although one *sui generis* and irreducible to its vital layers. This novel register does not cut its primogenital ties with the vital logos. It keeps them alive in the transformative forge of the soul. This is now a human soul, the crystalizing agency of the Human Condition. The new register of sense — that of the Dionysian logos — has its birth in the creative forge of the human soul. This register in accomplishing a revolution in sense undergoes a corresponding revolution in the ways, means, foundation stones that make order. As a system of order sustaining the continuing process of the specifically human, that is, creative, constructive, advance, it calls for and exhibits in its manifestation ordering principles or points of reference that should serve as guarantees of relative stability within the turmoil of gregarious/societal life comparable to those of the synergies of the intellectual code. The latter bring forth an appropriate selection of forces with propensities that serve as footholds and directional forces for the process.

With the advent of the Human Condition, which for its own specific sake breaks away from preordained footholds in a code and goes forth, there still must be something specifically human, forceful, lasting, active, and ordering for human commerce to hold on to, to rely on.

Anticipating the genesis of the Dionysian logos within the human creative forge, we will propose human virtues to be these footholds for maintaining order and the unity of sharing-in-life.

CHAPTER TWO

KNOWLEDGE AND COGNITION IN THE SELF-INDIVIDUALIZING PROGRESS OF LIFE

I. THE SETTING OF THE INQUIRY

First, I intend in what follows to introduce and to establish an essential differentiation between the concept of "knowledge" and "cognition," or rather between the notions that they express; secondly, I aim through this differentiation to clarify the self-individualizing constructivism of life.

In fact, these two notions coincide to some degree, which leads to pervasive confusion in their usage. As will be shown, their distinction, which may be clearly drawn only with respect to their basic functions within the unfolding of life, yields new insights into man's situation with respect to Nature and Life, since it throws a clarifying light on the progress of this latter. And, as I have been declaring for some time, no perspective within which to situate philosophical issues is more urgently needed by contemporary philosophy than the thus far completely ignored perspective of Life.

Consequently I propose to treat knowledge and cognition within the field of my now fully elaborated phenomenology of life.[1]

Indeed, though Occidental philosophy has produced in its long history a great wealth of approaches, methods of inquiry, and insights into knowing, cognizing, feeling, sensing, and thinking as functions of human life specifically, and though with each of them there has been introduced a new bias in the treatment of these particular functions, we have now at last reached a point from which we may approach them full-fledgedly, in an unprejudiced fashion, obtaining a definitive clarification.

We have, in fact, only now, at the junction of philosophy and scientific inquiry, found the clue to the labyrinth of the intertwined puzzles that the universe of man poses to the human mind. The creative act of the human being — from the realization of which all the human faculties flow, and which binds together the spread of their constructive operations as the specifically human circumambient world of life is constituted — *is the Archimedean point from which at once the human world, the perpetual making of life, as well as all the human reflective approaches to life can be unfolded.* Indeed, as I have voiced in my phenomenology of life, the creative act of man gives us the long awaited access to not only the springs of human reality but also to the field of the generating and unfolding of life at large.[2]

As a matter of fact, as I have presented previously,[3] the creative act of the human being deals ultimately with the rational profiles (filaments) of generating objectivities. Thus, as the fulcrum of the entire generative functioning of the human being, who not only pursues constructively an existential course but also endows it with a uniquely human significance (meaningfulness) of his/her own invention, the creative act delves into the life generative functioning preceding the specifically human circuits and partakes of them while establishing the specifically human world of life, human reality. Hence it offers us an access to both the field of life at large, the immeasurable spread of the all-engulfing turmoil of the unsurveyable cosmic drama, so elusive and otherwise inapproachable, and the specifically human universe of vital-sociocultural reality with its own innumerable, infinitely transformable differentiations.

As I have been showing in my previous work establishing the phenomenology of life as a new, pioneering field, *it is with the creative act of the human being that we discover the key factor of all ordering: the self-individualizing progress of living beingness.*[4] It is through the self-individualizing progress of living beingness that the surging Logos of Life expands itself through endlessly varying rationales, streaks, streamlets, conundra that join the elements of the gushing life forces and channel them into segments of processes; these infinitely differentiated rationales establish links among happenings, connections among synergies, taming the wild impact of erupting forces and so serving life's progress.

In this interplay of life energies — which in a continuity of discrete moves expand in all directions, like the spread of a spider web, in the constructive advance of life through self-individualization — both knowledge and cognition play an essential role. And therein their respective functions are sharply differentiated.

In the above-outlined context, in establishing what roles knowledge and cognition share and what roles are particular to each, I aim to bring some clarification of the progress of life's expansion of the Primeval Logos.

II. THE CALL TODAY FOR A NEW INVESTIGATION OF THE ISSUES CONCERNING KNOWLEDGE AND COGNITION

1. *The Issue of Knowledge*

The issue of knowledge with which we are dealing here emerges recurrently at different stages of Western history. Its various formulations express the phases of the cultural development of humanity and belong to the sociological study of knowledge as well as to philosophy proper. It has challenged several great philosophers to seek a new foundation for knowledge by reinvestigating man's cognitive powers, the very possibility of cognition, as well as the nature and role of science, while reconstructing explanatory edifices. The most striking instance of this stimulus crystalized in Leibniz's efforts to establish the notion of a "universal character" as the ultimate rationale present in all types of cognition: rational, empirical, spiritual, ethical, and religious.[5] When once conjoined with a "combinatorial system," this universal character was to function as a "universal science" — *mathesis universalis* — and reconstruct the primordial unity of all types of human cognition, knowledge, and scientific endeavor. The enormous growth of modern science with reference to strictly rational — mathematic — principles may be seen as a fulfillment of the Leibnizean enterprise. But the development of the sciences since the time of Leibniz remains a recurrent challenge for philosophy. It is enough to mention Kant's avowed concern with obtaining the rational legitimation of empirical science. A similar concern survives in Husserl's early preoccupation with the foundations of mathematics in principles of cognition. Responding in his later development to the so-called "crisis of the European sciences," which had become overrationalized and had lost their ties to human life, Husserl attempted a large-scale investigation of the lifeworld as the transcendental source of human cognition. In this he competed scientifically with another attempt, that of neopositivism, a school of philosophy led by thinkers like Hempel, Nagel, etc. whose efforts culminated in the *International Encyclopedia of Unified Science.* Both of his attempts at seeking foundations — the former proposing that they are in eidetic cognition and the structure of reality, and the latter that they are in the interweaving of the lifeworld and its transcendental source, upon which ground all scholarly and scientific pursuits were expected to rediscover their unifying links — were meant as two complementary aspects of a *mathesis universalis.* They have inspired the thinking and culture of the Occident (and beyond) in the first half of the twentieth century. However, the unexpected differentiation of modern science and technology, which results in the singularization and specification of the cultural lifeworlds, lays bare the roots of their differentiation. The Husserlian project fails to fully account for the new vistas opened up on the primordial human condition. On the one hand, it circumscribes too narrow a territory to be able to reach the entire spread of singular differentiations, a spread vaster than the one that Husserl's universe of thought encompasses. On the other hand, in his ever renewed attempt at classifying his approach, Husserl set too rigid a framework for investigation to do justice to his own insights.

The basic insights and principles of Husserl remain valid, but call for a new, deeper, and ampler analysis.

In fact, since Husserl's days, that is, since World War II, the immense growth in the technological sciences and the growth of knowledge generally — which has led to a narrowly rational specialization in methods, approaches, and outlooks — seems to have so completely severed the disciplines' ties to each other that no communication appears viable. Consequently, the human universe that the various sciences mirror appears to be broken into pieces. The human being, who refers to scientific and scholarly knowledge for his personal views of life, the world, the meaning of his existence, and his destiny, and who from the coherence of various types of knowledge weaves for himself a lifeworld texture — which is the texture of his existence (the texture in which he finds the meaningfulness of his undertakings, of his triumphs and defeats) — has lost a stable and consistent foothold within a universal lifeworld pattern. The lifeworld pattern itself appears to be disintegrated, chaotic, and consequently devoid of meaningfulness. The search for a basis on which to reestablish communication between and among the fields of human knowledge and to interpret that knowledge in its universal human significance (as shared by every culture in its distinctive way) becomes increasingly urgent. The search for a cogent human world is of paramount significance for contemporary life. Although the terms of the argument between neopositivist empiricism and Husserlian classical phenomenology will remain recurring points of query in philosophy, yet, with the progress of the social situation of knowledge, they at present retreat in importance before impending new formulations that the social situation of knowledge makes pressing.

In fact, these developments compel philosophy to attempt a turn particularly profound and rich in promise. The disintegration of the traditional image of the world and man does not leave him in an empty void or on sterile ground. On the contrary, against all the voices of pessimism, we must assess the appearance of vigorous germinal forces. Technology exerts an irresistible influence and attraction over all present-day cultures and challenges each to raise within itself an entire network of questions concerning the meaningfulness of man's existence within his lifeworld. Each of these cultural lifeworlds meets this challenge in its own distinct fashion. This confrontation opens a singular perspective on the universal human condition.[6]

In the first place, we are compelled to seek a clarification of what we understand by "knowledge" at its very source, namely, its origin and role in the constructive system of the world of life within which the human being establishes a specific type of existence. This world of life is envisaged as the field originating together with living beings of which the human being is just one species. It is the appearance and development of technique — technique uniting the forces of the cosmos, the system of life as such, and the creative apparatus of the human being — that calls for such an elucidation-in-depth.

In the second place, we must keep in mind that knowledge has also a specifically human aspect and source — even if that be one among many. This source, which plays a special role in the constructivism of the world of life, is human cognition.

2. *The Need for Clarification*

Our proposition that we seek knowledge in the virtualities and mechanisms of the organic processes of life appears startling. Do we not always see "knowledge" as being embedded within the significant context of a "knower" and the "known"? Does this context not imply a knowing "subject"? And yet we also have reasons to distinguish between knowledge and cognition. While knowledge is characterized by universality, communicability, objective validity, etc., human cognition is, in contrast, uniquely singular, incommunicable in its singularity, most intimately performed, and the factors that it constitutes as a state of affairs are, as it were, personally owned, strictly personally valid. It is true that these striking differences between what we understand by "knowledge" and "cognition" point to the aforementioned crux of their seemingly radical distinction. In current philosophical usage, cognition, for its being accomplished in *subjective* processes of the human mind, is in its source and in its performance a process going on in and accomplished by a single mind, by one human subject, whereas knowledge is the *objective* yield of the cognitive process, a yield comparable in all human subjects and, thus, universally valid.

We cannot dispute the subjectivity of cognition. What we will submit, however, in the next section of this chapter, is that what we call "knowledge" is not restricted in its essential features to the harvest of cognition. On the contrary, it is pervasive within the entire spread of the life system's constructive individualization. As a matter of fact, already accomplished life functions deposit what we understand to be "knowledge" and store it in readiness for virtual activation should an appropriate situation occur within the life process. Given this, knowledge is the crucial factor working within the universal progression of life. It is indispensable and present at all stages of unfolding life. Not only does it mark the actual or past progress of life, but it also represents the capital gain of life's constructive advance — upon which further steps in its progress may be based. Ultimately — in a metaphysical perspective — we may see knowledge as the rational filament through which the Logos of Life projects its main paths and insures their being blazed.

In the perspective of the constructivism of life, cognition as we will argue in the next section here, appears in its essential nature as life's specifically human vehicle; through it the Logos of Life expands into the societal/cultural constructivism. Cognition's source in human singular experience — experience that is by its nature uniquely intimate to the subject performing it and that presumes the subject's self-enclosure and separation from everything else — poses again special issues.

It is the present-day disconnectedness of the scientific disciplines and the call for some grounds on which to reestablish communicative links and the urgent call for cultural communication among nations that brings the question of the universality and communicability of cognition — and of knowledge that it yields — to prominence. Where should we seek the principles of universal validity for the specifically human knowledge obtained in the singular unshareable experiences of singular subjects?

To answer this question we will in the next part of our reflection delve into the nature of the "originary experience" of the human being, which lies at the heart of the specifically human self-individualizing progress. In doing so, we will set in relief the way in which knowledge obtained through human experience expands the self-individualization of beingness into self-interpretation-in-existence, into a unique human script.

III. THE ROLE OF KNOWLEDGE IN THE ADVANCE OF THE LIFE INDIVIDUALIZING PROCESS OF NATURE

1. *Knowledge in the Individualizing Progress of Life*

With the progress of contemporary science and technique the essential characteristics of knowledge seem to break out of the narrow "knowledge, knower, and the known" framework. We discover the presence of knowledge seen in its essential features as "spontaneous evaluation," recognition of the "fitness" of matter to be absorbed or ingested, for example, "know how," the triggering of "appropriate responses," etc., within nature's life processes.

These and other essential features that we attribute to knowledge constitute guarantees that life's processes will advance from one stage, circuit, or phase to the next in the functional progress through which living beingness delineates its self-individualizing course. I have already anticipated this argument in my study of memory and reason ("Memory and Rationality in the Ontopoiesis of Beingness," *Phenomenological Inquiry,* Vol. 13 [1989]), which pointed out that the three main features that we usually attribute to memory — recording, depositing, preserving and retrieving — are not necessarily functions reserved uniquely to the human mind, or to any mind for that matter. To the contrary, the deposition of a datum of significance — life-significance — emergent in the occurrence of an event or of a process as well as the preservation of this newly emerged datum in such a way that the next process or event will profit from it in the form of an intrinsic *proficiency* for taking a step forward, for advancing the constructive process a segment further, are present at all levels of life within the self-individualizing process of life itself and prior to the advent of the specifically human phase of life.

Now, let us consider in what the nucleus of what is deposited at each functional step of life's constructive progress consists. What is, for instance, deposited in the surging of the sap of the maple tree at winter's end as the first step towards taking up the active life process again after dormancy? The tree does not exhibit the central directing agency of a "mind," and yet its growth obviously depends on deposited proficiencies that have been preserved throughout the dormant state. In the growth system of the tree there is a trigger that at a certain point of its functioning unrolls a most significant driving force. This driving force is simultaneously informed as to what stimuli among circumambient conditions (e.g., moisture supply, climatic conditions, etc.) to respond to. It is informed by devices that will orient it appropriately toward the stimulation of renewed activity in the entire growing system, ending the system's hibernation. This force is also endowed with virtualities contributing to this stimulation; it surges then in a measured way so that there is such and no other quality of sap, and such and no other quantity. This proportionality in endowment indicates that the entire growth system of this individual tree is informed about the complex interlinkage of a tree's functional proficiencies, on the one hand, and about its own existential working system with its innumerable and most complicated existential dependency on and interdependency with the universal workings of nature, on the other hand, in such a way that it can *communicate* in its functioning with these workings' stable and changeable factors.

Strikingly enough, the terms that I am using in this simple example of a single vegetal operation to talk about the "know how" intrinsic to life's proceedings involves already terms like "communication," "information," "response," which we normally use within the knowledge, knower, and the known context. This crossing of the border between the subject related realm and that of "objective" Nature will be justified in our further reflections. We have now to substantiate more this usage as we move from a single operational complex to the consideration of the entire self-individualizing progress of living beingness.

In fact, when Aristotle, struck by the fact of continuing individual development in the life process, investigated "coming to be and passing away" through "generation and corruption" and formulated the knots of passage from one stage to another in the metaphysical concepts of "potentiality" and "actuality," he was dealing precisely with the phenomena of life progress that we are dealing with here: the discreteness and continuity of the constructive advance. This advance calls for the virtualities and proficiencies intrinsic to the living individual's unfolding. This brings us to conjecture a self-promoting, entelechial schema.[7] In fact, it is through a rational profile (filament) that a living being manifests itself within the field of life. And this profile, this entelechial schema of individualizing beingness, projects from within its very course, in virtue precisely of the virtualities and synergies through which it actualizes the self-individualizing progress, information for further stages yet — in storage and ready to be triggered on an appropriate occasion.

The maple tree before being ready to perform the developmental step of releasing its sap has to germinate from a seed and only in progressive steps and stages does it attain the developmental complexity that allows it to hold in readiness the trigger for its vernal awakening. Life is characterized by this developmental continuity, which projects itself through the discrete steps of a process, each of them accomplishing a fragment of the entire constructive project and each depositing an item of information to be used at the next stage — all of which is initiated by a germinal informative complex surging from the "prelife" level. In this germinal complex the categorial differentiation of living types that is the cornerstone of life's ordering is already deposited.

Modern and, in particular, contemporary embryology and genetics bring to light and scrutinize the significant data that are stored within the life promoting complex (endowment) and that direct and partly predetermine the course of life's self-individualizing.

2. *The Human Condition, an Information Complex Deposited within the Evolutionary Progress of Life and Marking Its Turning Point*

The increasing complexity of types that mark the evolutionary progress of life culminates with the deposition in the life system of a complex most strikingly different from all preceding ones, the complex of informative, regulative, and directing data, devices, and virtualities that I have called *the Human Condition.*[8] It appears to have been spontaneously released within the advance of the life system, and yet its radical novelty constitutes one of the greatest mysteries that the human mind confronts.

This mysterious state of affairs is due precisely to the fact of cognitive knowledge, of which the human being seems to be the unique beneficiary.

In the initial stage of the human life course and throughout its progress, the spontaneous life operations, and even the processes that select means toward ends operate — in spite of the specifically human flexibility of choice — according to the availability of means in the individual's specific circumstances and by determining their of natural fitness. Nevertheless, the specifically inventive virtualities of the Human Condition expand under the stimulus of Imaginatio Creatrix and endow the natural operations of life's forces with new proficiencies. Hence, the life apparatus of what is now human beingness not only invents unprecedented means for meeting life's ends but also transforms life's very needs. It transforms life's very circumstances, indeed, creates new circumstances that Nature did not provide. This is owing to man's unique and, in the evolution of life, unprecedented functional apparatus. The human being is capable of the most astounding accomplishments, which single him out from the rest of living beingness. Instead of blindly assuming his part in nature, he is capable of "cognizing" life's regulations and laws as well as the modes of the interconnectedness of life's processes; he is capable of both discovering and grasping them. This "grasp" of the rules that govern the play of forces within which the human being coordinates his own efforts to stir and maintain his own developmental continuity of existence has three noteworthy aspects.

First, this grasp yields an overview of the territory of life within which the human individual comes to grips with the challenges of existence; it yields as well an appropriative projection of this overview into a realm that he experiences himself dwelling within.

Second, this grasp is an appreciative, an evaluative grasp that measures the proficiencies and modalities of the interplay of life's forces and the person's own forces in dealing with them — and measures all with a view to how they may be used to serve his life interests.

Third, the human being does not merely passively register or evaluate these forces; to the contrary, in virtue of the creative powers particular to the Human Condition, the person is capable of using his acquired knowledge to devise mechanisms meant to help him in his life struggle. These, he constructs and manipulates according to his needs and aspirations, which reach beyond natural constraints on his enactment of life. This means that the human being takes the initiative in depositing the knowledge he has discovered within nature's workings and that what he has gained through the creative exercise of his powers — his very own self-generated knowledge — is now harnessed into operational schemas that work within the processes of Nature itself. The human being enriches, as it were, the ingenious workings of life with devices invented for the benefit of his very own avenues of life. In short, by his very own initiative, the human being first gains insight into the primogenital knowledge deposited within the processes of life and then he expands that insight through his very own inventive faculties and so reaches through his own "artful" devices into the primogenital ordering of life at large.

With this we arrive at the central task of our argument: the differentiation of knowledge and cognition and of their respective roles in the unfolding of life.

3. *Knowledge, Cognitive Knowledge, and Intelligence Distinguished*

We can see clearly at this point that although there are essential analogies to be made between the significant data deposited in the life system and the knowledge obtained in human cognition, i.e., cognitive knowledge, there are also sharp and transparent differences between them. These differences manifest themselves in the different parts each kind of information plays in life's progress and in the differentiation of the Logos of Life that sustains that progress.

We have already pointed out that the human being in the course of evolution gained cognitive knowledge about the operations of Nature and about the processes of life and that he inventively uses this knowledge for his own benefit. The human being has continuously expanded his ability to work with life's workings. First, the invention of the simplest instruments expanded directly his physical skills (e.g., the axe, the pitchfork, the spade, the wheel). Then came the invention of more complex machines that go beyond physical utility and are the means of discovering Nature's workings (the telescope, the microscope, etc.). Thereby, knowledge of the cosmic forces is deposited in an operational schema and added to the universal processes of the cosmos. And if he can in this way undertake to himself enter into the progress of life, it is first because he is capable of obtaining knowledge stored within the system of life, of retrieving it. Second, he may retrieve it because among the singular instances of life's operation he is cognitively outfitted to grasp it. By virtue of the creative orchestration of his faculties, he is capable of surveying a number of life's elementary constituents, of grasping them in a unique synthetic fashion "in his mind," and then of extrapolating this synthesized survey into coherent and "objectivized" — that is, isolated from the empirical and intentional processes in which the above operations of the mind have been performed — universalized statements about states of affairs in his circumambient sphere, life, world, and himself, that is, he is capable of "cognizing," which means positing a meaningful universe of existence, a vision of the All, constituted by his own powers. This vision circumscribing his vital and inventive reaching "out" of and "inside" himself, with his vital/inventive/cognitive powers as its fulgurating center, firmly anchors the human being in the world of life.

We have arrived here at the radical difference between life deposited knowledge and cognitive knowledge. While it is the role of knowledge deposited by the life system to launch and maintain the course of life on preestablished tracks, and while the projection of new steps in that system still follows initial guidelines, the role and function of cognitive knowledge is to transform the life system. For the knowledge that carries the organic, vital, and psychic processes of life's progress reposes in these operations alone and does not double itself into a partial mirroring of its proceedings or extend to the furthest intelligible "reasons" and "implications" seeking explanations, while the creative spirit of human cognition does launch the concrete, singular life performance out to the furthest horizons of the cogent universe within which the human individual/person establishes his experienced anchorage.

It would seem that we have already arrived at an understanding sufficient for our present purpose of what is analogous and what is distinct in the knowledge intrinsic to the life system and in the cognitive knowledge. However, there is still one more unique accomplishment of the cognitive powers of the human being to be considered, one that strangely brings both of these realms of knowledge together again and then draws them apart. Here is the last leg of our inquiry.

We have so far been using the term "knowledge" in a rather vague sense. In a strict sense, what we usually call "knowledge" is precisely an objectified statement about a state of affairs that is posited through cognition; that is to say that in principle "knowledge" denotes the result of the specifically human capacity to perform a set of psychic operations, those of focusing attention on, observing, sensing, surveying, and synthesizing and objectivizing what happens around and within us, and thus extrapolating that into the intelligible form of a meaningful statement. With this last operation we abandon the intimately subjective sphere within which the psychic-cognitive process has taken place and reach the circuits of intersubjective, universal meaningfulness.

Indeed, the meaningfulness of life's, the world's, the cosmos', and the human being's rationalities is specifically cognitive. In point of fact, the individualization of life is rational in its outline, but it becomes meaningful only when this outline is retrieved in human cognitive processes and projected into statements about the respective states of affairs. With that the communicability that in its primogenital form resides already in the rational segments of life's progress takes on a new form.

Furthermore, with the establishment of meaningfulness as the harvest of cognitive activity, the human mind unfolds a new sphere for its creative/inventive power: thinking. Relying exclusively on meaningful data, the human mind develops new skills that expand the specifically human universe of life: calculating, computing, reorganizing, forecasting, planning, etc. Thinking, as the specifically universalizing function of the mind, deals only with meaningful, fully intelligible data and in a strictly intelligible fashion, a fashion that we call "intelligence." Out of this activity of human intelligence come innumerable inventive findings as well as ways of manipulating them — with consequences that pose the crucial test for our argument on the respective prerogatives of knowledge and cognition in the progress of life.

Through millennia the human being has been applying cognitively obtained knowledge about the rules, laws, and concrete proceedings of life's processes — the knowledge deposited within the life system — to his inventive purposes, to improve his own life progress; simultaneously, he has been unfolding inventively his own powers, of which intelligence constitutes the culminating point thus far. In our times a most striking development in this course of discovery/invention took place with man's invention of a machine that, going far beyond the knowledge operative in Nature, is capable of reproducing the workings of human intelligence. Considering the import of the creative powers of the human being in the workings of human intelligence, it is already a question as to what degree human thinking, the speculative system of the circuits of the mind, is a work of Nature and to what degree it is the work of human creativity. Calculating, translating, etc., computers invented by the human mind operate according to the rules that govern the higher functions of the mind itself and perform these operations on data proposed by the human mind too. These artificial, nonindividualized, and asubjective mechanisms, which may be infinitely multiplied, handle the givens of human intelligence with skills proper only to it.

Not only has the human being used the technical means he has acquired to interfere with the natural laws of life's progress, to regulate according to his own devices the most intimate individual life processes (as in organ transplants, procreation technologies, chemotherapy, etc.), to outwit space-time limitations (supersonic flight, space probes heading toward the stars), and to facilitate the handling of the necessary tasks of everyday life down to their minutiae, but he has accomplished all this referring at base to the rules laid down by Nature. However, he seems to have now reached an unprecedented point, having accomplished a feat that seems to make him the equal of Nature itself. Not only does he enter into the schemes of the rationalities governing the life system, but he also is establishing such self-directing schemes himself.

There are some basic realities that should be kept in mind, however. First, the establishment of self-directing operating systems is accomplished by intelligence understood as a self-centered agency. Second, these invented operative systems or mechanisms obtain their entire blueprint from the intelligence of the human mind. They prolong its work but do not invent on their own anything in a line different from that of the directions that were deposited within them in their programming. Third, the human mind, although it seems autonomous in its intelligent circuits, is nevertheless existentially embodied within an empirical psyche; its universally valid, intersubjectively communicable yieldings are ultimately embodied in the uniquely singular, unsharable, and incommunicable empirical operations of the human subject. Husserl, who gave the most thorough attention to human intelligence, showing it to be the highest circuits of operation of the subject's intentional system, never ceased to emphasize that all cognition and intelligence is at base the empirical, individual experience of the subject.

Not only are the most abstract rationalities of human intelligence processed through the empirical acts of the singular subject, but knowledge of the workings of Nature is retrieved and universalized into meaningful states of affairs by innermost empirically embedded acts as well.

The borderline between knowledge and cognition is thus drawn by the subjectivity of cognition, by its sources within the human agent on the one side and its obvious informative entanglements with the system of life on the other. The great question that remains to be considered is that of how subjective experience so uniquely singular in its performance and so evanescent can generate lasting, universally valid and intersubjectively communicable knowledge.

This question concerns also the problem of communication between different cultures and historical periods. The question is the more pressing as human intelligence now seems to outrun the natural system of life and escape its control.

In the brief statement that follows, we will attempt to trace the main lines of an inquiry that will yield an answer to these questions. It will shift to the subjective circuits of Life's progress.

IV. THE "ORIGINARY EXPERIENCE" IN A NEW ANALYSIS

In the context of the above basic insights, the answer to the above questions is already anticipated. Neither the level of the intentional structures of the objectivity, fruit of the cognitive rational faculties of man, nor, on the other extreme, the ever deeper probing into the genetic forms of their origin following backward the constitution of these very forms in the line of the transcendental genesis of man's rational consciousness — on the model of the Western, Cartesian mind — can claim to present the full-fledged meaningfulness of the personal life process or of the intersubjective lifeworld. This meaningfulness is that of the articulations, entanglements, and intergenerative modulations of all human functions as they take part in the process of man's lifeworldly existence — and "to exist" means for a human being to "individualize" himself from within his own, Nature's, and the lifeworld's conditions by working out his own unique existential route in terms of his operations: organic, vital, passional, sensory, emotive, intellectual, moral, aesthetic, spiritual, and religious. What else is this route if not the devising at each step, with each operation, process, choice, etc., of meaningful elements or of networks of interrelated elements that have become meaningful owing to their operative, constructive cooperation? What does this route amount to if not to the establishment of the individual's meaningful existential script?

With the notion of the "source experience" we may now probe the origins of this script and seek the unity of types of experience in the way in which they all enter into its texture.

1. *The Notion of "Source Experience"*

Source experience emerges in the investigation of the genetic descent to the simplest forms of the constitutive unfolding of human consciousness, simultaneous to and correlative with that of the human lifeworld. Pursuing this genesis from the complex structures of man as already fully developed, actually given, we reach its incipient point, the point at which what is primordially "given" to experience and the experience of it itself — as such — emerge. This incipient point appears as a *tabula rasa* upon which the individual, prompted by his surging spontaneity, inscribes progressively his existential route; "human nature" is being devised in this process. Here the system of individual life is "decoded"; it spontaneously emerges and works as a system of reference, as it functions as a watershed distilling and filtering the unbound virtualities of the initial spontaneities.

Husserl failed to account for the individual being's "inscription" into the universal pattern of life and living Nature. His penetrating analysis of the modes of the experiential unfolding of the human being identified restrictively with the unfolding of his conscious functions, was from the start vitiated by the *a priori* precedence he gives to rational cognitive experience. In Husserlian phenomenology all the other types and modes of experience have been approached through this channel of investigation and been treated as ultimately subservient to the sovereign rule of intellect or reason. Although ethical experience has been recognized in its own right as the decisive factor in the intersubjective world, it has nonetheless been, in the last analysis, subsumed under the rational laws supposedly governing all objectivity (reality). The entire experiential network of the genesis of human consciousness and of the lifeworld has in Husserl's investigation been reorganized, according to the implementation of these rules.

When we impartially envisage the originary point at which the human individual sets out to unfold his vital beingness, that is, how his life promoting reworking of the virtual givenness into a life schema proceeds — to be precise, how "experience" that expresses this reworking originates and is accomplished — we are seeking the ways in which the given (not a ready-made datum, but unbound impulses, pulsations, drives, strivings, longings, etc. stimulating the unfolding of the individual's sentient faculties) triggers the surge of the principal operative life-significant factor: attention. Following its direction, we seek the directives for its orientation. We are then led to the question of the origination of molds for the reworking of the given stimuli, as well as of the schemas according to which the sensoromotor functions come together in order to promulgate vital operations. Last, after the patterns into which the diverse sentient elements are coalesced, such a formulation of these questions already entails that the experience toward the emergence of which these basic vital and sentient orchestrations lead cannot be restricted to one, privileged mode (e.g., the cognitive), but is in its very nature plurimodal. In fact, the awakening of life consciousness in its simplest form — primitive and undeveloped in its functions — calls for the apparatus that would mold the given. This apparatus, however, can be neither too hastily presupposed as the system of rational ordering nor identified with the already developed life system of one single human lifestyle developed in a specific culture, singled out as the starting point of the genetic investigation. Least of all can it be identified with the system for which Occidental rationalism has been the leitmotif.

Moreover, the origin of plurimodal experience within the context of the individual script of man's self-interpretation in existence brings to light the emergence of experiential modes: operational, vital, operational-sentient, sentient-emotive, emotive, passional emotive, objectifying-cognitive, evaluating-cognitive, imaginative-emotive, imaginative-yearning, imaginative-objectifying, etc. They emerge as the first significant signs of a progressively unfolding alphabet through the play of which the human being devises his meaningful vital script while enacting his individualizing progress.

2. *This Succinct Analysis Leads Us To Distinguish Two Distinct but Complementary Understandings of the Notion "Source Experience"*

While one of them expresses adequately a stage in which the experiential schema of human life experience unfolds, the other is but an expression of a hypothetical limit case from which the constitutive genesis of consciousness would begin, but that could not amount to a full-fledged instance of experience. Thus, leaving it aside for the moment, I propose to turn to the first. What we may call "experience" is a specific configuration of experiential modes coalescing together into a unique type of experiential pattern. The emergence of this patterned configuration represents already a decisive step for the human life significant self-interpretation phase of the person's constitutive unfolding. Through the discriminatory selectiveness of its modes, it differentiates the simplest basic types of experience. From each of these patterns evolve, then, correspondingly different genetic spectrums of experiential variations on the given type. The profusion of complex, nuanced, cognitive or spiritual or aesthetic or moral experiences, with or without sharp contours, have in their respective patterns their originary and representative basis, with reference to which they may be identified. Source experience, so understood, fixes in the flux of the genetic unfolding of man the point at which the functional development reaches a decisively regulative, constructive stage — a stage at which, in a spectrum of basic functional patterns, it lays down an operational basis and outlines the specific guidepost of the meaningfulness of man's self-interpretation in existence. These specific experiential patterns are at the origin of the differentiation of the moral, cognitive, aesthetic, spiritual, and religious types of experiences. Each of them filters an assortment of the given elemental spontaneities and simultaneously, by the constitutive preponderance of one of them, establishes its unique specificity (e.g., yearning spontaneities carry spiritual and religious experiences, whereas, in agreement with Husserl, I see in sensory perception the originary experience of all cognitive experiences).

Inasmuch as, in agreement with Husserl and Merleau-Ponty, I see sensory experience as an outcome of an intricate and complex set of operations, each of them rooted in turn in a complex network of vital, organic, and life processes, there arises the previously raised question of the constitutive descent into simpler phases of the conscious genesis of experience. In retracing the constitutive stages with Husserl, we seek after something that we consider a germinal experience — as if experience were something already prefigured in its kind even at an incipient moment of the conscious genesis. From this standpoint we must presume as a "source" or "germinal" experience the very first instance of a conscious act surging with the activation of human awareness.

We could conjecture such an incipient instance of "originary experience" to be the instance that Husserl called the "Ur-impression." He sought in it the most primitive — that is, simple — instance of givenness that the activated consciousness encounters as an "object" of its activation in starting its own self-constructive progress. Although the conscious receptivity may be considered at this stage "originary," inasmuch as it initiates the stimulus-experiential coalescence and differentiation between the diffused, "blind," "meaningless" vital operations and the meaning initiated product of focused attentiveness, yet we cannot consider it "experience."

In fact, the passage from diffused, "mute," "blind," motor operations and vital processes — that surge within our being as undirected and undefined strivings, drives of "instinct," pulsations, yearnings, longings, anxieties, and feelings — to experience, consists in their entering into a specific functional pattern. Different configurations of these patterns, consisting of a different assortment of elements playing various functional roles, operate as filters and molders for the elemental forces that they process. Thus the experiential pattern becomes a discriminating agency that makes selections, adjustments, intergenerative associations, and, in doing so, establishes objectifying and orienting directives. This pattern is tripolar. As the instrument of the all-embracing meaningful network of man's self-interpretation in existence, it has to coordinate the universal conditions of the life system, the postulates of the entelechial individual existential genesis, and the universal and specific conditions of the lifeworld (its constitutive conditions qua lifeworld as well as its conditions for a specific cultural situation that the given individual enters and evolves within). Ultimately, however, it has to account for the creative strivings of the personal human destiny.

The elemental spontaneities that we have distinguished cannot in themselves be an object of self-evident cognition. Only experience in its full-fledged rationalized pattern can. It is precisely the role of the experiential patterns (which we may thematize with eidetic certainty) that allows us to thematize these spontaneities by conjectural inference.

3. *Man's Self-Directed and Self-Oriented Agency as the Point of Reference for the Unity of Knowledge within the Human Condition*

From our analysis of the originary experience, it clearly appears that what we call "human nature," which in the history of philosophy was expected to account for knowledge, is itself not given but merely virtual. It consists, first, of a bundle of vital and specific virtual spontaneities; second, of the life system code of the living individual; and last, of the virtual experiential patterns that carry them constructively on, while devising the genetic advance of the living human individual within his lifeworld. However, even the "decoding" of the life system with respect to the human conditions within which the operational system exists indicates a self-directing agency at work.

Without entering into an analysis of the modes of operation of the self-directing agency (which I have considered elsewhere), it is enough to point out its entelechial scheme. The entelechial scheme presiding over the functional system of the self-directing agency projects constructively the existential route of the individual and orients it toward the actualization of a specific type of beingness. It functions as the system of reference for the patterning of types of experience, for the spontaneous coordination of the emerging specific functions, and thereby for the interrelatedness of the fundamental types of experience.

Each full-fledged experience brings into the common reservoir of man's knowledge new meshing with which to weave further the meaningful script of his individual, personal, and social existence. The weaving is then rooted within the conditions of the vital existence of many individuals, in which it finds its perfect coherence and unity. If such a coherence and unity are disturbed, as in pathological situations, human life and existence are distorted. Only at the level of man's existential confrontation with other adversely oriented life scripts or with sclerosed, inflexible conceptual systems, do we encounter radical discontinuities or seemingly irreconcilable adversities. Within the universal existential script expressing the progress of the individual and social life, experiences operating in their originary modes function in perfectly harmonious cooperation, like the loops that link into each other in a woven fabric. (We may also mention ecological research, which shows a subjacent interrelatedness between the rational and the empirical, the sociological research of Levi-Strauss, which shows the role of religion in structuring basic social order, and the anthropological inquiries of Mircea Eliade, which exhibit the interchangeable aspects of reason and spirit.)

On this analysis, experience, as has been pointed out earlier, constitutes the letters of the alphabet with which the script of human existence is being set; it also projects the order of the letters emerging from their functional interrelations as they enter into expressive formations. The meaningfulness of the concatenations into which experiences may enter, that is, their syntactic rules, stem, first, from the rules and laws of the universal schema according to which a virtual individual being works out his existential route beyond and in cooperation with the anonymous forces of Nature and the rules prescribing the possibility of living being's emergence. Second, this alphabet of experience and the rules of its meaningful formation into a full-fledged expressive/communicative system allow for several systems to emerge, each of them expressing a specific type of script by which man inscribes himself into the general schema of life and the concrete world. No matter how we intellectually diversify this common human world, the human being is basically a holistic unity of experience; thus the cellular unity of the human world. The person gives it its meaningfulness and chooses its significant factors from among others according to the experiential significance that he has established in his own life script.

Third, since already at the level of the source experience and then with the progressive unfolding of his inventive and creative powers, man is not only a self-centered, but also a self-oriented agent, it is in his own judgement, decision, and free exercise of his will that the significant points of his meaningfulness are selected. Thus his life script is not a mere "decoding" of the laws of Nature; on the contrary, his lifeworld is continually being changed and transformed according to his own judgement and tendencies.

Being simultaneously in uninterrupted contact with the primogenital laws preserving life's progress and capable by his synthesizing and universalizing powers of surveying life's overall situation, the human being has his hands on the steering wheel. Though false prophets tell of impending doom, the human being has the future of humanity in his own hands and has all the necessary means for providing for its further glorious unfolding.

CHAPTER THREE

THE CREATIVE RISE OF THE HUMAN SPIRIT

I. IMPETUS AND EQUIPOISE IN THE DIONYSIAN LOGOS

The play of sentient/emotive/passional forces prone to sudden outbursts, to surreptitious twists of sense, to flooding the entire experiential field and coloring all its circuits outside of the reach of our will's influence — all these impetus driven moves of the logos, call for appropriate measures if they are to be balanced against each other, to be negotiated in their pluridirectional tendencies. To prevent tearing the individual soul apart, to bring the living being through the chaos of its existential intertwinings, the logos of life secures selective functional patterns capable, assuming appropriate conditions, of reaching a stage of harmony, of equipoise.

We have first seen its harmonizing work in the subliminal soul, where the subliminal molding of brute vital moments is accomplished. The initial spontaneity must first differentiate, be differentiated, and be endowed with specific virtualities in an intermediate zone — which postulated territory we call the "subliminal" territory — in order that the functional progress and articulation of life and human being may come about.

However, the first fruits of the logos of life fully ripen here in the impetus pushing the living being toward an other and in humanizing contact.

Here is the release of the moral sense that, as we have seen, accounts for the sympathetic encounter between human beings; it is the humanizing transaction that makes the sympathetic impetus springing forth from the depth of the human soul the fundamental measure of social equipoise, allowing the social exchange of goods, and cooperation in serving social interests for the common good. It is in the subliminal realm of affections and passions that the roots of Ethos lie. The notion of Ethos is intimately linked with the subliminal factors of man's functioning that we in our naive attitude identify with "passions," "affectivity," "sensibility," etc.

The sense of the "common good" is, in the first place, the result of the logoic harmonization that surpasses diverse individual interests by bringing to them a measure of equipoise that is shared — or to be shared — by all.

In fact, it is the sense of the common good that grounds the rights of the individual in a social group. Social institutions grounded in the sense of the common good are the balancing response to the impetus surging from individual interests. Banks, markets, corporations, security exchanges, schools, factories, international agreements, etc. are measures for social equipoise amid the diversity of changing plans, attitudes, etc. and anxiety over the uncertain future.

These provisional measures guaranteeing relative durability, security, and stability in the lives of social groups, nations etc. in their onward rush and resulting turmoil have their counterparts in systems of government, of courts that dispense justice, of constitutions and bills of rights.

From the logoic prompting to seek defense against abuse by others, syndicates and labor unions emerge. Provisional measures for the uncertain course of life offer insurance, savings, pension, and retirement plans.

Lastly, but foremost in significance for the projection and maintenance of social equipoise, are the codes of ethics either secular or religiously inspired that are meant to outweigh the always present vital impetus to seek self-interest at any cost to others. These foster benevolent sentiment and promote the subliminal molding of the human psyche and prompt logoic forces counteracting human egotism, greed, ruthlessness toward others and keep alive moral consciousness with its highest criterion: righteousness.

However, the entire human social/moral universe would collapse without the pillars of the moral virtues so laboriously installed by the Dionysian labor as the depository of its achievements, as the groundwork as well of the most perfect manifestations of life.

II. THE CREATIVE RISE OF THE SPIRIT

The Passional Self-in-the-World As the Feeling Heart of Beingness — the Dionysian Logos

1. *The Logos of the Subliminal Passions*

The human universe, the fruit of ever new radiating forces, is not projected and maintained by the intellective synergies and the constructive network only. It receives its strongest impetus from the passional forces of the soul. In fact, while the intellective synergies find their outlet in constitutive/cognitive productivity, the aesthetic and the moral synergies that yield sense, the two other main lines of the virtualities of the Human Condition find fertile soil in the primal forces of the soul. In modern philosophy reason has been hastily contrasted with feelings, emotions, passions. But these feelings, emotions, and passions are properly contrasted with primal moves of the soul, desires, sensations, dark strivings, impulses, since they themselves represent the ever refined modulation, molding, modalization of the soul's primary and very own virtualities — molding, and modalization that follow from the introduction of the uniquely logoic synergies of the aesthetic and moral senses. The blind strivings of the vital energies as well as the virtual synergies that are the soul's very own flow into the creative forge of the human condition and on encountering the dominant logoic synergies are molded into subliminal passions. Generating energies and synergetic forces of their own, these flow into the arteries of the intellective apparatus making it vibrate with the pulp of life.

What would the intellective skeleton of the structuring operations amount to without the fulgurating power of these subliminal passions that not only give it the innermost impetus of human life — sharing, trust, compassion, devotion, love, beauty, sublimity, righteousness — but that in their intergenerative modalities are the vibrant ever renewed medium of human personal and societal experience.

It is marvelous to survey the ungraspable wealth of human personal/communal life as it is partly concealed and partly manifested in deeds and works of creation. This novel feeling modality of the logos of life has its source within the human soul. It is the human being who becomes the greatest processor as well as the greatest receptor of the inventive/creative outburst and proliferation of the logos of life.

It is within the human soul, its innermost, that this fulguration of understanding, com-passionate, and solicitous rays takes wing to fly outward toward the entire world of life, and it is within this innermost of the human subliminal soul that the moves of the entire world of life find an understanding, felt resonance.

Indeed, the human-self-in-the-world, the transformation station crowning the unfolding impetus of the logos, is the feeling heart of beingness. This *felt* modality, unprecedented in the unfolding of the logos, we call the Dionysian logos.

2. *The Dionysian Logos' Manifestation of the Vertiginous Currents of Human Action/Interaction. Their Inventive Networking and the Societal Edifices*

The intelligibility of human communication, its passional promptings and modulations of the background of vital survival needs, on the one hand, and that of creative strivings to advance toward ever more advanced forms of life, on the other, and most of all the passional promptings to share, to commune, to preserve and continue — all these, and innumerable other bearers of sense that are introduced by the triadic as well as passional logos — manifest themselves in the personal/interpersonal/communal spheres of the human universe of life. In fact, the intelligible universe of spheres of objective appearances, in which nature, the biosphere, other living beings, and human beings all appear as our sphere of life with the cosmos as its horizon, would remain an empty shell without human action/interaction. As it is, we stand spellbound for these networks constitute our existence and the essential strings upon which our functioning is suspended. These strings are like those that make puppets move, and we have both to respond to them and master and command them through the efforts of the inner life of our soul. The communal/societal life in which we are lodged is the main subject of the soul's inner discourse, the motor of its development.

The societal universe, if we may call it thus, that emerges with a human group, that unfolds, collapses, and is resuscitated from scratch after some immense interval by another group, is the very pulp of the human universe of life.

3. *The Action/Interaction That Crystalizes in the Harmonization of Attracting-Repelling Propulsions*

The modalization of the passional synergies leads to an innumerable diversification of the main passional propulsions that ultimately then fall into a spectrum of prompting-retaining, attracting-repelling, joining-holding back tendencies. These contraries fall along a vast and qualitatively diversified and nuanced range of fluctuations. How does the harmonizing of these diverse tendencies that enter into the attracting-repelling moves, their joining, cooperation, acting together, or interchange of action occur?

We deal here with several elements and perspectives on the inner logistics of the logos of life. First, in order to see how it is possible to harmonize such opposed passional tendencies as sharing and holding back, solicitude and envy, generosity and greed, forgiveness and vindictiveness, etc. as they spring forth from the passional synergies, suffuse judgement, and even dominate it, we need to recognize that the passional logos, the Dionysian logos, brings the judgmental apparatus to a vast expansion and transformation in its emotionalization.

4. *Moral Judgement and Its Triadic Pattern. Interest: My and Your Communal Interest; the Moral Sentiment That Harmonizes Oppositions*

To anticipate in some detail our analysis of the origin of the societal spirit, of its passional synergies and the foundation stones of the societal moral order, all of which brings to prominence the human transaction as the logoic key to the entire societal life as well as to the moral order that is its nervous system — and of which virtues are the foundation stones and righteousness the criterion of measurement — it has to be brought out that here as elsewhere the societal universe emerges from human transactions and reposes upon their harmonious deployment. Nevertheless, the harmonizing of the passional interests keeps that universe in a passional turmoil that is in perpetual quest for equilibrium. While such social equilibrium seems to be gained when we have a prosperous society at peace with its neighbors, further developments will show that it has, even in this seemingly harmonious state, already been corroded by deep-seated factors that will at an appropriate moment erupt with violent force.

And yet one of the strongest types of synergies that the unfolding of logos brings about is the societal type of synergies emerging from conflict situations. Competition — concurrent attempts to attain position, secure benefits, obtain power and influence, insure individual, family, group, and national preservation — prompts ambitions and galvanizes human passions, all of which brings about uncommon energies, endurance, perseverance, the withstanding of hardship, and the making of sacrifices.

It is of the nature of the social universe that passional conflicts of interest undermine the innermost process of the social universe with perpetual crises, upheavals, turmoil so that it is during short periods of "peaceful" equilibrium that the logos of life harmonizes interests and conjures up its most powerful engines of creative advance.

What are the foundation stones of this unstable social/moral line of progress that is ever on the verge of collapse — with individuals and societies being ever challenged by the possibility of annihilation, sometimes in natural cataclysms, but most often in conflicts in which people annihilate each other — and yet can and does pick up again and continue on its shaky but steady advance?

There is an underlying pattern upon which a judgmental decision is made by which the multiple pairs of opposites are harmonized, a decision to surmount the tensions in the harmony of an agreement and subsequent societal action/interaction.

In fact, underlying the selective process of every and each societal deliberation we find the logoic triad of interest. Deliberation and selection is intuitively directed first by consideration of one's own interest in making one or another decision on a matter. If it is a matter of choosing between two or several possibilities, it seems simple for the person involved to consider the choice's positive or negative sides. But when it is a matter of deciding on an opinion, we have always to consider that our first tendency is to form an opinion according to our own views, tendencies, convictions, and, lastly, even remote interest.

However, in both cases my own interest cannot be fostered without consulting the interest of another, either another involved in the matter or someone who contributes to my deliberations by having another possible point of view. Here a harmonizing logoic principle is needed: *the moral sense.*

Lastly, however, after due deliberation over one's own and another's interests and an agreement, the moral/societal decision cannot be made before the interest of the community of which we are a part is probed and taken into account.

5. *Moral Conflict and the Specific Synergies of Societal Action/Interaction*

In consideration, deliberation, and decision making, there enters the entire functional system of the individual human-self-in-its-world-of-life, and this with all its vitally significant relevancies. That means that not only our vital/survival interests but also our empirical desires, predilections, dislikes, animosities, etc. enter into the attitudes that we assume in "consultation." Furthermore, considering the scarcity of goods, opportunities, privileges, liberties, honors, etc., on the one hand, and the similar or frequently even identical interests prompted by need, desire, ambition, greed, on the other, that individuals, groups, institutions, and nations seek to further, there arises more often than not a variety of competing tendencies and hence individual and social conflicts. Therefore, the basic pattern of the interest triad is by no means a strictly intellectual/rational pattern. Since the interests for which it stands and that it means to distribute stem from and are carried by vital needs and desires, and by subliminal passions as well, and since both of these enter into the deliberative and decision-making considerations as inclining and appetitive tendencies, there arise unavoidable passional conflicts among individuals, groups, and societies that are bound to create upheavals and turmoil. With all passional forces playing their respective roles in them, all social conflicts are ultimately moral conflicts; the basic pattern of the interest triad is ultimately a morally charged passional pattern. There cannot be a morally neutral social distribution of interests since when conflict erupts passional tendencies surge that bring in the moral sense and the entire moral apparatus to deal with them.

This triadic pattern is instrumental, first, in human transaction's taking off and developing innumerable societal collaborative networks (even though they are disrupted occasionally by failures to harmonize the antagonizing forces).

But most significantly this basic logoic pattern at the foundation of human societal inventive/creative genius prompts societal institutions: governments, associations, religious institutions, educational systems, economic institutions, scientific organizations, recreational groups, etc.

In the midst of the conflict between attracting-repelling forces, the societal sphere of life is prompted further and further by the human genius, and, deprived and/or freed from the action of the entelechial regulative system; it seems to float upon the waves of often conflicting interests among nations, institutions, and ultimately individuals. Yet it thrives amidst recurring crises and through defeats. It thrives on the wings of the creative human spirit. It thrives creating, evolving, and maintaining its specific significant features, its peculiar style through an incessant companion effort, the effort of cultivation. It is this latter that brings with itself the virtues that are the foundation stones of human freedom in moral life.

6. *The Intellective Platform for the Generative Works of the Human Logos*

The striking generative proficiency of the logoic triad discussed above manifests itself in the way in which the incipient move of the intellective act concurrently prompts three major lines for the surging of the entire logoic apparatus of the human mind. In fact, without prompting them and being able to avail itself of their services, the intentional act, while it could perhaps get off the ground, would not reach its concrete destination.

First, there is the mind's constructive/constitutive apparatus for establishing objectivity in foundational experience, which we have already detailed in the main.

Second, there is already inherent in perceptual structuring activity, the indispensable inference that another way by which logic may function is called for. Indeed, it is necessary for the objectifying enterprise that there be available appropriate constructive principles and regulative rules. Furthermore, the exfoliating activity of the intellect may well imply analytic procedures of dismembering, discrimination, and differentiation, but distilling from that a distinctive structured content calls for the selection, choice, and decision making necessary for a synthesis. The apparatus of judgement surges to fulfill this role.

Moreover, we have to emphasize that while initiating a novel modality of life the act of intellection brings in a new system for spacing and scanning, the spacing and scanning of the intelligible universe of the human being; the spacing and scanning of the human spirit is further corroborated by the judgmental work of the mind.

III. THE AXIAL ORDERING OF THE HUMAN UNIVERSE-IN-ACTION: OPENING THE HORIZON OF THE HUMAN FREEDOM OF SPIRIT

1. *The Emergence of the Judgmental Apparatus in the Mind's Triadic-Axial Processing*

To accomplish the constitution of the intelligible universe, the intellective act prompts a new constructive set of synergies: the judgmental apparatus. That ramifies into three kinds of ordering principles — categories, ideas, and concrete primogenital sense. It is through the latter that the judgmental apparatus is intertwined with the primal/primogenital circuit, while it is with the categories and ideas that it brings forth with itself that the apparatus projects a novel specifically intelligible organization of the human universe.

In contrast to the rigid code prescribed by the entelechial schema for the organic/vital individualizing progress of life, the judgmental apparatus introduces great flexibility in the way in which the three kinds of ordering are applied in this novel way of concurrent singularizing and universalizing while shaping the building blocks of the intelligible universe. Furthermore, out of the cooperative functioning of the vital primogenital/constitutive-judgmental apparatus, a new intellective synergy surges: the synergy of the human will, which is indispensable for prompting the operations of judgmental activities and leading them to their final destination, their passage into full human action. The will is the crucial guarantor of the judging subject's innermost and uniquely own decision making in judgmental deliberation, selection. It determines itself by this judgmental function following the promptings of the will — a uniquely personal factor. It is owing to judgmental activity that, first, the universe of life is universal for all human beings and uniquely its own for each of them and, second, that this new self-individualization/self-interpretation of the living individual in his ways and avenues of existence is not only performed by itself but also by his/her own innermost choice, decision, and self-determination.

The judgmental apparatus performing the essential functions of the mind differentiates itself into three modes of functioning that follow the specific factors of sense, three virtualities of the human condition that infuse sense: the intellective/categorial, aesthetic, and moral factors of sense.

Whereas the intellective factor is implemented by the categorially informed judgmental function as it singularizes (individuates — which is not to project) intelligible objects and introduces the order of objectivity, the aesthetic sense prompts the judgmental function to explode space and the intelligible objects of the mind, and its procedures account for the spirit's soaring toward a seemingly limitless horizon. The moral sense allows the judgmental function to foster a system for the moral ordering and deployment of its own synergies of societal (and intimately personal) interaction.

Thus, the judgmental apparatus of the logos operates the major move of the intergenerative passage of the mind from the organic/vital to the inventive/intellective circuits of the living being's life course and projects in ever new realms the specifically human universe of life.

2. *The Mind As the Watershed between the Organic/Vital and the Intellective/Spiritual Significance of Life: Judgement, Categorization, and Ideas*

As we have already pointed out, the judgmental apparatus of the mind carries with itself its structuring and regulative principles. While the categories direct the structuring of constitutive objects as well as of intellectual meanings, it is ideas (transcendental ideas?) that regulate the singularization of objects in the constitutive process itself, endowing them concurrently with universal meaningfulness.

It is with the categorial deployment of the judgmental function that the passage between the organic/vital and the intellective/spiritual significances of life — with no denial of the first for the sake of the second — occurs. Here the energies, forces, synergies of the first flow into the work of the second.

Having sketched the origination of the primogenital transcendental experience from within which the synergy of the act of intellection bursts forth and then having traced the basic primogenital triadic subject-object act that constitutes the crucial processor of the triadic-noetic logos, we have now to go back and apprehend the passage from the primogenital transcendental experience in which only a specific intellective sense is distilled from the otherwise vitally significant manifold to the full-fledged intellective experience in which an "objective" meaning is formed. Husserl in his most extensive studies of this passage stressed the nature of this meaning. Whether we consider the passage in either his static or his genetic phenomenology, emphasizing either the constitutive or the contemplative aspects of the experience, there is still a constructive/structuring work here on the part of the acts of intellection that can be seen in the generation of this meaning, and at base this passage that fixes the intelligible universe of the human being occurs along with so-called "empirical perception" (either inner or outer). In this passage an entire apparatus unfolds that accomplishes the differentiation, discrimination, and selection of elements that previously were only spontaneously, hazily, and passively sensed and in a way that replaces the phantasms of animal perception with a "clear" and "distinct" form and yields intelligible sense, a clear-cut presentification within the emerging field of consciousness. Not only does the constitutive apparatus then call for the precise means for performing these functions, for the framing of the intelligible sense and its structuring, and for a specific mode of scanning ("inward" to the subject), but these means in turn call for structuring principles for particular objectives, for an overall set of principles for differentiating objects, that is, specific principles for spacing them within their own realm. With this we have outlined the platform of primary intellective experience, which yields objective (eidetic), permanently valid meanings. But it remains to be seen, first, how these meanings — intelligible bricks — may be differentiated among themselves, and second, how they acquire their permanent (or quasi-permanent) validity. Further, it remains to be seen how they can be the building blocks for new avenues of life, which amounts to seeing the manner of their spacing/scanning in the advance of life. However, where in primogenital experience the distilling of the intellective sense occurs through a subjacent vital spontaneity of the central agency that is already on the verge of transformation and there is there no properly intellective synthesizing effort involved, we may say that the synchrony of the various moments significant in the intellective sense's emergence is in part an effect of the stimulus that comes from the virtualities of the human condition under the aegis of the inventive/creative imagination, and that, in part, it proceeds from the factually relevant vital instinct. This, however, is not yet a clear and distinct mode.

The passage from this sense that merely flashes by to clear and distinctive meaning calls for deliberative activity on the part of the subject, for constitutive acts, on the one hand, and constructive principles, on the other.

In the bringing together of selected appropriate moments of objective constitution — that is, of valuation, selection, and decision making, in short — the act of synthesizing must follow in order to complete the formation of meaning.

In the history of philosophy, this valuating, selecting, and decision making activity has always been called judgement. Already at the very first steps in which the logos of life enters its human phase, the perceptual activity in the subject-object intellective act triad involves the emergence of the judgmental apparatus as the key factor of the human mind. The judgmental apparatus in molding its acts draws on the three above-mentioned sense giving factors of the human condition. It is the categorial, that is, the intellective sense giving factor, that provides the key element in the basic platform for the new enterprise; it brings forth its own means for framing perceptual, and hence all other, full-fledged intelligible meanings, what Aristotle calls 'categories.'

a. *The Categorial Triad: Substrate-Property-Relation*

In the constitution of an objective meaning a fundamental categorial pattern of mutually contrasting and yet intrinsically complementary elements, moments that form an organizing pattern from the manifold of primogenital experience, emerges with the judgmental functioning of the mind.

Three moments that imply the necessity of their mutual completion initiate judgmental functioning.

The repetitiveness of the experiential content of primogenital experience gives rise to the move to grasp it. This move entails already the act of intellection's tendency to analyze, to differentiate and discriminate. It calls for selection, articulation, and a form/axis around which objectivity may be crystalized. To this call responds the surging categorial functioning of judgement, which is oriented along the lines of the categorial triad. By attributing properties to the substrate of experience, an objective structure framing the objective meaning of the thus distilled experiential content is articulated. Selection, the valuation of relevancies, and attribution are synthetic operations indispensable for the formation and positioning of objectivities.

With these primary moves of the human mind, the judgmental function takes off.

b. *Regulative Ideas in the Singularizing of Objectivities*

While categorial judgement operates the synthesizing function on the lines of the substrate-property-relation axial pattern that forms objectivity as such, it effects the singularizing/concretizing of the activities of the intellective act in the constitutive process by bringing forth ideas that constitute a system of relevance and by setting them up as a constitutive a priori that yields an attribution appropriate to a given experiential constant.

It is in relation to an appropriate idea that the acts of the experiential/perceptual constitutive process discriminate and select the constitutive moments that yield an intuition of essence in the judgmental synthesis that results in a meaningful object: a singular objectivity that is supported by the subjacent concrete manifold of experience and yet endowed with a universal structure of meaning.

c. *The Infinitely Flexible Modalization of Consciousness As Such: the Modalization of Intellective Acts*

As the mind deploys its intellective apparatus, it not only projects novel structuring and connecting networks but also molds the acts involved. Elsewhere I have discussed the subliminal/poetic molding of the soul's passions. Here we broach the question of the molding of the acts that belong to the establishment of the intelligible universe of human beingness. The modalization of these acts gives to the workings of the mind the vibrating flexibility of conscious operations; it feels, as it were, full three-dimensional space; it accounts for "conscious life."

In the first place, we have to emphasize the already mentioned lived evidence that gives to the primogenital act of surging consciousness its "validity." As completed by categorial judgement, lived experience is, on the one hand, molded into a positional modality that makes of it a spaced and scanned "mark on a living map," a foothold in reality thus being ascertained. But the ascertaining activity of the judgement involved in the positioning of an object differentiates into contradictory, mutually opposed modalities in which the existential reality of the object positioned is ascertained. From the range of modal variations in which the positional act ascertains the status of reality, be it dubitable, probable, certain, possible, etc. along a vast range of gradations, only one is actually ascertained (although vacillation among them in the ascertaining makes for the vibrancy of the reality of experienced life). Further modalizing of the positioning acts of consciousness/mind open on different estimations of the realism of human reality (empirical, ideal, fictitious, actual, etc.). The modalization of conscious acts prepares the means for the deliberative weighing of scales of measurement, a matter to which we will return later.

d. *The Predicative Judgmental System that Generates the Intellectual Universe of Thought and Promotes Communication and Human Interaction*

Although the primary move of the mind brings in categorial judgement, the indispensable completion of the constitutive apparatus, and accounts for the positioning of the objectivities of the intellective platform, it does not stop at that. It is actually just preparing the groundwork for its next conjoined step, one that springs forth from the first one. This next step consists, first, in the analytic exfoliation and explication of the meanings obtained through categorial judgement. In taking it, the judgmental apparatus expands into multiplying, modulating, intergenerative, and inventive acts and schemas for acts.

In fact, the inventive logos has antecedently installed in the categorial triad its subsequent move. The triadic pattern of the categorial performance, the substratum to which appropriate attributes are assigned for the constitution of a meaningful objectivity implies another modality whose performance is triadic, that of subject-predicate-assertion. The assigning of the appropriate attributes to a given substratum is performed already by a predicative assertion. That is to say, a logical mode of a synthesis asserting the predicative relationship is indispensable already for the accomplishment of the categorial synthesis of the substratum and its predicates. In other words, each categorial judgement underlying an intelligible meaning is convertible to logical judgements about its composition and structure that are asserted in terms of a subject and its predicates.

With this new turn the judgmental activity releases its implicit purely judgmental, or "logical," sense, which has existential autonomy and is independent of experiential sense. "Logical" sense stands entirely on its own, independent of "subjective" connotation or reference to a singular experiential act. Furthermore, it moves from judgement about intelligible objects to that of the "state of affairs" (in the Husserlian understanding). In short, it gives objects complete universality and becomes a vehicle for framing experience in universally asserted meanings that are communicable.

Cognition obtained through the intellectual intuition of the intellective act turns into communicative knowledge.

e. *The World of Life, the World of Communicative Interaction-Interchange among Living Beings in the Unity-of-Everything-There-Is-Alive*

The constitutive activity of the constitutive triad is completed by the work of judgmental activity and extends its own constructive paradigm that frames empirical, one-time experiential content into constant building blocks, varying networks of intergenerative and connective patterns, links in the form of processes, events, etc., and then, with reference to this platform, the judgmental apparatus sets out with its own generative powers to expand these networks of generative and communicative linkage into an entire universe. The perceptual intellective act is thus completed by the activity of judgement. Consciousness now moves beyond perceptual constitution to a versatile state in which its intentional thrusts are multidirectional. They spring forth like the rays of the sun infusing light into all the arteries of the mind, into all minds around. With its own "logical" sense, the mind, informed by the inventive/creative synergies of the Human Condition, generates an ever transformable plethora of objective meanings, networks of interrelations, etc., which are projected by individual human persons in their existential communication, prompting and promoting interchange and interaction among themselves. An intelligible interhuman universe thus blooms.

This involves also existential attitudes and interaction with the animal kingdom and vegetalia. The inner-outer world of life thus emerges in the unity-of-everything-there-is-alive.

f. *The Inner Universe of the Soul*

Even as there is constituted from within a world of life without, there is built up the conscious/reflective inner universe of the soul. This universe of the life of the soul, to which we return, unfolds its own innermost virtualities; it captures the creative virtualities of the Human Condition, setting up a "creative forge" for processing its very own innermost yearnings, dreams, transcendent strivings into the web of a "subliminal reality" where the creative imagination spreads its wings freely.

Then, in yet another move, the soul initiates its unique move toward the "understanding of All."

g. *The Vortex of the Centripetal and Centrifugal Orchestration of the Vibrant Intelligible Universe As It Manifests Itself; the "Ego-complex"*

Just as the living branches of a tree spring forth from the trunk and keep close to the juices and energies supplied by it, so the functional segments of the agency of the mind grow from and keep close to the mind's "trunk." The mind is nothing without its supply of energies. They emerge from the fertile soil of the soul, engendered by the virtualities of the Human Condition. We have outlined the "branching" of the mind and indicated how in its "branching" the mind operates the radically new phase of the logos. However, in order to bring to light how the human spirit spreads its wings and soars toward the free sky, there are still some major turns of the new logoic, creative transformation to be assessed. We have to pinpoint the most significant transformation of the inward/

central agency of the living beingness that is now human into the vortex of the "ego." Here is the emergence of a novel and most powerful synergy. Furthermore, we have to acknowledge the criteria for a new valuation of the human significance of life. Lastly, a radically novel set of criteria for the spacing/scanning of human existence relevant to its personal interpretation has to be regarded, these criteria being those that aim at the "meaning of life."

h. *The Creative Human Agency*

It is fascinating to note that the logoic passage from one stage of evolution of living types to another is never a radical jump; on the contrary, it is always operated in a discrete continuity of processes and the new ensuing forms always assume what pertains to the preceding one.

So it is with the overall organization of living functioning. We have already distinguished the intellective triad of the primogenital act that harmonizes the manifold of experience in constitution, as well as in judgement. These two branches of the intellective apparatus usually called mind have now to be completed, and the completing factors will come to light as we try to grasp the novel creative orchestration of the functional agency of the mind.

Indeed, the branches of the intellective apparatus processing and generating the new avenues of life are differently orchestrated than are those of life in its organic/vital significance. That is to say, the central orchestrating of the latter, its centralizing entelechial agency with its rigid prescriptions is now transformed by the flexible system of intelligence. Yet this latter also requires an orchestrating factor.

While the organic/vital synergies and forces are distributed into various cooperative functions by an entelechial agency empowered by its own synergies, we must wonder whether the intellective transformation of the organizing agency does not release a uniquely intellective synergy to the same effect, one that corresponds to the valuative criteria of the vital significance of life and their role in the spacing/scanning of vital life. And we must ask whether there are not new intellective criteria for spacing/scanning in the novel creative significance of life that accounts for the human self-interpretation of existence.

To begin with the first question, a crucial transformation of the directing entelechial agency does indeed occur. The new inventiveness in constitutive procedures as well as the great flexibility of judgmental functioning and its unlimited invention in the proliferation of meanings do not allow of a rigid organization; on the contrary, there is now a most inventive orchestration among the various functional branches of the mind. This orchestration refers to a pivot, which in modern philosophy we are used to calling the "ego." The intellective constructivism of life does not follow haphazard fragmentary directions; on the contrary, it is itself organized around this pivotal complex, the ego. Ignoring all the disputable points concerning the constitution of this pivotal center, we see it first manifested in its concurrent drawing on all the functions that it centralizes — even as it distributes itself in them. All functions are "empowered" by the ego in their performance; all enter into collaborative intercourse referring to the ego's pivotal role; all represent this pivotal complex in their enactment as if they were governed from this central source of command. In this relation the ego appears as a vibrant, constant, and ever nearer undefinable moment that is translucent in its undetermined nature as an absolute unifying power. However, it draws its power from the synergies and forces of the performing functions themselves. Its unifying role is not to gather the circuits of operations unto itself. On the contrary, it in a centripetal fashion sprays them "all around" like rays. Indeed, we first stretch, expand, reach out, tend out of ourselves with our acts, and thereby we constitute our universe "all around" the ego that becomes its center. We live in a universe in the shape of an infinitely expanding globe. But the ego also lives in all of its acts; thus, they centrifugally "return" to its core complex.

The ego's main characteristic, though, the one that endows it with undetermined translucency, is that in dealing with the triadic intelligibilities it acquires full awareness of their significant moments, awareness also of a triadic mode: the significant moments of the intelligible are received first by the ego-complex in their meaningful content; second, these become an intentional object; third, they take shape as the ego-complex's "own" enactment. In this fashion the ego gathers and distributes awareness throughout the intellective apparatus; it is the "light" of this latter seen as "consciousness." It is only by abstracting from the concrete functional system that we may with Husserl speak of the "ego pole," or with Kant of the "unity of apperception." In our perspective on the logos manifesting itself in the full-fledged constructivism of life, the ego as the pivot of the intellective advance of life appears to be the central "power station," dispensing power as much as it draws it from life performances. However, there would be a major difference between the phenomenological (and transcendental) conception of the ego and that which emerges in the logoic manifestation of life. The ego as the pivot of intellective construction — the central factor of the mind — carries in full the inventive/creative synergy of the Human Condition, which it gathers from the performance of the mind, crystalizing it, and it is from this source that its crucial power flows: the synergy of the will. Its power and its synergetic propensities are informed by the development of the entire functional schema, but it springs forth from the creative forge where brews the forthcoming functions that crystalize in the Human Condition, and it unfolds with the deployment of these functions.

It is as it is empowered by the emerging will that the pivotal role of the ego complex takes off putting the entire system of the constitutive-judgmental-acting schema into motion. It is the will that loosens the hold of the vital/organic system and its inertia. The sphere of human action takes off.

i. *The Will's Introduction of the Creative/Inventive Valuating Criteria of the Human Significance of Life and Its Kairic Spacing/Scanning*

The will as the specific motor of human self-interpretation/self-individualization-in-existence introduces through the ego-complex novel creative/inventive criteria for the valuation of human life and consequently a novel set of principles for its spacing and scanning.

Without abandoning the vitally significant criteria for the valuation of life, the will, as the innermost arbiter of the human mind in the manifestation of the human spirit, sets as criteria for valuation the will to undertake, to understand, to believe, and to accomplish.

We may see how the entire span of specifically human existence consists in man's continually repeated efforts — springing forth from his very innermost beingness — to follow the promptings of the will in these directions; we see this also in how the human being evaluates his or her life by these criteria; lastly, we see how undertaking, that is, understanding, believing, and accomplishing, space and scan occurrences, events, processes, and their intertwining in a truly kairic mode in the course of the human personal interpretation of life.

We measure the allotment of time for activities according to the standardized time unit that is universally valid in society, but we scan our personal interpretation of life in terms of our efforts to undertake, our understanding of life's filiations, our belief in ourselves and others, and our accomplishments in our personal ventures.

j. *Truth: Measure and Equilibrium*

The question emerges of whether the apparatus of the human mind that promotes the human differentiation of the logos "spontaneously" unfolds the human universe, that is, of whether this universe blazes its own evolutionary path, shaping itself as it goes. In other words, does its order flow from the performance of the apparatus once it has been put in motion? If so, what would guarantee that it not go "out of order," that it not lose proportion?

Without a measure we could not envisage how the systems of relevancies that specifically human constitution obviously maintains with those other circuits of existence in which it is suspended could be maintained. We know that over the millennia of human experience and again and again in the earlier history of life on earth, cataclysms occurred and that they still occur, but that through it all the life system has not gone awry. It maintains its existential equilibrium. And so does the specifically human intelligible universe. The collective experience of mankind makes it manifest that there are several circuits of the human intelligible universe to which criteria of measure are being applied; each of these circuits is subjected to its own specific criterion since it manifests its own specific type of ordering and the criteria of measurement are applicable only for particular types of ordering.

The foundation stones of the intelligible order are intelligible objects and the acts by which they are constituted. It is their clear and distinct crystalization in constitution and predicative judgement that provides the fundament for all intelligible order. This clarity and distinctiveness is also the best criterion for the measurement of the intelligible manifestation of the human universe-in-action. But this criterion does not suffice to guarantee either the universal validity of singular perception or assertion — its agreement with the collective experience of mankind — nor does it suffice to guarantee equilibrium in the tortuous and tumultuous onward rush of the constructive stream of logoic dynamisms or equilibrium in the edifice of life meant to emerge from them.

We have pointed out above that the logos of life differentiates itself into three modalities that are most intimately interwoven in their enactment and yet promote three essentially different factors of sense proper to the Human Condition that subsequently account for the three main spheres of the human universe. Although each of them has for its ordering its own distinctive cornerstone and consequently requires a measure of its own with its own criterion as well as its own proportioning and deciding agency, yet it is the criteria of measurement and the proportioning agency of the triadic logos that provides the indispensable vector for their performances. The criterion of truth and compliance with it traverses all the deliberative, selective, and asserting functions of the mind and includes the constructive progress of all circuits of human reality. The ego-complex to which all circuits of life's functioning refer functions as the center of ontopoietic equilibrium; it reads, as it were, like a seismograph, all the disturbances and deviations from good proportion and then dispenses appropriate or equitable portions for each life function. As such it is a scale upon which all is weighed. It performs this function ultimately through the judgmental apparatus. Thus, the basic criteria of the clarity and distinctiveness of objects and "ideas" (that is, of the intelligible meanings distilled from experience) as objects of predication and thought have also to comply with the first judgmental criterion of measure, the proportion of correspondence to ontopoietic norms. These latter are to be drawn from the sets of ontopoietic relevancies to the laws of the biosphere and the cosmos. To be considered still are the criteria of the specifically intelligible measure and equilibrium in what guarantees the specifically intelligible human-universe-in-action, human interacting. The collective experience and practice of mankind has recognized truth as such to be the pivotal criterion for human experience and action.

The criterion of truth on which the main emphasis falls in the human understanding of the world, life, other human beings, and oneself, as well as in interpersonal communication and collective affairs, has in its search for a measure been given several interpretations in the history of Western philosophy, which are too well known to be covered here. All of them have been concerned with the concordance of different kinds of intelligibility. In spite of this variance in understanding the concept of truth itself, philosophers and logicians alike seem to concur in seeing the "truth" of a singular intelligible instance as being decided ultimately by its being concordant with the entire context of the intelligible experience of the world and life praxis.

This concordance, however, has to be weighed, balanced, proportioned, and asserted by the constitutive and judgmental functions of the ego. Thus, the equilibrium on which this measure depends, the equilibrium of man's self-interpretation-in-existence (self-individualization) comes from the center of beingness.

The harmonizing/repelling elements of human life — in empirical experience, emotions, thought, and human interaction as well, which are so unsettled, constantly at odds in some way — necessitate a harmonizing equilibrium that is at one instant attained, at another lost, in order for each living beingness to keep itself afloat in the stormy seas of life. That truth pervades all of the types of measure that human intelligence employs is one of the great prodigies of the inventive logos: it rises and falls with the freedom of the human spirit that guarantees its dominion.

3. *Cognition/Knowledge, the Medium of the Inner/Outer Universe of Action; Knowledge in the Individualizing Progress of Life*

The double face of intelligence, its constitutive/cognitive logoic versatility is the crucial medium for the emergence and unfolding of the intelligible universe around us as well as of its construction in cooperation with other living beings. Appreciation and communication emphasize the cognitive/knowing face of the triadic logos. To bring into focus the triadic accomplishment we should, however, place cognitive knowledge in proper perspective.

IV. MEMORY, THE DYNAMIC AND FLEXIBLE BUT PERSISTENTLY "LYING-THERE-IN-WAIT" BACKBONE OF EQUIPOISE IN THE TEMPESTUOUS PROGRESS OF LIFE

1. *Introduction*

Bedazzled, like Kant before him, by his discovery of pure reason, Husserl indefatigably pursued its intentional system of constitution in its ramified structured acts and their concatenations. In this focusing on pure reason as it extends in one way or another through the entire functional system, the traditionally acknowledged "faculties" of the mind, if not lost from sight — Husserl but marginally distinguishes imagination, memory, and will — no longer had their proper roles within the great task of constituting the human being within its lifeworld.

Husserl stressed primarily the nature and structure of transcendental consciousness at large as if it were a global organization of human functioning that assigned some subsidiary role to imagination, memory, and will. In my phenomenology of life, in which not the cognitive but the creative performance of the human being is the point of departure, and in which not the already accomplished stage of life's evolutionary development culminating in transcendental consciousness but, on the contrary, the evolutionary stages of life's ontopoiesis are the object of scrutiny, we discover that the individualization of life is not an accomplished function of pure reason — not even of some a priori rational functioning agency. To the contrary, following the progressive stages of life's functional complexity, we cannot avoid differentiating functional traits, the performance of specific roles within the entire functional system of living beingness. Within the culminating phase of this complexity, in the creative act of the human being, we encounter these aptitudes in their fully developed proficiency. It is then fascinating to see how the specific functional complexes carry with them quite distinctive forces that they process from natural spontaneities and how these interact, interplay with, and sustain each other in the great task of individualizing life. But the individualizing life simultaneously draws on the networks of rationalities through which the Logos of Life manifests itself. So rationality is all along essentially present in the great game of life. However, if the specific functional complexes that we traditionally call faculties — which are being retrieved and reaccorded their proper status in our perspective — be clearly differentiated, then "pure reason," understood in the strong sense of the term (Kantian or Husserlian), appears as just one among them. I have differentiated the faculty of reason from imagination, will, and memory, calling it the faculty of intellect. It is then most interesting to see how in this great game of self-individualization or self-interpretation-in-existence of living beingness the different faculties assume respective roles in the common task. Because of the various points at which we encounter reason-rationality in the ontopoiesis of beingness, I will, while attempting to probe into it further than I have done so far, focus on one aspect of this interplay: the role of memory in the projection of overall rational consistency in the constructive progress of life, *and* on its encounter with the faculty of the intellect.

We always associate the rationality of the constituted world with reason seeing it as the human faculty that is meant to process it. Within the perspective of the phenomenology of life, reason, seen as the functional organ of the mind, appears at only one phase of life's individualizing evolution. I have discovered that life's formal constructivism is, in its progress, conducted in a diffused way, and does not as Husserl maintains in his analysis of constitutive consciousness flow from a central directing agency, which would be constitutive consciousness. This amounts to showing that modes other than reason lead to linkages, interaction, interfusing between, and the structuring of, operations and organs. As a matter of fact, I disavow the sovereignty of reason (as mind or pure rationality). I have established in my basic investigation "Moral Sense in the Foundations of the Social World."[1] that the operations and processes operative in the self-individualization of life are not directed by "pure" reason, that is, by rational deliberation and choice, but by valuation. I trace the role of valuation in the progress of life's establishment up to the complex agency of the mind that directs human self-individualization in existence and show that it is the originary factor of morality and of social interaction. Valuation, and not rational calculation, is the source of morality, therefore, inasmuch as it is in simpler types of life the decisive factor for constructive progress in growth.

However, this contrasting of the tasks of reason and valuation by no means isolates valuation and its workings from rationality. Valuation is in the progress of life's self-interpretation-in-existence the factor accounting for linkages, the interconnectedness of operations, of interrelations, fusions, etc. It projects the linkages of life's constructive progress. Are not connections, interlinkage, etc. the basic structuring elements? Do they not by bringing varied and yet relevant elements together project rationality as such? It is then obvious that the various modes of valuation that we encounter in the primogenital ontopoiesis of beingness introduce the modalities of rationality through which the Logos of Life manifests itself.

Anticipating our exposition, I will submit that memory in preparing the steps of the constructive progress of life's individualization is constantly working in tandem with the factors and forces promoting the crystalization of what we may call in a large sense "reason." It constantly deals with reason's elementary functioning. It is therefore unwarranted to draw a hasty conclusion, as some recent authors do, and to oppose valuation to reason per se. Such an oversimplification leading to this false conclusion comes from the improper perspective in which they situate their entire query.

While I reject this opposition of faculties, I do differentiate them, however, and I have found scholarly confirmation of my own differentiation of them in Edward Casey's work. His extensive and thorough investigations, first of imagination, then of memory, is in line with my own tendency to differentiate the faculties — which also goes against the spirit of contemporary philosophy.

2. *The Life Context within Which Memory Appears*

The metaphysics of life that I have expounded is grounded in — or translatable into — a protophenomenology, that is, an ontopoiesis of beingness that underlies any phenomenology of the name. The aim of this study is to probe deeper into the genetic progress of the living individual as it acquires the forms through which life will unfold and transforms them attaining greater complexity. I seek to discover junctures in this ontopoietic progress that I have left unacknowledged in previous investigations and to recover from the fragmentary evidence a complete — if at this stage, rather sketchy — outline of the inner functioning that carries living being throughout the entire range of its progressive phases, that is, from the incipient point at which life asserts itself through self-individualization in its simplest forms, through the progressive development of valuative flexibility, to the phase that culminates in the creative human condition of man.

In my chief treatise (*Logos and Life,* Book 1: *Creative Experience and the Critique of Reason* [1988]), I stated, not without surprise, that memory is present in the work of the ontopoiesis of life from the start. What better leitmotif for this present investigation than to follow memory's own development and role in this constructive progress?

We "discover" memory as a function usually when we are concerned with our cognitive processes. When we try to "re-member," to "re-collect," or to "memorize" an item or a fragment of something already "known," a datum of cognition about an event, a date, some piece of information, which we have "learned" and which we naturally assume has been "recorded" and "deposited" in some sphere of our cognitive apparatus and is there in readiness for retrieval at the appropriate time.

Three features, then, we naturally discover in and attribute to what we call "memory": recording/depositing, preserving, and retrieving. And since in focusing on memory, we are concerned with our conscious behavior, memory belongs to the philosophy of mind and its investigation is circumscribed by the functional circuits that this latter traces.

In whatever form we meet memory, in our cognition or in our behavior, it always involves these three basic functional moments: the *deposition* of a datum of significance that emerged in an event, process, act, etc., and that would vanish with this act were it not in some way recorded; the *preservation* of this significant datum, which makes the process from which it surged a step forward, an accomplishment; and, crucially, the *retrieval* of this significant datum from the passive state in which it was "lying in wait," as I have put it elsewhere. The proficiencies of the subliminal soul emerge then into the field of life processes. The role that these three features are obviously meant to fulfill is that of allowing for the connectedness of processes and acts, for the continuity of life.

A starting point is to be acknowledged. Do we not encounter such a significant deposition of data in the phases of self-interpretation-in-existence prior to the advent of consciousness? Within our new phenomenology in which the point of departure is the investigation of the tentacles that the human creative act spreads and maintains and that reveal the intelligible network of the living being's self-individualization-in-existence as the ultimate foothold of intelligibility, we see how this self-individualizing progress is advancing step by step by processes and events with an innumerable variety of modes; and yet if any of them is to contribute to the constructive progress, it is as a result of a significant datum that it "deposits" before vanishing, a datum that then may be — and in favorable circumstances is — a foothold for a new stage in the constructivism of life's progress. From the simplest processes satisfying the least complex cellular mechanisms for the intake and outgo of substances needed for life, through the complicated operations that lead to expanded flexibility in the life promoting schema, segments of operations, and processes that lead to the formation of functional systems and organs, the differentiation of organic types, and, ultimately, to their mutations, each operational step forward is made possible and safeguarded by the deposited life significant data yielded by the preceding step. It is through the retrieval of this significant data that succeeding steps of the constructive progress take off. This happens in a line of growth; in the reverse course of the decay that is the extinction of life, it is precisely the dismembering of the significant data stored within the significant life script of the individual and their dislocation and progressive loss of significant articulation that leads to the weakening of operations, the dissociation of organic functions, the vanishing of functional proficiency and of the dynamic flow that prompted it.

So it is with the organic circuits of functioning. However, when we consider the vital circuits, we see there an expansion of valuation, given the number of possible relevant responses to challenges as well as the range of relevant data that can meet vital needs. And we witness an expansion of the role of the significant data that surge and are deposited within the framework of the life system and become there the foothold for a succeeding step; they acquire a progress promoting role. These circuits are the basis for increasing flexibility in the selection process. Without their role in preserving significant data, the selection process, which needs a basic register of items to choose from, would not be operative. But for selection to take place the preserved significant data must be "retrieved" and reactivated; then they can be combined with additional relevant data.[2]

Advancing to more complex living beings, we encounter at the threshold of conscious behavior — and even earlier, already at the phases of the higher vital/psychic types of functioning — the valuation process that, as we have established, carries on the self-interpretative individualization of living being as it expands first into multiple choices and then advances from merely mechanical "instinctive" inclinations to some measure of felt relevance, preference. We encounter here a simultaneous "sensitization" or "adjustability" of the deposited significant data as it is retrieved by *inclination/preference.* In this phase of valuation, in which the mechanical schema of response hitherto made on the basis of the *fitness* of substances for contact/absorption/ingestion is loosened up and the relevance of the items offered for selection is ascertained by vitally-psychically significant processes, the deposited significant data that the selective processes operate with seem to be retrieved in a way analogous to the operations of human conscious memory. Furthermore, owing to the new flexibility of the selecting process, they seem to have also lost the rigid contours they had in the primogenital mode and, in a way similar to the conscious remembering of items of experience, they are revived and adapted to the variety of moments inherent in preference and to the relevant composition of the available items.

With one further step, we advance to the specifically human phase in which deposited significant data are of a determinate conscious nature and in which the conscious apparatus that we call memory in the strong sense is engaged and active.

Nevertheless, this differentiation of the stages of the entire course of life's individualization makes it manifest that, in its three basic features, what we call "memory" is present at all phases. Through its very presence in every circuit of organic, vital, vital/psychic, and conscious processes, what we call memory itself projects its context. Instead of being confined to the sphere of its highest accomplishment, which is the human mind, memory stretches its activity throughout all spheres of unfolding individual life and makes its context that of life's progress itself. We recognize memory, in fact, in the active function of the mind that we call "re-membering" — how appropriately this word is formed, the bringing together of disconnected members of significant experience — or "re-collecting," the collecting of dispersed moments. "Memorizing" is an active effort to bring together an entire segment of data and by conscious effort to deposit it and record it so firmly that it can be re-called into the active field of consciousness at will and in its integrity. We try to re-call significant data of experience from the oblivion into which they were sent to connect them with the present and to create a consistent pattern of continuity of experience and the world. Here, however, a crucial distinction has to be made, a distinction that will bring us directly to the heart of the matter.

The depository role of memory is, indeed, essentially related to the proficiency of deposition as such that allows this "in wait," passive, static, existential status of data, which can then be readmitted into the active flux of life. However, it would be an unwarranted mistake to restrict the roles of memory to the radius of the eminently active, conscious process of re-collection, re-membering, and memorizing.

As we have established, the meaning acquiring processes of life proceed by more modes than those of lucid human consciousness. Undoubtedly, it is in the conscious processes of memory that its active modality reaches its peak. Nevertheless, this highest degree of proficiency in man's self-interpretation-in-existence is preceded by and prepared for by several stages of the self-individualization of life, beginning with the primogenital condition and then a series of significant steps marking life's constructive unfolding. This self-individualization, as I have established elsewhere, proceeds on the basis of the significant data at hand. Their succession, their discrete continuity, carries life's progress. What is here at stake is the guaranteeing of the succession of appropriate significant steps and thus of constructive continuity.

With the recognition that it is the entire spread of the human functioning that establishes individual significant existence — self-interpretation-in-existence — and that this is not restricted to the works of consciousness, as is usually assumed in philosophy, we can find throughout all the phases of life's constructive progress the three features that we attribute to memory: deposition of significant data, its preservation, and its eventual retrieval in the active flux of life. We discover, in fact, that in its prototypical nature memory is the primogenital guarantee for the continuity of life's unfolding, for the constructive progress of life as such.

Having now gained the new perspective of this enlarged conception of memory, as a matter of fact, conceiving of memory as being as differentiated and pervading as life's self-interpretative individualization, we must admit that in its prototypic features (the deposition of data, its preservation even though it may never be called up again, and its retrieval in propitious situations), memory is, *before all else,* performing a crucial life function. It saves significant life accomplishments from being haphazardly dismembered and annihilated in the rushing, recklessly advancing currents of life. These threaten to tear apart all contexts and to engulf all in chaos. Not only does the salvaging role of memory counteract aimless, blind forces by installing the means for channeling them into constructive routes promoting life's unfolding and progress, but without this harnessing of the striving that the individualization of life requires, the Logos of Life could not unfurl itself and spread through the arteries of self-interpretative beingness. This salvaging also counteracts the effects of ruthless struggles between the individualizing beingnesses that in their self-centered striving would destroy the routes of life that others delineate for themselves. By salvaging, by installing channels for life's unfolding, this prototypic memory is meant to provide for this unfolding as such. By projecting the connecting links, it accounts for life's progress. It contributes in an essential way to the unity and continuity of becoming. Each step ahead is possible only if there is already installed a significant linking datum that can be activated and used in the operational process. Without this "discrete continuity" there would be no constructive progress, no life, and no manifestation of the Logos.

V. MEMORY IN THE CREATIVE ACT CRYSTALIZING THE HUMAN CONDITION

After this review of how memory in its three prototypic traits establishes itself within the context of individualizing life, from the first manifestation of the Logos of Life and through its entire progress, let us come back to the creative act of the human being, which was our starting point. As pointed out before, the creative act of man is the culminating point of life's constructive unfolding; here the Logos of Life surges in the invention of an infinite variety of rationalities; here the automatic decoding of the life script deposited by the system of Nature is relaxed and we as human beings may inventively devise our individual scripts. With our own creative powers, we may project our uniquely personal significant script.

The specific phase of the Human Condition dawns in the play of life. The full register of human functioning unfolds under the aegis of Imaginatio Creatrix and the three other faculties that emerge: will, intellect, and memory; and of these, as they work in tandem, memory is not the least important.

In this connection we have still to differentiate — although in actual performance they are to a certain degree interchangeable, or are even identical, even as they diverge at crucial points — between two ways in which the now full-fledged memory as a faculty of the conscious apparatus of human functioning, or of the mind, contributes to human self-interpretation-in-existence or, in other words, to the specifically human constructivism of life.

In our extensive investigation of the creative act by which the human being transmutes the vital forces into the specific functions and faculties of the creative orchestration, we have already pointed out some of the junctures in that act at which memory plays a decisive role. We will now present these succinctly while advancing to some further considerations. While memory's role in the creative orchestration as such is of capital importance, it seems to have thus far been ignored. Instead, the main investigations of memory have been in the realm of its workings in the already established and automatic performances of the human mind in the so-called constitution of the human world of life. I will here bring forth, then, and dwell on what I have newly discovered, the role of memory in the creative act and process. This perusal of the nature and role of memory will only then take up its contributions to the constitution of the human lifeworld and to the everyday conduct of life's business.

In my previously established system of the creative orchestration, I differentiated four faculties: imagination, will, memory, and intellect. These are indispensable to each other as each assumes its respective role within the "creative orchestration" of the emerging human type of self-individualized being (or self-interpretation-in-existence). While the will processes the vital spontaneities that provide the initial impetus and prompts and, what is more, dynamically directs the progress of the creative operations keeping them on a unique track, and while the intellect provides an ever varying assortment of structures for the possible formation of the creative idea besides assuming the function of structuring the experiential acts and their objectifying directedness, it is, nevertheless, another faculty that completes this work by keeping the result from vanishing with the fleeting moment and by keeping it "in front of us."

It is the crucial function of imaginatio creatrix to, first, explode the rigid codification of the organic vital/psychic beingness and open it to infinitely varied "images" as "possibles" to select from, and, second, to open up the rigid, mechanical vital/psychic valuation hitherto prevalent in the animal sphere and to give complete leeway for flexibility in valuating and putting in proportion whatever is relevant to imaginative striving by drawing on the deliberative functions of the intellect.

And yet, this mutually supportive apparatus could not function without the absolutely essential contribution of memory.

To begin with, without memory, the bursting out of imaginatio creatrix in the midst of the automatic and restrictive vital/psychic system would for all its fulgurating, seemingly unlimited, imaginative hints, prospects, images, etc. be so much combustion left unharnessed by an engine. These energies if not captured and maintained will immediately die out again. As it is, we are often given a start by new prospects, glimpses of ideas, and imaginings and are energized for only an instant because the vision is not sustained and vanishes. It is the indispensable role of memory to sustain this energy by capturing the host of imaginings that appear in the focus of consciousness — which surge together with the functional apparatus under discussion — and by keeping them in focus throughout successive experiential processes more or less vividly, but sharply enough so that the deliberative functions of the intellect can dwell upon them. Memory retains some imaginings vividly; others have a vague halo around them. But it also manipulates images by bringing them from the background to the forefront of the deliberating faculties. It retrieves original vividness or adumbrates an image when appropriate for deliberation, giving the deliberative functions space. First, and foremost, however, memory saves from oblivion imagination's first fulguration, sustains the energies released, and by the flexibility with which it proposes images and revives them in varying intensities, it contributes to the directional work of the emerging creative apparatus. Second, it is indispensable to the deliberative work of the faculties, for it maintains and retrieves the data to be deliberated. Thirdly, it is instrumental in the establishment of the primogenital temporal "horizons" of human experience, as Husserl put it. In fact, the advent of the archetypal creative act of the human being is simultaneous with that of the lived present — the triggering of the specifically human intense lucidity of consciousness in which the roving focus of attention marks varying intensities, gradations in fullness of experience, etc. Then opens the horizon of the past, the depository of what the roving eye of the lived present witnesses and makes translucent — for which witnessing the entire functional system has been mobilized. It is the role of memory to accumulate data and inventory it, so to speak, according various principles, one of the most obvious being the order in which things appeared. And so memory sediments the already fading and ever more faded moment with each new instant of the incoming present. But moments only seem to vanish from our sight because they leave behind the constitutive sequence in which they occurred. This sequence will be made operative with the recalling of some of its data when it proves to be relevant at some future moment. It is the role of memory, however, to not only maintain sequence in its inventory of past events, but also to revive them when the proper moment comes. Even if they will never be revived, memory holds them completely "forgotten" with their sequence and their thematic interlinkage with other moments — as an extension of our actual experience. This extension of experience in which the present may reach backwards Husserl called the temporal horizon of the past. And is it not in a countermove that memory of the past or of the just passing present serves as the springboard — along with the atemporal horizon of objective possibles — for the spontaneous projection of the future?

To summarize this portion of our analysis: It is with the coming into operation of the specifically human faculties that, with the human creative act, the vital-psychic powers of life acquire the lucidity of human consciousness; the now conscious system acquires both flexibility and a self-projecting capacity. The mere registering of the succession of events and processes by memory in prehuman spheres of functioning operates on the basis of satisfaction. This registering now acquires the translucent expansiveness of experience in the temporal horizons of the lived present, the past, and the future, horizons within which each present experience dwells.

The importance of the role of memory within the prototypic creative act of the human being as a signal event in the unfolding of the specifically human significance of life cannot be stressed enough. We discover also that memory is operative in the creative orchestration of human faculties. I have discovered and shown in an analysis of the creative or inventive process in all its phases that memory is indispensable: first, in maintaining the fulgurations of imaginatio creatrix, and, second, in re-proposing past images, ideas, etc. for deliberation in the forming of the idea of a creative work,[3] as well as in sustaining the entire process of the transition from the already formed creative idea to the realized work of creation. Memory is not only the sustaining factor of the trial and error search for forms — of its haphazardly unfolding continuity — but is also a prompting force in the progress of the search as data are reviewed for their possible significance for the creative construction of an object. Memory revitalizes the vividness of what is recalled for scrutiny; working within the constitutive networks it acquires subsidiary energies that allow it to scan the sedimentations of past experiential data for a datum appropriate to and required by the creative process. Without all this, the creative process would not take off.

We are now prepared to ascertain the manifestation of memory's capital role in the emergence and maintenance of the human lifeworld.

1. *The Role of Memory in the Establishment and Maintenance of the Lived Presence of the World*

Considering memory as merely a function of the mind is the traditional and current practice, without any envisaging of its manifold involvement with the significant processes of life's self-interpretation-in-existence. We seem to always attribute to it the past as its domain. Memory is meant to safeguard for us that which has already taken place, that which once occupied our attention in the spacious present and then spontaneously left the present field of consciousness as ever new phases of the living present succeeded each other retreating further and further from the luminous focus of attention into the darker zones of our conscious system. It is memory that is credited with keeping the receding data in readiness more or less effectively to be, again by memory, recalled when needed and to re-appear within the lived present of our conscious life. There, though it is an item or fragment that comes from past experience, it in some or other way may still compose with the present an extended world of life that is significant to us.

We are so bewitched by this maintaining, preserving, and retrieving role of memory that we naturally relegate memory to a role dealing with the past. We are then naturally led to attribute the giving to past moments the intense lucidity proper to the lived present to the faculties and functions of experience other than memory. As we know, Husserl and classical phenomenology stressed in particular the constitutive role of the intellect, attributing to it an extensive structural apparatus in which the time sequence of present, past, and future were under the intellect's constitutive aegis. In this perspective the human lifeworld, which Husserl revealed to us as the all-embracing milieu of the human being, was conceived as the fruit of the intellectually governed constitutive system.

Our new protophenomenology or phenomenological ontogenesis avoids the anthropocentrism that takes consciousness and the role of the intellect as its starting point. On the contrary, it takes as its starting point the protogenesis of life. It follows the phases of life's unfolding until the human lifeworld emerges and only then raises again the question of the specifically human lifeworld. In this essay I will limit myself to examining the living present as the pivotal functional moment in which human constitutive powers attain the completion of their current, automatic performance.

When we direct our attention to an object, we focus on it for an instantaneous moment. And when we reflect on this focusing, we discover that to maintain the object in focus we instantaneously repeat our focusing for as long as we desire it to be maintained. With each singular focusing of attention we may take in a different aspect of the object. We may move our attention from one part to another, and, finally, we may constitute the entire object at one glance. However, even though, as we know, there is a sequential order to the constitutive process, the focusing of attention by which we perform it is itself instantaneous. Focusing can only be repeated. It cannot be extended. It is not without reason that Husserl and Husserlians consider human consciousness to be by nature composed principally of acts and the stream that they form as they burst forth and pass away. However, the living present that manifests our conscious beingness is a "spacious present," as it has been called. Although its focusing is carried out by a series of instantaneous acts, nevertheless, it perdures. We experience ourselves in a lived duration of acts. Bergson grasped this as the phenomenon of "la durée vécue." Yet this lived duration is clearly only a phenomenon because once the instantaneous acts of consciousness are brought to a halt, it vanishes. The question emerges of whether this phenomenon of lived duration, of the spacious present, would exist were there not a particular way in which any experienceable gaps between the instantaneous acts are precluded so that no matter in what slow motion we would put our conscious activity, we would still experience the undisrupted duration of the lived present.

The point to be brought out here is that in this spacious present each of our instantaneous acts endures — just as the luminous circle projected by a lamp extends to darker but still illuminated zones till it dissolves altogether in darkness — so that the aspect of the object on which it focuses expands into further and further objective regions that are the realm in which the object dwells together with innumerable other objects, familiar and consistently unified regions that belong to our experience in such a way that we consider it altogether to be our "world." It surrounds us, but in it we also dwell, and to it we ourselves as living experiencing beings belong. This world is not only that which at any present moment we could clearly encircle in our imagination, but that which we syncretize from all that we have experienced, know, and can imagine. And although we could never at any instant of our lived present embrace its vast expansion even in imagination, still we feel hazily that in its contours and in substance it belongs to the instantaneous acts through which we carry out our conscious existence from moment to moment, existence in which, despite the utmost fragility, since each instant could be the last without appeal, we feel firmly established and secure.

Indeed, we are so firmly established in each surging and perishing instant, despite having no force sufficient to retain it, only because each of them is enmeshed in the canvas of the lived world into which it extends and that it revives and carries within us in its emergence. Classical phenomenology devotes scrupulous attention to the ways in which the constitutive consciousness — which within our enlarged context of individualizing life is the consciousness established by the creative orchestration of the human faculties and made automatic in order to perform the universalizing constitution of the lifeworld — structures and constructs the world simultaneously with the unfolding of its own potentialities. Memory certainly plays an important role in this work, which role we have anticipated in discussing the specific function of memory in the creative process.

What should be brought out here is the thus far not duly emphasized and yet striking role of memory in establishing and maintaining the lived world for us as our most intimate, familiar sphere, the inseparable, ever present milieu into which we extend, as the sphere into which our spacious present expands. In fact, I will submit that it is memory that, by keeping open the horizon of the past, organizes and maintains the lived world guaranteeing it its proficiency for taking part in the present and the future, for which it serves as a springboard — all of which accounts for the presence of the world within the spacious present. For, and this is of primary significance, it is memory that maintains the instantaneous acts within the objective sequences of experiential — constitutive — patterns, and in such a way that despite their surging and vanishing no experience of a gap between them is possible.

I have analyzed in detail the moment of the individual's setting off on the journey of the creative undertaking.[4] However, the "moment" of the individual's transition from the vital/psychic functional stage to the actualized Human Condition and its radical overturning of the hitherto active self-interpretative system of functioning still remains to be envisaged with any acuity. Within the new system that the actualization of the Human Condition — of its virtualities and their proficiencies — brings about, *I have considered the thus triggered Creative Act to be the Archimedean point for life's unfolding progress.* The long brewing functional mutations of every kind leading to this Copernican revolution within life itself elude concrete analysis. We may, however, on the basis of that revolution's effects approach it by conjectural inference as an atemporal moment, as a central moment gathering up the threads of the most significant developments that in their full-fledged unfurling become the specifically human functioning, bearing and distributing/bestowing the human significance of life to vitally significant situations, events, acts, etc., which are observable.

With the surging of the creative act as such — not this nor that creative process or act of any human individual, but of the primogenital, original creative instance — the human was brought into life's orbit. This was the moment at which the ever advancing in complexity living beingness came into its own, the instantaneous I crystalizing the human type of beingness in which all the respective functioning faculties assumed their essential roles: will, imagination, intellect, and memory. We may see from this schema that the faculties together conduct the course of the expanding Logos of Life, "Universal Reason." The new products of the specifically human individualization of life are three — will, imagination, and intellect — but memory has been at its work from life's first bursting forth.

TYING POINT FOUR

THE LOGOS OF SUBLIMINAL PASSIONS — THEIR CRUCIAL ROLE IN HUMAN SELF-INTERPRETATION IN EXISTENCE

CHAPTER ONE

THE PASSION FOR PLACE AS THE THREAD LEADING OUT OF THE LABYRINTH OF LIFE

I. PHILOSOPHICAL ANTHROPOLOGY, AESTHETICS, METAPHYSICS REFORMULATED

How intriguing it sounds, when we think of it: "inhabiting" the earth, "dwelling" upon the earth — as if the living being, the human being too, and earth were two separate spheres, the one of individual life, the other of an indifferent "entity," a planet. It is as if the living being could also inhabit or dwell elsewhere, on some other planet, in an unknown neutral sphere. We ask ourselves, indeed, whether there are other planets favorable to the existence of life, and other intelligent beings there. Does this mean that human being needs to "occupy space" for its existence? Does human being need to "have a place" in which to establish itself? Would then "space" and "place" be a middle term between the earth and the living being? Or, rather, would "spacing" and "placing" as such constitute one of the primogenital modalities of life?

This juxtaposition inherent in the current and old expressions ("inhabit" and "dwell") that put the living being — the human being — in contradistinction to its sphere of existence, its territory, its soil...raises, indeed, some basic questions of life. Here is the very question of life's foundation, grounding, elementary conditions, on the one side, and its most complex manifestation, freedom from them, on the other. We will attempt to enter the subterranean labyrinth of this intricate and enigmatic situation through the investigation of the human passion for place.

It is a labyrinth truly. We could say that the Minotaur that we meet on venturing into its meanders is another human passion, that for philosophizing. As if it were not enough to approach the issue of this most intricate phenomenon of inhabiting, dwelling through the elemental passion for place that human being exhibits in all its dealings with life and that comes to have striking manifestations in our own reflection on it — feelings expressed in science, literature, the fine arts — it appears that we cannot in our quest find repose in any one of them; to pursue it, we are drawn into the heart of the labyrinth by our philosophizing reflection, of which the three fields usually called "philosophical anthropology," "aesthetics," and "metaphysics" reveal themselves to be the proper fields for reflection. In them the entanglements that seem otherwise to be loose ends are tied together and the elemental passion for place may be adequately approached, grasped, and clarified.

In saying this, we are obviously indicating that there is a "center" for this inquiry. In current postmodern thinking all centralized organization is denounced; there seems to be no center to anything; it seems that we could venture into any investigation from any place, at random.

Yet the evidence of experience contradicts at every point this lighthearted assumption. In our journey into experience or the world we do go astray unless we find a starting point that is either a center from which the main lines of our inquiry flow or a center into which all otherwise dispersed lines of research gather. We ourselves at the outset of any physical or mental journey are such a center around which the world and universe of questioning is grouped. In the investigation of the human significance of life as well as in the investigation of the universe around us, it becomes obvious that in the human approach to the world, of life — and for that matter, in that of all living beings — we envision everything as lying either "within" or "around" us, we, living persons, being the "centers" of these inquiries. It is as if through the wishes, desires, appetites, appreciations that surge from within us, from the center, and by which we bestow significance on the "objects" that lie "without," we establish all that is without as a world "around" us. This is by no means a haphazard throwing out of hooks for a catch; the "within" is establishing itself as a "center" of our lived, experienced, envisioned universe. The hooks thrown send out circles around themselves like those that expand from the points where pebbles hit the surface of a body of water, circles that are always at a certain distance from each other, always in a certain order beginning with those closest to the break in the surface and spreading in ever further circles till the vibration of the water diminishes, the waves become broader, disperse, and disappear altogether.

This order of circles spreading over the surface encountered by our hooks seems to be the real principle of mutual exchange between the two substantial factors of center and periphery, although it varies as the waves are of different thicknesses, widths, depths, and even number according to the differing modes of the two factors and their respective conditions; these modes may differ but the order of the waves that they respectively set in motion and receive enacts the vibration that encircles them. It is from their center thus projected in exercise that the living beingness reaching out establishes its existential realm.

There is no point in following a passing fashion or in beating around the bush about repudiating it. Inasmuch as the human being is a center from which its individual and personal concerns spread, establishing it within the unity-of-everything-there-is-alive, it is also a center into which lines from innumerable other centralizing life-processing agencies gather, maintaining it in existence and allowing it to advance.

In point of fact, the passion for place is to be pursued, investigated through those lines of reflection that flow from what we could consider a "central" philosophical focus, namely, from the Human Condition. This focus corresponds to so-called "philosophical anthropology," but this inquiry gathers lines proceeding from the many circuits of what we call the ontopoiesis of life, which circuits sustain life, reacting to the circuits of the cosmic laws that meet in man, and it manifests too the aesthetic factor, which is crucial to the Human Condition in its lifting its significance to its specifically human level in the human creative act with its intellectual, aesthetic, and moral sense giving factors. It is the aesthetic factor that comes to the fore in our inquiry into the elemental passion for place. Our inquiry lies then within these three intertwined philosophical domains: the Human Condition, metaphysics, and aesthetics in the novel formulations they acquire within the phenomenology of life.[1]

II. THE PASSION FOR PLACE AND THE METAPHYSICAL CENTER OF AESTHETICS

As the center of aesthetics is metaphysics, so the aesthetic center of metaphysics is the elemental passion for place. This remains, however, to be shown, and I intend to demonstrate it in a succinct fashion.

First of all, the passion for place — as do all subliminal passions — lies at the heart of the Human Condition in the unity-of-everything-there-is-alive from whence this very condition is mapped out in all directions. However, while all the subliminal passions — except that for light — emerge within a territory, a frontier, so to speak, territory that the Human Condition with its creative forge draws on and encircles within this unity in which it partakes, the passion for place in leading us has the unique role of traversing this entire territory and does not stop at that, for it mediates all the steps of unfolding bios, of individualizing life, to life's furthest frontiers.

Among the elemental passions for the air, infinity, the sea, light, etc. that spur heroic deeds, elevation of the soul, human communion and expand the spheres of life in all directions into specifically human endeavors, the passion for place, as we will see, has no match other than the passion for light. It reveals itself upon scrutiny to be the basic driving force of the spacing/timing function of the logos, in particular of the network of life being projected by the living beingness initiating its course, with specifically human beingness drawing from it its full-fledged inspiration.

This far-reaching role of the passion for place gives us a precious thread allowing us to situate in a definitive way our inquiry into the creative source of the specifically human significance of life, an inquiry to which we have devoted two decades of work in the International Society of Phenomenology and Literature. This inquiry has been focused on the elemental passions of the soul proceeding from the creative forge of the Human Condition. In the specifically human adventure, the passions are the bearers of the sublimation or "humanization" of the soul, promoting the life of the spirit. Thus, our work has, on the one hand, discovered and elaborated the mysterious workings of the creative mind and this mind itself (what belongs to what is usually called anthropology/philosophy of mind), and has, on the other hand, though these are inseparable and intermingled aspects of our quest, pursued the sources and the unfolding of the beautiful, the sublime, and the elevated, what would usually be classified as aesthetic/moral inquiry. Thus, our hitherto accomplished digging into the sources, nature, and unfolding of the elemental passions has been a study of the Human Condition with a highly aesthetic bent.

The inquiry into the passion for place will, on the one hand, make a breakthrough: from the human condition we will be brought into the spheres of bios and cosmos; this will be the situating of the human condition itself within the vast network of life as such and of its cosmic contingencies. But simultaneously, the passion for place as the driving force of life in all its circuits effects its most essential impact through and upon the creative, subliminal footholds of specifically human existence on earth, which has its roots in the soil and its hands, head, and spirit raised above toward the sky, the furthest horizons of which are those that only the soul may guess.

It is precisely the aesthetic elan, the aesthetic/cosmic breath, the aesthetic/creative sense of the Human Condition, that is at the center of the role played by the passion for place.

As we have shown in our previous works, the aesthetic significance of life is found in exemplary works of the fine arts, literature, technology, architecture, etc. These exemplary works of human creative genius, cannot be approached at random, not in just any perspective, however ingenious; it cannot be adequately understood through its already accomplished and reified products — works of art, poetry, prose literature, theories — or through a focus on its expressive medium, what is fashionably called the "text," alone.

Nor may we expect to discover the pervasive role of aesthetics in our grasping the life significance of the human universe through an emphasis on man-the-creator alone.

Aesthetic inquiry, that is, inquiry into the emergence and manifestation of imaginatio creatrix, has thus far been situated in human works, that is, within the sphere of the human condition, which inquiry falls into the field of philosophy called "philosophical anthropology." Although I have challenged the usual conceptions of philosophical anthropology, in which the accent falls on anthropos, the human being singled out from the rest of living beings, in its own "nature," and although I have broken with the unwarranted artificial isolation or demarcation of the human being from the web of life and have emphasized the Human Condition as a unique territory in the evolving web of life, as a unique, novel, vitally significant station in the evolutionary advance of living forms, I have not thus far gone to the end of the road in this aesthetic pursuit of the origin and genesis of the specifically human significance of life accomplished — launched — by the elemental passions of the soul. With all of the passions so far investigated, we have held back at a halfway point.

What has been brought out in the main is that the human condition that springs forth from the unity-of-everything-there-is-alive and constitutes a creative novum does not isolate itself from the web of life. We have seen that through the soul it remains a link in the chain of life's forces and forms as a transformer of the vitally significant forces and energies into specifically human, creative ones; we have brought out that this creative human station functions as the final accomplishment of the vastly diversified degrees of an increasing flexibility in the workings of life, having greatly extended the rigor of the entelechial preordering of life's rules and regulations. Here life, in fact, passes into a system of creative freedom; human being invents its own routes of life.

This creative freedom comes, again, from a "center," and so it may be seen as a novel, specifically human "creative entelechy."

Within this framework, the study of the elemental passions of the soul has meant for us the pursuit of the origin and genesis of the specifically human creative endowment, of the new freely unfurled significance of life over and above its vital significance, though continuing it. It is the specifically human course of existence and the human dominion reposing in the human spirit and its origin in the creative forge that we kept in focus.

But with passion for place — as with that for light — we withdraw from this dominion. We still locate the center in the creative forge of the human condition, but we penetrate much further into the vital territories of the poiesis of life, into metaphysics and ontology in traditional terms. With the passion for place we reach the nest that the human condition finds in the womb of life. This nesting of life leads us further — through the spheres in sequence, through the further and further existential horizons in their referential networks in the system of life, with its cosmic contingencies, and on to the cosmic laws themselves.

The proto-aesthetic element accompanies the passion for place throughout. Hence, through this meandering way we arrive at our initial submission: as the center of aesthetics lies in metaphysics, so the aesthetic center of metaphysics is the elemental passion for place. The specific significant factors that imbue the human significance of life with aesthetic sense lead us through their dispersed rays to the center: through the human condition to the metaphysics of life.

CHAPTER TWO

SPACING/SCANNING AS THE FOUNDATIONAL FUNCTION OF INDIVIDUALIZATION WITHIN THE TERRITORY OF LIFE

I. THE NESTING OF INDIVIDUALIZATION

We have to begin our inquiry with a clear recognition that we do envisage things from a "center" within. By attributing to elementary spacing/scanning an aesthetic quality, I am by no means giving, as some thinkers do, deliberate priority to the aesthetic function of human being in establishing its "humanness." Recognition of the inquirer's position as the central position in mapping any territory of mind whatsoever allows us to avoid the one-sidedness of that view. We must disavow as well the priority that several contemporary thinkers give to moral experience and ethics, just as much as we would have to do if priority were given to the senses, etc. The starting point of the human condition that we have adopted prevents any such one-sided and limiting arbitrariness and makes it obvious that in the making of specifically human beingness all of the dominant factors are indispensable and play an equal role; they need to be harmonized at the very surging of their respective faculties in order that the human significance of life as such may surge and unfurl. If with respect to the passion for place we have to bring out the aesthetic function of the human condition as being "first," it is because it is the *poetic* factor in human functioning that accounts for the infusion of psychic acts with the "breath of life," which brings expansion, a "spacing" of the acts such that they may spread their "content" into experience, which then spaces itself as our own intimate domain, our domain within, our inwardness, our innermost self.

The "breath of life" is already the expression of experiential acts, introducing a novum into the empirical dynamism of the vital acts and operations of the psyche. It resides in the empirical vital operations and yet it spaces and scans itself in an "intellectual" fashion with the intellectual act of experience. As a matter of fact, the intellectual act, as the specifically human act, is carried by the aesthetic breath of life through which its continuing performances coalesce into "inwardness," the "sphere of the self."

We are here in a borderline territory between the ingrownness of living beingness and the spread of its tentacles through the circuits of "nature-life" with its system of relevancies that stretch to the cosmic laws, on the one hand, and to the specifically human expansion of this living beingness, on the other. The groundwork does not change, it remains always empiric nature-life; however, with intellectual acts, the specifically human significance of life takes a new turn; it expands through the aesthetic breath and spaces itself, so that "inward space" is expanded through experience, an inner realm that breathes the specifically human significance of life.

Thus, when we pursue the rays of this significance back toward the center from whence they come or into which they are gathered, we find first in every sense — vital, sentient, emotional, affective, intellectual, etc. — a basis for the outer/inner placing of individualizing life within the universal system: a nesting of individualization. A bird builds the station in which it lays its eggs in relative safety, a cozy place in which to nurture its young; in the security of its circumference, they may grow. The wolf, the boar seeks out a lair in the wild for comfort and security. For a human being who seeks it out or builds it, the nest is more than a place to be nurtured or to seek retreat from danger, cold, tempest, etc. The human being for his self-individualization in existence unfolds the aesthetic, moral, and intellective senses.

1. *The Human Nest: Home*

What for animals at large is a "nest" in which the individualizing life grows and from which it inserts itself into the network of nature before it is ready to enter on its own the avenues of its lifeworld, a place in which, as at a hearth, the animal family keeps mutually "warm" — and that is abandoned until conditions, inclement weather, perhaps, prompts a return or the construction of a new one — is an enduring reality for the human being who unfolds from within a much vaster area of meaning, of existential self-interpretation-in-existence. The human being makes its nest "for keeps," does not abandon it after seasonal use, or use it intermittently. A human being endows objects, materials, events that pertain to his building his "nest" with the entire spectrum of his uniquely own and personal experience. Thus, his nest becomes a hearth into which ever new experiences are thrown, keeping alive the flame that throws a glow all about, for the sake of a continuous flow of existence that ever renews itself. In fact, most friendships we have kept and keep, books we read, pictures we like become tightly interwoven with the innermost spacing of our experience, which becomes a place where we surround ourselves with all that we enjoy, where we retreat to refuel, to find repose, security, healing — to retrieve our very own selfhood if it be faltering owing to gales without. For the human being this nest within the situation of life and the world's schemas becomes a "home": a place not merely for retreat from the pressures of the outside world, but a place in which we expand our very self, with which we identify in our tastes and tendencies, a place for intimacy, a center of our very own in an alien world, a place for intimacy with ourselves, one to relish and share with others.

This atmosphere of intimacy we attempt to carry with us to the workplace, the vacation spot, the hospital by installing pictures, posters, pots of flowers that will recreate some of its atmosphere. In brief, nature's nest aestheticized by the human being becomes his home.

Our nest-home itself does not hang on some tree in an indifferent, neutral space; on the contrary, it is "spaced," that is, situated within a specific circumference spaced by itself, "marked" as a special place by the natural unfolding of the individual's functioning. Animals instinctively mark their vital territories in a forest, on a meadow, in the sky, etc. as their very own areas for hunting, pasture, repose. Human beings who extend their vital/instinctive life involvement by developing their own means for developing their territories — innumerable means, to begin with tools that transform nature's resources in their vital significance to realities that correspond to newly crystalized life modalities — also expand their nest-homes. But, what is more, human beings seek to anchor home in nature's groundwork: a "field," a "garden."

While the animal anchors the nest in which the individualizing lives of a new generation are nurtured in the nature-life system with its web of symbiotic ingrownness with everything-there-is-alive, the human being besides expands his spacing by tying together all of his experiential personal and interpersonal reach: affect, appreciation, taste, will, judgement, all contribute to the establishment of his nest-home as a "relatively stable" station that has to maintain a balance between his nature-life grounding and the waves of societal interaction. Our outlook on life, our bents and attitudes, our very personality contribute to the "aesthetic atmosphere" of the home; a home breathes and exhales an aesthetic atmosphere; could we not say that in constructing a home it is an atmosphere of our very own self, of our soul in its tastes, habits, convictions, beliefs, allegiances, and traditions that we aim at?

Home is, in short, a "place" that is established by the spacing of the specifically human aestheticizing function of our experience: morals, thoughts, reflections, lyric inspiration, all interwoven with the necessities of vital existence in a unique sort of convergence. All is prompted toward coming together in this convergence, toward the weaving together of a consistent web of personal significance, first of all by the elementary animal instinct to weave a nest as the center from which to confront the crucial moments of existence. Then, however, this elementary instinctive urge is simultaneously blown, like the surface of a soap bubble, into myriad colors by the specifically human "passion for place," a passion for setting a cornerstone for the working out of our very own interpretation of existence, for finding a place to anchor our very own means, talents, desires in the fully creative crystalization of the Human Condition, for having our very own fulcrum with which to probe our environment, get our bearings within life's expanse, and so gain self-confidence.

Yet from his home the human being digs deeper, seeking to excavate and rejoin his roots. First, he moves into the nearest ground, the garden, but then, ever on the move, he seeks what is beyond.

2. *The Garden*

It is in what we are used to calling a "garden" (in contradistinction to a "field") that we would establish our nest, our home, as a site upon which we unfurl our inhabiting/dwelling life modality, our springboard as well as retreat, our inner sphere among living beings, wherein we encircle ourselves in our very own territory, "on the land," as it were, and differentiate it from without, that is, a terrain the spacing of which is measured and apportioned according to our own personal appetite and rhythm of life.

It is not objective, surveyor-measured space that determines the rhythm of life in our garden; the smallest objective space may manifest a life exuberance, dynamism, and appetite for life that a large city park spread over large acreage of objective space will lack.

The reason is that our garden is ours insofar as we throw our very own vital as well as spiritual functioning into its life — a garden is alive, is prompted and maintained by our vital functions, and its modality of life basically translates our vitality, the pattern of our experiential system; it carries the stamp of our taste and even of some of our dispositions. In brief, what we call a "garden" is nature-life's dispositions and resources as admitted by the human being into his dominion — his womb — by specific human means in order to fashion it as his very own grounding. Gardeners live with the growth of their plants, their blossoming opens a new space in their vision of the world, their maturation and fruition fills their very being. Theirs is nature's harvest, nature's accomplishment or failure. Thus, the paths radiating from the center to the corners and bounds are sectioned and measured according to the "spacing" of the cultivation project; the garden and the gardener are both spaced; the measure is that of the vital spaces for growing that plants, trees, bushes call for, it is the space of walks to be enjoyed at leisure in the moving light of the sun or of the moon. Concurrently, there is the synchronization of the gardener's life occupations and the organic demands of the plants, trees, bushes, etc. In gardening, the human being enters into the spacing/timing rhythm of organic life and that life becomes part of the very circulation of the blood in his veins.

3. *Our Cosmic Breath*

Spacing our desires, intentions, projects in relation to our measuring, apportioning space, and the placement of our vegetable and flower beds, sowing and planting them in coexistence according to their vital requirements, we discover in our movements and calculations the scheme of contingencies of nature-life. Apportioning the objective "geographical" plots in size and shape according to the requirements that determine the places in which the various plants will flourish, we discover reaching further the relevance to nature-life of the cosmic laws. In fact, we situate the members of our body, their positions, their reach, their efficacious maneuvering of the tools that measure surfaces in a neutral "objective" way common to the measurement of everything without distinction within the system of cosmic laws, and we draw from their intersection with nature-life.

Tending the plants, the soil, watering in view of the conditions beneficial to growth, we discover and enter into in our day to day preoccupations the contingencies of individualizing life in the schemas of the forces of nature-life first and then those of meteorology, planetary motion, till we arrive at laws of spacing/placing/timing that reveal the ultimate structural condition of life, laws equaled only by those of light.

4. *The Cultivation of Nature-Life*

It is touch, the touch of soil vibrating with life, that makes our fingers, our hands, our entire body vibrate with life's pulsations. By planting and weeding with our naked hands, by digging the soil and crushing the clumps of dirt in our fingers, we pulsate with the moves of organic life itself.

Gardeners feel "grounded" in nature's life in their garden. The soil is the soil of gardeners' very existence when they themselves become rooted in its depths. The human being dwells in his home but keeps his roots in the soil of the earth: in gardens, fields, forests, etc.

Yet, as we have mentioned, in the garden the entire human being is involved, present, manifested. The garden fulfills some practical functions: it produces vegetables and fruit for food; it may — and almost always to some degree does — serve a "disinterested" pleasurable function, offering delight to eye and nose. A pleasure garden, be it a tiny flowerbed in front of a house, is an aesthetic creation par excellence. We speak of knowledge of agriculture, on the one side, and of the art of gardening, on the other. Agriculture and gardening intermingle, of course, but the emphasis differs.

The main features of human temperament, character, taste may manifest themselves in the harmony of a garden or in its disorderly arrangement, whether it be in an orchard, a patch of corn, or a flowerbed. Artistic taste for either formal perfection (in the arrangement of lines and the choice of flowers, plants, colors) or for a random mix and spread will distinguish the so-called formal garden with its structuring of space and plant selection, and even the determination of timing, from a romantic garden inspired by fantasy, in which the garden spontaneously follows the natural lay of the land and the flowers are mixed, etc. and, with only some intervention from the gardener's hand, are left free to adapt themselves and proliferate, as in a meadow.

In the garden we encounter in an immediate experience the soil within which ultimately the basic roots of our own organic/vegetal place among the play of vital forces in the unity-of-everything-there-is-alive is established, our basic organic/vegetal sphere. From thence one's own core bios sprouts and spreads further into one's psyche in its varied spheres within the Human Condition, and on entering the psychic sphere it brings about creative virtualities that revolutionize the entire ontopoietic design. Our elementary bios and the empirical psyche/soul, which has tentacles that spread in all directions, are together the source and main processors of energies, forces, synergies.

In everyday practice I am not aware of my ties with the universal play of vital forces in which I am "located" through my own energy processing system; we seem, on reflection, to lie within the dimension of "civilized," technical modes of societal existence, far away from where the elementary game of life takes place.

Gardeners, however, vicariously enter life's womb. It is not their own existence that is played out in the vegetable beds, but that of tomatoes, parsley, corn, lettuce, etc., but with their naked hands they enter into life's workings. I prepare the ground for life to sprout from, digging, fertilizing, smoothing the earth. I plant the seeds, and then water until the seedlings throw moisture-seeking roots into the ground and become strong enough to stand on their own against the sun, no longer needing moisture from my hand.

I look anxiously for the sprouting of the seeds and follow, with a breath of solicitude, their first moves on the earth; with fingers that amorously handle the soil, I weed around them as they grow and, taking the most tender care not to cut their roots, stems, or leaves, I will transplant them in an appropriate location, that is, I will prepare with my hands soft, loose soil in a new bed and plant them one by one, preparing cautiously a place for their roots, a place for their delicate adjustment.

II. AESTHETIC CREATIVITY

Yes, it is the basic sphere of life's progress that I enter through work with the soil and through putting into it the indispensable participation of my most delicate and tender feelings, expectations, and following that up with the thought that is care. Thus, I renew in myself the primogenital and primordial affectivity — the vital/moral affectivity that springs into life with the emergence of the human condition, entering its game. These are the creative virtualities that distance me from my absolute allegiance to my individual interests and spread into affective networks embracing others. The care — the moral involvement — will not break but serve the freedom of expansion within limits.

Personal character, disposition, taste in enjoyment, and beyond that our aspirations for beauty, for free expression, for manifesting in the world those of our inward dispositions that no words can express, and, what is more, our inward inclinations, whims, fantasies, nostalgia, desires, in short, the most intimate pulsations of our subliminal self, are all developed in the garden.

So, while, on the one hand, we work out the growth of our roots in the visceral sphere of nature-life, on the other hand, we find in our garden the proper climate for the unfurling of our innermost subliminal dynamisms. Indeed, from the soil of a garden that we fashion according to our intimate selfhood, we swing with our yearnings toward further subliminal placing.

1. *The Rhythm of Life, the Rites of Life and Death, and the Garden as Sacral Space*

In the garden we are involved in the cycle of life. Plants spring forth, and on meeting with favorable rains, warmth, sunshine they flourish. But droughts will make them wither away. Thriving life and death are constant inhabitants of a garden; to ever begin and wither away is its usual rhythm. Entering into its rhythms we experience it as a ritual: the ritual of cyclic growth, death, and return to life. Life's law is renewal. We follow the ritual of life from spring to late fall and through a waiting space of seemingly inert but in fact burgeoning life.

Into this rhythm of life into which we enter with the vegetation and ants, bees, earthworms, mice, rats, woodchucks, frogs…(the innermost rhythm around which spacing/timing/placing revolves, the rhythm that lies at the heart of our vital and subliminal concerns, the rhythm to the measure of which our imagination moves), are swept up the resources and energies of our entire soul, empirical and subliminal.

The human being who catches this rhythm scans his specifically human life and transforms it into a life ritual. Then his subliminal nostalgia releases life transcending yearnings. He endows the ritual of life with transnatural significance; it is within the precincts of a garden that he establishes special places, "sacral space," in which the rituals of life are performed in an elevation above pedestrian vegetal or simply animal significance, that of a fully human sphere, one that answers to soaring human yearnings. These may be an arrangement of rocks, a spring, a grotto. The acres where people are laid to rest are a type of sacral garden to which generation after generation goes to revive remembrance of forebears and loved ones, to reverence, pray for, and commune with them.

CHAPTER THREE

THE RELEASE OF SUBLIMINAL YEARNINGS

I. THE "IDEAL" PLACE: EDEN, EL DORADO, SECRET GARDEN

It is on the land that humans enter into the works of life itself and share in them. Working the fertile soil serves to expand the innermost resources of the soul, its longing, its striving, its innermost passions.

As we are attempting to show, the search for a place is not only the principle requirement of life as such but also the most fundamental and forceful passion of the soul.

Indeed, the passion for place in all of its esoteric variations, such as a passion for a sacred mysterious site, for a temple, a spring, a cemetery, a hidden labyrinth, subterranean caverns, ocean depths, a place where wealth is stored, where the secret of life, eternal youth, genius, etc. would be deposited, answers to an all-pervasive human thirst for the marvelous, the extraordinary, the miraculous. It is no wonder then that the garden has offered humanity fertile soil on which to nourish these yearnings in a full spectrum of deeply felt imaginary significance. It is a springboard from which to take flight and a center on which all can converge.

We seek "another place" where life would be redeemed of its shortcomings. East of Eden, there was a garden of life without end in which all vegetation and animalia happily coexisted with humans without the weaknesses inherent to the present material order with its servitude, danger, conflict, and death, a place of undisturbed felicity and innocence of life. This "paradisal" place we would like now to reach as we navigate our tiny craft over the waves of existence. It remains always on the far-off horizon of our dreams — an "El Dorado," a place where gold, opportunity, and happiness await the adventurous. This dream recurrently haunts humanity. Only its location changes. We always believe that by going to a new place our life would radically change.

Human beings endowed with these yearnings may not, cannot immerse themselves entirely in the ritual of life with its ever turning wheel of renewal. They cannot become absorbed in the cycle or fully identify themselves with its laws. As the ancient myths of every culture testify, human longing to overcome the necessities imposed by the laws of life, the inexorable turn of the wheel from birth to extinction, together with all the other limitations that life implies, has found outlet in innumerable expressions. The longing to overcome their confines is as deeply ingrained in us as are the laws of life themselves and haunts the deepest quietude of our absorption, complacency, enjoyment, fulfillment. The human being looks to a far-off horizon seeking total transcendence over the servitude that life imposes, over its unbending requirements.

1. *Homeland*

I look now from the shade of a tree at the pond, at the sprawling meadow beyond the sheep fence that separates the territory of the grazing sheep from my garden, the plants and flowers of which have to be protected from their voracious appetite. They graze contentedly and yet throw inquisitive glances beyond the fence: "The grass is greener on the other side of the fence." The sheep stand ready to sample that other vegetation the moment the fence should lose its effectiveness. The surface of the pond, into the depths of which my thoughts plunge, reflects serenity, repose. The blue lubin in blossom on a hillock before the fence separates the pond from the field beyond — a closer realm of intimacy, of shared roots and life cycles, one that expands, however, throughout the field, which is for the most part left to itself and to the sheep. But this vision of an intimate place expands still further — trees, another hill, a forest, and a mountain upon the horizon. And beyond that stretch other fields, ponds, creeks, pastures, gardens, and houses for sure. The delicate bluish haze hanging over the forested mountainsides tickles my imagination and spurs reflection.

The horizon opens from my home, my garden, and myself incorporated within it — from the center. It moves with my changing positions in this my very own milieu, the milieu of my self. My musing reflection extends far beyond visible contours, proceeding from the depths of viscera that are centered in the very soil of life. It runs through the entire span of my existence and comes back to the center from whence it flew.

We humans participate in the rhythm of life, know its cycles, but unlike plants and animals we do not vanish without a trace on our passing. There is another rhythm, that of generations, and in the line of generations we all leave a trace after we are gone, a trace of the work of our spirit on others. This work — the cultivation of the logos of life — is shared by families, communities, nations. We share in the spirit that was deposited as a common trust by past generations, we maintain some of its seeds and cultivate them in our own existence. Thus, those with whom we share these sedimentations of past generations cultivate similar passions for a place of their own: house, garden, meadow. They cultivate the same ideal of a home, of its habits, mode of life, mode of human relations, and of language, which expresses it all. Hence, there is an extended place — around my home and garden — that reaches as far as those sedimentations of culture have been deposited by succeeding generations. This is my homeland.

We are bound viscerally to the "native soil," the field, garden, forest where we have our roots, "native" not only because we like any animal first saw the light of day on a particular site, but because our forebears had their roots there and left behind the culture of this "land." No matter where we physically move within that land in order to establish our very own place, to put down roots, we can participate in and cultivate a way of life, the tastes, ideals, and virtues that are those of the land's people.

When my gaze wanders along the horizon past my meadow, my memory wanders too to a faraway homeland that I have left behind to make a new nest in a foreign land. I have found here a new intimate place, but memory takes me back to my ancestors, to their valorous deeds remembered, recorded, and passed into legends that molded my mind, then to the cycles of life of people I knew and cherished, to their fates, their heroic virtue, magnanimous hearts, their weaknesses as well as strengths, who — whether victors through their selfless dedication to others or victims of history's visitations — have passed to others the torch of endurance, of perseverance, of hope against all hope. Some have died, some continue to cultivate their legitimate places in the homeland, rendering full allegiance to its spirit. I cultivate my own spot in far-off parts.

2. *De Patria Mea*

Where is my Home? So my idly wandering spirit asks. Are my roots in the soil, in my very own place, my garden and my home, not merely my receptive organs — receptive to the forces of Nature-Life as well as to my creative virtualities deriving from the Human Condition? Is not all the outgrowth that begins with my deepest involvement in life-cyclic growth and extends to personal desires, tastes, cultural habits, virtues, and ideas of human destiny not situated at the center where they brew and from whence they spread, at the center of my heart? There I may embrace the past and the present of my innermost experience in one harmonious whole. There is where all converges and bursts forth.

But do I with my deepest longings repose in my heart? The heart joins all the arteries through which the life significance processed by the vital organs, the subliminal forge, practical life enactment, etc. pass to the center, but it does not retain them. Just as the natural horizons beyond my garden, my meadow, beyond the sea recede ever further as they are approached, so the webs of significance that emerge from our naturally processed experience, recede on reaching their zenith as new waves of experience come in. The sedimentations gathered by generations are adumbrated by incoming material, and, at best, these may advance with some impetus for a time. But they may also be disrupted and vanish. Social pressures may transform them beyond recognition. The natural existence of humanity for all of its creative inventiveness in surpassing nature has a horizon that ever recedes, that offers no definitive halt.

Carried by my subliminal longings for the final repose of fulfillment, the fulfillment of my life's course in life as such, I muse about a place beyond any place, a place of repose that is not caught up in the repeated rotations of the wheel of life, a place that my soul alone will recognize, for it has been seeking it from the beginning.

II. ESOTERIC "RATIONALITIES-IRRATIONALITIES" — A VITAL TENSION OF OPPOSITES

Through cracks, crevices, any openings, from all sides, vital forces penetrate the surface of the tight network of intelligibility into which the human endowment is woven, forces that have not been harnessed in the ordering of the life course. Just as these unharnessed vital forces may surge in violent outbursts with respect to the ever renewed accomplishment of harmonization or its failure, the remnants of the vital forces that manage to pierce the sieve of the intellective network penetrate as if mute and blind the entire texture of the living world with respect to the ever fluctuating and ever to be projected and newly enacted harmonization, keeping it, as it were, on alert. Whenever we withdraw from the business of current life, life itself is made the object of various types of questioning. Among the quests for objective answers, and the intellectual and philosophical queries, these otherwise mute and blind vital forces spur a questioning about life that is not oriented toward rational answers apprehended in the lucid clarity of the mind but seeks secretive bypasses, issues, and powers that would be equally secretive in their effects and workings, yet would resolve the questions.

To the fullness of the intelligible act of intellection that sets up the synchronizing coming together of the functions of what in its final operational composition we call "mind," there goes in the final analysis the differentiation and processing of all rationalities that the human being manifests. Yet, the clearness and structural cohesion of the intended objective varies in relation to varying factors.

In the first place the mind's highest degree of clarity and structural cohesion is reached in principle in the purely intellectual/constitutive acts that deal with the so-called "ideal" objects of thought and mathematics.

Second, the intelligible rays of the act favor a focus on the objective in immanent as well as sensory perception; thus they attain a complete constitutive schema of objectifying intentionality.

This is much less the case in perceptions of imaginings, fantasies, and so with experiences of emotions, sentiments, and inspirations, the intentionally constitutive focus may be, of course, present, but it is sidewise so and only accompanies the passional complex. Focus on a precise objective, direct or sidewise, breaks down even more in the processes of invention and creativity where secondary and random constitutive processes enter into play. Nevertheless, in the fullness of the intellective life of the soul, where consciousness itself becomes the intelligible field, experiences surge to the surface of intelligibility that do not exhibit the distinctiveness of intellective acts, that are dim, undifferentiated, having an intelligible core apparently without objective focus (e.g., foreboding, free-floating anxiety, felt intimations); these invite a search after their "reasons," their significance as well as causes. Along these lines, there appears throughout human existence a vague but most seductive tendency to seek "reasons" in between or beyond the reasons that can be legitimated by the clear system of the logos of life. These are "reasons" that would in a hidden way govern those reasons that are obvious but are at their rational joints and fault lines imbued by ponderous wonders and "extraordinary" phenomena. We would venture to say that in our rational life dim "invisible" uncertainties and unclarities prompt at times the surfacing of what we call "esoteric" questioning of the reasons for occurrences and events.

"Esoteric," "telluric," "magical" emotions, foreboding, anxiety, felt intimations exercise a profound attraction upon the human psyche; they intimate secrecy and exclusive initiation into it, they intimate hidden powers.

Because of the discrepancy between them and the life carrying manifested rationality, they are considered "irrational." And yet they contain an essential grain of reason, a virtual reason. Irrationality as the opposite of reason cannot be conceived without this virtual reason. We cannot assume anything "irrational" — magical, esoteric, secret, occluded — without visualizing on its far horizon the virtual possibility of its "clarification," "unveiling."

It is in foreboding evocations of the irrational powers that surround us that reside the springs of natural, primitive sacredness and religiosity: magic, sorcery, shamanism, etc. There lies the power of secrecy that gives *raison d'être* to esoteric societies, belief in magic, sorcery, etc.

Between the extremes of reason and irrationality there lies a vast gamut of differentiations that progressively pass one into each other. But what would lie beyond their tension is also beyond the scope of life.

PART FIVE

THE PROMETHEAN DIRECTION OF THE LOGOS OF LIFE IN QUEST OF ACCOMPLISHMENT

THE DIALECTIC OF EMBODIMENT AND FREEDOM

CHAPTER ONE

THE HUMAN SELF IN THE COMMUNAL FABRIC

I. OVERVIEWS OF THE MANIFESTED FIELD IN ITS PHENOMENAL SPHERE

The Living Fabric of Communal and Societal Life

When we focus on communal and societal life, we seem to be moving along a "nervous system" of interhuman relations, with set rules and regulations that vary from one communal and societal group to another. And yet these seemingly stable rules are in practice flexibly applied in individual situations. They are only relatively stable, displaying fluidity; they emerge, are maintained for a time, and dissolve within the living fabric of communal or societal life.

This fabric is a conundrum of all types of experience within innumerable cross sections of sedimentations from the past, experiences that go back to commonly shared even if truncated or elusive or dimly felt beliefs. These are in turn nourished by myths having their own deep-seated sources as well as by even more deeply nested archaic religious relics lurking in our hidden anguishes, reflexes, expectations, foreboding. In the last analysis, these reach the subliminal forge of human creative functioning — a watershed between survival oriented instincts, drives, etc. and specifically human, aesthetically modulated emotions and moral sensitivities.

All this enters into some or other configuration dependent on the particular history of the human group, into the living and ever in transformation communal fabric.

II. THE SELF-IN-ITS-WORLD-OF-LIFE

From our concrete analysis it follows that although the living human being is incessantly engaged in multifarious experiences and interactions, their movable complex making our concrete existential pulp, yet the interpretation giving sense to each and all comes from a center.

Human existence unfolds, grows, changes, undergoes transformations, is appreciated, lived, and enjoyed in the inner life of the self-in-its-world-of-life. The intentional strings projected by the mind toward the objects of its perceptions, experiences, wishes, fears, hopes — toward the outward world of life — mark only tentative, fragile points in the territory of the soul's life involvement; these involvements are in a state of constant fluctuation; they are objects of constantly renewed negotiations.

While the "inner horizon" of the self comprises the experiences and operations of the functions of the soul as its uniquely own space of inner action, drama, struggle, victory and defeat, there is still an "outer horizon" of the self's territory. Indeed, strange to behold, events, processes, beings that, measured by the standard stick of spacing and scanning, lay at noncontiguous points at vast distances from each other and are separated in their individual life courses in the vital territory of a human being, are through the intentional strings of the mind pulled together and become in the territory of the given self objects of concern, pondering.

1. *Awareness of the Self-in-Its-World-of-Life*

With intellective acts, that is, acts that are "transparent" in their occurrence and procedures to the subject from whom they proceed, the human mind not only "brings to light" its own circuits, but radiates some illumination on the vital regions from whence it ultimately arises and in which it remains firmly rooted through primogenital experience. Awareness of this "inner spacing" in the various degrees of dependency on the central agencies of the mind and on the ego, does not remain fragmented in the manner of the singular instances of awareness found in particular acts or their segments. On the contrary, the passional soul of the human being becomes aware of its self in what we call the central articulative reference of the individual acts that entitle us to experience the "power" of my "I," the "ego," in the inner spacing of our passional soul in its innumerable life involvements, its participation in the outward stream of life with its turbulent course, multifarious interests of concern to the soul, solicitations and rejections, etc.

Awareness of the self-in-its-world-of-life encompasses the entire inward territory that is outlined by the functions of organic/vital, intellective, societal, and cultural significance in which the individual living being is engaged in its inner/outer oriented course of life and is projected as well as circumscribed by both its vital and its intellective moves and acts. The soul/mind performing the living being's individualization/differentiation of its course in the midst of the stream of life intertwines with, encroaches on and intergenerates with, and promotes the growth or decay of other living beings; just as a stone thrown on the surface of water sends out ever expanding circles, so the soul/mind ranges outward and through the intentional resonance of its circles expands a corresponding inward spacing. It encircles itself within the horizon that it itself projects "all around." Indeed, in a manner fascinating to behold, we experience, and think, and envision "all around" us the vast extension of the world of our existential involvement, which extends out to our existential situation, other human beings, animals, things, nature.... It is within this ever extended self-spacing world of my intending that I find my "self."

My ego manifests itself in articulating my will, my judgement, my acts, and in the feeling of "power" over them. Through these acts, throughout their reach, acts "substantiated" by primogenital experiences in their kinesthetic extension, the primal moves of the soul, and the reasoning in the sensory elementariness of our bodily functions, with their system of relevancies to the rules and laws of bios and cosmos, I differentiate myself in my full vital/intellective/spiritual beingness and project a vision of myself in my uniquely own spacing in the world. While the ego is the fulcrum of the logoic apparatus as it administers, projects, and controls the inventive/creative avenues of the individualization of life, it is in the vision of the self that its manifestation culminates.

The ego can then adopt the interests of the other, join in the other's passional interests, spurning the tendencies in the soul that run counter to the interests of others in the outside world. There is in the soul not only latent turmoil that may at any instant break out from among passional elements but also a constant tension between the innumerable interests binding the individual to the interests of other beings within the life stream. Contrary tendencies, attractions with opposed aims, conflicts of interest, powerful stirrings with negative aims, all these have to be harmonized.

Awareness of the self-in-its-world-of-life is, indeed, the result of the harmonizing efforts of the subliminal soul.

Placating the storms of the inner passions and averting destructive confrontations with the powers of the external world by negotiating the attunement of the individual beingness to external constraints are two constant projects for the soul. Awareness of the self-in-its-world-of-life is suspended upon harmonizing the inner with the outer, the human being with others.

Thus the self-in-the-world appears as the network of equipoise in which the tensions between the person's inner agency of intellective and judgmental ordering — the so-called transcendental subject — and the constraints as well as postulate of its outwardly directed enactment are harmonized and dissolved. With commitment to interests, ideas, goals and ideals held in common, human persons advance from awareness of "self" to awareness of "we," and the societal spirit arises.

III. THE HUMAN PERSON AS THE FRUIT, THE FIELD, AND THE ACTOR OF THE CULTIVATION OF THE LOGOS

The human individual inhabits both the organism whose workings function on the level of vital significance and the human soul, the mind that grounds specifically human significance; here through the progressive cultivation of the logos there appears specific patterns of molding, of acquired and retained habitual responses, of revealed or strengthened bents and unfolded dispositions forming a motivating core of experience, volition, and subsequent action unique to each human being. This unique complex is the fruit of a progressive historical unfolding of specifically human propensities in which the logos of life acquires peculiar forms. The molding, transforming, and developing of the entire field of the human psyche/mind then becomes the main factor imprinting the further molding of the significant life course. Since this patterning encompasses the entire field of the soul/mind complex, it is also the field of the entire cultivation of the logos, which molds the faculty of the intellect and motivates the will. The human person is both the field that the cultivating moves and processes work and the main propelling force behind the initiation of a course of action and perseverance in it.

The creative cultivation of the Dionysian and Promethean lines of the logos within the inventive structures forming the intellective act, deposit deeply ingrained molds/habitualities in the human psyche; these molds of feelings, emotions, passional tendencies, aspirations, evaluations are then instrumental in inspiring the will.

These specific modalities unfold under the influence of educational emphases as well as the impact of the milieu on our appreciation of life experiences; they are the footholds of the human person's cultivation of the logos within a group, community, culture.

The functional pattern of the person, which surges through the enactment of life, is founded in elementary dispositions that are carried over from the vital orbit and reappear in the elemental/subliminal circuits of the soul. Even though they acquire the molds of the inventive/creative apparatus surging in the creative orchestration of the human faculties, they yet remain the basic synergies derived from the vital zone and are to be counted among — whether our subsequent aesthetic/moral molding is in accord with them or not, whether we attempt and struggle to tame them — the synergies of the dispositions to anger, resentment, vengeance, or their opposites, and they are to be taken into account in the specifically human turn toward life enactment.

Our motivations and subsequent actions derive originally from our dispositions and emotions, for which we manage to "cultivate" significance; in part, they also derive from strictly utilitarian and practical considerations, which, nevertheless, are to a great extent molded by our aesthetic/moral "taste" for types of experience as well as by our own "feeling about ourselves."

Here, with the question of taste, we reach the spinal column of the interactive, inner/outer molding that the person both operates as well as is subject to in its ingrownness in the societal/human world. Much of the cultivation of the logos of life through persons, human groups, communities, societies, the world at large is a matter of developing and transmitting "taste" for such or other objects, qualities, emotions, appreciations, creating a "style" in art, clothing, food, dwellings, soil cultivation, etc. in the group's individual members. Hence stem the styles by which people express themselves as they enact the cultivation of the logos. The person crystalizes itself and its potentialities in the making by acquiring tastes and perfecting styles.

In a fashion typical of the logos, we have here a cultivation that proceeds amid a tension between two opposed tendencies. Here we have the specific bents and aspirations that set a person apart and make him singular; there we have the modes, tastes, styles of life acquired by a human community. Each conditions and implies the other; each is cocreative with the other in the progressive cultivation of humankind — in the creation of culture.

While, at the one extreme, we find human societies acquiring various phases of practical approaches to life, tastes, and styles, which define them in their specificity, so on the return of the boomerang we find the human person unfolding through education and interchange with the milieu of the era a design for its own cultivation of the logos, one peculiar to itself. Such a relatively stable functional design defining the person's peculiar modes of life enactment — taste, style, and motivations — we call "personality."

IV. THE PROMETHEAN ARDOR

The human being's spontaneities flow far beyond the script of natural existence. They flow beyond the closed cyclic curve of the script that lifts him above amorphous matter but then bends him down again to fulfill his natural role within the universal life schema as he exhausts his natural spontaneities and vanishes from the scene leaving space from which a new individual and novel segment of life may take off. Unfolding himself within the pattern of the natural script, he passively submits to its burden; he undertakes its fulfillment as a continual labor. And though he be inspired to go far beyond the pregiven line, he never goes so far. With every attempt to go further, man has to be satisfied with initiating just one routine loop in the weave of things.

But is man content with this submission to his natural doom? Is his aspiration, striving exhausted with his life? Or, on the contrary, might not some small piece of his inward promptings, his specific spontaneities be passed to another being who will carry on the task?

The specifically human spontaneities do prompt him to reach further. They intergenerate and intensify in their dynamic forcefulness. Finding no ready canvas on which to find expression and no script of a higher, independent order to enter into, they brew an ardor. They cannot follow the arc of the cycle of life; they are alien in their rarefied nature to the conduits of man's toil. They are carried by the yearning to be free from the natural canvas and submission to the inevitable. They would open the unforeseen but promised. They prompt man to rise from the anonymity, the mediocrity of his bent posture and shoot upwards.

Indeed, the human being seeks in the significance of his natural life script the opening of a promise. The script of existence cannot stop at the calculable, foreseeable, possible. The human being seeks the "impossible" promised by his innermost being. The quest for this promise is the innermost thread of the weave of human existence. It is as a silver thread interwoven with each loop, with each mesh. It is present as a point of interrogation within every articulation of the meaningful unit. The human script is held in midair, lifted up by the brewing spontaneities that do not allow the human being to submit himself to the system of meaning limited to survival. We are prompted to rise above survival concerns even though we are again and again pulled down from each momentary escape, each fling upwards by the heaviness of life's weave. Hence to be human is to be rising up and falling, exulting and suffering. If, as with Prometheus, chains run eternally through our beingness and suffering devours our flesh, it is not because we have dared to reach too far and have had to be punished for our daring but because to be human is to be devoured by an ardor to reach for a promise that lies within our very being but cannot be incorporated in the natural script.

The ardor has to fail, but does this mean that the promise is false? Is it, as Gide seems to have shown, that man tries in vain to free himself from the chains of the everyday torpor, repetitiveness, that there is no exit from the domain encircled by the natural pattern? On the contrary, while man has to fall onto the ground each time he lifts his wings, this does not mean that he has raised them in vain. He can reach the sun and convert his ardor into a lasting substance, even as the substance of life is doomed to vanish. It is the ever repeated rising — even if followed by the fall — that may bring about a new weaving of the uniquely human script, which then can maintain itself through all the spasms of quest and defeat.

CHAPTER TWO

FROM HUSSERL'S FORMULATION OF THE SOUL-BODY PROBLEM TO THE DIFFERENTIATION OF FACULTIES

I. THE SETTING OF THE TRADITIONAL PROBLEM

Our here taking up the theme of the soul-body problem in Husserl's phenomenology is not arbitrary. On the contrary, this problem presents itself to us inevitably as we carry out the research objective of our symposia and congresses, which is to explore, step-by-step, the fundamental questions that phenomenology poses, by means of criticism and then to refashion them. The soul-body problem reveals itself, astonishingly, as Husserl's deepest concern and thus as an integral part of the foundation of phenomenology as such. In the course of my Husserl/Ingarden critique, the problem presented itself to me compellingly because Ingarden was quite apparently stumped by it. At the end of his monumental opus, Ingarden, who adopted all of Husserl's acquisitions in this area, simply stopped short at the soul-body problem and could not get beyond it. We could then say with T. S. Eliot: "And the end of all our exploring/ Will be to arrive where we started/ And know the place for the first time."

In his *Cartesian Meditations,* Husserl relates his work specifically to Descartes, to whom the dichotomizing that is the origin of the soul-body problem in modern philosophy is ascribed. It was the overcoming of the idealism/realism antithesis arising from this position that constituted the basic intent of transcendental phenomenology, as has so often been pointed out. Husserl expressly recognizes and affirms Descartes' "cogito ergo sum," which seeks the certainty of cognition in the awareness of having made the breakthrough to immanent consciousness despite the necessity of the completion of cognition by pure transcendental consciousness to guarantee this assurance of cognition. Husserl devoted his studies to this purpose, especially the second volume of his *Logical Investigations*[1] and *Ideas* I.[2]

Pure consciousness, which is the domain of the certainty of cognition, correlates with the assumption of the intentionality of consciousness, which has, in Husserl's view, the essential task of sense bestowing, establishing the priority of the intellect.

But for Husserl it was not enough to pursue the Cartesian concern for the certainty of cognition. With a grand sweep he undertakes to bridge the gap between intellect and empirical psyche and to make good the loss of actuality/real world that resulted from the Cartesian undertaking. *Ideas* II can be regarded as the attempt to regain reality by catching it up in the same net with pure consciousness. Examined more closely, the investigations contained in this work constitute a coming-to-grips with Descartes. Point by point, Husserl works out, in his own way, the intuitions that are contained in the Cartesian "cogito" or that are merely suggested by it. He demonstrates, in this fundamental essay, that the full development of the system of immanent consciousness — which Descartes lacked — extends to indubitable cognition by way of the connection to the empirical and to reality altogether; he charts a new course. Yet, insofar as he is attempting to regain the organism and reality that eluded Descartes' immanent consciousness, this is no longer preeminently a matter of the manner of cognition, but one of the existential connections between the human functions that can be differentiated by the phenomenologically objectivating consciousness.

The continuous existential unity (in which the body/organism plays out its existential autonomy) of the functional provinces would also include the human "I can" in all its fullness.

Husserl's efforts to set forth the unity of the human being will here be treated from many standpoints in serious explications and discussions.

Husserl's differentiation (which has not been given sufficient attention) of functional realms in the human being ("domains," as I have called them in my Ingarden/Husserl critique), i.e., body/organism, organism/soul, soul/mind, constitutes an undertaking that I consider to be most important and interesting. I shall also begin with the distribution of the functions and not with the customary constitutive/cognitive orientation. But I maintain that, in spite of Husserl's frequent treatment of this differentiation, his way is not the way to achieve full access to the faculties underlying the functions — the "I can" in all its manifestations: I think, I value, I know, I will, I do. *But now it seems that Husserl's overall conception is not only to be found in those of his analyses oriented to researching cognition, i.e., of the objectivating subject, but is present also in, and indeed must be sought first, in his explication of the faculties engaged by the ethical, that is, practical subject* — the design of the collective functions that, in his view, makes up the unity of consciousness.

In Husserl's view — as appears already in his lectures given before the first volume of *Ideas* was published — the acting subject extends not only from the cognitive, i.e., logical functions of reason with which the feeling functions of the affective mind are tied but also to the valuing, willing, and doing that stem from both functions. Nevertheless, following Husserl one finds that it is only in the intentional founding of objectivating, valuing, and willing, whose unity is accomplished as, and comes to an apex in, the unity of the "total consciousness" in human action, that one could find an essential correlation between actuality and present consciousness as a presupposition for an absolute cognition.

It is universally assumed that Husserl identifies consciousness with intentionality. Alphonse De Waelhens emphasizes this in his study "L'idée phénoménologique de l'intentionalité," but asks whether Husserl understood the presuppositions of his own theory. In his opinion, intentionality, on which the whole structure of consciousness rests, instead of being "un point de départ facile et évident," must itself be grounded "dans le cadre d'une théorie complège touchante les rapports de la conscience et du réel." Now it is in the idea of intentionality that the attempt to overcome the alternatives of realism/idealism rests (also in the differentiation of the functional domains). Does intentionality really go to the root of the interrelations of consciousness even if the latter were grasped in the whole breadth of its functional operating, and does it really go to the root of the real? As Levinas emphasizes, intentionality has nothing to do with the relations between real objectivities; it is essentially "...l'acte de préter un sens" — *Sinngebung* — sense bestowing. But I wish to emphasize that sense bestowing should not be limited only to the functions of the intellect.

In Husserl himself one does not see the relationship of consciousness to the real treated only in the intellective, fully constitutive, or primordially constitutive perspectives. The question is, however, whether the unity of "total consciousness" in Husserl does not ultimately rest on intentionality thus hindering the explanation of the emergence of the other normative functions.

Here I shall undertake a short presentation of Husserl's conception of the human faculties that are said to make up "total consciousness," and then I shall proceed to my own conception of sense bestowing, which I call "ciphering," in its imaginative form giving.

II. TOTAL CONSCIOUSNESS AND THE DIVISION OF FACULTIES IN HUSSERL

Husserl proceeds from the basic faculty of the subject "I can" and ascribes to reason all the functions of consciousness that in their unity activate the human faculties. There is a fundamental distinction made between objectivating, valuing, and practical consciousness (or attitudes of consciousness). Corresponding to these distinctions there are, respectively: the object of judgement, the object of value, and the object of will. But this differentiation refers (as Diemer also shows) to fundamentally different faculties of reason: understanding, which is altogether responsible for the logical structures of cognition; and affective reason, which is, on the one hand, the source of affective valuing and value measuring and, on the other hand, the source of all affective conscious processes. This means that in every conscious process, in every really immanent act, we "are-there-too-with the object" ("mit dem Gegenstand dabei sind") and that we "live" in the act of experiencing.[3] This also implies the corresponding division of kinds of objects according to their predicates, relations, and states of affairs (e.g., in the objectivating orientation of consciousness objects are constituted by means of logically predelineated, objectively valid predicates). Yet the value-objects and practical objectivities that arise in the affective mind and are supplied with predicates of the affective mind (referring to purely subjective valuing acts of the subject) are half-subjective.

Although there is extensive interweaving between the objectivating understanding, which is directed toward rational structures, and affective reason — interweaving that is constantly changing in the changing life of consciousness — and although values stemming from affective valuation are supported by rudimentary objective references, the two faculties coincide, in Husserl's view, most sharply not in the higher or originally intellective spheres but in practical consciousness.

Decision making, which, as is known, distinguishes the intellect as an autonomous domain of functioning from the psyche, is to be related to the intellect, i.e., understanding, only in its objective aspect; but it derives, as willing decision making, from the active volitional orientation that, proceeding from the valuing done by the affective mind, actualizes itself in willing and doing. Although it is the objective valuing, i.e., the valuing explicated by the understanding, and the volition that radiates from value that constitute the ethical Subject, yet it is in the feeling faculty proper — of the affective mind, which enjoys, and accordingly brings forth valuing and feeling — that unique subjectivity comes about through the purely subjective predicates of the object that carries value. These predicates signify a unique subjective sense bestowing that takes place above all in the psychic-mental realm.

The lifeworld as "harmonious social living," which Husserl ascribes especially to the effect of the unity of the functional domains of body/organism, organism/psyche, psyche/mind, reaching their apex in the mind, is to be traced — on the part of the faculties — to the ethical Subject, i.e., to the valuing and acting Subject. As Husserl himself expresses it:

Gesetze des Handelns in einer sozialen Gemeinschaft von Personen, die auf eine gemeinsame Welt von möglichen Werten bezogen sind, die sie realisieren, in deren Realisierung sie sich kennen und fordern können...sind in der Obereinstimmung von den aus subjektivbedingten Lebenssituationen hervorquellenden Wertungs-und Willensakten der einzelnen singulären ethischen Subjekte untereinander in den durch die Vemunft-Einsichtigkeit-Objektivation intersubjektiv....[4]

Ultimately this unity of reason — i.e., of understanding and of the affective faculty — operates in a combined way in the personal Ego, which promotes the whole range of the "I can" in the human being. As Husserl expresses it: "das Ich ist der identische Brennpunkt aller Aktionen und Affektionen, das Ich des 'Ich kann', 'ich tue', 'Ich denke', 'ich werte', 'Ich will'."[5]

I have tried to show how the functional domains of the human being — which in *Ideen* II are differentiated only from the standpoint of the objectivating consciousness, in their existential links — in the perspective of Husserl's total thinking, first of all derive from the division of faculties. Second, I have tried to show how by the major weight being shifted to practical consciousness, i.e., to acting and doing, and proceeding from this standpoint, there exists still another kind of existential interweaving of the functional totality: body-organism-soul-spirit. In fact, it is apparent that without taking these into account one cannot expect to find a solution to the difficulties that Husserl's view of the unity of the soul/body entails.

Without occupying myself further with these perplexities of Husserl's, I turn now to the constructive treatment of the problems — first by proceeding from what we have gained from Husserl's view and continuing, then, with my own research.

III. EXISTENTIAL SENSE-BESTOWING AND THE NEW DIVISION OF THE FACULTIES

At the outset I would like to throw open, two questions. If one takes some distance from the objectivating direction of the immanent processes and turns to conscious living itself, the life of consciousness appears to us in a unique unity and fullness of conscious living and of living-through. The question arises: "is this fullness and unity a matter of the affective mind alone?" As a matter of fact, if one turns to the social living of the lifeworld where we see, with Husserl and also beyond him, how the individual acting subject effectuates itself, this self-effectuating being not merely self-enclosed but also extending over the individual span of time, so that the question arises as to the possibility of continuity in social living. If it were *only* a matter of sense-bestowing by the understanding and the affective mind linked together, how could one account for the undeniable continuity of living through one's life within the intricacies of intersubjective relations?

1. *Enjoyment and Creative Orchestration*

As far as the first question is concerned, I wish to assert that the unique fullness of the mental life process in which the Ego is rooted is by no means only the fullness of feeling of the acts of mental life, whether of knowing or of valuing. Nor is it a result of all the constitutive and preconstitutive acts that bring about the "identification" of the conscious I. On the contrary, the fullness of the mental life process is the basis of this constitutive process of identification. It is a uniquely peculiar fullness of conscious life that not only fills out everything but also encompasses everything and that leaves no gap between the self-singularizing mental life processes; it is, rather, a *self-renewing* fullness of mental life processes that, living itself out in all acts, always means more than these; the living subject lives and rests in it even though it also contains it. It does not allow itself to be taken up in cognitive, valuing, and acting consciousness alone. In it the consciousness of the child, not yet broadly developed, finds an incomparably rich fullness of the lifeworld that by far exceeds that of the fully constituted consciousness of the adult. While objectivating reason aiming at a logical order and objective distinctions, sets up rifts and cleavages; while the affective mind produces valuing distinctions and the will produces action, causing sharp breaches in the continuity of the individual's lifeworld constitution — all directed to objectivating distinctions — what else but the imagination[6] could bring about this unity that is so fundamentally essential to individual mental living? It seems that what I call "imagination" is a particular faculty that has a unique role between cognitive, objectivating reason and the affective, valuing sphere of emotion and volition — the role of filling in, from within, the emptied life of consciousness, which is continually impoverished by the differentiating functions, taken apart, and weakened, of filling all in with affective phantasma that hover freshly before it. In this way it enriches what has become impoverished and transforms what has become unenjoyable with a new enchantment. In this way 1) imagination would prevent the constitutive composite image of the conscious functions from being torn apart by producing the fullness of immanent life and providing a bridge of invention; moreover, 2) the possibilities, on the part of consciousness, for coalescence and adaptation, which form the basis for its alterations, and otherwise remain incomprehensible would be ascribable to imagination; and 3) *when imagination enters the scene in a completely unfolded creative context of the total functions that unfold together with the imagination, the possibility of making a moral decision would be ascribable to imagination as a morally unconditional decision.* (As we remind ourselves, it is in the moral decision that Husserl sees the autonomous standing out of the spirit from the soul and the ground of social community life.)

2. *The Orchestration of Existential Sense-bestowing*

With this we come to the second question thrown open above. How does sense bestowing occur in the social lifeworld? It is there, in the fully developed acting and doing that employs all the human faculties, where full sense-bestowing, i.e., the preconstitutive and supraconstitutive activity of all the human functions taking shape in man's inwardness, is exercised. It is there, also, that we can observe all the faculties that activate these functions, since here it is no longer a matter of a conscious identification on the part of the conscious subject, but rather one of *existential* "self-individualization" of the human monads in action and work. *The question as to the unity of the functions is sharpened here to that of the continuity of the process of the existence of the human being that is being carried on by means of the constitution of the lifeworld.* The being of the lower animal is rigidly bound to its surrounding lifeworld, whose vital, meaningful net has been cast forth by the animal, with the result that when the sense links in its environment are broken, or changed by external circumstances, its sense-bestowing organs of sensation are thus also prevented from functioning. The human being, on the other hand, when he meets up with the inadequacy of the domain of his lifeworld, will immediately search out a way and undertake to accommodate the latter to himself once more. He develops inventive powers for reshaping this meaningful network of life, or even for casting forth a new one. This is the distinguishing mark of the human mind and of the human decision. *Thus human existence stands out as historical.*

When Eugen Fink says that "the constituting life of transcendental subjectivity is in itself historical," one can see in that much more than the continuity of the course of the processes of consciousness cast forth logically and as an object, and of the universal lifeworld constituting itself in them. What we mean by human "history" is precisely the continuity of the sense bestowing human existence that must continually be cast forth anew, through and above the drastic reshapings of his lifeworld that the human being must undertake again and again from one epoch to another, from one revolution to the next. The fact that there is a continuing weaving of the connections in spite of the continual emergence of the new and in the midst of radically opposed tendencies and currents indicates that there are connections that cannot ultimately be explored. It is this that brings about the sense of wonderment and pondering about history and that gives human life deep and unexplorable roots in the common fate of mankind. But this connection is shaped through interweaving by inventively and enterprisingly oriented sense bestowing on the part of the existentially self-individualizing human monad. This brings to light the fundamental role of the imagination, which — as a third faculty together with the intellect and the affective mind — sets human sense-bestowing, from its origins, on a unique plane. It is only on this plane that one can see how the functional unity of the human faculties, which come into play together, brings about a specifically human totality (i.e., of body/organism, organism/soul, soul/spirit) in a venturesome, active nexus of meaning that carries out its effect on human existence.

Just as reason is oriented to cognitive identification and the affective mind is oriented to affective valuing and willing, imagination is oriented to action. It is only when imagination emerges as an autonomous faculty of the human being that one can grasp the psyche-organism relation by its roots.

This is the image that I propose, on the basis of my attempts to look at these things in a new way.

CHAPTER THREE

TELOS AND DESTINY

I. THE QUESTION OF THE TELOS IN CLASSICAL PHENOMENOLOGY AND ITS OVERTURN

At first sight, what could be more unexpected than to raise the question of a telos of the human being within the framework of Husserlian phenomenological research! Is it not the first aspiration of this type of philosophical approach to provide for the exfoliation of the constitutive system in which man and his universe emerge by direct intuition — both certain and indubitable — which permits "things" to speak for themselves, by offering them, as it were, a tongue by which to speak? Remaining on par with the nature (structure, essence, eidos…) intrinsic to things, this intuition presupposes their exhaustive self-interpretation. On the basis of this double premise and in the context of man's self-interpretation, would there be a need to seek principles and external causes parallel to classical philosophy?

Starting from this double premise, therefore, what reason would there be to raise the question of a principle of finality in the transcendental system of man within a lifeworld assumed to be self-explanatory?

Of course, in giving a key place to the progress and development of this philosophical reconstitution in a discourse that is identical, in the thought of Husserl, with the philosophical discovery of the universe of man (but understood at the same time as presenting the constitution of man by himself within his lifeworld), it is natural that this flux of the transcendental constitution itself refers to a principle of orientation, to a purpose, a telos. It is not surprising, then, in the conception of the transcendental genesis of man-in-his-lifeworld, to discover the idea of a meaning that it tends to accomplish; first of all, the direct telos of the singular constitutive act, which is the intended object, and then of a telos of Western man or even of human history as such. Moreover, as we know, phenomenology itself is seen by Husserl as this telos underlying the genesis of Occidental culture. Thus transcendental genesis is conceived as a teleological system.

However, the question arises: "Would the intuition of this teleology, insofar as it emanates from the genetic reconstitution, be elucidated in an adequate manner by this reconstitution itself?"

Husserl himself uses the term 'telos' in several ways. He uses it to denote the aim of a particular constitutive process, notably, the final outcome of the constitution of an object, which is the basic segment of the genetic process, limiting it in itself. The term telos appears in the current of Husserlian reflection again in the sense of an ultimate finality of all transcendental genesis as such. It is in this regard that the paradoxical question arises: either the double methodological premise does not suffice in order that things reveal their intrinsic nature; or this eidos, which at the same time presides at every genesis and remains intrinsic to the constituted object, does not contain in itself its ultimate telos. And yet this genesis requires it.

Leibniz proposed that in order to exhaustively define a thing, it is necessary to have recourse not only to the knowledge of its intrinsic nature but equally to the mode of its origin.[1] And yet it seems that classical phenomenology in its two lines of development, namely, the eidetic ontology and the transcendental genesis of Husserl, having pushed this demand to an extreme, reveals this demand to be yet insufficient. It remains also unfulfilled by the thought of both thinkers. But if pursued further, it appears that phenomenological reflection taken as such does not stop at any point of its genetic development but refers further to the postulation of the *ultimate* telos of things being constituted. Whether it would be an individual object, the Man-in-the-World, or humanity, the disclosure of their respective teloi seems indispensable for founding a definitively phenomenological edifice of philosophical reconstruction. Attempting to grasp in its fullest extent the object, whatever it might be, the real autonomous individual, the Man-in-the-World, we will necessarily encounter the question of its "final cause," its telos.

Within the phenomenological perspective the question should then be asked whether objects do not each possess an intrinsic telos in themselves (for example, a telos proper to the real individual or to a purely intentional object, to the Man-in-the-World, a telos proper to the constitutive cycle of individual or social development). According to the differentiation of the universe of the phenomenological discourse, there would be a corresponding differentiation of the telic principle. Husserl and Ingarden differ radically on this point.[2]

It is apparent that Husserl's need of a teleological principle arises from the genetic nature (that of a process in progress) of the self-interpretative method of transcendental phenomenology. Meanwhile, in the eidetic ontology of Ingarden, which consists in a chain of ideal objects, this need stands ignored. Ingarden, who is opposed to Husserl, categorically refuses to attribute any meaning to the notion of telos.[3]

Ingarden's formal ontological research, which provides us with access to beings in their rational structure *par excellence,* aims at fulfilling a twofold role by means of differentiating their composition. On the one hand, the intrinsic structure of things and beings accounts for their possible *actio et patio* and, on the other hand, it is meant to serve as a foundational principle for the universal order of the human universe. The question arises, first, whether this research has provided us with structural differentiations sufficient in themselves to fulfill these roles. In particular, do the differentiated structures contain the indispensable elements for fusing the particular links of the hierarchical scale of beings that Ingarden's investigation differentiates in accordance with the continuity of the objective world, on the one hand, and what is more, in accordance with the continuity of human experience in its fullest extent, on the other?[4]

Second, does this formal ontological research offer us at least a basis of unity for these two types of beings: man and the real world? (As we are well aware, the existential claim of the autonomy of the real world was the principal object of Ingarden's preoccupation in his controversy with Husserl on the subject of Idealism/Realism.)

As we have attempted to establish previously, a basis for such continuity either of order or of experience is lacking in Ingarden's vision.[5] Indeed, although Ingarden succeeded in moving beyond the reductionism of Husserl's transcendental monism by his ontological pluralism, he was unable to arrive at or discover the continuity of existential links and failed to establish a unified ontological territory on which empirical reality and transcendental consciousness would meet.

It is necessary, then, to ask if the ontological structures, the ultimate foundations of being, which on an Ingardenian analysis remain definitively separated, are not therefore perfectly incapable of founding the universal order of things, as we know or understand it in the continuity of human experience. Perhaps eidetic analysis alone is incapable of revealing the intrinsic unifying factors that would account for continuity.

After having critically examined the Ingardenian analysis, we shall see that, in fact, the ideal structures are not the ultimate factor of the universe of things and beings. They offer merely a system of reference for its rational articulation. It is rather within the existentially prior context of actual existence that the actual individualization of beings is established together with the arteries of life.[6] By stating this, we have moved from the ontological to the cosmological approach; from an eidetic approach we have proceeded to a conjectural one.[7] These two are inseparable in our inquiry insofar as, on the one hand, the ideal structures provide indispensable orientation for the conjecture while, on the other hand, only within the context of actual existence may the ideal structures prove to be valid as well as show their flaws.

Shall we not, following the progressive individualization of the real individual within the world process, see that this latter postulates not only an ontological mechanism of motion and self-sustenance in existence, but first of all, an intrinsic nucleus accounting for the individual's self-governing agency? What else could account both for the intimate transformation of dynamism to structure, and for the intrinsic linkage between Elemental Nature and the system of the hence emerging world of things and beings, if not an entelechial principle?

Putting emphasis on the real individual being, we bring together the subjective (Husserlian) and ideally objective (Ingardenian) biases. By the same stroke, distinguishing the individual's entelechial principle, we propose to solve the problem of the existential continuity of beingness.

And yet one further question has to be dealt with: Are we going to find the real individual being self-explanatory in his own progress, and in the progress of the real world of which he constitutes the foundation, or must we search further in order to account for progress?

From this brief outline of the phenomenological aporetic, which appears once phenomenology is confronted with the issue of the telos, there clearly emerge the great controversies in which the phenomenological project is entangled. Phenomenological investigation also appears ripe to confront them and to revise its classic position.

First of all, when I focus phenomenological self-reflection on the question of its intrinsic teleological orientation, that entails questions concerning "reasons," "causes," and "principles" that the encompassing, hitherto unquestioned framework of analysis has hidden but can now no longer disguise. Second, *teleological investigation lets us break out of the closed circle of man-within-his-life-world and open the gate for an integral phenomenology of Man-and-the-Human-Condition.*

In what follows I propose to approach the issue of the teleologies within the phenomenological program directly, that is, to outline the *three stages* of phenomenological inquiry at which, as is shown in my previous work, the question of the telos emerges and is implicitly treated in its own way.

In the first place, searching below *the intentional act - eidetic structure correlation* that represents the narrow sphere of the objectivity of the lifeworld, the real individual being will be approached in its actual existential origins as it individualizes itself within the network of the elementary forces of Nature manifesting their own regulative telos.

Secondly, breaking the closed circuit formed by the transcendental, specifically human lifeworld with its complementary zone of the preconstitutive, we will seek to retrieve, through phenomenological discourse, the hitherto lost subliminal resources of the human being. The creative telos will emerge as our indispensable guide. Will it prove itself self-explanatory? Far from it.

The creative approach to the subliminal ground will lay bare the thread of the specifically human telos: the transnatural destiny of the soul-monad. Only in it can the other two telic principles of the real individual find their outcome.

II. THE REGULATIVE TELOS OF THE REAL AUTONOMOUS INDIVIDUAL: TELOS AND ENTELECHY

The eidetic intuition instrumental in both ontological and transcendental phenomenological analysis gains its clarity and apodictic evidence by focusing on the object in its intentional, that is, strictly rational, ideal skeleton, in which it "appears" with constitutive consciousness. Although we may retrace the steps of its transcendental constitution, that is, the progress of its origin with intentional consciousness, yet these origins — or "appearances" — are themselves repetitive. This "repetitiveness" — as opposed to the uniqueness of the actual origination in existence of the real individual being — is the very condition on which the phenomenological retracing of the intentional constructive progress of the object rests. Indeed, it seems, first, that we may intentionally follow only the genesis of an object preconceived as having been already accomplished in its constructive process. Second, if in doing so we find an intersubjective consensus, it is because what we do follow is beyond the *hic et nunc* of the concrete existential genesis: what the intentional analysis pursues consists merely of the rational itinerary of its articulations insofar as it retrieves segment by segment the phases of its structures as *reproduced.* In this perspective it becomes even clearer what I mean by claiming that the phenomenological analysis of the transcendental constitution is but a formal "reconstruction" of the universal structures of the constructive mechanism, insofar as the construction itself has already reached the clearly objectifying level.[8] The actual existential genesis of the concrete individual in its material process of construction does not reach to the surface of the formation that "appears" as, or is embodied in, the intentional objectivity. It is, then, the repetitiveness of the object, in its ontological "universality," preconceived as already completed that, on the one hand, allows for the eidetico/intentional analysis and imparts to it its methodological privileges. On the other hand, however, taking into consideration merely this present phase of the actual genesis of the existence of the individual being, at which its genetic progress has already taken the shape of the structure "appearing" objectified before consciousness, means paying the price of missing the principles of the actual existential origin of the being.

These origins are not to be found in the final stage identifiable by a repetitive, universal model that is representative of the role of human consciousness in assimilating a givenness, categorizing, and establishing universal interconnections within the intersubjective lifeworld. Even if, as seems undeniable, the individual real being progresses by ultimately falling into a universal category of objective structures, yet his origination in existence, his individualizing himself amid otherwise anonymous Elemental Nature is a unique and unrepeatable instance.

Indeed, if instead of severely narrowing the field of intuition to the sphere in which the already categorized forms appear, we open it, we need not plunge into a precritical attitude. The indubitable evidence of the consciousness that was instrumental in their final shaping belongs only to a stage at which they are already individualized from the whole of the elementary processes of Nature, and their appearance with intentional reflection completes the process of severing them from the same elementary processes. *Where peering into the expanse that intuition opens would mean losing this absolute evidence of consciousness, we may thereby gain in our understanding of being.* In fact, in the attempt to follow the real individual being in its emergence from the primitive forces as it progressively stands out as an autonomous factor of the whole that then becomes the "world," we gain the evidence of its beingness.[9] With his beingness in focus, we discover that the real individual makes his route as an autonomous agent. That is to say, he differentiates himself as a singular being from the rest of Nature by virtue of his own regulating devices.[10] He enters the self-constitutive progress with the nucleus of his own material endowment; together with it goes a virtual operational equipage ready to be triggered into action should external conditions allow it. In this way he may direct from within his constructive differentiation, unfolding his potential material as well as using to this effect the matters external to his very own field.

This inside/outside direction of his inner workings not only establishes him progressively as a self-directing agent but also makes him assume the role of a constitutive factor with respect to the entire realm of Nature. It makes him exercise an impact on the circumambient forces of Nature and to become a center in which these forces of hitherto anonymous Nature converge. While unfolding in his outer/inner directed constructive processes, the agent of the real individual develops his operational mechanisms, and in dealing with external elements strengthens his very own, unique position. In short, he projects his own itinerary, establishing himself in existence, countering or evading some of the crosscurrents he encounters and assimilating others for his own benefit.

Accomplishing in this way, step by step, the constructive individualization of his being from within his own inwardly unfolding rule and on the basis of his own initial resources, he establishes himself within his context of actual existence, which reaches as far as the radius of his operations.[11]

Our perspective on the constitution of the real individual being, which we have labeled elsewhere "cosmological," retains as its system of reference both the eidetic and the transcendental systems of interrelated rational networks, but it rises above them on its own wings. As Eugen Fink has observed, the philosophical significance of the phenomenological analysis postulates from its findings indications of how to proceed further, leaving the analysis itself behind. Verified by these systems of references, the context of the actual existence of the real individual yields, indeed, evidences of its own. They corroborate partly those of the classical phenomenological inquiry; and they partly enlarge on them and bring some essential corrections. First of all, the real individual within the context of his actual existence bridges the gap, otherwise insurmountable in Husserl's thought, between the empirical functioning of the human being and intentional consciousness. As we know, Husserl struggled in vain with this problem until the end of his quest and seems to have been ready to give up the initial project of the total hegemony of intentionality. Without it, however, that is, without bringing the empirical zone of functioning within the intentional system, no unity between the two realms can be maintained.

Secondly, by starting with a focus on the real individual in his hazy, changeable, blurred contours, from which we may nonetheless "conjecture" the outline of his actual existential process, we avoid with one initial stroke the Ingardenian predicament.

In fact, Ingarden had, in his ontological analysis of the fundamental types of things and beings, given an eidetic skeleton of the real individual. He attempted to supply the Husserlian transcendental conception of the real individual being with an ideal intrinsic structure. In the Husserlian conception we see the real individual constituted in his character as a transcendent real object by a series of noematic glimpses (*Abschattungen*) with which the perceptual process of consciousness fills the spaces appropriately left empty within a preposed model. As close as this investigation may come to the object's ways of coalescing and to accounting for the features in which the object "appears" via transcendental constitutive consciousness, yet it does not reveal its essential intrinsic structure, showing what it is in its own right. On the Ingardenian analysis complementing that of Husserl, it is then the complex ideal structure governed by the nucleus of an essence containing a constitutive nature that should account for the real individual's existential distinctiveness from other types of beings and for his autonomy. To the degree to which the intrinsic essence of the being would indicate the necessary coexistence of structural mechanisms that allow for exchange of externally conditioned properties, internal processes with external outlets are founded in the structure itself. Thus motion, action, and interaction are "ideally possible."

The differentiated structural mechanisms explain the existential status of the real being. That is to say, they allow him to be a center of action and the foundation for the causal network of the domain of the world. In other words, the temporal spread of his existence is accounted for. However, there is an important difficulty that Ingarden's analysis entails without solving. The individual cannot remain in a stationary stage; in order to maintain his identity in the temporal spread of existence, the individual can not enter indiscriminately into action undergoing haphazard transformations. To exist as one and the same throughout a succession of temporal occurrences, to be an "individual," means to organize *actio* and *patio* into a developmental process. The question then emerges whether the Ingardenian ontological structure of the real autonomous individual provides us with the principle of his intrinsic development.

This question has to be answered in the negative. Neither the question of the progress of the real being's development nor of his itinerary is raised within the framework of Ingarden's ontology. Hence it is easy to see why the idea of the telos seems to Ingarden altogether irrelevant. But can the conception of the real individual do without it?

Furthermore, I consider the incapacity of pure eidetic analysis to find the internal principle of the development of the real individual being as the reason that ultimately prevents Ingarden from establishing existential bonds among the various types of objects that he categorizes. Disconnected as it were, they cannot function as a base of the universal order. The existing world asserts itself in perpetual evolution, and its process-like progress in its continuity has to be accounted for step by step. Without a principle not only of order but also of the dynamic progress intrinsic to the individual — the foundation of the real world — its systematic progress cannot be grasped; progress and systematic development then fade away, and the origin and point of extinction of the individual being goes unexplained.

Could Husserl have provided a plausible formulation of the problem of teleology after having assumed that, while the eidos does not hold an intrinsic principle of development, it yet serves as a conduit of the individual object's transcendental constitution? Would the framework of research that he has set for himself allow an adequate formulation of the issues involved? Although he with time expanded considerably the methodological framework of his research, he never abandoned the eidetico-intentional assumptions that determine the analytic possibilities of the investigation.

In contradistinction to these two approaches of phenomenology, Husserlian and Ingardenian, I have proposed elsewhere that it is the element of *entelechy* that asserts itself as a decisive factor in the cosmological perspective, within the individual structure of the living individual's autonomous being. The entelechial factor of the living individual being appears as the material nucleus of his existential mechanism, as absolutely his own, and as carrying out the progress of his being's particular identification.[12] While containing the whole project of its cyclic development, the entelechial factor presides over the overall orchestration of vital processes that allows an individual being to unfold, on the one hand, and over appropriate ontological structures and the corresponding biologicopsychic "programs," on the other. *It is only on the assumption of an entelechy inherent to the development of the real individual that we can give an account of the emergence of the individual real being into existence, as well as of his cyclic progress,* which leads not merely toward an unavoidable extinction, but primordially aims at a constructive attainment. In fact, while both the eidetic and the transcendental perspectives emphasize either the universal type as such or the universal progress of the constitutive function, it is this latter approach of mine that implies an orientation. Without this orientation, would there be a life unfolding that advances from one constructive step to the next? Without progress would there be an order of events? And without order, would life and real existence be at all possible?

At this point, however, a new question emerges: "In order to achieve its fundamental existential role with respect to the real individual, should not the entelechy comprise a foreshadowed principle of its final attainment, the telos, the development of which the entelechy is planning and to which it aspires?" With this question, which pinpoints the conjectural postulate of individual becoming, referring it to its intrinsic prerequisites, we project the common network of the existential agent of real being as extending between his entelechy and its telos.

Concerning the nature and the significance of this established telos, we must ask: "Does it comprise the final explication, the sufficient reason for the being of the real individual object?" When I approached this subject in the Leibnizian fashion elsewhere — that is, assuming the bias of the question, "Why is there something rather than nothing?" — I noted that, conceived in the cosmological perspective, the real autonomous individual does not contain in the framework of his being his sufficient reason for being. His individual telos is custom-made for dealing with temporal, contingent, and natural-existential conditions; therefore it is subjected to various circumstances and is partly predetermined by the total life process. The reasons ordering his origin, those accounting for the emergence of specific means to deal with his existential conditions, and those of his having been selected as such a being from among other possible beings, remain foreign to him. Asked to account for them, his telos refers to the concrete universal order of things. This, in turn, refers to the architectonic plan of the objective constitution of the universe of things and beings.[13] Once there, the thread of the cosmological analysis, that is, of conjecture established on the basis of the eidetic and intentional data, breaks down and we are not led to any terminus.

These questions remain unanswerable despite our having begun with the consideration of the primitive initial dynamism; we must accordingly approach them from another angle.

III. MAN THE CREATOR AND HIS SPECIFIC TELOS

Consideration of the real individual's development as it is oriented toward a telos of his natural accomplishment encounters a particular problem as regards the progress of the specifically human being. But then we may wonder whether there is something "specifically human"? However, in contrast to empirical-reductionist tendencies, any philosophy of phenomenological inspiration seeks to claim and establish a specifically human aspect of this real and autonomous individual being that is man.

In this connection it suffices to mention Dilthey, who had already defined Man by his capacity of self-experience; Jaspers, who saw the specifically human element in man's ability to transcend himself; Husserl, who perceived the specifically human factor in the full expansion of transcendental consciousness; and Paul Ricoeur and Ingarden (in his last years), who seem to point to moral consciousness.

Aside from Ricoeur, Husserl as well as Ingarden conceive of man's freedom as an integral factor of moral consciousness, which is on their analysis confined to the exercise of human faculties within the intentional configuration. This position made us wonder whether, if restricted to the basic constitutive system, man would be capable of a complete moral life. Would he be altogether capable of the sociocultural development recognized in the phenomenological conception of the lifeworld? Furthermore, does not the progress of this development presume a renewal of structures? If man remained closeted in the categorial jurisdiction, all the possible variables of forms would be quickly dried up; as progress becomes stifled, all life would be snuffed out. The successive stages of the constructive progress, according to the variability of the preposed set of categories, arrive at a limit at which a new reshaping is necessary. In the postulate of a new reshaping is included the request for the renewal of the types of objects. As to moral freedom itself, if limited in its exercise to the framework prescribed by an unchangeable set of categories, it would be governed by corresponding preestablished ideal values. What meaning would it have to the ethical man to move solely in the closed circle of preestablished ideal values? The central question here is: "Could man-in-his-world unfold completely through the productive and reproductive function of intentional constitution alone?"

And yet, throughout the entire span of phenomenological reflection, the very possibility of lifeworld constitution assumed here, that is, the possibility of the total renewal of forms undergirding this reflection, has been ignored. Indeed, this school appears to have neglected, literally without exception, the genuine originality and exemplary uniqueness of the masterpieces of artistic creativity and scientific and philosophical theory that have manifested themselves in the present phase of the constituted lifeworld and slowly gained recognition and established themselves, although at first they were met with distrust or lack of understanding and often had to wait a long time until man shaped crucibles of the right shape for their assimilation.

Although in his aesthetic investigations Ingarden established varied forms of "transcendence" of the structural forms of objects with respect to each other — in contrast to the all-embracing homogeneity of the constitutive process in Husserl — he did not acknowledge the unique status of the "exemplary" Work of Art in the world. On the contrary, following Husserl, he accepted explicitly the total domination of the categorial system as "die oberste Gattung," and insisted that no new and original object could be introduced into the world. Husserl, in contrast, reserved the privilege to bring forth and introduce into the constitutive genesis a new and original proto-element, but to the *Urimpression* alone. Once introduced, the constitutive genesis molds and reproduces it into endless variations in its preestablished and universal forms.

Thus, neither of these at least partly correlative conceptions of man-in-the-world takes into consideration this need for the complete reshaping of the types of objects on which the very existence of the lifeworld in progress seems to depend. On the contrary, both of them seem to have erected obstacles to its recognition.

In approaching the world through these two channels, we are bound to ignore the issue that appears to us concerning the main condition of man's existence in the world. It is obvious that we must then seek to enter directly the path leading to it, that is, into analysis of the creative function of the human being.

We cannot forget that Bergson was conscious of the necessity of the renewal of the forms; yet, having identified too quickly this need of the constituted world in progress with Nature's innermost need, he perhaps overlooked some conditions found at both levels of phenomenological consideration. In the final analysis, the two should undoubtedly rejoin each other, although in appearance they rest on two opposed poles: the one on elementary conditions (blind Nature), and the other on the human freedom to go beyond the limits of objectivity. We will now attempt to investigate these poles separately and bring them together precisely by means of an analysis of the human creative process.

Our first step involves the discovery that the creative function, guided by its own telos, generates Imaginatio Creatrix in man, as means, par excellence, of specific human freedom — that is, the freedom to go beyond the framework of the lifeworld, the freedom of man to surpass himself.

In the second place, by following on the opposite side the networks and elements of its synchronization, we enter into Elementary Nature. While recognizing the creative function as a basis of the fundamental human condition, we shall discover, then, that to be human means to be creative.

The creative act, which we have analyzed elsewhere, emerges within the human-being-in-conflict, in man's rebellion against the interpretation of the Real that the present phase of the constituted world gives us. Then during his first new set of operations, as he orients himself to the direction to be taken in giving a new shape to his personal and intimate vision of the Real, man finds out that his perception of reality has through habit become the surrogate of experience and is disintegrating. Whether it be in poetry, painting, visual art, etc., or in the pursuit of an idea, a thought, a color in response to our aspiration for a metamorphosis, the constitutive perception of the object of our attention disintegrates in all its aspects. It falls into pieces, leaving, however, a half-open door allowing us a glimpse of the unshaped searching impulse that is rallying all our being's faculties, the passionate, emotional depths wherein all is fused together within a spontaneity that puts to work all that is in us. This spontaneity offers the vision of a new universe.[14] Indeed, upon this disintegration of constitutive consciousness, our efforts do not fall into chaos or inertia as in the case of the pathological disintegration of consciousness. On the contrary, all our energies are mobilized for the discovery of means, ways, and materials, and for channeling them into a constructive apparatus capable of concretizing a new vision. A line of progress is thereby delineated, a path of constructive advance. This path must emerge from a gathering of our faculties for the universal and anonymous constitution that we have unveiled as such. And yet, meanwhile, we pursue a unique and personal objective, within a new orchestration of our faculties, which become organized under the pressure of a spontaneity directed to its own telos as foreshadowed by the vision. Is it possible that this spontaneity, comprising the blind and dynamic passion of our being, may be itself blind — as Husserl argued concerning the "passive synthesis," that is, the unfolding of the intentional constitution about which man does not *decide* but with which he flows *nolens volens* onward, or as Kant sees the *Einbildungskraft*? In such a case, could it be capable of giving direction to this progress? In what way then could the telos of the vision toward which the creative process aspires be attained?

As we have shown elsewhere,[15] on the one hand, it is in the work of art, of scientific invention, philosophical doctrine, etc. — that is, in an intersubjective object — that the vision will crystalize. On the other hand, this new orchestration would not be able to fulfill its role without acquiring its own constructive rules and directives, that is, rules and directives that establish it in its own right, outside the categorial system that controls the eidetic and intentional structuring. Thus, it must create its proper forms.[16]

In this connection it is appropriate to consider the nature of the vision, emphasizing those features that allow it to fulfill its function of aiming at a specifically creative telos, on the one hand, and to have intimate bonds with the particular type of spontaneity that it may prompt from the specifically human condition, on the other.

In fact, the vision of a "new universe," in contrast to the shape that has been granted to the Real throughout the present phase of the constituted world, glimmers on the horizon of every creative impulse. We identify in it again our nostalgia, disabused by the actual lifeworld, our yearnings for the "authentic" in contrast to the "fictitious" in this world, and our desire to surpass the world while undergoing a corresponding transformation of ourselves. This vision, nurtured and carried by the very passion, the emotive forces, and the inner pulsations and impulses of the heart of our being, crystalizes our volitions into a definite objective. It aims at finding or inventing a new form by which to grasp the Real more adequately and embody it as an object that can enter into the present phase of the constituted world — provisionally suspended — in order to transform it. In this way the vision foreshadows the telos of the creative process, a process that it guides while prompting it into action.

Prompting the process, the vision appears to stand for that elementary spontaneity carried silently by the elementary forces we have mentioned. A doubt still arises: While rising from those blind depths, could not this spontaneity generate itself in a totally "neutral," that is, aimless, outpouring, which may be mute and blind, such as *Einbildungskraft* or the *vital impulse*?[17] If that were the case, how could we explain the nature of the emerging vision, which not only seeks an outlet for a constructive fulfillment but also proposes its goal?

Indeed, the deep-seated remaking of our operative faculties in the creative orchestration, which the vision initiates and brings about, and the nature of the progress of the creative process into which its dynamism flows and that it leads, bear witness to the contrary. Already at the point of its taking off from the initial spontaneity, the vision already carries in it the seeds of this creative tendency and outlines the scope of the creative process and its significance for man: *man through the quest for a form embodying his vision is seeking to acquire his own form for his self-explication in existence.*

Thus, man's elementary condition — the same that Husserl and Ingarden attempted in vain to break through, by stretching the expanse of the intentional bonds as well as by recourse to prereduced scientific data — appears to be one of blind nature's elements, and yet at the same time, this element manifests itself as possessing virtualities for individualization at the vital level and, what is more, for a specifically human individualization. These latter virtualities we could label the 'subliminal spontaneity.'[18]

However, does the specific human telos embedded in the mechanism of the creative vision rest in itself? Does not its own genesis indicate other conditions that have to be fulfilled in order that man's creative self-interpretation may acquire its specific significance? As to creative freedom, which appears as the source of specifically human freedom, is it not necessarily bound to another form of freedom, equally inherent to the human being? The question arises: Is it enough for man to go beyond the boundaries of the lifeworld (*Lebenswelt*) in its present preestablished phase, surpassing oneself by self-explication within a work unique and original in relation to this world, in order to satisfy his primitive yearning? Does creative self-explication bestow upon human existence its final significance? Is the transcending creative impulse carried by the vision fulfilled in its concrete completion in a work of invention: art, theory, social life, world development?

We may conclude this discussion of the telos that presides over the enactment of the creative function by stating its crucial role in three major points:

(1) It allows the human agent to break with his imposed survival oriented patterns and advance toward the orchestration of *Imaginatio Creatrix*. Thus it opens up the exit from the closed horizon of Nature (and of the transcendental circle, for that matter) for specifically human freedom.

(2) Leaving behind the preestablished regulative principles of the intentional system (subservient to Nature), the creative telos guides the origin of new forms as fruits of human invention. Thus it leads man from natural determination to creative possibility.

(3) Finally, the creative vision offers us the much sought system of reference for thematizing the preintentional dimension of human functioning as its "subliminal" resources: they appear as a specific endowment of the real, human individual basic to his human condition.

IV. THE TELOS AND THE SACRED: THE SOUL'S TRANSEMPIRICAL DESTINY

By asking whether the creative telos provides the final direction for man's individualization from within Nature and the circumambient lifeworld, we have in fact sought the key to the ultimate significance of specifically human existence. The various stages of the individualization of the real individual in his specifically human form may be considered, as we have proposed elsewhere, as man's self-interpretation in existence.[19] Endowed with a unique meaningfulness, the creative pursuit may be seen as its crucial stage.

With creative interpretation, the possibility of choice being opened, the question emerges: "What are the ultimate reasons for the principles of choice?" That is, what are the final reasons of man's progress in existence?

It is the specific privilege of men, that is, in being human and in being able to deliberate, select, and invent, to pause for a moment and to contemplate the enigmatic position that man has himself outlined in the process of weaving the texture of his existence within the otherwise alien forces of the natural universe and the social world. *He will search for the end toward which the ever refining meaningful pattern of his progress tends.* This pattern in itself appears to be advancing within its fragments of intricate plots, itself always fragmentary, without revealing the secret of "what it is all about." He will wonder: "What is the point that the flowing concatenated meanings intend to make? What is the final aim that glimmers enigmatically foreshadowed and that our spontaneities prompt and tend toward accomplishing? Shall man pursue it by himself alone or have recourse to the Other?"

Indeed, when we pause for a moment to consider it, in a way detached from the flux we flow with, we see that our dynamism will be spent; the works of our invention might outlast it but will last at the utmost only as long as humanity. And yet, caught within his genetic progress itself, man must seek for its final significance.

To have failed to find final significance in the creative telos does not mean — as is sometimes concluded — that there is no answer to these questions; it means merely that the thread of continuity of the meaningfulness of human existence does not stop there but continues at a further stage: that of the sacred. The investigation of the primordial role of the creative function in man's existential unfolding has prepared the ground for approaching it.

Indeed, once the absolutism of the constitutive function is disclaimed and contrasted with the creative function, the constitutive genesis of the human being appears as just one basic functional system of the real individual promoting the stereotypic channels of life; that is, *when we can clearly see the subliminal ground of the initial spontaneity, the territory is ready for the quest after the transempirical destiny of the soul.*

Where the existential life oriented processes in which the real individual unfolds come to the end of their course and meaning acquiring, it is the quest for the significance of this course itself that engages man. Having scanned the possible scope of the two previous lines of unfolding and stated that within their telos no key to man's self-interpretation in existence is given, we can now search for an ultimate aim of this differentiation of the human being from all beingness.

The question arises: Is this accessible to phenomenological inquiry? Our answer is already given: we may approach it with reference to the creative function.

The process in question shares with the creative function first of all the initial situation that both emerge from a reaction against the constituted world. Both of them surge in the form of a quest. Both of them, working through and with the given elements of the lifeworld, carry beyond the lifeworld as it is experienced in its present phase. However, in contrast to the creative process, the search for the key to the enigma of human existence does not mean a rebellion solely against the present phase of this world. It emerges only when the validity of the human world as such is disclaimed — when we arrive at questioning the purposefulness of its changing, precarious, never completed course, when, at last, its lack of sufficient reason is recognized. In short, this search presupposes loss of natural belief in the validity of the lifeworld as such as well as of *any* objectivity. The frenetic quest into which the human being enters on discovering the contingency of natural life in the world does not contain any positive scope. In opposition to the creative impulse, it does not seek a new form in order to incarnate the Real.

Since he has already abolished the outer world, it is toward his inner self that the inquiring subject will turn. Yet the line of questioning he follows, in which he scrutinizes also all the objectively constituted aspects of his experience, relies in this negative fashion on intentional constitution and its product, the lifeworld. But while leaning on them, this questioning runs a different and original course. Likewise, it uses the creative function; and yet, while making use of all the operative elements of its orchestration, this quest avoids the crystalization it aimed at in the pursuit of a creative work. Thus, lacking reference to a definitive objective, it proceeds tentatively, in the "darkness of the senses." Yet, we are capable of tracing its progress from the landmarks it establishes on its way.

To be sure, phenomenology cannot ignore these landmarks of experience, once it has retrieved the subliminal depth of the human condition.[20] Indeed, they appear in their self-givenness authentically absolute, i.e., with evidence that is not relativized to the system of objectivity.

This quest, which we recognize as the road to the experience of the sacred, does not progress without *any* aim. It reveals an inner element of direction, its very own telos. If, following the Husserlian tendency, we call "soul" everything in the human monad that represents the mute element of passion, affective pulsations, internal urge, yearning, etc. — that is, the material ontological element that flows from Nature's depths and that represents the empirical resources of human existence — then we can, in parallel as much as in contrast, call this telos of the expression of the sacred the soul's transempirical destiny. As can be foreseen from our earlier discussion, it is within that telos that the two teloi previously discussed find not only their completion but also a necessary condition.[21] I have indicated briefly that man's two basic functions, the constitutive and the creative functions, are at the same time utilized and surpassed in molding it. Now it becomes necessary to ask: What would be the purpose of this quest? Of what does human destiny consist? In which way could it be the fulfillment of the specifically human genesis in all its arteries? Lastly, how would it offer an answer to the great problems of phenomenology, those with which we have been concerned from the beginning of our inquiry?

V. THE PURSUIT OF DESTINY IN CREATIVE SELF-INTERPRETATION BETWEEN THE SELF AND THE OTHER

The initial question here is: Do we pursue the quest for our destiny within a soul alone, in isolation from other beings? Or, on the contrary, can we find in the "short waves" transmitted between the self and the Other a meaningfulness in our lives other than the struggle for survival — by entering with him into a rapport that may be called "inner creative reciprocity"? The study of the various forms of communication in which human self-interpretation in existence proceeds shows that, apart from organic and vital interactions as well as strictly utilitarian types of involvement with other human beings, it is in particular in the human relations that have classically been termed *Eros, Agape, Philia, Storge* that the specifically personal quest for the meaning of life proceeds. It is carried out in a dialogue between two persons, which is essential to the relations just mentioned. There is, however, more to this dialogue than what Jaspers has so well described. It remains still to be shown what in the human functioning and condition allows it to take place, that is, in virtue of what the metaphysical horizon of human existence is founded and opened. We propose, in line with our argument, to show that interpersonal dialogue is founded in an ontological state of the human person, that of "inner creative reciprocity." This state itself constitutes but a further stage of man's ontological individualization or self-interpretation in existence.[22] Yet, unlike in other phases of self-interpretation, it does not take the Other to be merely a means.

Furthermore, and this merits emphasis, it is a creative process. We will consider from among its innumerable manifestations only the innermost profile of the search after a transempirical destiny — a search shared by two persons.

In the common pursuit of destiny, self-interpretation has two reciprocal poles. The "reciprocity of love," for example, consists in a simultaneous search after the meaningfulness of events for our own existence, which results in "ciphering" their significance into the weaving of the fabric of our own existence, as well as in trying to find the meanings that the beloved gives to them with reference to his own thread, that is, in "deciphering" its meaning with reference to our own. Such an interrogation, intimately unique to each, is shared by both members of this relationship. Already man's self-interpretation in the creative work was directed (in its search for new forms, rules, and principles) toward the other self — the spectator, reader, listener, etc.[23] The message that creative work is conceived to convey is also formulated as the creator's very own meaning of human existence and with respect to another self. In fact, the creative process remains the fundamental vehicle of both.

Indeed, the profile of the self-explication of destiny may be clearly recognized as the thread running through the above-mentioned dialogical relations only in the perspective of the creative function of man, because it proceeds using all the ways and means of creativity. Nevertheless, they differ on several essential points.

First, in opposition to the interiorization and total self-absorption of the artist in the creative process leading to a work of art, in the pursuit of destiny we seek for clues to the meaning to be given our interior existence in a radical turn toward the Other approached as caught within *his own* identical quest. Second, although evidence of our inventive and constructive ingenuity is present in this search, we do not seek to construct, discover, or project any object. Neither do we aim at prompting an event or occurrence of any sort in the lifeworld. In fact, self-interpretation in destiny is not oriented toward ciphering a rational, intersubjective message.[24] In that inner urge to forge a significant thread of our existence in response to the ultimate question that at the point of our discovering the contingency of lifeworld existence can no longer be postponed — the question of how to salvage from the fleetingness of life something of lasting value — we direct our scrutinizing interrogation to the Other. We seek to discover whether, and how, *he* proceeds. What significance does *he* attach to *his* own existence? We provoke him to reveal *his* secret thread by revealing, without any reserve, our own interior existence. The Other receives the revelations of our quest in his own inwardness, in a spontaneous reaction in which he responds; while receiving it he makes clear to himself what he himself is striving for. Inasmuch as he receives our manifestation as the revelation of our virtual self — the one that we tentatively project but have not yet crystalized — he is copresent with us and yet remains absent, because he remains within the network of his own quest for the final meaningfulness of things. From our self-revelation he grasps this virtual self of ours, fluid and fleeting inasmuch as he may compare, confront, identify, or assimilate some of its aspects with *his own* tentative and undefined tender substance.

In fact, what else are we after in this mutual examination of each other, in which we tend to penetrate into the most secretive tendencies and intentions of the other self and the way in which he appreciates their significance confronting it with our own, if not to dig deeper into our virtualities? We seek to scrutinize our resources for feeling, comprehending, and digging deeply enough in the never ceasing current of experience for some clues by means of which a thread of a perduring validity may be sustained through its course. This thread would sustain the perduring validity of our very self; nevertheless, it is to be wrung from the fleeting segments of existence and spun by their means. Progress here requires that we advance to the discovery of the virtualities of the very undercurrent of our being. In order to transcend oneself — that is, our natural, empirical, everyday self of the present stage of existence in which we are constituted and seemingly stabilized — within the scheme of a new meaning of the current of existence, we have to reorganize even our vital functioning at its elementary state.

The spontaneity of our quest, which is continuous and carries us relentlessly onward while undergoing innumerable metamorphoses, crystalizes in this new weave by trial and error in the course of mutual self-revelation. My own new self becomes "other" without it ever stopping in its tracks and without identifying itself with a form or mold. Indeed no definite form may grasp it. It flees from and evades its own eye as much as that of the other self who would try to maintain it in vain in any fixed shape, be it for an instant. We would be tempted to consider that self almost a "nonbeing," because it may not be subordinated or paired with any definite mold of experience into which our constitutive apparatus would like to force it. And yet, the self "exists" because it manifests itself as a *thread* running through our innermost concern, and it is incessantly "molded" and "forged" in its own way in relation to it.

As we have already pointed out, the processes of creative reciprocity emerge from a situation comparable to that from which the creative process *sensu stricto* surges. As we have shown already, their orientations not only diverge on many essential points, but first of all, the quest after destiny decisively addresses the *final* concern, one no more to be postponed on account of present-day concerns or tasks. Where the creative process stops each time a new proposal for the answer to the enigma of the Real presents itself, the quest for the ultimate significance of human existence never stops. And yet it is essentially creative in the ways of its progress.

We address this interrogatory quest simultaneously in a twofold direction: toward our innermost self and toward the Other, while we attempt to scrutinize the most intimately personal experiences, convictions, and attitudes in the foundations and reasons of each. Hence we draw clues for the meanings that we should give to the state of things present, and also past. But in neither of these essentially conjugated steps can we expect to find a status quo, an already present solution to our perplexities and uncertainties.

Also, it lies in the very nature of the quest, in which it is our own being in existence that is in question, that we do not expect to find within the Other a "sense of life" already established — and one that we could accept. It is from within our *own* resources and on our own evidences that it has to come. But none of them has ready material to offer, waiting to be discovered. We would reject any such ready-made finding as not being our "very own." We would not stop at any assimilation of anything we did not ourselves deliberately choose; that would be to grant ultimate significance to a merely haphazard, circumstantial occasion, the reasons of which remain hidden to us. And, first of all, no already available form, or answer, will ever fit the purpose. It is precisely in awareness of all reasons that we have to acquire this final answer to our interpretation of existence as giving the key to its complete script. And in order to know "all the reasons" for our choice, we have to invent the key ourselves.

This interrogation is thus not a passive flow; but in the process of scrutinizing and questioning just what one's own and the Other's attitudes toward one's own existence are (attitudes that remain hazy, undefined, in the twilight of the consciousness of both), we have to probe them by inventing ever new attitudes according to particular biases that we also invent — as fitting our own, and the Other's presumed, circumstances — and according to the clues that we have already half-established, half-projected. In our tentative approach to the Other's quest, we provisionally assimilate, through our interpretation, his own answers to ours, checking the validity of his presumed stand over against our own.

We address ourselves to the Other also as to a *witness* and a *judge*, seeking his approval or acknowledgement of our deepest concerns and dilemmas. We try to introduce him into the very heart of our perplexing attempts to interpret the given, by scanning one by one the possible meanings of the given by which we may approach it. We show him the reasons we envisage, weigh, or reject outright, the evaluation of the meanings that our imagination proposes, probing thereby ever more deeply their intrinsic virtualities as they suggest themselves to us. We then introduce the Other into the very heart of our creative investigation as a *second self.* We face him simultaneously as an "other self" insofar as we expect to receive from him, in the common current of interrogation, the opening of his own being-on-a-quest. We face him as a being-on-a-quest other than ourselves, over against whom we may measure our own quest. He is an "other self" sharing our own self insofar as he receives, observes, and retains for himself the essentials of our own quest, and also develops *his* quest by in the same way reciprocally checking its progress with ours. Each of us, in order to reveal to the Other the meanings that he might propose for facts, events, feelings, decisions, acts, longings, and deepest urges — and to along with him seek to understand them and check their relevance to understanding — scrutinizes one by one all the available and possible systems of interpretation. Trying them each in turn and rejecting each. None of them, in fact, appears capable of transmitting, or of holding, this unique significance we seek to establish.

This reciprocal inward quest develops and maintains itself through innumerable instances, fluctuating, instantaneous, and yet sustained by the thread they forge. This thread seems to sustain itself on its own. It does not have any definitive foothold in empirical reality nor a preposed set of regulative guidelines to follow: it proceeds by trial and error. Relying essentially on human invention it would be accessible merely in disparate instances of spiritual experience, if it could not be understood and accounted for as the process of man's ultimate phase of creative interpretation. This quest of the self and the Other for a transempirical destiny is seen as a quest for the final telos of man's self-explication in existence only when we discover that it proceeds by means of creative self-interpretation.

To realize this, let us summarize the positive points of our description.

In the first place, as the dynamic thread of our communication with the Other's self, the quest for our destiny is — like the creative process — constructive. Although it does not construct an intersubjective message, it carries on a unique and unprecedented transformation of the state of our own being. Second, like the creative process, it advances by breaking, one by one, all the joints of the functional intentional network oriented toward the constitution of objectivity. Furthermore, as with the creative process, self-interpretation in destiny employs for its own end the means of human functioning and all the points of reference that the constituted lifeworld and ourselves may furnish for its use — and yet, it is by no means subject to the lifeworld's organizing regulations and projects. On the contrary, self-interpretation in destiny is worked out through a functional system devised for its own unique "purpose" by the interrogating process itself. No principle for its interpretative articulation or of its meanings is preposed. All are essentially tributary to our invention with reference to our deepest self in its communication with the Other.

Like the creative process, this search that self and Other make for our destiny in personal interiority is carried on in all functional dimensions and by making use of all the operational mechanisms of the human being, considered as a real individual as much as a person. The questions that the self addresses to the Other, and the answers that he gives in turn appear ambiguous, just as do the questions and answers of the creative interrogation — because, although they bring to light the most profound and secret longings of our innermost being, yet they are merely *tentatively* proposed. Their meaning becomes more precise and nuanced as they coalesce in their encounter with the response of the other self and in our own probing appreciation of that encounter. In this bringing forth, reaching out, and turning inward movement, the reciprocal quest suspended between the two poles establishes, like the creative process, a specific orchestration of all human functions. It draws directly on the subliminal resources of the human being, which in the creative context become liberated from their intentional constitutive ties.

In the search for its own script, the soul lends to this creative dialogue all her means, down to those most rarified, which otherwise cannot be partaken of. She addresses the other self with *transparent sincerity,* a sincerity that she is not capable of projecting toward even herself when alone. She addresses him beyond the reach of any objectivity, leaving life pursuits, concerns, and values aside, that is, beyond the vicissitudes of time. Since this address does not aim at any particular object, its objective being to conjure the ultimate significance in which everything comes into its own, it dwells in the supratemporal, the abiding, the eternal.

And yet, the communication in which its being is crystalized seems to be in danger of breaking down at any moment. The ambiguities of this reciprocal interpretative creation are so extensive that we may even ask: Do the encounters ever come to pass? If so, they are a moment's games. We would have no objective evidence — intuition could not bring it about, since there is no objective content to be sighted — had not each encounter meant just a new clue that we discovered or invented, weaving a further mesh for our transnatural destiny. Our quest seems to abandon all of life, and yet it salvages its lasting significance.

But what about the other self upon whom — as it seems — we have suspended our whole being and existence? Do we really ever meet him in *his* truth? Each self progresses, in fact, separately. Each of them, in the quest after the new interpretative system that would be final, has scrutinized all available cues and rejected them. Nevertheless, it has accomplished the essential task. In the quest for a destiny uniquely his own, each self discards interpretative schemas one by one, like the leaves of an artichoke, despoiling himself of explications. The quest advances through the invention of new ones to be discarded in their turn. We believe in the process that "encounter" with the other self, the communication to him of this ultimate concern of ours, would occur if we could only find together a meaningful system in which both his search and ours would be explicated. Unreserved radical commitment on the part of both offers the unique possibility to the self and the Other of transgressing the closed framework of the self, with each advancing toward that which he is not in this impervious elan that becomes more and more intense as our creative interrogation becomes more demanding and refined. The progress of its "positive" accomplishment, that is, of the "ciphering" of our very own exfoliation of existence, is, however, constantly disrupted; we lose the loops of the thread — the meaningful forms proposed; we lose the meshes that were expected to bring our experience together with that of the other self into the same net. While we believe we have thrown a hook for him to grasp, we see that it has no meaning for him.

If for the accomplishment of our endeavor we would expect the genetic "ciphering" of a lasting meaningful system, and a "text" — a task to which our repeated efforts seem to be applied — a text in which the other would recognize himself, this is never attained. At a certain point in this seemingly constructive advance, when we are convinced that our self and the other self is firmly established within the same meaningful text, meaning that we have constructed a common "universe" that we both share, a universe that is meaningful with reference equally to oneself and to the Other, and when we believe we have grasped the other self in his identity, at that point the entire edifice collapses.

To answer some of the questions that force themselves on us concerning the significance of this attempt or the nature of ultimate self-interpretation in existence, we must state, first, that although no definitive meaningful system can be applied to interpret the progress and attainments of this quest, that does not mean that all its scanning, discriminating, and renewed adjustment were fruitless. On the contrary, it has served several purposes.

First, although we fail to meet the Other in any rationally definitive, objectively "lasting" way, yet the instants at which we communicate together in our ultimate concern for existence are extremely rich in "substance." They expand their significant ways into all hidden dimensions of our being and reveal thereby that our virtualities may serve as the cornerstones of progress in our quest. Second, although our quest does not lead to the definitive establishment of a system of meaning for our route, it fulfills the crucial role of transpiercing the objectifying strata of the lifeworld structures that keep us closed in on ourselves and in contingency.

In fact, it is in this progress of liberation from ties to empirical, life oriented structures and processes that run through our being that we do actually uncover our being's virtualities. In the common search with the Other, the intimations are found that may be worked into a thread of destiny.

And yet with broken communication, this thread itself breaks. Every discourse, every interpretative process, every communication fails. The virtual sense of their genesis, for the sake of which we have employed all our resources, escapes a finalization, a completion of its meaning. Does that mean that we face a void on reaching the breaking point in this pursuit? That would undoubtedly be the case in an isolated, self-centered search for the meaning of life or of human existence — as presented by Kafka, Camus, and other contemporary writers. In an isolated search, one missing the creative reciprocity of the Other as engine and support, the initial impulse does not crystalize into a constructive process; it does not unfold the inventive imagination that would nourish and stimulate its advance; without it, the search becomes sterile and dies.

The elan of the reciprocal inward quest to decipher the ultimate significance of the human life course, on the contrary, is prompted and intensified as it proceeds. When it breaks down in its weaving, the soul, on the one hand, is left free from empirical ties; when, on the other hand, the Other detaches himself and vanishes from her horizon, she finds herself lost "nowhere" and with "no one" to turn to. Cut off from the world, all her spontaneities flow into the urgent impetus of her quest. When that abyss opens, the soul finds herself face to face with the Radical Other on the other side.

In retrospect we come to see that in this quest that works through but leaves aside the empirical dimensions, we have been uncovering and elaborating *the inward ground of the sacred in the human being.*

VI. RECAPITULATION

Thus, we see how the telos of the soul's transempirical destiny is the key to the significance of man's specifically human individualization (or self-interpretation in existence) considered within the whole scope of experience. It is ultimate; but this telos is not constructively achieved. The transcending impetus of man's quest stops at the boundary of human experience. All significance that the progress of destiny might have acquired dies in it. However, the impetus goes beyond and does not fade away. On the contrary, it reaches its apex. Its encountering the Absolute Other, the Ineffable, does not need any further explanation or interpretation. Its significance is final and as such appears as the key to man's self-interpretation — so much so, that without it, the specifically human element would vanish. Here, the philosophical discourse ends. If we would try to pursue it further, it would be necessary to perform a total "conversion" of our principles of self-interpretation (or in the words of Kierkegaard, a "radical jump").

I then propose that man's triple telos contains an intrinsic interconnected network of bonds of which man is the creative agent. These bonds, at the one extreme, have their roots in Elementary Nature, and at the other extreme, tend toward Transcendence, the Radically Other as their ultimate reference point. The emphasis on his telos, then, places the human being within the total expanse of his condition.

TYING POINT FIVE

INTRODUCING THE MEASURE: CHRONOS AND KAIROS

INTRODUCTION

LIFE'S TIMING ITSELF VS. THE HUMAN ESOTERIC PASSION FOR ACCOMPLISHMENT

The impetus of the logos of life never stops at its singular catch, at a concrete reach. In its fulgurating outburst radiating in all directions, it nevertheless tends toward a relatively stable post within the self-individualizing process. Those processes relate to an ontopoietic design whose functional and generic rays radiate in all directions within its circumference. The ontopoietic line of life's unfolding gives life an appearance of temporal succession; life flows, we flow in ever new events; it flows inexorably onwards. Although its continuity is interrupted by innumerable events, constructive/destructive/arresting etc. events, we refer to their constructive succession along the ontopoietic course of individualizing beingness. We measure this succession as "temporality," time in the "substantial," abstract form given to it by our reflecting mind, that is, "chronos," which flows ever onward, without stop, halt, repose, return, or retrieval.

But much of the ontopoietic process by nature attains a measure of repose, of achievement in the respective tasks of its operations. As small as those tasks may be, without this constructive proficiency of the logoic impetus, no world, no life, no "something" would be possible at all.

So in the web-platform of the Dionysian logos there appears a reinforced strategy of the logos of life in carrying out its pattern of impetus and equipoise: the entire system of moral life. Its foundational schema comprises a direction taken at the prompting of an ideal, the turmoil of new interactions and interrelations, and then a relative repose, halt, found in the radius of forces gathered in the effort to accomplish the ideal.

Indeed, within the Dionysian sphere there is no predominant direction in the flow and passage of events. To be human means to act, to invent, to create. The flow of our human acts into the accomplishment of an enterprise however small is of great, paramount significance for the inward play between the Dionysian impetus and equipoise.

With these few words of introduction we may now pass to the nodal point (junction) from which the numerous logoic rays that enter into the measuring of life may be delineated.

CHAPTER ONE

CHRONOS AND KAIROS: ORDERING ON THE ONE SIDE AND RADIATING ON THE OTHER

The question of time preoccupied philosophers already in Antiquity, and its interpretation in history has known several culminations. Strangely enough, time does not fall into any ontological category, for all of them depend on it. In fact, the various modes of existence (e.g., reality, ideality, fictive existence, absolute existence) are to be distinguished fundamentally by their relation to time, which their respective ontological structures indicate as being either necessary (real existence, which allows change and action) or inadmissible (ideal existence, which does not compass the possibility of change), etc. The question thus emerges — in virtue of what does time play such a basic role with respect to ontology?

In a different way, we may presume the same crucial role of time with respect to metaphysics and philosophy of mind. But what about the role of time in culture and social life, that is, in "practical" philosophy, on the one hand, and the specifically human vocation of seeking a transcendent personal destiny, on the other?

From the end of the nineteenth century to our day, we have witnessed a distrust of speculative thinking, its dismissal as the spinning of overall theories or "systems" of philosophy in which illegitimate connections are made between and among various facets of questions for the sake of projecting an all-encompassing context for philosophical inquiry.

And yet, when we turn to the interpretative approach prevailing at present, we find that no single piece of the "text" may repose in itself; interpretation is referred to its "deeper" layers of meaning, and in the last analysis, we should pursue it in a regression through the entire development of culture.

This shows that philosophical issues cannot be formulated or pursued adequately other than within a context that, if this be at all possible, does not call for further justification or explanation. Thus, we propose to here treat the question of time, that of chronos and kairos, as it appears in its original mode, that is, squarely within the context of my "phenomenology of life and of the human condition." This is a context ambitious enough to repose in itself without turning to further points of reference.

I. CHRONOS AND KAIROS IN THE ONTOPOIESIS OF LIFE; SETTING THE ISSUE

Contrary to the established phenomenological tradition, according to which we begin by asking questions and then through painstaking effort seek an answer that may yield a definition of the factors involved, I will begin my inquiry by proposing a definition of time and will only then reveal the analytic context within which it receives its determination and proceed to its exfoliation.

I propose that time, in both of its modes — chronos and kairos, as differentiated long ago in Antiquity — is the grand, infinitely complex, flexible artery-in-progress of the constructive advance of life, encompassing its relevant cosmic links, on the one hand, and its reach for a portal to the transcendent aspirations of the human being, on the other.

This definition situates time amid all the infinitely changing modalities with which the simplifying human mind categorizes the innumerable ways in which processes, events, functional units, etc. advance life (chronos) toward its nodal points of accomplishment (kairoi) as well as amid the specific circumstantial, as well as essential, strivings (and the crowning significance of their accomplishment) at the very heart of the poiesis of life. Time is the main artery through which life's pulsating propensities flow, articulating themselves, intergenerating.

By situating time and temporality (together with space) within the framework of the phenomenology of life, or, more precisely, of the ontopoiesis of life, we are radically shifting away from the classical and modern contexts in which time has been envisaged and within which its conception has been situated.[1] And it has to be remembered that a philosophical conception always occurs, whether deliberately or implicitly, within a framework, within the "context" of a larger conception that determines its bearings and sets guidelines for its treatment. With the notion of time, it seems that it has oftentimes, owing to its pervading presence within the realm of givenness, assumed a frontal, primary position and become absolutized. But our encountering of time and temporality within the orbit of life restores it to its proper framework.

Let us recall that the context of the phenomenology of life is circumscribed (projected?) by a matrix: a) by the human creative condition as the Archimedean point for initiating inquiry; b) by the self-individualization of life's being then the thread inviting and leading us through the labyrinth of life; and c) by the existential tentacles that if followed reveal the unity-of-everything-there-is-alive. And no longer are any of the objective and subjective aspects of time drastically isolated; endless succession and endless duration are no longer so distinct. When time is no longer absolutized, time and temporality are situated in their original, primogenital setting. The genuine perspective that had been speculatively lost is regained.

The difficulties of assuming an absolute status for time were seen already by Aristotle when he set forth the notion of time saying "it is impossible to conceive of time when time was not" and recognized an organic bond between time, movement and place, and the soul (*Metaphysics* 1071 b7).

Indeed, Aristotle saw time as being inseparable from motion and movement as the essential feature of physical reality (*Physics* III, 201 al0). With these two points made, time is already immersed in physical processes and generated together with them. The view that time is also inseparable from the soul (consciousness) completes the picture of its bearings (*Physics* IV, 223 a27). According to Aristotle the reason for which time cannot be separated from the notion of the soul is that time is envisaged in now moments involving a "before" and an "after." Here is the "number of motion," and in order that there be a distinction between "nows," there must be "someone" to measure or count the lapse of time.

Thus, consciousness, the soul, (an observer) is conceived as being the indispensable synthesizing principle for the measurement of time.

We see from whence there emerges the Aristotelian context of the physical body as the context of the notion of time: the context set up by the time-motion and space-soul complex. Because the role of the soul appears to be central, it is puzzling why the good Aristotelian Leibniz, who brought out forcefully the monad as a living crucial factor in the world and stressed *vis viva* as the motor of the world of nature, would exclude the soul from among the principles that he proposed for "every possible world," predicating only the space-time matrix. This omission could be explained somewhat only if we assume that he interpreted the notion of the "soul" exclusively in terms of consciousness, that is, of a measuring observer.

In our own account of time, we will emphasize the Aristotelian apprehension of a "synthesizing" principle that is indispensable to the space-time matrix vis-à-vis the physical world. Yet we will give it a fresh interpretation, one acquired within our context of the ontopoiesis of life, for we deliberately expand the physical world into that of life. Thereby the system of relations between the three moments of the Aristotelian matrix will undergo a transformation of roles, a reversal of priorities in the roles they play. The soul will receive a fuller meaning (as Aristotle intuited it). The notion of Being will recede from the horizon as a new metaphysical matrix of Beingness emerges to take its place.

Let us anticipate our itinerary. Within the ontopoietic context we will a) move from a neutral, universalized notion of "motion" to constructive moves within the cosmos and to constructive, telos oriented "movements" of bios; b) go from an understanding of the "soul" as a counter of movements unrolling before it, as it were — "nows" involving "past" and "future" — to a notion of the "soul" as an agency of life's constructive progress, a living agency that distributes the articulations of life's movements in operations and processes and generative moves; and lastly c) by distinguishing in classical fashion "measurable" and "qualitative" time, chronos and kairos, respectively, show how they come to emerge in their unique roles within the exercise of the specifically creative virtualities of the Human Condition.

II. THE REORIENTATION OF SIGNIFICANCE WITHIN THE NEW LIFE CONTEXT

Aristotle, in *Physics* I-IV, 219 b1, establishes an intrinsic relation between time and number, saying first of all that time is the numbering of movement according to the prior-posterior order. But he later specifies that time is a "kind" of number (ibid., 219 b5). That means clearly that time is not in itself a number, and is not a means for numbering either, but is itself numbered. Time and motion remain in reciprocal relation to each other when it comes to measurement: "we measure not only motion by time, but also time by motion, because they determine each other mutually: because time determines motion, of which it is the number, and movement determines time" (ibid., 220 b14-22). If it is the soul that counts, several questions emerge: if the soul is the measurement of movement and of time too, does that mean that it creates the measure? Or is it the case that it merely performs the operations of measuring and counting? Or is the soul, consciousness, itself time — as some contemporary theories of consciousness would have it? And what about number as measurement itself? Is it created by the soul or is the soul the occasion for the numbers' being applied in ideal form to the concreteness of motion and change? If so applied, the soul would be at the source of the intelligibility of time. What is here at stake is the understanding of the soul as such and of its involvement with concrete motion not as an intellectual observer, but as an integral factor of all that moves, that is, of living beings and the life-processes in which they are entangled. As a matter of fact, we may discover in the full conception of the soul of Aristotle himself the same intuition that carries our phenomenology of life. This is an intuition of the soul and its role with respect to the processes and progress of life, of "coming to be and passing away," that we share with Aristotle in our conception of the self-individualization of life. We will reinterpret the above-mentioned time-motion-soul relation in the perspective of this conception.

If we consider the fact that, according to my conception of the self-individualizing-in-existence of life, it is through a crucial specific existential/ontological device that differentiates all life from nonlife, that is, through the inward/outward oriented central "agency" of the individualizing beingness, that life's ontopoietic processes are carried out, then we see that it is not "universal" or "universalized" motion that is at stake, but the movements of bios, which numbers time. Aristotle saw in this central agency of each living beingness a "soul," a system of articulation with a glimmering of consciousness. And indeed, even at the simplest level of complexity, as Leibniz also maintained, the "animal soul" is seen as having a synchronizing capacity. This is not a mere "observer" but an articulator of motions into their successive, concurrent, and telic knots, a configurator of phases and culminating points of accomplishment. In our context of the ontopoietic unfolding of life, the soul plays, then, as in the Aristotelian perspective, a "counting" role, but this "counting" is understood as introducing the distinctiveness of the phases of temporal progress: it involves an organizing, articulating, and dynamically operating function, a living function.

The soul, which we have conceived of as stretching over the entire territory of the self-individualizing processes of life, comprises then all the being's circuits of energy and forces, all stirrings, moves, operations, and processes, from the simplest vital ones, through those psychic and gregarious ones orchestrated by the entelechial principle of vital significance, up to the emergent Human Condition, which with its creative orchestration ushers in freedom and the specifically human significance of life. It is the soul that gathers all the life processing functions and through them articulates the constructive progress of life. We share here the Aristotelian intuition, and, in agreement with Aristotle, as well as with Leibniz, we see the presence of consciousness, be it a glimmering of "selfhood," in even the simplest forms of life. We, however, take a further step, and going beyond Aristotle, see the soul, as the concrete principle of life itself, as promoting through life's functioning the movement through which time is numbered.

Time, being originally and primordially coeval and congenital with the constructive deployment of life, and hence to be distinguished among life's relevancies, both those cosmic and those abstractly rational, participates in two lines of a combined strategy of life, and hence has one twofold modality, that of chronos and kairos.

CHAPTER TWO

CHRONOS AND KAIROS SEEN IN THEIR ONTOPOIETIC ROLES

I. TIME AS THE MAIN ARTERY OF THE PULSATING CONSTRUCTIVISM OF LIFE

Time and movement are inseparable, and, indeed, there is no relation of priority and posteriority between them; they are both functions of a constructive progress of life. Time appears, in fact, in the crystalization of the motility proper to life's functioning; it appears in manifold modalities and is as ungraspable in its entirety by the simplifying human mind as is life's progress itself. In contrast to the classical and current identification of temporality with succession, it seems that, in the main, the constructive phases in which life crystalizes, namely, in the phases of generation, growth, fruition, decline, and extinction, it is the modality of *confluence* that characterizes the measurement of the lapse of time. That is to say, in these phases of growth and diminishment the vital processes of bios do not proceed through what we can only abstractly distill from the entanglements of the functions that carry them on — a direct cause-effect nexus in which the lapse of time is to be apprehended as a sequence in which each occurrence or event would have a clearly traceable cause and be in turn the cause of the next occurrence, change, transformation. The ontopoietic course of life, as I have been showing for some time, proceeds in contrast through the confluence of numerous operations, processes, moves, stimuli, etc., each of them either belonging to some or other conundrum of forces or motion, or surging forth spontaneously. This confluence performs the transformative or sensoromotor operation. However, this confluence itself crystalizes in a multiple *motio*. Hence, it crystalizes in time, which lends it a "moment" of fulfillment, the measure of the step onward in the process of growth or decline. Each constructive advance of individualizing life (e.g., the opening of the petals of a flower, the rise of the sap of a tree in early spring, the cross-pollination of flowering plum trees effected by insects...) is a result of a bundle of results — of numerous operations and processes, each of them crystalizing segments of time that flow together to work a change, a transformation, a moment of constructive progress. Each advance in life is neither the effect of a single cause, nor does it single-handedly contribute or effectuate another change. On the contrary, each occurrence in the course of bios' unfolding is significant in various inward/outward radiating directions (*inwardly,* the opening of a flower is a phase preparatory to fruition; *outwardly,* it is the opening of a source of nectar that nourishes bees, wasps, hummingbirds, etc.).

Thus, whether intelligence registers it or not, life-bios is timing itself. It measures itself, and thus its temporal spread, by its natural constructive advance in the cyclic cosmic order.

It is not time that bring in this order; it is not that time is added to the constructive operations from the "outside" as if it "existed" or is "just there." The moves of the vital operations beginning with the lurking organic forces that emerge from seemingly nonorganic ones, or the moves of a virus entering a cell in order to replicate itself, all these time themselves from "within," and their advance marks and measures temporal progression in itself. Thus, it is from the inward moves of the sentient and living soul of the living being that there stems the measure of motion — organic/sentient/psychic motion — and the time that crystalizes itself in it.

II. THE INTERPLAY OF CHRONOS AND KAIROS IN LIFE OPERATIONS

Life operations simultaneously display a twofold orientation: "chronic" and "kairic." That is to say, we cannot envisage chronos as a linear sequence of events. Events are end results of operations and processes within a conundrum of moves drawing on and restricted by the environmental complex within which they occur. Some events are oriented by definite factors. Others occur at random, result from processes of mixed direction. It is only by simplifying their run that we may speak of "causes" (and not one cause) and "effects" (and not one effect), considering that the causes are directly or mediately multiple and so are the results. A snowfall, for example, is the cause of a spectrum of "effects" on nature as well as on human existence. To seek a clear-cut sequence of succeeding events is simplistic. All that leads to the "kairic" moment of the flux of becoming — the event as such — is to be sequentially organized as chronos only with respect to the end result. In the very distinction between these two modes of time there lies the strategy of the logos that congenitally conjoins them.

Time, temporality, chronos is thus not linear; there is no linear flow of time; the linearity of time is a pure abstraction of a geometric nature. What "there is" is a flux and reflux of forces and synergies charged with their own dynamic propensities that come together operationally in various functional sequences and fall into a constructive pattern centered on the fulfillment of an end result. Their flowing together into the same segments of functioning and their articulations bring about knots of fulfillment (kairic moments), each of them marking a constructive step "onwards" and a constructive progress for which some external influences had to fall away while other marginal moments concurred. It is this step of progress that punctuates the advance of life processes; abstractly we see it as the tracing of a "line," which we then artificially divide using a universally accepted measure. The motion of the sun, its linear progress through the sky, suggests to us such a conception of time.

Then there is the inward/outward universe of the psyche/consciousness with its vibrating sentient, emotive, intentional, and creative strivings — passional forces having their very own streak of significance for the construction/constitution of the specific universe of human being-in-progress.

Scanning and spacing assume a mode of their own in this progress. The conscious universe appears with its very own expansion, its very own "inward" spacing of objectivities that are presented outwardly in stabilized form. Time appears as the very course on which our outward acts are embroidered, but also as the measure of these acts of embroidery themselves, as chronos. But the inner process that brings these acts to the world involves will, judgement, decision, and initiative. Their purposive orientation makes the case for a uniquely human strategy of the logos, the surging of the kairic passion.

There is no need for intelligence, or for any other observer, to register the lapse of time. Life simply proceeds and temporalizes itself. We must consider that in addition to the universally constructive crystalization of time and its contrary, destructive tendencies as seen in the whole cycle of individual life going from generation to extinction, there is that which departs from the pattern, a particularly vital tendency of bios, its self-constructivism.

As I have been pointing out throughout in my treatises, what makes life is its particular, constructive tendency, which is decisive for its unfolding and its spread, nay, for its very taking place. And this constructive tendency is embodied in a self-individualizing progress. It is only by consistently deploying forces and directing them from within that a living beingness may establish itself and maintain a course of progress within adverse, neutral, or merely propitious outward conditions. This constructive tendency is then appropriately "embodied" in a life principle intrinsic to the emerging life process — to the emerging living beingness — what I call the "entelechial principle of life." According to its direction ontopoietic constructivism unfolds, unfolds while it constructs with each move. There is an intrinsic ontopoietic "agency" of life directives. This amounts to the unfolding of an "entelechial schema" presiding, as it were, already in its germinal form over life's constructive advance to a self-fulfillment in accord with a "model."

In fact, in the ontopoietic unfolding, each juncture of development, each process, and each step of the advance participates in a vast schema, a larger fabric, an intertwined, interconnected, cooperative network flowing from different functional sectors. Each operational confluence marking a segment of the constructive advance is constructive because it appears to be oriented by a telic principle, an aim to achieve, a task to fulfill with respect to the previous stage of the constructive advance, and to a future one that is being prepared as an even further step of constructive advance.

However, the confluence and positive constructive run-on of these operational moments is not to be taken for granted. All depends, first, on the coming together of innumerable inner/outer oriented factors. Or, in other terms, since the confluence to be accomplished is to be a moment of time, it gives to time, temporality, the confluent moment of the constructive mode; it projects a special significance upon the preceding temporal moments, the timings of the indispensable conditions that come together and allow for the confluence that in turn makes its temporal moment share in the significance of the conditioning: that moment acquires a special significance and together with the confluent occurrence appears as a "propitious moment."

The routes of bios are ceaselessly punctuated by "propitious" and "unpropitious" moments for its forwarding itself; they have to unfold in favorable or unfavorable circumstances; there are climatic, ecological, gregarious, as well as cosmic transformations and cataclysms that weigh in.

If we identify the long winding operations and processes of timing with the notion of "chronos," we then subject them to the simplifying effect of human intelligence's "impartial observation," which reduces time to a succession or continuity of occurrences and events. The multifaceted timing of the constructive achievements that punctuate life's progress may then be seen as instances of "kairos." Chronos is the order and sequences of life. Life-constructive fulfillments that mark ontopoietic progress, their occurrence within the play of favorable and contrary conditions, are the moments of kairos.

III. TIME EXTRAPOLATED: THE COSMIC RELEVANCIES OF BIOS

With this basic clarification we enter directly into the puzzling issue of "cosmic time," as well as that of "mechanical" temporality. As I have pointed out before, life is, indeed, not only the point of a specific convergence of forces that emerges within a conundrum of specific cosmic conditions, but life in all its forms "implements" the specific laws of the cosmos itself. Although these laws (e.g., gravitation) are not directly "present" within the structuring of the forms of life, they are re-presented within its specific forms and processes through what I have called the system of "cosmic relevancies." Thus when we ask about the nature of cosmic movements and their temporality, we are already raising and formulating these questions "from within" those very cosmic relevancies in accord with which we are constituted as living beings, even as we participate through our very endowments in that system. In short, we approach the cosmos with our life-established notions of and devices for measuring motion and time. Our estimation of cosmic developments over billions of years is made by transposing onto a sphere unknown in itself our own very specific, uniquely specific, life factors. We do not discover in the cosmos — at least we have not so far — other forms of life. Should they exist, how would we be able to recognize them from within our own form of life and specify their means and relevancies? We can identify the motions of the stars, the falling of meteors, the streaming of cosmic forces, etc., but only partially the constructive moves, processes, and operations that we distinguish in the sphere of bios as they orient and surge from "within" and work a constructive achievement.

Thus, having no evidence of "movement timing itself" in the cosmos, we approach these moves and processes as stripped down versions of the timing of life and so posit for it an abstract line of succession, seeing only lifeless mechanical motion. In this fashion, we arrive at both "cosmic time" and the "time of mechanical motion" as well as the uniform abstraction of all of time's qualitative life coordinates in the measurement of time by clepsydra, clock, metronome. We empty the prototype of living time of all its genetic content, leaving a mere skeleton.

IV. HOMO CREATOR: THE KAIRIC MOMENT AS THE MOMENT OF CREATIVE ACCOMPLISHMENT

1. *Freedom: Kairos between Fancy and Arbitrariness*

The realm of bios is timing itself, poised as it is between the two extremes of necessity (the constructive entelechial principle) and hazard (the external conditions for the principle's deployment). Chronos and kairos, which are life's arteries, take shape in this oscillation.

However, this situation underwent a gradual change as the animal species emerged in evolutionary progress, in the complexity of their forms and their ever growing flexibility in matching inner demand and external supply as they counteracted adverse conditions and availed themselves of opportunities. This flexibility reaches its culmination in the surging of a new category of life, namely, that of human freedom. In fact, it is between freedom and arbitrariness that human self-individualization oscillates. In indicating these bounds, we point to creative endeavor as the vector moving within them, using them. Within the context of the phenomenology of life, I have singled out, within the evolutionary progress of constructive types of living beings, the Human Condition as the Archimedean point of the ontopoietic unfolding, that is, as one of the three main factors in the metaphysical matrix of life. Having focused above on the unfolding of bios and the animal forms of life, now is the moment to move on to the ontopoietic crowning point of complexity reached in the creative Human Condition.

The timing of life's self-unfolding, its self-interpretative course, here undergoes several transformations, with time as kairos coming into its own. Kairos here assumes original, uniquely significant roles with respect to the specifically human significance of life, its projecting of new avenues, history, and the personal quest for transcendence.

In the crystalization of the creative Human Condition, the hitherto relatively rigid entelechial decoding of life's enactment seen in the progressive unfolding of species in evolution attains the flexible, inventive, endlessly transformable progress of "free will" or whim, of human beingness. Chronos, and more particularly kairos, now take on novel modalities and roles.

Cutting through innumerable questions concerning freedom, constraint, arbitrariness, as well as the universal human round of life — order and turmoil, monotony and revolt, peace and crisis in personal and societal life — is timing itself. At one hand, are our everyday regular activities, gestures, and, at the other, there is the ceaseless flow of our personal conscious psychic life of stirrings, sensing, emotions, motivations, and conscious acts variously concatenated. From this ever vibrating conundrum, in its variety ungraspable, I will single out the main artery of specifically human self-interpretation in existence, of human self-individualizing, namely, the intrinsic timing of the modalities of chronos and kairos as they work in tandem in the implementation of the unique prerogative of the human being, the creative endeavor.

To cut a long story short, human self-individualizing progress defines itself by striving for accomplishment. The accomplishment of tasks, aims, ideals stands in contrast to the vital organic sentient movements of life's constructive progress heretofore. These are now projected inventively by the human being, the creator. The tasks themselves emerge as fruits of a human creative reorientation within the circumambient world. The creator's innermost tendencies and desires as transmuted by imagination are tied not into the knot of the appropriate natural response, as with animal life, but by a free deliberated decision and impetus to seek an accomplishment, to undertake the projects that ensue. Natural promptings are transformed into those of the surging will, igniting an initiative to undertake an accomplishment. Due credit must be given to the progress of the evolution of species from entelechial flexibility to the exercise of freedom, to the point where the creative orchestration of emergent specifically human functioning comes into its reign. But I see the crystalization of the creative function in the Human Condition as the instance of the emergence of the unique timing that is kairos. Human will seems to soar on the wings of imaginatio creatrix through open and unlimited skies, "free" from all constraints, and seeming to enjoy all "possibilities" in a play of fancy and the "free" whimsical selection of them, that is, nondirected, altogether arbitrary choice. The human will to undertake, initiate, and institute — which means to move freely in the self-individualizing progress of life — is caught in between fancy and the arbitrary. It is within the new functional creative orchestration that the virtualities of the Human Condition bring in, first, the primordial moves of the human spirit; second, the thrust toward the other; third, the will to undertake; and lastly, the deliberative inventive quest. These functions are not performed alone, however. On the contrary, they may be activated only within valuative schema that install the human creative condition, within the aesthetic, moral, and intellectual perspectives that introduce the specifically human valuation of life.

Furthermore, these factors of human dynamics informing human action operate within the bounds that the world of life projects in structures and rules. Ultimately the human being, although aiming beyond through a creative swing, keeps his inventive creative spread within the human world of life that he unfolds, steering a course basically within the bounds of the system of life itself. Hence, as much as the human creative prerogative makes surge and guarantees the exercise of freedom, the same creative spirit does not let freedom be carried away into a wild chase after phantasmagoric possibilities of fancy or to fall into the accidentalness of arbitrariness. It remains within the open and yet not absolutely unrestrained boundaries of the human.

2. *Kairos as the Propitious Moment; Accomplishment and Measure in Human Creative Self-Interpretation in Existence*

The specifically human life operations draw on the energies, forces, strivings of our animal circuits in the self-individualizing process that extends their orbits, maintaining a balance of freedom between arbitrariness and fancy. The excesses in either direction lead to crises and upheavals in personal and social life. Both of them, however, belong to the "nervous skeleton" of self-individualization, the self-interpretative progress of the human being, and weave the fabric of the creative orchestration of the system of human functioning.

These specifically human operations, then, are "informed," imbued, molded by the significances with which the human being as creator endows life. Thus their timing is their partaking of shades of significance that vary in the infinite modalities and qualitative nuances of the aesthetic, moral, and intellectual sense giving factors brought into the progress of life by the Human Condition.

The timing of these fulgurating and constant moves, operations, processes is fully imbued by their modalities, tendencies, aims, successes, and failures in striving for accomplishment.

Since the confluences, fusions, and resultant steps of successive progress occur within the creative milieu of human interactions, individual personal encounter or avoidance of encounter are essential factors of human transaction. When chronos diversifies into innumerable streamlets of occurrences, there emerge conundra of propitious conditions that allow the tying of the knot of accomplishment. It is by the tying of these knots of accomplishment that the human personal — and social — self-interpretive course of existence not only proceeds but is also measured. Indeed, along the path of human creative self-individualizing, kairos is concurrently the timing of the propitious circumstances and forces leading toward the realization of constructive projects, their accomplishment, and finally their measure.

The specifically human itinerary advances essentially through two intimately inextricably interwoven and intermotivating factors: the inner realm and the external world. The entire web of inner experience, with its feelings, thoughts, deliberations, judgements, is woven partly by the universal requirements of life and partly — and ultimately — by the innermost striving for accomplishment, for advancing, for self-realization in the attainment of ever "higher" aims in the domain of outward life (from there to be transposed into the inner self). It is through action directed toward others, through activity with others that our intimately personal life within advances. Hence, the accomplishment of an undertaking, the culminating kairos, is in its outward realization prepared in the interior life. And, indeed, with the advent of the human condition we witness the emergence of an extraordinary sphere of expansion of the inward/outward, self-individualizing and self-interpretative agency of living being.

However, the inner life — considered to be the life of experience or of consciousness — manifests a timing of its own; chronos and kairos become more sharply defined in our sensations, feelings, thoughts, judgements, and their concatenations. Despite the complexity of each state of the field of "consciousness," these experiences remain in constant flux. We speak, then, of the timing of all this experience for a task, an aim, striving, a move. We do not experience stagnation and inertia. We experience a "continuity" of the "present" of our actual state in our preoccupations, steps forward, striving, all being taken over by receding into the past, as we simultaneously have the experience of projecting ourselves into future moves, steps, projects, tasks, aims. Thus, our consciousness of existence, of even closer acts, acts of being alive, acts that are unqualifiable, times itself onward and never stops while we are in a waking state, not through the whole of life's itinerary.

As pointed out above, although inwardly processed, this inner conscious life is, within the creatively orchestrated functioning of the person, oriented outward toward interaction. The measure of it, then, is essentially external. Hence, the timing of the inner, conscious self-interpretative course with its ungraspable complexity falls into the pattern of an inner chronos insofar as acts not only succeed each other but in their advance form new modalities of experience, falls into modalities that may in the case of routine, repetitive performances in the external world qualify as being monotonous, falls into an experiential chronos, then. However, when animated by hidden inner stirrings toward ventures, projects, and aspirations that are to be actualized in the external realm of life, inward acts acquire the "exciting" rhythms of kairic striving for accomplishment. It is the innermost striving for accomplishment with its entire kairic concurrence and modalities, both circumstantial and essential, that constitutes the characteristic that distinguishes human beings from other living beings. Thus, human life's inner experience and external activity is carried by two interwoven arteries; that is, it is timed into chronos, the everyday tacit carrying on of repetitive assignments for life's maintenance, and the kairic rhythms of urgency, promise, expectation, ecstatic hope, and final attainment of goals. Such attainments require the right proportions, the right measure, the right occasion. They progress in the constructivism and creative advance of people, social groups, cultures, nations.

3. *The Kairic Passion — the Passion for Accomplishment*

As the original moves, operations, processes introduced the primary scanning and spacing, continuity and discreteness, identity and alterity, ordering and measure laying the groundwork for becoming, they also laid down the means for it to be flexible, changeable, and yet expansive as the wealth of diversification increased with every turn, with every new phase of transformation of the circumambient environment and its seminal riches, with every change-transmutation of type. The scanning and spacing of life in this becoming operate as basic arteries of this advance — and retreat. In becoming and its stationary transformatory "mechanisms" of the moment such as organisms, individual beingnesses, objects, human groups, nations, inventions, institutions are articulated and ordered. These assume as many ordering modalities as there are main types of beingnesses processing life in their unique relatively stable "stationary" modes.

The scanning and spacing — abstractly called "temporality" and "spatiality" — takes then a specific inward/outward or outward/inward turn with respect to the inward/outward processing of the individualizing life. We distinguish, first, the spatiotemporality of "outward" objective reality as seen at the "surface" of life's seemingly objectively fixed presentification. We distinguish as well the spatiotemporality of the inward/outward processes and mechanisms or of our inwardly embodied organs, which follow their own rhythms, which rhythms we attempt to synchronize with the outward objectified and artificially universalized rhythms. And then, we have the scanning, spacing, of the specifically human life constructivism.

V. FROM SELF-INDIVIDUALIZATION TO THE DELINEATION OF HUMAN HISTORY

1. *History in Its Interpersonal, Societal Circuits*

In talking about specifically human self-individualization in life as "self-interpretation in existence," we have brought out that thread of the human being that has become so much in vogue in contemporary philosophy, that of the "historicity" of the human being. In my previous writings I have emphasized the point that each human reaction/action proceeds from and is sifted through the creative orchestration of the faculties of the human being, whence it acquires a specifically human life-significance. The life itinerary of the human person is the significant design for its life, a design that issues from the person's very own fabric of tastes, tendencies, vital strivings and, on their finding their voice, from subliminal passions, whether unfolded freely or subdued, and from the will and its selections as balanced by the demands of circumstances, etc. This design of the life course, which the human person works out in the interplay between its endowment and circumambient situations, forges the uniqueness of the person as well as the route of this forging, reality and its history.

Unamuno, following Spinoza, emphasizes that there is without question a compelling innermost "drive," a conatus, that spurs life's progression toward something perduring. We will specify that this innermost drive that threads through the constructive processes of life, being itself a constructive drive to advance, to undertake, to initiate, to accomplish, is a kairic drive, for it proceeds by synthesizing and synchronizing the entire inward/outward functioning of self-individualization in existence from vital functioning to conscious functioning and the creative will that prompts it onward. As such, the unfolding of human action through kairic accomplishments carries with it the specifically human valorization of facts and acts: through moral circuits that imbue personal and public life with rightness and wrongness; through aesthetic circuits that bestow the taste for beauty and lift the human spirit by ideals; and lastly through circuits of intellectual valuation that endow experience and action with formal structures, articulations, and constructive devices and measure the logical underpinnings of experience and judgement. These sources of individual valuation are common to human beings as such. Their practice elevates the individual to a distinctive personhood; it makes him or her concurrently different from any other person, unique among others, and makes him or her partake of the entire human community; it is instrumental in binding individuals into a communal chain; it accounts for the modalities of constructive sharing in life; it advances the communal spirit in which all members partake. In sum, it is through this set of circuits that human history, as the history of a society, a group, a nation, and humanity as a whole evolves.

Human history (e.g., the history of a nation or of a given culture) is made up of all the factors that enter into personal self-individualization, into self-interpretation in existence, and yet it is interwoven with some outstanding phenomena: peace and cultural flowering, social revolutions, wars, internal societal crises, natural cataclysms. In short, it can be ventured that the history of a society advances through the succession of day-to-day business, of "regular" tasks, aims, while kairic accomplishments punctuate indispensable everyday progress — that is to say, chronos and kairos advance human history in tandem. Historical progress is made by communal effort and acquires a significance in essence shared by all. Once a cultural model of experience and understanding is established by common accord, it becomes universal. And yet, to understand it is to go back to strictly individual experience and appreciation, back to the individual play of passion, taste, tendency from whence it took off.

Recalling the Crocean distinction between action and thought, it may be said that while kairic accomplishment — the final "act" — is carried out by human action, its final significance (or interpretative account) pertains to thought. Yet, in contrast to Croce, I do not see division between actions and thought. On the contrary, although interpretative and appreciative thought distances itself from the facts and action, in the process of understanding the origins of historical facts and their significance for the communal and personal life of people, facts/actions are elevated by the subliminal passions to the aesthetic, moral, and intellectual life of the human spirit. Facts/actions become moments of human history only on attaining specifically human significance. Human history is, indeed, made of kairic moments that find their full significance in the spirit and when passed on to posterity with that significance. Great national victories and defeats, political and economic accords, advances in the welfare of all, stretches of peace, great cultural events that transform or elevate the life of the spirit directly, as well as the failure to achieve the same despite a great tradition of concentrated effort and expectation, in short, all that punctuates the progress of history, are kairic moments that obtain their significance in the human spirit and in their rootedness in facts.

When they have withdrawn from the actuality of life into the shadows, these facts can be retrieved by succeeding generations in their significance by reviving in thought the individual, personal experiences, passions, tendencies from which they issued.

These revivals naturally illustrate a clear distinction that I introduced long ago between "primary" elementary passions and "subliminal" passions, the latter being the fruits of the creative functioning of the human being.

It could be said that since it is the occurrence of kairic accomplishments or their failure to crystalize that decides the progress of human history, the historical interpretation of facts is performed with kairic direction, from the kairic point of view, through an assessment of the kairicity of events.

2. *The Kairos of the Transnatural Genesis of Sacred Providential History*

The timing of the human undertaking, of the nodal points in the progress of culture, societal life, and specifically of creative endeavors, manifests the interplay of chronos and kaïros in extraordinary fashion. In the timing of a personal or social crisis, of turmoil, revolution, the interconnective networks through which chronos is timed become distorted, broken, deviated, and in order to pick up the threads of life's continuity, of the broken patterns of the spirit, new aims, tasks, etc. have to be devised. Unless totally annihilated, the dynamic stream of life pushes onward calling for the restoration of the broken loosened continuity of the spirit. Thus, restoration, on bringing about fresh germinal streamlets of forces that will recapture the dynamism of novel undertakings, manifests the timing of the networks of life, of kairos at work, but a kairos that is both in and out of the established arteries of chronos. If we would consider the established nervous system of life's timing of itself through recurrent intervals and regularities, the basic progress of life, we would see kairic timing at work in the picking up of the broken threads and working "in" and "out" of time. We would be considering the timing of critical situations, of great undertakings. Kairic timing works with and works beyond the constitutive intentionality of human affairs. Thus, by definition, the role of kairos in human affairs is played on the borderline of the constitutive networks of timing and its breaking points, at its limits.

With this in mind, we may, with Evanghelos Moutsopoulis, the philosopher of kairos, approach the most difficult issue confronting the human spirit, namely, the quest for the ultimate sense of life, for personal destiny, the life-transcending quest for the Divine.

We may see, in fact, first, how already in the creative accomplishment of an exemplary work of art, societal construction, on the one hand, and in the various turmoils of life, on the other, kairic timing, that of the creative process, draws from numerous lines of intuition a creative idea that is concurrently implicated in and distant from them all. Thus amid the discontinuities, the breaking points in the timing of chronos, a restructuring of elements occurs; they are broken apart from their intentional (if psycho-intersubjective) or vital-rational (if processes of nature) moorings. Secondly, this timing that is both in and out of the basic time of life's advance lends understanding to the specific out of time continuity through discontinuity seen in the individual genesis of the sacred within a human soul. Indeed, the "instants" in which the human being experiences the sacred emerge, as it were, out of nowhere in the person's "natural" stream of experience. They do not belong to the otherwise intentionally subtended flow as links in a chain of occurrences. They differ in quality. In their *sui generis* nature, they break into the flow and disrupt it. And yet they manifest accomplishments that could have been prepared only by a subterranean work in the soul. Their occurrences within timing manifest the psychic flux of natural life, of experience, on the one hand, and constructive steps that follow one on the other indicating a progressive transnatural hidden work of the soul, on the other hand, a work that progressively ties the kairic knot. These transnatural moments manifest themselves as interruptions of the intentional schema of our natural psychic processes. To the intentional schemata they constitute irruptions of discontinuous accomplishment. They express the kairic "right time" as well when they interrupt, they mock the "right measure." We may thus say that the unique transnatural nature of the instants of sacral experience is best grasped in terms of the kairic timing of its subterranean genesis and as the equally kairic surging of sacral moments within the regular continuity of intraworldly processes. But though, as we have already seen, human history as such proceeds through the kairic tying of new knots when the lines of intentional significance are disrupted, it is clear that that kairic timing has no relation to the genesis of the sacred.

As emphasized above, the kairic irruptions of sacral "instants" are not themselves devoid of interconnective links as discontinuous as they are in the schemata of life. Their surging forth at unpredictable intervals in the intentional schema of natural experience and their actually standing over against it indicates a punctuation in accomplishment performed in a hidden network of interconnected progress — a genetic network of "transnatural history," which the soul's progress in sacredness projects. We may call this "providential history," the history that in religious contexts would be considered the historical unfolding of a covenant between human beings and the Divine.

In the religious perspective, divine interest in human lives, Divine Providence, does not belong to human history. The Kingdom of God is not of "this world." In the perspective of the natural intentional schema of this world, Divine Providence proceeds in a continuity of discontinuous acts.

Through this kairic timing of Divine Providence, we may also understand the interventions of God in worldly affairs, God's rewarding and punishing human beings right in this world. We may also view the Christian mystery of the messianic Incarnation of the Divine in time in this perspective. In grasping the dynamic of the kairic timing of becoming, we see how two different orders may remain parallel and yet irrupt into one another at propitious moments and introduce a measure and sense of proportion appropriate to the workings of the level broken into and yet of a different order of rules. This is particularly helpful in grasping the Son of God's taking flesh to Himself. Here is an utterly unique kairic intervention of the Divine in human affairs.

The timing of chronos and kairos here, while intertwined, proceed from two different orders: the one of human history, the other of redemptive history. This contrasting distinction calls for a specific modality of freedom, one in which human beings are called by innermost stirrings of their souls to meditate in action, proportioning their direction and pace within the ontopoietic boundaries of life so that the designs of Divine Providence may be accomplished.

PART SIX

THE STRATEGIES OF IMPETUS/EQUIPOISE IN COMMUNAL SHARING-IN-LIFE

CHAPTER ONE

THE FULGURATION OF THE LOGOS IN THE "OVERT" STRATEGIES OF THE EXISTENTIAL INTERACTION

The Communal Significance of Life

I. DELIMITATION, BEINGNESS, AND THE UNLIMITED

With its very first constructive impetus the Logos of Life engaged in a primogenital play of opposites. This play from there on subtended the entire logoic constructive project, taking a specific modality within the surging sharing-in-life intentionalities of the Dionysian logos. Lastly, it led to the astounding new intentional dimension and reversal of direction that is artificial intelligence with its still in process expansion of the reality of the media of communication among human beings, which is the intentional redoubling of the human world, human reality. In philosophical discourse focusing on the human sphere there is a new reality that human beings project by their own intention upon cybernetic space, one that not only cannot be ignored, but that enters into our very conception of the human living being.

This "new reality" doubles the "space" and "time" cross section (axis) of life in such a way that the spacing and timing of the fundamental, basic, originary world of life is being transformed. It seems to open the way to freedom from the constraints of this primogenital axis of life and reality: the spacing and timing of becoming. But before we enter into this crucial point, we have first to sketch the fascinating dialectic between the delimitation of beingness in that very becoming and the unlimited from which it is carved. Second, and most significant, following this dialectic in reverse through the entire foregoing exfoliation of life's progress, we will conclude, that the logos of life from the start, that is, within its intrinsic proficiencies, carries the law of this dialectic, though with a vector — an ontic necessity to delimit encounters and an opposed Promethean ideal of liberation from constraints, of self-devised freedom.

As a matter of fact, beginning with its primogenital moves in installing the cosmic conditions of life, the logos of life carves from the thus far unlimited cosmos, laws for heavenly bodies, which amounts to establishing frontiers, confines, delimitations. It is only as the significant opposite of these carved out "limits" that encircle and so define the "out of bounds" that the "limitless" may acquire a sense.

Going through the emergence of the womb of life, the "biosphere," toward the entire process of embodiment in the emergence and unfolding of the ontopoietic process of the becoming of beingness, we see how each step in the constructive moves means a specification, that is, a separation of a particular element from the "rest," in this case from "life forces" that then become an "outer realm" in the very carving of the new being's dominion. It establishes its very own identity precisely by carving out constraints all its own within the already partially delimited.

In short, the emergence and the subsequent unfolding of beingness-in-becoming is a progressive delimitation, a progressive setting-up of constraints on the unlimited — to begin with on the untamed forces and play of energies. The becoming of beingness is the self-imposition of ever new, ever tighter constraints. With each new type of living being, with the ever more refined complexities and functional twists, the delimiting role of self-individualization becomes more precise.

However, *mirabile dictu,* the logos of life simultaneously pursues a reverse course. Along with the progressive functional complexity of living beings, there occurs a progressive opening toward novel spheres of life. The manifested world grows with the moves from the soil-fixedness of plants to the ambulating of animals, to the upright posture of the apes and the human beings. As living beings in their ongoing development of new features distinguish themselves more and more particularly from other living beings, the limiting constraints this places on them in terms of their range of existential conditions coincides with an expanding freedom to select, to plan, to self-devise existential routes. The delimitation goes together with a constant striving to open possibilities and increase self-determination. In the human sphere we call this self-determination "freedom."

We witness this ontopoietic self-determined drive toward freedom with unprecedented clarity. The human ontopoietic course has now even developed artificial intelligence — a strategy of the logos that we are not ready as yet to fully understand. But undoubtedly we are here facing a specific turn of the logos, and I will attempt in what follows to sketch the meaning of this feat, which clearly aims at striking the final blow for the "self-liberation" of beingness-in-becoming, which of necessity has progressively been bound in chains, beginning with the primogenital chains of cosmic laws and the life system.

Shall we conclude that the logos of life purposes the creation of freedom out of constraints?

II. THE DIONYSIAN LOGOS IN ITS INTENTIONAL NETWORK OF COMMUNICATIVE LINKS

1. *Communicative Links in the Dynamic Fulguration of Intentionality*

Intentionality has been widely discussed and analyzed, not only in the transcendental phenomenology of Husserl and the Husserlians but also in scientific inquiry and art theory. Even when the concept is extended from the acts of intellectual, transcendental consciousness to the kinesthetic interactions of bodily movements, it is always considered as a specifically human prerogative; it seems to be reserved to human conscious organization.

It is time that we consider it as a logoic life device. There can be no doubt that the intentional act is in its fullness the implementation of the Dionysian logos that surges with the human creative condition, and yet if we look closer, consciousness is also the prerogative of animals, even if it be in degree less developed or more rudimentary. And so the intentional network that lies within the human constitutive system fully accomplished already in different degrees subtended the dynamic unity of the operations in all animal life.

The role of intentionality is to secure a primary level of equilibrium through establishing unifying links in the overt manifestation of life amid the passion, action, reaction of the struggle for existence. This entire phase of the manifested life is a function of the awareness-conduct of living beings enacting their specific life, their existence.

First, it is the means by which the individual, singular beings to throw out their very own vital hooks so that they may obtain a presentation of the world around and within.

But this presence of the surrounding world simultaneously indicates the interrelating links that each singular living being in virtue of its intentions projects with the external sphere of existence while situating itself among other beings-in-progress and establishing its own hold on the whirl of life forces. In a simultaneous, secondary way, it is through this most ingenious twist of the logos of life that the vital forces, desires, needs, sentient operations are brought into a sharing-in-life that provides mutual reinforcement of life situations within the entire realm of existence and a station of repose for individual overt efforts to remain in existence, to advance, to flourish. This sharing-in-life as such — the throwing out of ties between and among individual ontopoietic life courses for the establishment and maintenance of each in overt, manifested existence in mutual interrelations — is a capital feat of the logos of life. This is not just a generality. On the contrary, these interrelating ties provide existential support as well as foster the development of sentient, affective, emotional bonds among individuals. As is obvious, such sentient, affective bonds are the flesh of coexistence among living beings. This flesh unites everything-there-is-alive, making sharing-in-life not only possible but also a living experience.

In this sharing-in-life, as we have already mentioned, developing intentionality plays a further role: a signal, a sign, a voice uttered. It is as if the mute intentional ties of sensing and affection free themselves from a cocoon and open another dimension. This sign, signal, voice serves sharing-in-life through a further function of intentionality, namely, communication. Animals — and humans — communicate to signal danger, to indicate better pastures, easy prey, accessible and desirable water....

Although all the previously mentioned functions of intentionality play an essential role in the ontopoietic course of human individuals within the human sphere, it is in this last function of communication that the Dionysian logos excels and attains the greatest heights of logoic achievement.

Let us, first brings out that beyond the sentient, affective, and emotional ties among humans there is what Karl Jaspers so beautifully described as human communion. This is a level beyond human communication, that is, beyond the emotional/affective level. The logos in its Dionysian swing projects human ties serving common societal goals and then at the peak communion in faith and ideals.

However, the swing of the logos in its intentional sphere, in the sphere of sharing information, knowledge, and of participating in communal, societal, and cultural affairs has now reached a dizzying development in "artificial intelligence."

The impetus of the Dionysian logos does not stop at any step reached with technological invention. So-called technological progress is nothing other than the impetus of the Dionysian logos in its Promethean aspiration to set the human being free, to make him master of not only his destiny but also of his very own ontopoietic course as set down by the system of life. We wonder how far it may go. It is the vertiginous level being reached by the intentional creative/inventive pursuits in establishing ever new means of communication among human beings that greatly accounts for this swing.

But more will be said about this later. Let us now continue to examine the societal significance of life and elucidate the measures of equipoise that the logoic impetus brings to individuals, groups, and societies together, to nations and international organizations, etc. We look into how this swing toward conglomeration finds equipoise in what basically are "social contracts," and how the human individual preserves individuality within the social conglomerate in which he or she unfolds personal existence.

Because at the human level there is self-awareness, knowledge of cosmic laws, and higher aspirations, there is a particular existential striving to protect once individuality and to gain a measure of independence within the circumambient ties of life, to be free to accomplish one's dreams.

2. *The Communal Significance of Life; Rights: Individual or Communal?*

Sharing-in-life is the basis for the formation of the network or fabric of communal existence among particular types of living beings. It is observed in ants, bees, monkeys, apes, and human beings. Communal existence manifests an equipoise between individual needs, tendencies, characters, criteria, and ways of behaving and the interactive and cooperative links between individuals that are indispensable for the maintenance and promotion of appropriate individual scripts, for the proper apportionment of opportunity and freedom in the devising of individual self-interpretation-in-existence. There must be a measure of equipoise between the diverging interests of individuals in their exercise of personal freedom, existential autonomy, on the one hand, and the common good and its requisite cooperative spirit, on the other. Communal existence is possible only when there is a distribution of roles in interactive tasks. Responsibilities are offered, and rewards of appreciation and material benefit are given in such a way as to satisfy at least basically the existential needs and expectations of all. In the cooperation of common endeavor there surges a "communal spirit" that transcends concrete individual benefits. In this shared effort there is taken a deeply ingrained satisfaction in communicating existentially with the meaning of life itself.[1]

The communal spirit differs according to the nature of the communal effort. Those belonging to a military detachment entrusted with an important mission share a self-transcending commitment to "duty." They will then expose themselves to the risk of losing their lives. Others may be transported above everyday concerns by a shared social ideal, to work for the improvement of humanity, to raise morals, to refine tastes, to increase earthly happiness by reducing life's contingencies, to assure eternal happiness by beseeching surrender to the divine will.

The communal spirit is the lively, vigorous watershed between the tendency to be self-centered, congenital greed, animal self-interest, on the one hand, and our equally weighty existential need for consistent cooperative effort. On the basis of our untamed individual tendencies the achievement of that cooperation may appear altogether impracticable. But then we see situations in which nations fall under totalitarian rule and people acquiesce to following rules they are not disposed to favor and do not in fact accept. The communal spirit that naturally holds common efforts together then dissolve, as people become passive even apathetic and discontent deprives them of the joys of life.

To attain true equipoise, one always vibrating with change, one that is at every turn being negotiated anew, and to maintain a discrete continuity in the integral identity of a given community — nation, town, parish, etc. — that is, an identity that goes beyond its intellectual rational "purposes," points of reference have to be recognized and accommodated. This is a deep concern, insuring basic communal coexistence, arranging matters so that all may benefit and harm to any be minimal, given all our elementary drives, passional stirrings, outbursts of emotion.

The fact that the natural linkage between the circuits of individual existence and communal coexistence involves a tension onto which equipoise has to be projected indicates a basic situation of conflicting interests. The common good certainly does not benefit all individuals, and individual "claims" for freedom in the spacing and scanning of their existential course may go against established canons. Hence there is the eruption of revolutions, social protest, etc.

The measure of equipoise to be found shifts from one side to the other imperceptibly in both interpretation and actual practice. Hence the social order incorporates laws to protect individual freedom. Constitutions incorporate clauses on "human rights," which are then vindicated in special social groups that unite to this effect.

Part of this search for equipoise is the provision of communal protection of the individual from abuse and harm inflicted by others. There is a deep-seated tendency to violence to be protected against, some of it purposive, some of it random.

But when we deal with the effects of pathological influences on individuals and society, we are concerned with another kind of equipoise, one that has to be addressed each time an outburst of violence disrupts communal ties: the equilibrium between excess and penury in vital/psychic/societal life dynamics.

CHAPTER TWO

THE DIALECTIC JUNCTION IN THE LOGOIC STRATEGIES: MORAL LAW VS. COMMITMENT

I. MORALITY AS THE LOGOIC STRATEGY FOR SECURING A MEASURE OF EQUILIBRIUM WITHIN THE "WILD" DIONYSIAN GAME OF LIFE

1. *Setting the Issue*

"What is freedom?" The way in which we usually employ this term conceiving of it spontaneously is such that we can express it only poetically as the "breath of life" — only poetically, because what we mean by "freedom" does not fall under any ontological category. Freedom is neither an object, a thing, a living being (individual), an idea, a class, a quality, a process, an event, a force, a function...not even a relation. And yet freedom is undeniably, nay, primogenitally a modality of life.

That freedom is a constant matter of concern in human life is an obvious and striking fact. Humanity at large is locked in an unrelenting struggle for the freedom of nations, social groups, institutions, and human beings. What this really amounts to is that it is the freedom of the human person, as it is nowadays formulated, that is at stake. However, when we consider the freedom of the human being, the human person, the fact should impress itself on us that freedom is most intimately involved with life and consequently has to be considered with respect to all living beings — for higher living beings of the higher species of animals are also persons.

With this preamble, it is clear why in order to approach the question of freedom in a foundational fashion we have to situate it first within its proper context. Only situating it within the network of the generative life functions spurring freedom's emergence can we do justice to the role of freedom, a germ of which at least is essential to this emergence itself. It is by following that development that we may then grasp freedom's significance for human life. I propose that the appropriate context in which to treat and discover plausibly the conditions and parameters, the virtualities and functions proper to its emergence and its assumption of a specific role within the world of life, the world of living beings, is the philosophy/phenomenology of life, a philosophy of life that is integral and scientific. Freedom concerns life in a unique fashion. That is to say, within its context we have to seek to differentiate the modalities of life that bear the name "freedom," conveying it to living-beings-in-action in order that they may "act freely."

2. *Freedom as the Primogenital Manifestation of Life*

What is more obvious, and yet it usually eludes our attention, than the fact that the bursting out of life as such is an emergence of a novel, original phenomenon, novel with respect to its premises, original with respect to its source? It is an emergence into the "open" yet its premises remain self-enclosed; emergence into "light," yet its premises remain in the "dark." It is an exhalation out of an inert area. There is ignited a spark for a fire that spreads, a life engine that can pull inert energies into the process of becoming. Becoming is a factor at the presiding and carrying out of the constructivism of life and its spread into existential modalities unprecedented in matter. In brief, to breathe, to move, to feel, to think, to desire, and to struggle for existence is an exercise of freedom. How wonderfully free we living beings are in breathing the air, in thinking and feeling and desiring and moving and expanding ourselves this way, in projecting our very own tentacles outside ourselves. Indeed, we are free to project them *within and without.*

In this respect each manifestation of life as this expanding from within is a manifestation of freedom; we human beings share it with everything-there-is-alive: plant, animal. And yet there are — as hinted at above — marked differences in this primogenital freedom, differences in extension, in modalities, in inner control.

Indeed in line with great thinkers such as Aristotle, Thomas Aquinas, Leibniz, Kant, Hegel, and others after them, we discover ever anew that life as an eventful emergence as well as a process-like development that proceeds all from within and is worked out by living individual's functioning, is unfolded within the enactment of each and every move.

And yet, as has been pointed out often, there is, a great differentiation of types of living beings with respect to the variety of their inner systems of functioning and corresponding to that differentiation with respect to their integration with the circumambient sphere of life. It is precisely in the spread of the active tentacles that the living being throws out around itself that its range of possibilities as well as its power to control the inward/outward directed progress from within that life's unfolding expands.

Having recognized as freedom this primogenital expansion from within one's own virtualities and powers, *we will recognize the growth in complexity in the evolutionary differentiation of life as growth in degrees of freedom.* In fact, with the advance of the complexity of types there goes a larger range of possibilities for the living individual to match his innermost existential needs with the available means in order to satisfy them. From the rootedness of plants and trees to the simplest animal is a radical and most significant jump in mobility, be it the least mobility, yet it allows for a novel line of freedom, to "seek" and to "choose."

With the lowest species, those simplest in functioning, vital needs are satisfied simply by the "fitness"[1] of the materials available in the circumambient environment. With the higher sensitivity of more complex functioning, this satisfaction of needs is guaranteed by the sentient features of internal propensities. Accompanying the acquisition of greater mobility is a more fully developed "discernment," with sentient recognition of the nature of the needed as well as the available resources, recognition that paves the way for the exercise of an "inclination" toward such or other choice. However, it is obvious that there is a close and stable correspondence, a match between vital needs and the range of their possible satisfaction and the available resources of the environment. The crucial point is that both of these express the laws of life itself. That is to say, it is an inner code of the individual's functioning that carries with itself the traditional indications for this correspondence between the individual unfolding life and the environmental conditions indispensable to it.

The laws of life are represented by an inner code — what I have termed the 'entelechial nucleus' of a living being — by which the rules of life's unfolding strictly implement the laws of life itself. These laws strictly circumscribe primogenital freedom in its range of choice, selection, and deliberation.

The emergence of the Human Condition within the evolutionary advance of the forms of life is a radical novum — even with respect to the types of higher animals, the ape, for example, that have a seemingly vast referential system. It is the surging forth from nature of elements that are heterogeneous vis-à-vis that system that brings about a transformation of the entire vital functioning by expanding it into the human mind. As a result the living being loosens its bondage to Nature-Life, opening on a uniquely different sphere of existence.

This transformation occurs precisely around the human mind, the axis of a novel kind of freedom, a freedom that on the wings of imaginatio creatrix — which proposes ever new possibilities, informs with ever new tendencies, and prompts the faculties of the mind — soars far beyond the circumference set by the laws of aviation, often defying them. Imaginatio creatrix proceeds by always seeking its satisfaction within the specifically human significance of life, which the creativity of the mind brings about. Thus is the sphere of the human spirit coaxed forth.

The main question that I want to treat now is this: In virtue of which factors of the creative virtualities does specifically human freedom emerge and unfold?

But we have first to bring forth the innermost manifestation of this specifically human freedom beyond the existential web of life as embodied through the human person.

II. CREATIVITY AND THE DRAMA OF HUMAN EXISTENCE

Let us repeat that the Human Creative Condition involves a radically novel modality of freedom in which desire, selection, and choice are not circumscribed by the strict laws of the entelechial nucleus in its attunements to circumambient conditions, but, to the contrary, go far beyond those conditions with the entrance upon the scene of imaginatio creatrix and the new factor of invention, which proposes new avenues of life.

This new creatively expanded and projected network of specifically human life opens a new, most complex type of freedom, one that — not being strictly circumscribed by life's laws — becomes as it were the engine of a drama that inwardly and outwardly tears the human person apart, pulls in various directions, brings in new existential categories of defeat, disaster, and misery as well as of victory, accomplishment, and happiness, etc., categories that are absent from animal existence.

The point is that, as present day social life most acutely demonstrates, the possibilities for realizing one's own dreams, intentions, pursuits depend on the situation within the societal world. There is a limited societal space for realizing each individual's pursuits. As in the animal kingdom, so in the human societal world the existence of one individual means coexistence with others. This means a tightly spun network of interaction, interrelation in which (unless societal circumstances are favorable to a certain accommodation of numerous individuals) each individual is allotted within the common net its space of action in which to push his own intentions while renouncing others so that the social law's protection of one's "rights" infringes on the rights of a fellow man. One's self-instigated exercise of free selection, invention, promotion may, and often does, infringe in some way on the rights of others. While we individually and as a race tend toward "absolute" freedom, should not that goal be seen as a utopia in fact?

The drama of the human aspirations that raise the human being high above his frame and of the defeats that plunge him into despair, the drama of legitimate aspirations, heroic efforts to fulfill them and of how the indomitable forces of life cut their wings and annihilate them, this drama of existence is rooted in the scarcity of goods in the natural and social worlds, in the limited possibilities offered for the exercise of our imagination, and in the inner and outer struggle we engage in as we strive at some or all cost to express the force of the spirit as prompted by a yearning to accomplish, win, gain, reach something particularly significant in our existence.

Success and satisfaction, disappointment and distress, these are the extremes between which the pendulum of human freedom swings. Moving from the vital level of freedom as the primogenital engine of life to the level of human freedom, we discover in the latter the engine of the human drama.

The transformation of the modality of freedom from the vital to the specifically human significance of life is such that we have to raise *the question of what is the specific coincidence of elements in the human condition, of what is the original virtuality that moves the creative orchestration of the human mind toward that configuration of factors that makes specifically human freedom surge* — a freedom that is, on the one side, unbound imaginative/inventive prompting and, on the other, conditioned by the entanglements of the natural/social world of life. What are the reasons for which humanity denounces and struggles against the direction exercised by governments, cultural expectations, and social groupings? If left without direction society degrades into a spectacle of mutual abuse among its members. The law and the systems of ethics meant to protect in an even-handed way the rights of all individuals fall short in their implementation, and this reveals that they lack sufficient power of appeal to us as human beings to begin with. What would be objectively applied in a particular instance somehow misses the internal "measuring stick" within the person, if there be one. The indispensable resonance that would make the prescription work in each case does not come about. If that rule hinders or goes against our interests, law and ethics lack the power to convince that would awaken a desire and willingness to implement the rule of ethics or law even when they are recognized to be fitting. We are tempted to go around the rules, avoiding the proper resolution. We will just lull ourselves into thinking that to be right.

Having thrown out our question, we will now go deep into the roots of the free human person as it crystalizes itself from the virtualities of the human condition and so trace to its origins this conditional freedom in order to answer our question.

III. THE PRIMOGENITAL MOVE OF THE MORAL SENSE AS THE IGNITING OF THE SPECIFIC FREEDOM OF THE HUMAN SPIRIT

I have previously proposed the moral sense — as the "benevolent/malevolent" sentiment — as the factor of humanness that emerges from the creative virtualities of the human condition.[2] I see in it the primogenital experience that first "humanizes" the living individual; second, it simultaneously throws a hook towards human companions as partners in the game of life.

Indeed, as I have described before, there takes place — in the first stages of the human being's evolution from the sharing-in-life of gregarious herd existence — an encounter of the living individual with another.[3] This is an encounter in which simple animal herd feeling does not prevail, it being a "confrontation," a kind of measuring up of the other in which, first of all, the possible intentions of affect — distrust, fear, indifference — are superseded by the benevolent sentiment that the creative virtualities of the moral sense release from the Human Condition in this confrontation. It is like a synergy, a prompting force that carries the factors of significance in it. With the release of the "benevolent/malevolent" sentiment (it can be a negative event), there follow the *three founding moments* establishing the humanity of the person, the humanity that, although it means a transformation of the very innermost core of the individual in his inwardness, bonds or refers the individual to another, prompting the throwing out of links that create a new kind of "togetherness" in life.

In the first moment of the transformational experience that the benevolent/malevolent sentiment releases, there is a recognition of the being now confronted as having a likeness to oneself, that he is "like myself," or to put it more strongly is "an *other* self." (Paul Ricoeur's concept of 'un autre moi-même' was put before the public ten years after my monograph "The Moral Sense in the Foundations of the Social World," just cited above.) The second moment follows, which consists in the attribution to the "other self" of an equivalence to myself in the vital significance of life and beyond. These two moments give rise to a third, that is, they prompt *the gesture.* The gesture, an exchange of gifts, lifts itself to a higher level of significance; it acquires the significance of an action that transcends itself, of *a trans-action.* This act transcends its vital setting, transcends the pragmatic giving and receiving of goods by raising all to a specifically novel sense in which the exchange of goods bestows on the occasion "moral sense." The gesture means the freeing of the moral sense lying there in wait as a factor of sense among the virtualities of the Human Condition and now playing an active role in the advance of life.

The emergence of the moral sense's bestowing a specifically human significance on life is borne upon the three-moment primogenital establishment of the basis for the human equitable experience-of-oneself-and-the-other, the self-awareness of our sharing-in-life and of the equality of rights among human beings and of a measure for apportioning rights to living beings other than man. It finds a measuring stick in the ontopoietic design of life's individualization within the networks of interactive relationship in the unity-of-everything-there-is-alive, the basis of the differentiation between good and bad.[4]

Lastly, but foremost, the release of the moral sense accounts for the capacity to decide about oneself, which is the full-fledged meaning of decision.

With these three moments of the new meaningfulness of the reality of life, the moral life is established. The benevolent/malevolent sentiment, its carrier, is a prompting force of a nature entirely different from those of vital energies. It is not *only* that vital spontaneities and synergies carry only the vitally significant rationalities for the meaning bestowing that is dedicated to the promotion of natural life, but also that, as said before, their meaningfulness follows the schema of life strictly. The synergy of the moral sense brings in rationalities that transcend the circumference of the vital significance of life; they even contradict vital self-interest at moments of generosity, denial of self for others..., which means putting the interest of the other before one's own, the renunciation of self-interest for the benefit of the other. Here the sharing-in-life common to all higher animals takes a new turn. It is grounded in the specifically human rationalities that the creative orchestration of the faculties captures from the entire spectrum of the emerging new meaning bestowing virtualities. Under the aegis of the ever active Imaginatio Creatrix, human beings can in self-awareness of their interests project at the novel level of imaginative planning, using their intellectual knowledge on the platform of life interrelations.

Now owing to the synergies of the moral sense, this new platform sinks roots in the interaction among human individuals, which means a sharing-in-life on a higher level than that of the gregariousness of higher animals. This interaction is moderated by reference to individual rights in sharing, by applying the highest principles of Good and Evil, and by any of the variety of systems of human values that the intellect constructs at the instigation of the moral sense.

The understanding, the acceptance, the application of the moral precepts of ethics and law and recognition of the validity of moral values depends not only on intellectual analysis of the situation — the understanding of it in terms of the moral valuation established by the exercise of the moral sense in human transactions — but first of all on *the force of the moral synergy of the benevolent/malevolent sentiment* of human persons.

Here we are reaching the focal statement that emerges from the foregoing inquiry. With the emergence of the synergetic moral sentiment there came forth a specifically human freedom, springing forth from within the unfathomable depth of the human soul and nurturing all the human functions of the creative orchestration, a freedom to invent, to plan, to valuate, and to decide on sharing-in-existence with others in a common life, in life's fullness.

Human life is engendered in the freedom of this self-awareness with others. Hence, the unquenchable thirst for freedom. As I mentioned above, the freedom to act of one person may, and often does, encroach upon that of another. The game of freedom between my fellow man and myself carries, however, all the prerequisites for harmonious attunement with the proper understanding and exercise of the moral sense. The very root significance of my freedom lies in my recognition of the equivalence of the rights of other fellow men.

The last but not the least result of our inquiry, which did not come to my attention in my previous study of the moral sense, concerns the extravital origin of the moral sense that ignites specifically human freedom. I attempted above to emphasize that the appearance of this freedom is a unique emergence within the life system of a transvital moment, on the one hand, and that it transcends all vital necessities and determinations, on the other hand. Thus is brought in the inward freedom of the human being, who emerges self-conscious and self-deliberating and able to decide not only about himself with respect to others, but with respect to himself too. Personal self-valuation leads to the questioning of human destiny and opens the gate for transnatural longings, premonitions, foreboding of the supernatural, etc. The moral sense's transnatural ingression into nature may be seen as an ingression of the supernatural, which leads to the religious intuition of sense in nature and to the religious intuition of the Creator. This opening of the human soul to receive sense in a way transcending the vital sense giving network and the gregarious sharing-in-life conundrum of logoic life oriented interrelations in process as well is a point of encounter with the otherwise *sui generis* and transcending elan of the soul when it abandons the bounds of the logos of life, frees itself from its web, and rises toward the quest for the "transnatural destiny of the soul." In other words, it appears as if the opening within the soul accounts for a round trip traffic in which human spiritual longings find a door to a haven along a road that has been already paved by the entry through the same gate of the moral significance of human conduct within the societal sphere of life.[5]

Nota bene. This coincidence of the transcending origin/destination that we discover along these lines explains the mutual motivation of human moral and religious inspirations, as well as progress along their ways. It also becomes clear why morality with its principles and religion with its laws and precepts meet.

Finally, this opening toward the transnatural on reaching the fonts of the human soul allows it to draw on its own resources for its transnatural destiny.[6]

Summarizing, it comes clearly to light that specifically human conditioned freedom emerges in the wake of the moral sentiment that prompts it and then distributes the moral principles, emotional qualities, valuative standards and leads toward the intellectualization of this entire network of moral values. Hence, it is also clear that this modality of freedom is simultaneously the promoter of the individual modalities of expansion within the world of life, the unity of natural and societal coexistence, inasmuch as it is essentially mindful of the possibilities of the expansion of the other fellow human being.

The intellectualization of this system tends to take over. The intellectual calculus of values that we tend to make by applying strictly rational means for valuating our part in a transaction and then making decisions is easily twisted by our egotism and self-interest and we may evade the distribution of justice by false calculations of merits and dues. It is reference to the moral sentiment that is the source from which the distribution of the equal rights and responsibilities of the human person flows and to it human persons as humans should naturally have recourse.

In conclusion: no sterile debate about the respective validity of socially competing ethical theories will help solve the social disarray and perplexity of our present age. It is the revival in human persons and social groups of the primogenital moral sentiment alone that will give us true direction.

IV. THE LOGOS OF HUMAN FREEDOM

Within the human sphere of the Dionysian logos fall the bodily delimitations that the human being attempts in various ways to amend or circumvent and even with the development of the technological strains of the intellective logos to overcome by technical means, e.g., by changing the time a distance is covered by using a car, plane, ship, or by changing the course of functional developments within the body by various medical measures that restore its constructive course or that repair defective organs allowing their normal employment, etc. However the main weight of the logoic push toward liberation from the delimitations of self-individualization within the life conundrum falls on the social significance of life. The societal ties of sharing-in-life do not compass only mutually supportive strains; it is owing to them that the delimitations within the societal sphere of existence appear in human experience and play their games within the affective, intellective, and self-aware sphere. Our plans, our striving for accomplishments, and our feeling that we are members of the community and society at large are involved in the logoic inner strivings toward liberation from limitations; these latter signify abuse, wrongdoing, deceit, cheating, oppression, enslavement, extortion, tyranny, domination, etc. Laws have been set by various societies as far back as the time of the reign of Hammurabi over the Sumerians, some 3,750 years ago. It was followed by those of Draco in Athens in 621 B.C., which were improved upon by Solon shortly afterwards, the laws set up by Alexander the Great for Macedonia, the Greek states, and conquered parts of Asia and Africa, the codes of Roman law, Magna Carta in England, the principle of Neminem Captivabimus in Poland, the Code Napoleon, which endures till now. These are just highlights of the dynamic logoic search after principles serving as points of reference, of which with the obverse side is the call for a commitment to implement them in everyday societal practice (even under pain of punishment for contravening them). These principles simultaneously safeguard the independent status of the individual and impose constraints on his/her conduct in order to safeguard the independence of others. Again we find here the logoic play of the opposites of independence and constraint, this time in the societal sphere. But the point in this matter is the human logos' striving toward gaining a measure of independence, freedom from constraint whatever it be.

We will come to this logoic driving toward "freedom" after we turn from the principles regulating the play of individuation/delimitation and constraint/freedom in another logoic device: the device of establishing a platform of equipoise for this game, this strife, this struggle in society that goes back to what has historically been called the "social contract."

V. THE SOCIAL CONTRACT AS THE MEASURE OF EQUIPOISE IN SOCIETAL ORGANIZATION, INTERACTION

As discussed earlier, progress in self-individualization, in the attainment of ever greater self-promotion within the world among other living beings has followed the line of life's evolutionary ascent. But this logoic tendency is coupled with an innermost necessity of promoting ever more articulated ties of interaction and interchange, which are meant to further the advance of the logos' drive by enabling, facilitating the ontopoietic specification of individuals. The primogenital ties of the simplest animals in sharing-in-life, shelter, food, the avoidance of danger expands with the progress in complexity, first to the gregarious coexistence found in packs of wolves, herds of buffalo, and elephants, etc. Study of the "social behavior" of animals shows us how birds occupy their own territories, how they safeguard territory, and how they communicate giving individual signals to each other about impending danger. Jumping along the evolutionary line we encounter the much closer ties of coexistence among the apes of the African savannah. With their much developed mental proficiencies, they manifest a well-distributed division of labor within a system of social layers that are delimited according to the services that the respective members render to the group (grooming, food seeking and preparing, caring for the infants, etc.), with a governing individual ape at the top. These and others are well-known facts. Our aim in referring to them is to indicate that the convivium of individuals occurs at all levels in the evolutionary line as an intrinsic necessity met by the logos of life: individualizing/socializing. Along the different ways of this logoic thesis/antithesis tension, life proceeds.

We see this already at the lower stages of individual development, but it reaches its peak at the specifically human level — where it is to be pondered in the various stages of anthropological progress, each with its degree of development of the individual and corresponding articulation of social life. Underlying each schema is a logoic matrix with reference to which the articulation of the individuals in a group — with the group emerging precisely owing to the implementation of this articulating matrix — and the individuals' interaction with each other is outlined through the prescription of "good" comportment, interdictions, taboos, etc. This intrinsic logoic matrix subtends all the laws of the group.

This matrix of the social contract serves as the measure of equipoise for the singular implementation of rules, laws, prescriptions of interindividual convivium. In its progress, first in the human anthropological, so to speak, unfolding and refinement, and then in the subsequent and simultaneous need to devise new societal articulations, humanity has from the earliest periods known to us — Assyrian, Sumerian, Minoan — through all the periods of unfolding civilization projected constantly new forms of social order that call for the appropriate organization of interaction in individual and societal existence and provide a system of reference for maintaining relative stability in the constructive progress — whether progressing or regressing in quality — within an inexorable flux and turmoil of competing forces, influences, drives, desires.

Societal existence itself is a spread of coordination in several orders that works out the articulation and organization of the dynamic progress of individual human lives.

With various degrees of flexibility it is rooted in or called for by these several orders. First, we have the root order of nature promoting human individual existence in a primordial fashion. Second, we have to consider the system of individual personal faculties that emerges from the embodied human mind in its interplay with the human spirit. Lastly, we have to consider the natural spontaneous ordering of individual personal lives within the sphere of a communal sharing-in-life.

The classical theories of the social order (Rousseau, Locke) were based on certain assumptions concerning the human unfolding in existence. One of these assumptions was — in line with the times — based on the predication of a universally constant and continuous line of human development in individuals. But this was coupled with the achievement of a utopian ideal and definitely stable societal order. Lastly, again in accord with the spirit of the times, society as such was conceived as a self-enclosed, self-determining domain of life, the development of which was accounted for by a social contract.

Yet, as we have seen, and what is drastically borne out by the history of the Occidental world and culture, as well as by that of other civilizations, such a static view is very far from the mark. Not only is the ontopoietic unfolding an outward/inward progress, that is, not only does society's singular unfolding depend on the varying life conditions, but the bents of human psychological — i.e., embodied mental — life and character are unforeseeably mobile and varying. Moreover the extravagant and conflicting with each other tendencies of socially oriented individuals does not allow the grasping in theory of the order pertinent to a particular social group. The concept of the social contract must encompass open frontiers if it is be applied to the interaction found in particular groups. A reappraisal is needed of the intrinsic mechanism of the social function that the social contract is called on to play. To fulfill that function it has to refer to the root conditions of societal relations.

If we want to reach to the inward logoic mechanisms that subtend every particular "social contract" as such, we have also to take into account, recognize, and valuate the essential postulates of the Human Condition, namely, the fundament of sharing in life with other human beings, the drive toward personal fulfillment, the acceptance of responsibility for oneself along with a responsibility for others, that is, for the "common good" and for conducting oneself accordingly. Every social contract is thus based first on recognition by the individual of the proper attitude toward personal fulfillment, on consideration of the interests of others, of their fulfillment, and then on a commitment to comport oneself according to the rules that promote this interplay of individual plans and projects within a society. Yet the logoic matrix of what I call the primogenital social contract briefly described here, which contract underlies all other changeable social agreements among individuals and governments, changeable according to varying life conditions, cannot be conceived in the classical universal and static terms. Its constant and continual fostering of the common life has to be appreciated in terms of the dynamism and discreteness of life and the individual. Needed is a constant but moving referential system reflecting the dynamism and directionality of a crisis based conception of life.

As a matter of fact, the young social sciences are bringing out the dynamism underlying societal life, the crashes, catastrophes, revolutions that cannot be foreseen and are very difficult, if ever, to control. Economic fluctuations that have their repercussions on societies and individuals: crashes in the financial system — which carries the burden of human practical transactions — inflation, debt burdens so great that the obtaining of further investment indispensable to national livelihood is hampered. Then there is the eruption of nationalistic passions with their covetous ambitions to appropriate territories by force and cleanse them of others by terror. Ideological-religious antagonisms can reach such intensity that they disrupt the societal convivium. Lastly to be mentioned, and yet maybe the most disruptive, and the least amenable to a static model of the social contract, is individual turmoil, personal disregard for the most rudimentary rules of existential cooperation, devious and bloody mindsets that feed violence and undermine the stability of society, shaking the feeling of personal security, even in the home, the last refuge from abuse by external forces.

Throughout the history of the Occident and the Orient such violent disturbances were a commonplace in the course of things. It is not astonishing that humans sought fixed systems of societal reference to guarantee some relatively stable existence, some degree of control over the turmoil. What is surprising, however, apropos such systems of reference is that they do not take into consideration this dynamic and unruly flux of life in the first place.

If we want to do justice to this situation, to appreciate the hitherto unfolded system of the ontopoietic unfolding of life as such, the specifically human surpassing of life strictly subservient to survival interests through intellect and acumen, and the whirl of societal life with its own dependencies, conditionings, uncertainties, dramas as well, we will be not satisfied with any social theory or outline but will seek the nature of the societal equilibrium in the above-mentioned logoic matrix upon which all this hangs in one way or another, that is, in the mechanism that projects societal equilibrium amid the evolving current of turmoil and upheaval.

The original ground on which the logoic matrix works and from which it emerges is the communal existence of the human individual. Its existential and ontopoietic axis should be sought in the ontopoietic sequence of the individual.

As to the first, let us recall that although social equilibrium is supposed to be — and time and again is — orchestrated and maintained at all levels of the individual's existence by organizations large and small (governments, economic and social institutions, churches, schools, associations, clubs, etc.), the very functioning of these societal institutions is itself rooted in a basic logoic set of dispositions of the human individual/person — willingness to share and cooperate, commitment to others and to causes, trust.

These originary dispositions to specifically human participation in life provide the basic pattern of communal attitudes. It is within the nuclear community (parents, family, friends) in which the human person develops and acquires a specific mold, personal bents toward others and himself, that there unfolds the fundamental pattern of the person's "self with respect to others": social roles embraced, social responsibility taken, social restraint accepted, social trust given.

And here comes an essential point. This pattern is in itself movable but yet endows the person with an axis of communal attitudes that are inlaid in all possible modalities of emotive, intellectual, and spiritual life and mold the individual as a specific, nay, unique type of person among others, as well as a member of a community. Here, in contrast to the classical view of the individual as developing on its own, as well to Durkheim's views that society is prime in the lives of individuals, we steer a middle course. Although the human individual appears to be the cornerstone of the societal order, acting as such requires the acquisition of a set of essential propensities and features through the sieve of the pattern of community life that molds the individual's actions/reactions in interaction with other human beings.

Ultimately key is the most sensitive, delicate, and basic pattern of the person's own existential dynamic fabric with his bents and proficiencies for unfolding essential ties with other human beings, ties that have to be seen not only as a subsidiary factor of the person's drive for fulfillment, but also the foundation of the social contract seen in its logoic sense as underlying the whole of the societal convivium. This holds true as well with respect to the natural existence of the individual within the world, with respect to his highest strivings.

The rays of the implementation of the logoic social contract are countless. In surveying the interpersonal ties of the individuals at the core of patterns of social involvement and attitudes, we have to focus first on the originary nucleus of the genesis of interpersonal attitudes within the ontopoietic design of self-individualizing life. It is with reference to this existential genetic design that we must view the unfolding of societal attitudes: the forming of aspirations, the assuming of social roles, particularly leadership, the drive for personal/social achievement, the willingness to perform chores for the common good, to share in responsibility, to accept responsibilities, etc. This calls for the ideo-intellective formation of the concept of personal life fulfillment and the valuation of the life scene and the existential possibilities available for that fulfillment. Lastly, but most significantly, it calls for the eliciting within the genesis of the ontopoietic unfolding of the person of a crucial trust in others and a commitment to undertake tasks and be steadfast in the pursuit of common goals.

To sum up, let us say that the very formulation of the issues involved in societal agreements of all sorts most significantly reveals the logoic social contract presiding over them, offers an insight into the strategies of the Dionysian logos in giving impetus to individualization, impetus to the conscious, fully reflective, selective, and enterprising level of the specifically human deployment articulating the schema of equipoise. Apprehending this equipoise avoids views that oversimplify and deviate from reality by positing all to be either individually self-determined or societally conditioned, as well as any static and universalized conception of human nature, for this is an equipoise achieved amid the dynamic unfolding of the individual within the world of life, one that draws on the moral, intellective, and aesthetic sense giving factors of the Human Condition. Again the logoic strategy contravenes the constantly changing flux of the life struggle amid constraints, crises, and sheer uncertainty by endowing the Human Condition with the ability to project and follow a constructive course of life.

VI. HUMAN RIGHTS IN THE STRIVING FOR EQUIPOISE AMONG SOCIETAL FORCES

We witness the continuing unfolding progress of the logoic drive for individuation/independence in the specifically human sphere informed by the Dionysian logos of social existence. In rebelliousness, anarchy, libertinism, it leaves its impetus unquenched, evaporating within the ensuing societal disarray. The drive is then renewed for a safeguarding defense of the individual against unwarranted submission to societal forces with their political schemes and gross abuse of individuals, treatment that hinders their natural course, frustrates their achievements, and even disrupts the main course of their vital progress. Hence we have the urge to specifically recognize the unique status of the human being within the web of living beings, that of moral status. Humanity is urged to recognize that the human being stands out from the rest of life. Man is rooted in the ontopoietic vital unfolding, yet through the implementation of the Human Condition attains the unique sphere of existence informed by the Dionysian logos, the spirit of societal relations referring to common aims, concerns and ideals and the spirit of invention and the creation of new avenues of life, all this on the wings of the moral sense that inspires respect for the otherness of human beings, attitudes of duty and responsibility for others. In short, the moral sense as the main factor of the Dionysian logos prompts the establishment of the specifically human person throwing out a ring of Dionysian links binding human persons together in a moral network of existential solidarity.

It is on the wings of this solidarity, on the one hand, of an innermost conviction of one's union with the unfolding of the human course, a conviction that the person's intimate moral significance is rooted in the human significance as such, that there surges the idea of human 'dignity,' which is a moral notion. On this understanding the specific status of a human being as human is a step in the progressive evolutionary unfolding of life, just as much as it is a constructive logoic advance in human self-interpretation-in-existence. Human being endowed with the creative logos gives itself by itself a new "sense," or rather, a new "significance."

This significance of the unique status of the human being is, as mentioned above, that it is an expression of the Dionysian logos' impetus for singularization, for attributing a specific value, which is then recognized after the fact by society.

This new logoic tension between recognition of the specific ontic/societal status of the human person/individual and the concurrent claim made on society on its behalf for the special safeguarding of this status is the ontic/societal basis of what in our epoch takes the form of the assertion of human rights.

The concept of 'human rights' is in principle a political appeal to the apparatus of the state to respect this unique status the human person has within the web of life. We see throughout the history of mankind constantly renewed attempts to propose some social regulations for the ordering of human relations so that people may live in a group and contribute to the nation and the local community.

The laws of Hammurabi, of Draco, of Solon, and the Mosaic Law were mainly dedicated to preventing gross abuses in human relations. However, already in the Middle Ages in Europe there was enough logoic consciousness of the "value" of the human person that the law as a point of reference not only indicated and set punishments for misdeeds but also — and foremost — provided courts of appeal in cases of individual abuse. Jurisprudence then became the measure of equipoise amid individual recklessness. With the first universal charters of jurisprudence: the Polish principle of "Neminem Captivabimus," the English Magna Carta, a universal status was attributed to the human person, one that in "political terms" became valid for all citizens of the nation.

Yet with the human advance in self-awareness, in the Dionysian striving for larger and larger fields for fulfilling the individual's plans, on the one hand, and with the brutal disruption of individual lives and suppression of their strivings under various regimes, on the other, we enter a new phase of tension, that between the rising, already risen, awareness of the specific value and status of the human individual, of his moral-existential standing and "dignity," on the one hand, and the not completely extinguished, nay, even renewed forces of coercion, on the other. We have seen and see still massive political waves destructive of human beings, as if they were not worthy even of the consideration afforded lower living beings, animals. On the rebound, then, there is a stronger surging of the logoic higher moral striving for the equipoise to be attained in law, a striving to establish what are called "human rights."

Although the impetus of human rights claims pertains essentially to the civic, juridical, and political spheres of social life, it is not to be identified with any of them because its significance spreads to all sectors of human life. Although it does not express any particular right/obligation it pertains to all. Lastly, and most significantly, although it is not to be identified with any particular state, feature, virtue, or line of conduct of the human being, its claim is raised on behalf of the very Human Condition.

The human condition, to which all the higher strivings of the human spirit go back, communicates that the human being is fully human only insofar as he proceeds in his existence by his own devising and the enactment of his very own interpretive existential scheme.

Human dignity, in virtue of which the human rights claim is raised, resides precisely in the uniqueness of human self-interpretation-in-existence.

VII. EIDOS OR ETHOS? THE MODALITIES OF HUMAN LIFE

1. *Setting the Issue: Eidos or Ethos*

In order to advance, we have to regress to our springboard, which leads us back to the logos of life and the initial spontaneity. The task of the philosophical investigation of man, his faculties, his resources, and the basic conditions of his existence, is, in fact, to clarify the reasons intrinsic to the human condition, which we may then either despair over or glorify. The pessimism/optimism controversy tied to the image of man in literature is properly addressed on philosophical ground. Accordingly, the first issue that philosophy is called upon to resolve is the validity of those judgements that reason may render. More precisely, we have the doubt whether acceptance of the sovereign rule of reason in matters concerning the life and destiny of man is not an unwarranted reductionism.[7]

There is nothing new — it seems — in this doubt. Phenomenology emerged from and developed in reaction to the abuse of "scientific reason," which, phenomenology claims, distorts "things as they are." The phenomenological inquiry has attempted to vindicate alternatives to the scientifically "objectified" view of the world and of man, something that it has endeavored to achieve by trying through several approaches to reach back to the originary genesis of both man as a conscious being and his lifeworld.

The question arises of why we should here reopen the investigation of man within the phenomenological perspective when several major lines of inquiry into man are already fully developed; among them the two dominant trends in phenomenology, the transcendental-constitutive and the ontologico-formal investigations.

However, for reasons we have discussed previously, the core of experience that phenomenology proposes to retrieve in its genetic descent to the pretheoretical, naive lifeworld remains unaccounted for by these analyses, which merely retrace the steps of rational reordering, structure, and formation. By looking at the world only in its universal and ever recurring structures, classical phenomenology has neglected to notice that the cornerstones of its progress are radically novel and original objects for whose emergence the phenomenological constitution cannot account. In studying them we discover that *the primordial function of consciousness is not constitution but creativity.*[8] Before the reality of human creativity, constitution appears to be but its anonymous, automated instrument. Man himself appears in a new perspective. The concrete man in his uniquely personal growth is revealed to advance not by the automated universal routes of his functioning but through conflicts, controversies, deliberation, choice, and imaginative projection. In short, the referential framework of man's life, the life of conflict, strivings, aspirations, passions, victories and defeats, is not limited to rational survival games but is epitomized in man's basic aspiration to transcend himself and his limits, to reach far beyond the rational framework of intentionality.

Finally, in personal life as well as in scholarly investigations or artistic pursuits, the resources of reason so often come to a frustrating end —

or even if still projecting their schemes — fail us and then we make the "right decision" or take the "right step" by following an intuition or a "blind impulse" that runs counter to the rationally preestablished routes. In art, science, and any creative endeavor of man, the moment of invention — of a radical step forward — is brought about always through the breaking in of a sudden, unforeseeable *spontaneous* moment into the previously outlined rational system that may prepare its surging forth but does not account for it.

There seems indeed to be a crucial mystification underlying the pessimistic interpretation of man in the analysis prevalent in present times. This analysis of man seeks to discover the significance of human existence in an objective status of man's nature and his lifeworld. It seeks a feature, a rule, a relation, or a principle to be found within man's natural, passive course of functioning as the positive key present there. In this orientation we deliberately make two assumptions: 1) that we should remain within the rational-cognitive perspective and find this key in an objective state of affairs, in the rational meaning of its *eidos,* which is expected to yield it; 2) that this (*eidetic*) state of affairs is inherent to the rational man and is just to be found. As with Kafka's "assignment," it is assumed to be ready and prepared and that man need only make himself open and wait patiently in order to grasp it. This assumes passivity and passive receptivity toward a preestablished factor.

But in asking the question of the ultimate significance of life and the virtualities of human condition that may fulfill it, are we seeking an Eidos, the abstract, inert, indifferent, impartial principle of an object or a state of affairs, or an inspiring dynamic factor of experience, specifically, an *Ethos*?

The basis for our controversy lies then in an investigation of the nature and the complete resources of human experience. What is the status of an Ethos? Do we discover its mode inherent in the constitutive function of reason or do we forge it by deliberation and choice? What is the vehicle of its progress; furthermore, if it be directed toward transcending the empirical framework of man's finiteness, what framework of reference does it have in its advance?[9]

In our proposal to retrieve the dimensions of the Ethos in experience, we aim to grasp the complete human act. To this effect we have, however, to turn to the creative context for an appropriate analytic framework.

From this sharp differentiation it should become clear that the pessimism-optimism controversy is about the *significance,* and not about the *meaning,* of life. Accordingly, in the investigation of its foundation in the human condition we will have to explicate its axiological quality rather than the genesis of the formal structure of experience; and this axiological quality must be considered 1) as most intimately relevant to man's innermost strivings, 2) as it emerges and takes shape through various dimensions of man's functioning, 3) as it appears in the tissues of the modalities of human life, and 4) as it surges from and points to a subterranean source in an Initial Spontaneity.

VIII. SPONTANEITY AND PHENOMENOLOGICAL INVESTIGATION

The Initial Spontaneity as the Postulate of the Intentional, Ontologico-Existential, and Cosmological Contexts

Each age, in its own way, is concerned with the great issue of the significance of human existence; therefore the progress of culture — as exemplified in art, literature, historical interpretation, science, and philosophy — is characterized differently by the various differing attitudes towards the human condition. Moreover, the role of spontaneity, as exhibited in man's relation to nature, to cosmic forces, and to his own personal surging into existence and course, has remained a source of recurrent wonder to human reflection. In different philosophical systems spontaneity appears in different forms and under different termini. It is sufficient here to mention Heraclitus' perpetual flux of all beings, Nature in Kant's *Einbildungskraft,* Rousseau's concept of the feeling from which man's complete destiny may spring, Schopenhauer's will, and Bergson's 'élan vital.'

(1) The notion of a 'spontaneous unfolding,' 'spontaneous act,' etc., occurs, as we know, also within the framework of transcendental phenomenology in a twofold way and plays an important role in the description of the way in which the constitutive genesis of transcendental consciousness of the lifeworld proceeds.

First, in Husserl's thought the term 'spontaneous' denotes a prompt unfolding at the level of 'passive synthesis' qualifying the unfolding of the concatenations of intentional acts and processes. This unfolding springs promptly from a further undetermined source but is self-explanatory in its progress; in this respect it seems also to stand for the unfolding of its forms or segments and their articulations within a 'given,' that is, a preestablished system unguided by any outside factor or constituting conscious agent; that is to say, the "spontaneous unfolding of constitutive acts" would mean also their prompt self-promotion. In this sense, the qualification of the constitutive system by the term 'spontaneous' is coupled with the other descriptive feature that for Husserl qualifies the intentional progress, namely, its automated "passivity" of operations with respect to the human conscious agent, that is, without any active role being played by the ego. The human agent himself, the ego, unfolds himself "passively" through the course of a spontaneous constitution of the means and ends that in their emergence punctuate his growth. By "spontaneous" is meant, then, that the route of the person's unfolding follows step by step its own intrinsic design and that this stepwise progress is thus entirely independent of man's conscious will.

Further, since it is precisely man's differentiating function (ability to discriminate among different elements) and his selection that, on the one hand, constitute the vehicles of his will and, on the other hand, allow him to reach beyond the brute functioning of natural forces into the lucidity of consciousness, the "spontaneous" unfolding of his being insofar as it is independent in its forms and progress from any discriminating and selective interference is a "blind" unfolding. If man were to remain at this level in his individual unfolding, following exclusively the spontaneous course of his own — and his world's — constitution, he would remain a mere element of Nature.

From this viewpoint, the spontaneous aspect of the genetic unfolding of man and the lifeworld is a feature of the universal constitutive pattern insofar as it designs its operations, forms its constituting means and the intertwining of their operations, projects their concatenations and thereby promotes itself in its building of man's operating functions in synchronization with the development of man's milieu. In retracing its network of articulating functions and operations from the most developed down to the incipient simplest ones, in which the genetic origins are sought, Husserl stops at the borderline of the constitutive articulations seen in the instance of the simplest "originary impression" (*Urimpression*). Since it is the formation of significant cognitive-existential ties that is here focused on, the emergence of the "originary impression" as the first instance of constitutive activity and the *unique* contact with the preconstitutive (that is, "originally given") marks its incipient instant as the "source" and "reason" of the spontaneously unfolding genesis of man and his world. When Husserl, in late reflections published by Iso Kern, attempts to go into the empirical dimensions to seek the organic roots of the constitutive system, he lacks the means with which to introduce their naive, prephilosophical, and prereduced data into the phenomenological orbit laid down by his program.[10] We see Heidegger and Merleau-Ponty too reaching the threshold of intentionality and stopping at the unreduced, that is, the "brute facts" or "wild being" of Nature, which as they find it are obviously impossible to interpret phenomenologically.

The question that I have raised elsewhere is whether intentional genesis is truly self-explicatory.[11]

Does the seemingly self-guided and self-promoting "spontaneity" qualifying the progress of transcendental genesis explain more than the ways and means by which it actually proceeds? The decisive quest for the factors and conditions from which the incipient instance of the genetic flux is precipitated cannot remain infinitely postponed.[12]

Husserl sought an answer to this question in the "absolute" existential status of pure consciousness; in *Ideas* I "pure" consciousness was conceived as its own existential source and reason for its dynamic unfolding and order. However, as we know, its divorce from "empirical consciousness" could not be maintained within Husserl's ulterior reflections. Seeking the interlacing of pure and empirical consciousness in the expansion of the intentional system into the experience of the body, Husserl could, however, reach phenomenologically only as far as the referential radius of the pure ego can carry the intentional system. The pre-ego-referring dimensions of consciousness remained intentionally inaccessible, indicating that we cannot find an answer to the question of the source of man's transcendental genesis through an infinite regress into his origins.

Second, we also encounter spontaneities as forces proper in the Husserlian transcendental system, as the "active" promptings springing forth from the pure ego: spontaneities of judgement, of the will, of valuation, and of practical action. In radical contrast to the operations that unfold "passively," they represent the "active" role of the ego that triggers them. Nevertheless, on the one hand, they are seen as totally bound in their operations to the regulations of the entire constitutive system with its threefold temporal limitations. On the other hand, their immediate source, the ego, is — as we have emphasized above — constituted by the passive genesis itself.

But could not our argument be undermined by bringing out that "originary association" — the basic factor, if not the very vehicle, of the passive synthesis — can be seen to contain an element of spontaneity that has the essential form we are arguing for, namely, that of a "generating force"? In point of fact, as Elmar Holenstein has emphasized,[13] Husserl's conception of the 'originary association' (*Urassoziation*) — opposing the classical Lockean theory of the 'association of ideas' — though limited to a static coming together of qualitative elements according to their mutual affinities, contains also a qualitative "melting together" (*Verschmelzung*) and "mutual motivation" (*Motivation*) between and among the affective elements of the constitutive data;[14] as generic principles, they promote the dynamic advance. As generic principles, they could be seen as moments of spontaneous "force" (*affektive Kraft*) released by the "passive" genetic system itself to promote its own progress.[15] And yet their role, however large a scale we may attribute to their activity, is still played completely within the total project of a passive genesis limited to the constitutive generation of its own modalities and functions.

The reach of the intentional network carrying on the constitutive genesis breaks down in Husserl's investigation at the threshold of the "originary impression." The intentional continuum does not reach any further. Reaching this threshold, the limits of its self-explanatory power is revealed. Husserl does not envisage the possibility of an affective association that is preconscious. Also revealed is the limited span of the self-promoting capacity of the individual: this continuity does not extend beyond the initial moment at which the genetic process begins and the final moment at which it ends with extinction of the human being. We must say of the two functions responsible for the "spontaneous" unfolding of the constitutive advance, its essential, intrinsic constituents, namely, originary association and passive synthesis, that they both are again nothing other than vehicles of structuring and order.

But what accounts for the triggering of the constitutive activity that they organize? What accounts for the incipient prompting of the set of its operations?

Even in order to bring to light the intricacies of the role played by originary association (at its threefold level of constitution: memory, expectation, and the living present) recourse to the essential laws of the constitutive system with their temporal horizon is indispensable. But can this alone explicate the setting into operation of the total project? Can it account for the incipient spontaneous surging of an "originary impression" already loaded with all the constitutive potentialities needed for its unfolding?

Moreover, from another point of view — that which I advocate in this argument — the prerogatives attributed to the generative role of the originary association are again too limited to take care of the complete network of man's functioning and reach to its roots in the human condition. In fact, the generative spontaneities of this basic factor of intentional constitution may lead to "new" qualitative and emotive complexes emerging from mutual influences.[16] They lack, however, the generative moments leading to *invention,* to the projection of the new forms and unprecedented qualities that are indispensable for authentic "innovation." In order that these may be fostered, something more is needed, namely,

Imaginatio Creatrix, which breaks through the constitutive system.[17]

We may conclude as follows. To account for the actual fact that the generic unfolding of the originary association with its regulative coordinates (that is, all the regulations of the passive synthesis) springs forth into operation — that the "originally given" moment enters into the constitutive system of the "originary impression" — a factor of *spontaneity* external to this complete system must be postulated.

(2) In his disagreement with Husserl over the "absolute" status of pure consciousness, Ingarden outlined his investigation beforehand in such a way that it could be expected to open access to the empirical realm of human experience; in this approach he introduced major changes into his basically Husserlian conception of consciousness.

By means of his previously established distinctions of possible ontologico-existential ties between and among the elements of objective structures, Ingarden makes a deliberate attempt to bring both the "pure" and the "empirical" sides of consciousness into one and the same ontologico-existential complex.[18] Pure consciousness is first seen by Ingarden as a set of ontologico-existentially interdependent components that include a) the stream of consciousness, b) conscious acts flowing from it, and c) the ego pole at its center. Then, improving on Husserl's conception of the 'person' as being constituted by pure acts of consciousness, Ingarden introduces the more complete notion of the 'living person.' The living person, although in her unfolding formally constituted within and by the system of pure consciousness is in her *material nature,* seen by Ingarden as essentially partaking of the realm of passions and blind forces represented by the notion of the 'soul.' Furthermore, going beyond Husserl's conception of the 'soul,' Ingarden sees it as being articulated ontologically in existential ties of interdependence not only with the blind and otherwise inaccessible realms of brute forces but with the 'living ego.' The 'living ego,' seen as incarnating empirically the "pure" Husserlian ego, plays the role of an "axis" within this complex of existential ties; as a 'living' ego it is not only the pole toward which all the acts of the ego converge and from which they spring forth for as such it also articulates and forms the blind forces that stream from the 'soul' and enter into the existential ties of the 'living person,' or so

Ingarden supposes.

The 'soul,' conceived in this fashion, is then seen to extend through the ontologico-existential ties into the "body-as-experienced" forming what we could call a 'soul-body territory.'

Would we within the reach of the ontologico-existential ties be in position 1) to grasp the realm of "forces" and "passions" of the "soul" in what Ingarden calls their "essential state," and 2) to reach by the same ties beyond the "soul-body territory" into the physical body (and not only the "body as experienced within the intentional system")? Were this feasible, we could expect to gain access to the originary source of the intentional apparatus in the *elemental* realms of nature and to the human condition as well. However, this ontological attempt fails. Not only does the material ontology postulated to yield this "essential state" of the naively assumed "forces" and "passions" of the soul fail to present a basis for such development, but the "soul" as so conceived — the soul-body territory included — is accessible to Ingarden's inquiry only insofar as it may be reached through the intentional network. All the supposedly "empirical" moments of this soul are assessed with reference to the nature of intentional acts and the pure ego. They constitute, as with Husserl's analysis of body-as-experienced, an integral part of the intentional inquiry. No legitimate link between the "soul-body-as-experienced territory" and the body-as-an-element-of-nature is in sight.

When in his two last books, Ingarden sought access to the *actual* empirical forces by, on the one side, tracing the causal network within the real world that releases them and, on the other side, apprehending the moral action that runs through the various levels of man's functional "systems" and "subsystems," as Husserl had done before him, he does not succeed in assessing these "primitive forces" and dynamic "semi-closed systems" philosophically; his inquiry proceeds only as far as an *eidetic* apprehension of formal-ontological ties.[19]

And yet, is it not beyond the last outpost of the realms of eidetic ontologico-existential ties and the intentional genetic system alike that we strike at the *Initial Spontaneity* that stands for the "primitive forces," "passions," "physical dynamisms" of brute Nature?

Does not, on the one hand, the intentional network with its dynamic flux of operations point to an Initial Spontaneity that would bring it into operation, a spontaneity that carries germinal virtualities with it into the incipient moment of the intentional genesis? Does not the operative intentional network require an initial spontaneity that channels into it the blind, haphazard elements that it may then canvass to promote its project further, its structures of objectivity?

On the other hand, does not the network of formal-ontological ties require ontologically accessible *material* to bind into its schemes as it establishes a structural groundwork for existence?

Philosophers throughout history have attempted to introduce the spontaneous factors of existence into the framework of their view of the world and man. The question of pivotal importance for phenomenology is that of *how, beyond becoming naively aware of physical dynamisms (as subjacent to specifically human functioning), to integrally introduce these spontaneous factors into the universal schema that phenomenology has been attempting to establish as it applies the criteria of a rigorous methodology.* Within the proper perspective, they may reveal themselves to be the crucial factor of the grand phenomenological system of man's and his world's constitution.

(3) When we at last consider (in our own cosmological perspective) the real individual within the world-context-of-actual-existence in which he appears as an agent centralizing and distributing the forces of the universe, we cannot fail to ask after the incipient impulse that makes him coalesce into a self-governing agent.[20] As an inward/outward oriented agent, through whom the brute forces of nature pass in and out after he has used them for his own aims, he himself cannot be their natural outcome. In the mere play of primitive energies and their haphazard strivings could there originate an intrinsically cogent and self-supporting agent able to resist and control the impact of the elements? Even if that agent's actual concrete mechanism is built from elementary forces streaming together, how would they converge without an Initial Spontaneity that would in one stroke give to their chaotic strivings an impulse and a purposive linkage?

Within the complete phenomenological framework of inquiry, the Initial Spontaneity emerges as an authentic counterpart of our systems and subsystems — of consciousness, intentionality, constitution — in which systems it represents the *elemental* ground of the primitive forces and the subliminal source of man's passions, drives, strivings, and yearnings.

IX. SPONTANEITY AND THE CREATIVE CONTEXT OF PHENOMENOLOGICAL INQUIRY

To have vindicated the subterranean spontaneity as the ground factor of human functioning is undoubtedly in itself to have radically breached the self-encircled eidetico-intentional array. However, the notion of "initial spontaneity" has been retrieved from beneath this rational web as its dynamic undercurrent by postulation rather than by immediate inspection. Questions of paramount significance thus arise: 1) Is it legitimate and useful to accept this notion without substantiating its role within the concrete texture of man's and the world's existence? 2) How can the initial spontaneity be ascertained to be the intrinsic factor of the concrete human functioning that mediates the naively postulated primitive "dynamisms" and "forces" in the human condition? 3) Are there ways and means remaining in accord with the basic phenomenological criteria of validity by which the initial spontaneity could be investigated?

Using three metaphors we could venture to submit that the "initial spontaneity" we speak of might be related to the universal state of things in a threefold way at least and that in each of these three ways it can be correlated with a different but specific framework of ordering operations.

The initial spontaneity as a current might, first, flow into its bed by following "the force of things"; it might, secondly, be channelized into a bed; thirdly, its current might also flow into a bed that it cautiously prepares for itself using the means and possibilities offered by the given circumstances while adjusting to them its speed, its volume, and its course.

In the first metaphor we might see a reflection of the Bergsonian universal elan, which recurs by virtue of being a universal force of Nature. The second metaphor, that of channeling, might well mirror the

Initial Spontaneity's introduction of the primitive dynamism of brute nature into the rational order of the constitutive system. Does not intentional constitution mold the initial spontaneous flux — from the incipient moment of its surging in — according to the rules of man's constitutive apparatus of controlled motivations, associative links, chain reactions, etc.?

The human faculty called 'will' — the capacity for decision making after deliberation and for sustained action to carry out decisions — is, in the current phenomenological perspective, seen as an intrinsic element of the intentional web. And yet at turning points in our rationally ordered existence, we face the sudden surging of forces totally independent of the manipulations of reason. They gather momentum and, taking reason by surprise, break the dams set by it and depart from the designed course. They seem to run wild. In fact, this outbreak of primitive spontaneity can only be seen as blind turmoil unless the Initial Spontaneity actually makes for itself the bed it runs into.

This is the third metaphor. Its use is postulated on the Initial Spontaneity's searching for the appropriate course to follow. Here is no blind and wayward element. While seeking to prepare for itself an appropriate array of lines for a constructive course, the Initial Spontaneity is differentiated into various functional modalities. *But more than anything else the Initial Spontaneity has brought forth the will to create, to transcend the common, preestablished course of reason by lifting the real individual to the specifically human status.*

In fact, the conception of "spontaneity" — in contrast to order — is pervasive in philosophical reflection. In the thought of Thomas Aquinas, the concept signifies the moment of indeterminacy, promptness, suddenness, and unreflectiveness. In Duns Scotus the "spontaneous act" is opposed to the system of cognition that is subject to the universal order of things.

Historically, the notion of spontaneity is also linked essentially with the outburst of the creative act, either that of the Creator — as it is in Duns Scotus and Scholastic thought in general — or of the human being. Will appears, then, in focus as "free," that is, as the prerogative of the creating agent himself to select and decide according to principles stemming from within his own deliberations and invention. As such it becomes the center of controversies.

Phenomenological inquiry into the human creative process indicates, in fact, that it does not proceed passively following a predelineated route but, on the contrary, advances through the freely acting creative agent: the specific spontaneity through which it surges and is carried onward could not be a result either of an elementary force or a whimsical outburst.

The inquiry into man's creative endeavor to create original and unique objects — or exemplary objects, to use Kant's expression — reaches, as I have shown elsewhere, deeper into the original conditions of man's functioning than intentional analysis.[21] From the standpoint of the dynamic constructive extension of the *creative* process and with reference to the person as a deliberating and deciding agent, the creative process proceeds from the actual phase of the constituted world. It emerges, however, in revolt against its actually established forms, its automated constitutive procedures as predelineated by the objective horizon. In its dynamic surge it breaks progressively through the acquired constitutive molds and the complete passively settled system toward the establishment of its own network. Thereby it clears the ground for the exercise of the *constructive, freely acting human will.* Like constitutive, intentional acts, the creative process is also directed toward an "objective." In contradistinction to them its objective, however, is not proposed from within the reservoir of the a priori possibilities of reason, but is, on the contrary progressively formed through man's active deliberation and choice.

In radical contrast to the "free variation" that guarantees a margin of flexibility in the constitutive act and the "reproductive imagination" accounting for it, the creative progress may advance freely owing to the generating power of Imaginatio Creatrix.

Within this delineated framework of analysis in which the constitutive mechanism enters, albeit in a subservient role, what comes into the fore, are — as I have shown in my *Logos and Life,* Book I, *Creative Experience and the Critique of Reason* — the functional subsystems of affections, emotions, and passions: *the material of the subliminal life stream into which we enter* after having crossed the borderline of the intentional network and freed it from its straitjacket.[22] Within the dynamic network of the creative context there is revealed — this time no longer precritically but upon philosophically prepared ground — this pulp of experience, which indirectly, although concretely, we may now assess and investigate in its fluctuating substance at the point at which its elements enter into various configurations.

In conclusion, *the Creative Spontaneity that emerges from the creative configuration of human functions has carefully prepared its own bed.* In order that the creative impulse of human free will, the will to act and create, may spring forth, all the types of human functions enter into a novel configuration. New, specific roles are distributed to all the types of operations among the various ramifications of man's functioning. The elements of the creative configuration coalesce toward this aim.

By following the different modalities that experience takes within the creative orchestration we may find the hitherto blocked access to the subliminal realm of affections and passions where lie the roots of the Ethos and modalities of life.

X. THE EXPERIENTIAL MODALITIES OF THE ETHOS AND OF HUMAN LIFE

1. *The Experiential Modalities of Life and Spontaneity*

Let us recall that we have in passing linked together the notion of the Initial Spontaneity with that of the "modalities of life" conceived as molds through which the initial spontaneity not only "makes her bed" but is also prompted, silenced, or even extinguished in its course. We have also linked the modalities of life with the genesis of man's attitude toward his specifically human condition relative to the pessimism/optimism controversy. Finally, we have proposed that it is the originary role of the Ethos — in contradistinction to the Logos of the constitutive unfolding of man — that is at stake in this controversy.

Since the notion of the Ethos appeared indirectly as being intimately linked with the subliminal factors of man's functioning that we in our naive attitude identify with "passions," "affectivity," "sensibility," etc., we have proposed the creative context in order to retrieve its roots in the subliminal dimension of human functioning.

However, did not Max Scheler, on the one hand, already bring forth the notion of the "person" as incarnating the Ethos; and has not, on the other hand, the realm of affectivity been treated by Scheler, Merleau-Ponty, Ricoeur, and Husserl himself? The contributions of classical phenomenology in this respect cannot be ignored, and yet — as we have argued elsewhere — the generative sources of the Ethos in "passions," "affections" and the ultimately subliminal dimension of man could not have come to their own rights within its intentional framework.[23]

Consequently, the role of the Ethos in the system of human functioning was not adequately acknowledged. It is indeed a question *whether the axiological quality of experience does not belong to the very nature of the specifically human act.* It might well be that to discover the roots of man's decisive attitudes toward his condition as he develops in the modalities of life, *we have to retrieve the human act in its complete validating-affective set.*[24]

In fact, as we have already pointed out above, Husserl did not overlook the affective component in the human operational system. On the contrary, he readily acknowledged that every conscious act is an "experience" (*ein Erlebnis*); this means that in the progress of his research, while trying to move away from the notion of *Empfindung* as a real component of the perceptual *hyle* he was, as Holenstein has rightly observed, more and more attentive to the "experiential" aspect of conscious acts, which, following Kant's transcendental aesthetics, he called "affectivity."[25] This allegedly founds the activity of the ego.

This analysis stretches even further, to the acknowledgement of a level of "primal affect." In fact, going beyond Kant, in his *Phenomenological Psychology,* Husserl differentiates, on the one hand, Kantian "receptivity," which he acknowledges also in detailing the passive genesis into receptivity proper, that is, the already active attentive orientation of the ego toward the "affect factor," and "affect," on the other hand, identified as the "primary activating of the subject," which does not come from the ego center but belongs to the "pure" "originary passivity";[26]

Husserl sees affect as "an active passivity, an affective force" as opposed to an activity proceeding from the ego pole.[27] However, Husserl ties affectivity to originary association,[28] and the fact that in his conceiving it can be elucidated only conjointly with originary association again makes its interpretative analysis relative to the constitution bound analytic framework, which forces him to conceive of affect as being intentional.

Although Husserl — according to Holenstein — distinguishes in a later manuscript between *Uraffektion,* which contains no intention pointing toward, and affect related directly to the perceived or the objective, which affect is of an intentional nature, the preobjective and preintentional affective units are still understood with strict reference to the intentional system. They appear to be oriented of necessity toward the intentional world-constitution.[29] Indeed, affectivity appears again in Husserl's analytic framework only insofar as it is relevant to the associative rules of its rational intentional framework and not in its own right.

Husserl also recognized that human acts at a certain "lower level of passivity" refer not simply to the ego as a pole but to the complete human *person.* He writes in *Phenomenological Psychology* that "ich gewahre, ich erfasse, ich betrachte" refers already to the I, seen as the complete person.[30] However, to start with the last point, he concedes then that the appropriate concept of the "person" is still missing. In *Cartesian Meditations* he conceived of the person in a too narrow way, merely at the level at which the pattern of the person is already transcendentally (intentionally) formed or constituted by pure conscious acts; he did not consider it at the level at which acts are merely virtual, constitutively undetermined either with respect to the horizon of the past or with respect to the horizon of constitutive potentialities. Furthermore, he seems himself to recognize that this conception of the person lacks relation to the body.[31]

Consequently, in this acknowledgement of the "primal" level of experience and of affective force, we may see the elemental realm piercing through the intentional system even in Husserl's thinking. We may also see in his recognition of the complete act springing forth from the person an insight into the person as the source of the completion of the human act. However, his essential reference of primal affect to originary association and like essential reference of the person to the pure conscious procedure prevents him from advancing. He is bound by the assumptions of his framework.

Indeed, as we well know, it is from the originary constitution of perception that the whole constitutive system of the lifeworld is assumed to start and in which the more complex cognitive-constitutive operations are supposed to be predelineated. This is accomplished by a *simple synthesis* initially oriented toward the constitution of the spatial object and to that the analysis of axiological and practical comportment within the emotional realm remains ultimately tied. At the same time, the analysis of the affective elements and complexes in their reference to the originary association (*Urassoziation*) — the level at which they appear in the Husserlian analysis relative to the study of association as the fundamental structuring of intentional genesis — proceeds with essential reference to their role in the constitution of the material, physical object, as presupposed for its sake. Thus, the Husserlian analysis of the level of the supposedly preintentional "primal affect" falls within the consistently conjoined frame of the analytic context with 1) the grounding stream of consciousness referring to the ego pole, the center of the domain of investigation; 2) its orientation model: the structure of the intentional act consisting of the intention and its objective correlate; 3) the network of acts through which the stream flows as the genetic process constituting the correlative network of objectivity; and finally, 4) the horizons of the constitutive advance: the temporal phases, on the one hand, and the a priori objective structuring principles, on the other hand. Since the objective correlate of the genetic progress, as the aim pursued by its operations, is the structural model of the "object," the constitution as much as its analysis hold on to the object as a guiding thread (*Leitfaden*). Consequently the analysis itself is from its start pushed into the channel of intentionality: preconceived with reference to it, the level of "primal affect" in its essential state eludes our approach.

To bring to light the essential role of affectivity and its ways in human functioning, we have, on the contrary, to unravel the level of primal affect in its own right.

Is it not only by approaching it from the opposite side, the side of the initial spontaneity, that we may retrieve it? At the juncture of its wealth of virtualities (which then may enter into the constructive orchestration of the specific creative spontaneity) and the emergence of this orchestration itself, the *subliminal* realm from which at the crosscurrents of all human resources, the act of human experience springs forth, may come to light.

Indeed, only within a framework suspended not only upon the human function of rational ordering but, first of all, upon its source may we approach man not merely within his individual and collective lifeworld but within his basic condition of "coming to be and passing away" and in the full range of his strength and weakness.

2. *The Modalities of Human Life and the Ethos: Preliminaries*

Heidegger undertook the task of investigating man not in the Husserlian perspective focused on cognition-constitution but in the perspective of man's existence-in-the-world. In his conception of the basic set of *existentialia* we can see concern for what we would call "the modalities of life." However, what Heidegger presented us seems to be the most universal skeleton of man's basic "being there" rather than the modalities of his functioning that express simultaneously the ways and means by which man enacts his existence and the ways in which human existence springs from its source. The specific conditions leading to human existence, the conditions of man's coming to be, his growth, decay, and passing away, and also of the forces and weaknesses that man may command or contend with, lie beyond Heidegger's grasp. Emmanuel Levinas seems to have come closer to these realities in his conception of man's fundamental existential modi, seen as "fecundity" and "habitation."[32] It seems, however, that although concretely pertaining to actual existence, both of these modi are shared by all individual living beings. The question emerges, then: "What are the modalities of life that would not only in general but also specifically enter into the composition of human life?" "How do they come about from within the elemental conditions?" "How do they express this condition itself on which not only man's life and death but also his self-designed destiny, depend?"

We have to this point only prepared the ground for raising these questions, and in what follows we shall confine ourselves to a preliminary outline of some analytic elements; these elements allow us, on the one hand, to articulate the notion of the "modalities of life" within our complete project, and, on the other hand, to indicate one path to their investigation as it pertains directly to our initial concern.

In a preliminary way it could by suggested that we first distinguish those *orientations of the elemental functions operative in man's "coming to be" that manifest the existence and maintenance of the autonomous individual real being. Among these we can count Levinas' concepts of 'fecundity' and 'habitation'*; but we should also add to them individual and group solidarity — which expresses itself in looser or tighter forms of *intimacy and play.*

Second, however, when we consider the specifically human being — that is, the one who from the mere individual reworks himself into a person — his coming to be and specific growth do not proceed only along the above-mentioned channels; *in addition his functioning is oriented essentially along the lines of work, feasting, natural piety, belonging, and guilt.*

Each of these modalities spreads through more concrete modi of experience. Natural piety diversifies into the more specific functional lines of filial piety and of the sacred and of love (love of other beings as well as of "motherland"), etc. Human work spreads through functional channels of various modi divided between two basic lines: *life survival* functioning and *creative-life-advance* functioning. This implies basic moral attitudes presided over by the decisive antithesis of justice/injustice. Intimacy establishes itself also as a specific modality of human personal life along several functional paths concerning both the uniqueness of experience and its communicability. Feasting diversifies chiefly into threads of ritual, spectacle, and conviviality. The "celebration of life" and of "beauty" is essentially to be completed by the aesthetic determination of experience.

Modalities of life, diversifying and intergenerating in concrete experience with other experiential factors and elements and among themselves into infinitely complex and nuanced effective threads, carry into and throughout the unfolding of the experiencing man the initial as well as the subsequently generated spontaneities. Our living being is the expression of their interplay. The initial spontaneity is maintained and multiplied by the way in which the modalities of the person unfold; the human being, as he thrives, distributes them through joy, enthusiasm, exuberance, energy, enterprising spirit, courage, etc. If this is not the case, all shrinks within him, his "life" dulls, and he loses courage to endure it.

However, if the spontaneities that flow together in the genetic unfolding of man and those that are subsequently generated fulfill this role, it is owing to their specific virtualities, which mold the act of experience as a specifically human act into which they enter. In fact, we offer the insight that when we consider the formation of the experiential act at the point at which man's virtualities come together when he "enters into their possession," namely, in the creative act, we distinguish, in addition to the positing character of the intentional form of the act, the "validating" ascertainment of its material texture. While the first posits the functional mechanisms and principles of the objectifying structuring of man and his lifeworld, the second is the vehicle and content of the actual progress. Through validating ascertainment in the human act as it is actually being performed, we reach beyond its ontologico-constitutive skeleton of formal mechanism into the very pulp of experience carrying and promoting life. It appears that human life is carried on by acts that at any degree of awareness are already in their lived experience-texture performed in a validating mode.

Indeed, it seems that within the ontologico-intentional framework of a conscious act, there has to enter *a three-channel set of spontaneities* that not only expresses the subterranean stream that prompts the act's emergence but also establishes it as the vehicle of life, of specifically human life. We discover these three channels with their specific virtualities through analysis of the validating mode of experience in which they function. In point of fact, whether they flow into a constitutive or a creative configuration there seems to unite within an act, as an act of human experience — in contradistinction to a mere organic "operation" — *the dynamic, enjoyment,*[33] *and axiological spontaneities.*

Concerning the first, what we naively see as the "dynamism" of a conscious act, we grasp through the differentiation of experience into nuances, intensities, and degrees of vividness or dullness; also, there are the degrees of impetus with which it spreads itself throughout the field of actual consciousness with its horizons of associative evocation. "Coloring" the past and the future, it pervades progressively the primitive vital functions, drills into their depth, vivifies them. However, with what does it "color" them?

The "tonality" of life, which it "colors," "heightens" or "lowers,' "illuminates" or "dims," is produced and enhanced through its association with the second spontaneity, that of "enjoyment," and is determined through the third, namely, that of "valuation." Each act of experience is carried through a "vibration" of "enjoyment" that spreads "horizontally" through the entire spectrum between the two extremes, the positive and the negative (e.g., between "sheer delight" and "unbearable pain" and the qualitatively almost neutral vibration of "just being," in the middle). There is also a "perpendicular" flow of the vibration along a vector of "elevation" upward, and "abasement" downward (at one extreme we experience our whole being lifted above the "common doom," while at the other extreme we feel "lower" than any other living creature, worthless and insignificant).

It appears through these examples that it is the axiological spontaneity that infuses the two others with *affective* value, distributing among them their own "dynamisms." Indeed, this third basic spontaneity flowing into the act brings into it the virtual desirability or undesirability of its elements; it thereby anticipates the completion of affective valuation in the aesthetic, moral, and spiritual aspects of experience.

By releasing within us a commitment toward the elements of experience — as desirable or undesirable — instead of distancing them as "objective correlates" of experience for a disinterested spectator, the axiological spontaneity enhances the very nature of the emotive quality of experience. Not only is our being ourselves now validated through the spontaneous matter of experience — by its being "our own," manifesting our "self" — but even "objective" elements of experience presented along this line release in us different types of commitment; instead of "neutralizing" them with abstract indifference, we "cling" to them.

The emotive instances that we concretely experience appear and then seem to vanish. And yet their appearance beyond the threshold of lucid consciousness has bound into the configuration of the given modality of experience the otherwise blind (because remaining below this threshold) "forces" active within the human being, forces that we are tempted to too quickly relegate to the prerogatives of universal Nature and to reconstruct conceptually as "impulses," "drives," "instincts," "strivings," "yearnings," etc. Undoubtedly there is that to which emotions point as their existential source. However, "brute" Nature could hardly by itself explain in its universally designed progress the *differentiation* of the initial spontaneity as it flows into the act of human experience. This differentiation must be one prior to the act of life in general and that of the human being specifically if it is to come about. Once bound into the orchestration of human functions these blind energies of Nature may stir its apparatus from below and supply it with "forces." However, the initial spontaneity must first differentiate, be differentiated, and be endowed with specific virtualities in an intermediate zone — which postulated territory we call the "subliminal" territory — in order that the functional progress and articulation of life and human being may come about.[34]

The modalities of life that we seek to investigate are generated within a given functional configuration. Is it not from the vigor of the affective substance of experience crystalized within it that they draw their *existential matter*, which they themselves crystalize? While they distribute and infuse pivotal spontaneities, they bring into the functional system the initial spontaneity together with the resources of the primitive "forces" of Nature that promote and intensify its course. Depending on their own active or dull state, the functional articulations of the modalities of life (e.g., the innermost ties of man to the family, to his country, to a cultural group, and to his fellow man, which form the modality of belonging) either intensify or shrink. They either draw their juices or dry up depending on the influx of attitudes toward life they receive (e.g., enthusiasm or apathy, enjoyment or ennui, tenderness or dry-heartedness, faith or despair, confidence or insecurity, compassion or egoism, elevation or abasement, etc.). Man's decisive attitudes toward his existence and his condition are already set down at this elemental level. However, insofar as we may already by this preliminary probing gain insight into the human condition, nothing is presumed as to whether a pessimistic or optimistic attitude should be taken toward it. There are virtualities present justifying both attitudes. But is man free to command their growth in one or the other direction? This ultimate question can be treated only within a fully developed inquiry.

Indeed, thus conceived, the modalities of life are not only the texture of the dynamic unfolding of the personal existence of man; they also initiate it. Not only do they transmit the primitive spontaneity filtered already by the subsystems of the natural, vital functions of man, but they generate and create further forces and novel molds and, in one bound, carry us onward as agents of both our most intimately personal concern with ourselves and our ingrownness in the world. The modalities of life decide whether it glows or stagnates, spurning now passion for life and now apathy; they differentiate in their structural mechanisms with reference to their orienting values. But the very substance of our personal lives resides in the value imbued emotive complexes and configurations of our acts. Their objectives might appear illusory in the light of reason, but the valuative validation that our acts invest them with makes us believe that life is worth living and its pursuits worth undertaking.

CHAPTER THREE

THE CREATIVE FORGE OF THE LOGOS WITHIN THE HUMAN CONDITION

The Twilight of Consciousness and the Human Virtues

I. THE SUBLIMINAL ZONE OF THE SOUL

To get a focus on the origin, nature, and role of the so-called moral virtues in individual and societal life, we have to proceed in a roundabout way. I have elsewhere explored the genesis of the specifically human significance of life. Here we seek to trace its specifically moral vector. I have also proposed earlier that the humanizing trans-action fulfills the promise of the human condition, making the living individual "human." We must take up the question of the humanizing trans-action again to probe more deeply into its crucial role in getting the entire moral apparatus of the human being off the ground. In order to get at the specific role of the virtues we have to situate that role within the framework of the entire moral agency of the human being.

Our major contention is that the virtues are not the product of our reasoning powers and not a gift of nature either, but are the fruit of the subliminal passions of the human soul — wherefrom they receive their main impetus (which must be maintained in order to secure enactment) as well as their qualitative determination and telic principle. This indicates a roundabout course of inquiry. Our first step is to investigate the way in which feelings, inclinations, emotions, bents, tendencies acquire their moral qualities. Interwoven with this is a renewed study of the humanizing trans-action as a specific type of symbiotic sharing in *sympatheia* that founds the societal bond through the "intersubjectivity" that it installs and sustains. In doing so the entire moral apparatus of deliberation, distribution, apportioning, and measured judgement becomes operative in norms, laws, principles, and ideals.

The establishment of a system of societal-moral ordering would be futile, however, unless we can locate it and maintain it in enactment in the humanizing trans-action. In the ever advancing turmoil and fluctuations of the inner self of the human being, how may any order for moral behavior — that is, any societal order — be established and maintained? How may moral norms, principles, laws, ideals pass into the societal interpersonal linkage without being distorted on their way to real life enactment? And even if enacted could they ever be repeated in changing situations? That is to say, how is the moral-societal order at all possible?

With these questions we are *in medias res*, that is, into the study of the virtues. We will elaborate and stick to this middle ground that consists in inquiry into the genesis of virtues and vices (although the emphasis here will be on the virtues alone), and their role in implementing and maintaining a societal-moral order in the midst of changing conditions.

II. THE "MIDDLE GROUND" BETWEEN NATURE'S VITAL SIGNIFICANCE AND THE HUMAN CONDITION

The Sentient Communicative Logos and the Dionysian Logos Enter the Scene

1. *The Philosophical Bearings of Our Query*

The moral apparatus of which the virtues form an instrumental part is embedded within the entire spread of life. The philosophical conception of it depends on the approach that is taken to the great questions concerning the ontology, the metaphysics, the cosmology of life. To distinguish our "middle ground" within this vast expanse it is necessary to start by indicating, be it even in a cursory fashion, our bearings.

Emotions and feelings, as I have established in a minute analysis elsewhere, *mark the territory of the human soul that is the middle zone between cosmic nature, nature-life and the emerging human universe of the spirit.* It is in this middle zone that comprises both spheres that a preexperiential vital play of forces takes place, which forces are molded there into feelings and emotions. Now is the moment to see how in this middle zone of the soul, in the creative forge of the human being there takes place the radical shift from the vital-gregarious significance of life to the specifically human significance of life — i.e., to personal/societal and cultural significance. A consistent significant schema that serves as an axis for human conduct, the criteria for its enactment, and, most significantly, virtues, emerge from the subliminal forge of the soul.

It should be clear here that the traditional anthropocentrism of philosophical anthropologies has no place within the field of the phenomenology of life. In envisaging the progress of life's unfolding of kinds we intuit them from the start to be transformations of vital forces, the emergence of new synergies and virtualities — all of them caught in an incessant constructive fusion, interplay, and transmutation. There is no question of any possible separation or division of life's constructive phases, which are characterized by the engendering of novel types of living beingness. Thus, I think about the emergence of the Human Condition as the appearance of a landmark, novel, unprecedented type of life's constructivism. There is no way to see any split in the bundle of vital forces that gush forth with the initiation of this new phase and the virtualities that crystalize in the radically novel type of beingness that is the human being. Neither is such a split conceivable anywhere within the vast spread of life.

Within the radius of the virtualities that the Human Condition brings about there crystalizes the functional life system of a living human being; there we have to seek all the joints of the transformability of meaningful factors, modes, principles of valuation, cadences of vital prompting, urges, nuanced responses, etc., which account for the passage from the animal to the human type. What in traditional philosophies was assumed to be an abrupt hiatus calling for a bridge between "external nature" and the "human mind" (vainly sought after), or, in modern times, between "subjectivity" and "objectivity," a bridge to explain the marvelously differentiated and discrete and yet strict continuity of life's progress, phenomenology of life has revealed to be a series of transformation phases within the play of forces of nature itself, in which transformations the Logos of Life in its unfolding comes to reveal its further potentialities.

With this in mind we will now focus on the question of how in this transformation process moral feelings, emotions, principles, and ideals emerge from the merely vital-gregarious (or psycho-gregarious) significance of life proper through successive species up to humans. With this question we mark off the middle ground, asking how in the humanizing trans-action the meaningful vital promptings, forces, and synergies are molded into feelings acquiring a moral aspect. To discuss that we have to plunge into the human soul, the functional territory from which the moral apparatus takes off.

2. *The Twilight Realm of Consciousness Wherein Is Set Up the Forge That Produces Novel Senses*

I have shown elsewhere how the human soul as the crucial territory of life's self-individualizing progress extends through what we call the "empirical psyche" into the dynamic circuits of the so-called "conscious life" with its "preconscious" zone and also gathers unto itself all vital organic and physiological forces, spontaneities, synergies. I have particularly emphasized that the human condition with its unique virtualities breaks through within the most complex area of the pulsating energies of Nature. Under the aegis of Imaginatio Creatrix, its crucial novum in the evolutionary advance of life, there is set up a unique type of "workshop" that I have elsewhere called the "creative forge." First, in the encounter between the organic-vital-sentient forces that are rooted in innumerable life promoting and sustaining process complexes, the virtualities of the human condition establish a new organization of these complexes-in-process. Under the prompting of the imagination there surges a set of faculties that divide roles in organizing this field of energies as they enter into the survival oriented segment of performing proficiencies and purposes, a network of new operative functions with a vaster scope and with expanded flexibility in pursuing constructive goals. Thus life avenues are expanded beyond teleology and the laws of the life system and its individual entelechial codes. That is to say, in a generic interplay the organizing-operational faculties — the parceling and coordinating functions of life — differentiate themselves, undertaking respective roles; imaginatio creatrix sets all in motion, impelling all into a new swing projecting novel horizons, novel spurs and incentives and directions. In short, the sky is opened for a new adventure in life as the virtualities of the human condition, hitherto germinal and lying-in-wait, on the one hand, and the thus far undeployed synergies of organic, vital, sentient forces, on the other hand, are released and enter into play.

All the forces thus ushered into this intergenerative turmoil reciprocally prompt each other to unfold their respective proficiencies and so open a new theater for the genesis of life: the subliminal forge of the soul.

This is not the zone of full human consciousness, yet neither can we identify it with the zone of the animal soul — the vital soul, the sentient soul or psyche. It does not partake of the human mind — for the primal operations of Nature remain still in "obscurity," and only the full consciousness of the human mind "illuminates" through its very own intellection the natural circuits. And yet this forge does not work in complete darkness. On the contrary, it operates in the "twilight of human consciousness." It is subliminal. The creative orchestration of the faculties already at work in the midst of all living functions brings the rays of intelligence into the game. Thus, this primogenital sphere of humanness-in-origin in generating its very own intelligibility of existence is a play of "darkness" and "light," of the elementary forces of nature now brought together in new synergies and of the creative intuition that crystalizes the virtualities of the Human Condition. From this play emerges precisely the specific "realm" of functioning — now specifically human — that I have termed the "twilight region."

This is an intermediary set of circuits that hovers just "above" the true primal zone of the soul. The elementary organic, vital circuits, significant forces, synergies, promptings, which contribute impetus to the operations of the processes of life, drawing from the womb of Nature, is the meeting place for them all as the theater of the ongoing progress of the universal current of life, of living Nature at its expansive work and the emergent synergetic drive of the creative virtualities of the Human Condition. All of the constructive work to be accomplished in this encounter occurs here. The wild forces proceeding from the reservoirs of living Nature have to be "tamed" here through that human touch in which every gradation of life's circuits receives new significance.

When the human condition enters into the game of life at the prompting of Imaginatio Creatrix there is imparted to life the three new specifically human sense giving factors, namely, the moral, the aesthetic, and the intellective senses. It is in terms of these specific factors that new life significance is imparted to life's functioning in the human individual. The new orchestration of living functioning requires means by which to bring about the "humanization of life." In the midst of a wild turmoil of forces it has to this end to find, forge, distill some factors with which to achieve a new combination of energies and forces; therefore it seeks to find and establish some framework for the novel fabric of life to be woven under its aegis. This orchestration tries to invent and firmly fasten this new fabric to pegs on which a new order may be suspended — an order no longer permanently encoded by the elementary life system, but informed by human genius, which takes it into its own hands.

The new fabric of specifically human life is so different from that of primal life owing to its being "societal." Thus do we arrive at the question of how societal principles of order — that is, the ethical moral rules, laws, and qualities of experience — acquire their specifically moral aspect in the primogenital phase of the transition that transpires in the creative forge.

We have now to ask what is the occasion on which all three factors of human sense giving, the moral, aesthetic, and intellective senses, are released by the Human Condition. With this question we pass to the second part of our inquiry in which the entire apparatus establishing the moral significance of life in its societal enactment will come to light.

3. *The Humanizing Trans-action and the Takeoff of the Moral Apparatus*

a. *Symbiotic Sharing-in-Sympatheia*

On what occasion does this transmutation of the significance of life occur? Do the new sense bestowing factors enter spontaneously and of their own "initiative" into contact with the vitally significant propensities that are dormant but "lying in wait"? Our investigation bears out the contrary: it is on a peculiar occasion that this primogenital prototypic encounter occurs and bears its specific fruits.

When we speak of the "humanizing trans-action" we mean a prototypical occasion, an entire functional complex involving two or more individuals; we do not mean by this term — and analysis we have performed elsewhere bears this out — any type of what we call "action," nor any societal transaction in the current sense of the term, but a primordial type of symbiotic-sharing-in-sympatheia that is a special type of linkage, as it is a special kind of "sharing."

In fact it is a type of sharing among individuals of a more complex and differently significant kind than the gregarious or gregarious/societal sharing in goods, services, stratagems, or play. All sharing means the fundamental enactment of life's "symbiosis," "symbiosis" meaning sharing in bios, in life.

Symbiotic sharing forms one of the crucially vital links among living beings. We see a symbiotic linkage among individuals belonging to the same animal type at all stages of development. As a matter of fact we never find in the history of life an animal individual existing all alone; living beings exist in groups sharing in various ways in the same life conditions, and they maintain among themselves numerous existential sharing interrelations. At the lower levels of life's complexity there are grades of passive, semipassive symbiosis, (for example, sharing pastures, body warmth, etc.); in the higher animal species we find not only semipassive and semiactive, but also interactive symbiosis.

This symbiotic interactive sharing acquires in the gregarious life of the higher species a sentient-psychic moment of group solidarity. Its specific novum with respect to the previously mentioned kinds of symbiotic sharing is an affective component. Although lacking a clear and full psychic complex of consciousness, this affective component displays nevertheless rudimentary primal awareness of other beings in a primitive "we are together" animus that is a movement of the "animal soul." Hence there is a symbiotic sharing-in-the-animus.

Symbiotic sharing-in-the-animus is a crucial step in life's unfolding because it introduces a societal linkage system.

For example, the monkeys of the African savannah maintain a highly complex societal system, one centered on the domineering figures of powerful mothers. Monkeys who attempt to join a more powerful group apply themselves to rendering personal services to the members of that group (e.g., grooming). A strong feeling of allegiance to the group is visible: if a bantling is in any way mishandled in play by a bantling from an other group retaliation against some adult member of that group by an adult monkey of its family will follow. This manifests empathic feeling in a lower key, the empathic feeling of the animus.

We know that services rendered by members of species to own kind can be advantageous even to members of other species in the same life situations (e.g., the sounding of an alarm issued by some species of birds at the approach of a serpent, hawk, eagle, tiger, etc. is understood by other species of birds as well as by some other species of animals like squirrels, rabbits, monkeys. This is "symbiosis" in one of its basic senses. But the affective moment, the empathic moment of the "animus" is not to be found in it.

b. *Symbiotic Sharing-in-Sympatheia As the Fruit of the Moral and Aesthetic Sense Giving Factors*

In the passage from almost accidental symbiotic arrangements to what takes place in the humanizing trans-action, namely, with the appearance of that which we have just called symbiotic sympatheia, a radical change in mode of life takes place. To this change we attribute the passage from gregarious-societal life to full societal-moral human existence. To clarify this passage we have to refer back to our already established conception of the human mind in the unity-of-everything-there-is-alive, as it emerges from the virtualities of the Human Condition.

To fulfill the intention of Husserl to discover the very origin of the meaningfulness by which the innumerable lines, types, categories — the ungraspably nuanced proliferation through which the human universe of life is constituted and the avenues of human individual and societal life are delineated in the most radical way — I have already elsewhere established that there are manifested in the Human Condition three meaning giving factors: the aesthetic, moral, and intellective senses. Let us here emphasize that these are not at all faculties that are operative in their own right; on the contrary, they are virtualities "lying-in-wait," ready to be released when the creative orchestration of all human faculties is put into action by the performing living being. This is, let us recall, the "humanizing trans-action" in its prototypic mode, which consists in an encounter between two or more living individuals characterized not by repulsion, nor by a primitive allegiance or by a joint self-confined self-interest, but by full-fledged conscious sympathetic sharing.

The trans-action is this occasion's prompting the initiative of the three meaning giving factors, and in what may appear a circular way, these factors then make the trans-action the prototype of full-fledged human realization.

We will later take up the role of the intellective sense, which becomes fully operative within the moral apparatus. For now we will set forth the role of the two other senses.

The humanizing trans-action is, in fact, the crucial occasion that makes the moment of symbiotic sympatheia. While in the highest animal interaction of a gregarious nature there was, as mentioned above, already present a sentient-affective reaction between individuals, here with the surging of the moral sense into the psychic functioning of the individual a moral sentiment of clear affective benevolence/malevolence is released; this sentiment emanates from the individual, and in its very nature is more fully "empathic" than is sentient/affective gregarious interaction. In point of fact, the aesthetic sense at work enlarges the scope of psychic moves in such a way that room is made for the qualitative and dynamic expansion that this moral sentiment effects; moreover, with the structuring contribution of the intellective sense, these expanded moves now take the definitive form of psychic "acts."

In prompting the expansion of psychic moves, the poetic sense not only contributes its own quality to the sentiment but also causes it to radiate throughout the entire inner life of the individual even as it extends its rays essentially "out" toward the other.

This extension toward the other through inner rays, toward his/her beingness, goes beyond affective sharing in the animus; it manifests the imprint of the poetic/aesthetic sense; it carries on the empathic move — now in a full conscious act — toward the other.

The novel empathic mode of sharing consists, in fact, of various specific moments. First, as mentioned above, there is recognition of the other in his/her beingness; second, there is a recognition illuminated by a self-identifying realization that "There is an other like myself." This "other like myself" illumination awakens the entire psyche; in it one feels oneself to be feeling "with the other" and "for the other" as for oneself. The empathic act, one that recognizes in the other "flesh of my flesh and bones of my bones," here acquires its full measure, and opens the way for the next decisive moment. With the exchange of a gift, a service, a good, this empathic recognition, which is one-sided in its elementary animal mode, becomes mutual. Henceforth the ensuing mutual sharing is sustained by a corresponding empathy felt by the other like myself too. This mutual empathy that is felt on the occasion of an exchange and makes the exchange possible I call the "humanizing trans-action." It is the prototype of human moral-societal symbiosis with all its far-reaching consequences.

I adopt the term "trans-action" because in the current usage it is the most general designation of a human exchange; however, I give to this term, as is obvious, a very complex and elaborate meaning, which makes it an anthropologico-metaphysical notion. I hyphenate the word to indicate that it transgresses the individuals who enter into it and opens out a universal societal meaning and validity, just as it transgresses any particular instance of its enactment. Lastly, and most meaningfully as the focal point of our present argument, in its enactment it transgresses the factual event in effects that sink deep into the inner selves of the persons involved.

It is, indeed, of great importance to see how its enactment draws on, and then penetrates into the psyche. As I have pointed out before, the humanizing trans-action is rooted within the twilight circuits of the human soul. Furthermore, it is the pivot of the new type of activation that follows from the moves that lead to humanness. It surges in this preexperiential sphere where the entire human functioning of the individual human becoming comes together. It stirs new movements within the circuits of vital functioning, gathering and drawing on their energies and activating the burgeoning synergies deposited in them by the Human Condition. This means that it draws on the whole of organic-vital-psychic functioning and gives it a focal point for its reorientation as well as an axis for the new organization that makes it operational. Lastly, the humanizing trans-action throws out from this operational axis the new lines for symbiosis in the specifically human environment, the societal mechanisms for the construction of a unique world of life.

It has to be kept in mind that the rush of drives, impulses, pulsation, etc. flooding out of the vital circuits and invading the primal zone of the soul is not a chaos. They are somewhat organized by the life promoting telic principles of vital functioning; they are subservient to the aim of survival. They consequently bear a rudimentary vital significance for survival, which binds them, however qualitatively crude, raw, and wild it may be.

Quantitatively they surge from functional operations that, being narrowly fitted to restricted aims, do not reach the conscious amplitude of psychic acts. Entering into their midst, the trans-actional linkage stirs them, dislocates their hitherto prevailing interconnectedness freeing them from strict bondage to vital processes.

There are released, first, the three sense giving factors, but simultaneously all the virtualities germinating in the vital, primal zone of the soul — in organic-vital functioning — are activated around a new axis. Here is the germinal crystalization of the Human Condition. Indeed, at the moment at which sympatheia is enacted in sharing, the entire ensemble of the Human Condition comes "alive." With the operation of the moral sense all the vitally oriented functional circuits are pulled together. Either their narrowly focused interconnectedness is given new orientation, or operational circuits that thus far had remained isolated in their self-enclosed routine performances — in the self-enclosed indifference of one animal being toward another — are broken through and affective antennae are thrown out that allow "empathic awareness."

Here an important point has to be made. The point at which these affective antennae open a living being's narrow system of self-confining circuits toward the other, the moment when "awareness" as an affective-cognitive moment surges, awareness of "an other like myself," is also the incipient instant of the crucial constitutive moment of human beingness as such. Indeed, "self-awareness," the act of a beingness' identifying with itself, is the element of the new zone of the human soul that thus surges: the zone of full human consciousness that opens with the new operations and pulsations and gives them the focus as well as the extraordinary amplitude of human experience. Now, it should be emphasized here that this self-awareness may surge only together with empathic awareness of the beingness of the other.

On the occasion of the humanizing trans-action, the moral and the aesthetic sentiments perform in tandem this crucial operation. With it the primal zone of the soul becomes the workshop of the full-fledged Human Condition. With that an orientation center — the conscious selfhood of a living being — emerges, around which the novel experiential system will then be plotted by the creative orchestration into which the newly emerging faculties draw the infinitely simpler primal functional apparatus, vastly enlarging it.

This description all too briefly outlines the essential role in the humanizing trans-action played by the new symbiotic linkage that sympatheia effects.

What is given here in broad outline I have substantiated on several points in previous writings in which I surveyed the vast horizons projected by the specifically human faculties in the forms of the human world of life, that is, the world of life in the full societal-cultural form that we know and its infinite horizons. Therein I have found that the axis around which the entire human sharing-in-life revolves is moral agency. In the main probe of this inquiry, in which we will focus further on the role of the humanizing trans-action, we will be directly drawn into some of the moral agency's basic forms and functions.

III. THE MEASURE: THE MORAL APPARATUS AND THE ELEMENTAL PASSIONS THAT PROMPT IT

In order for the symbiotic sharing-in-sympatheia to take place, the benevolent sentiment has to be appropriately "cut" to the situation, the individual partners, and to the nature of the sharing itself. It has to be adjusted to their multiple demands, apportioned according to natural requirements. This calls for the cooperation of the appreciative and deliberative functions of the intellective judgmental faculty. And it is owing to its services now taking the lead that the required primordial measure will emerge.

There must ultimately be a measure. As I have pointed out in my previous study introducing the notion of the 'moral sense' as the founding factor of moral/societal existence, the practical maxim of the moral sense is "Everyone his/her due." This points to the measure that we seek. In point of fact, the appreciative and deliberative functions naturally seek points of reference for orientation, an aim and an axis around which sharing may revolve.

All the facets of the situation are to be envisaged precisely as they might function in the elementary "assent" mutually given by people when they share: you have to give assent in order to present and receive a gift, a proposal a mood, a memory, etc. Beyond assenting to a gift, one might reciprocate, and there is a sharing in mutual assent.

With model of the humanizing symbiotic trans-action we circumscribe a sphere preceding in a preparatory way the enactment of contractual sharing and reach behind that enactment (which is in its objectivized form in current language called "action") by transgressing its specific significant complex already appropriately established in the concretization of this one-time irrepeatable enactment. We reach also beyond by opening up all the significant rays that surge on the wings of imagination and illuminate the sentient, valuative, appreciative, and constructive moments that enter into the trans-actional complex, the new horizon for their unfolding.

Here lies a very important point in the transforming enterprise of life. With the codes and laws of the system of life now broken at a sensitive point, all the life forces and energies break loose from the forms hitherto binding them and stand ready to enter into an expanded arena. Yet here as elsewhere the rule of order is to be introduced.

In the harmonizing effort of the soul in all its elementary, vital, and subliminal circuits there is a constant search for equipoise among all the moments within their respective circuits and between and amongst the interplay of circuits. The breaking of that harmonizing equipoise or the failure to reach it, on the one hand, weakens the unity of the entire schema in its performance; on the other hand, it leads to discordance, emotional upheaval, the breakout of elementary bents, and the strengthening of their tendencies leading in some instances to irrationality's coming to dominate the disoriented whole; envy, anger, hatred then lead to significant expressions of violence.

An order of life — of specifically human/societal life — has to be found. This new societal order takes root through the apparatus of human sharing — a prototypic instance of communal bonds. A measure of order needs to be installed here, a measure for the innumerable moments of the ever new trans-actions of symbiotic sharing, one that can be used in the ever new and irrepeatable situations that come up among living beings and that have to be endowed with the human significance of life, the specifically human life promoting unity of full conscious existence.

As a matter of fact, the conscious agency that we are describing as emerging in symbiotic sharing-in-sympatheia consists, to begin with, of the matter at hand — of discerning, discriminating, deliberating, choosing measures and proportions. Here no sides may be taken arbitrarily; persuasions, inducements, side considerations may not be given way to; biases may not be adopted; inclinations that would tip the scales toward one side may not be passively submitted to. Already the vocabulary that I use here makes manifest that the moral deliberation required for proper apportioning calls for taking a righteous stand.

Righteousness, *rectitudo* in deliberation and proportioning, is the axis of the human trans-action in its moral character. The agency of the moral circuit of deliberative operations called "conscience," which emerges on the occasion of the humanizing trans-action, must, in order to accomplish its role, be oriented by reference to this axis. The measure it applies may be expressed in the dictum, "Everyone his/her due."

Only if societal life in its unfolding is capable of upholding this principle can it endure.

On the occasion of the humanizing trans-action a new order of life surges distributed by the moral agency of conscience by means of the novel forms in the differentiation of the logos of life, around novel axes of rotation, novel modes of intergenerating interplay, with novel problems to solve and novel points of reference. Thus is opened within the system of life at large a zone of circuits that involve not an automated or prescriptively circumscribed range of functional linkage but the specifically morally informed, aesthetically amplified, and intellectively structured and oriented ordering schema left to the individual conscience of human beings to improvise.

Through this new functional schema there now runs the morally-aesthetically-intellectively oriented impetus of the living being who becomes human in its exercise and establishes his unique societal life and community.

IV. THE HUMAN PREDICAMENT

We have already outlined the primogenital moves that get the apparatus of the subliminal soul off the ground. That is to say, we have circumscribed the platform that the new creative orchestration of human functioning projects. We have also described in brief outline the central mechanisms of the subliminal workshop culminating in the "measure," in the criterion for apportioning used by the primogenital moral agency in its emergence on the occasion of the specifically human linkage that transpires in the trans-action of sympathetic sharing. With the emergence of the new sphere of the Dionysian logos of work, a novel type of equilibrium emerges, that which applies measure.

Now is the moment we turn to the subliminal soul's inspirational forces and energies — which in their release prompt and regenerate and promote the human significance of life throughout its innumerable channels. These subliminal forces contribute the shading of the infinite qualitative appreciations by which our sentient moves are endowed with feelings, emotions, attitudes, yearnings — in short, the entire spectrum of what we call "experience." These new aspects of sentient life, in return, draw out of the synergies of life the soul's very own impetus, so that these subliminal forces and energies may radiate. On the wings of these impetuous energies our sympatheia soars beyond concrete individual solidarity or loyalty to the holding of shared ideals that surpass the narrow, pedestrian, petty concerns of common morality. As a matter of fact, we reach the fulfillment of the complete moral significance of life only with the complete unfolding of the soul with the release of the elemental or subliminal passions. *The factor crucial for this radiating growth of the soul is fabulation.*

Fabulation is not a primogenital energeia, but the fruit of the already completely operational creative orchestration of the human faculties. It proceeds from but rises above the concrete groundwork of the subliminal soil, the circuits of life inspired by the moral-aesthetic-intellective senses. On the wings of the faculty of the intellect and at the prompting of imagination, we are projected into an "ideal" and abstract realm of intellective, synthetic, categorizing activity. This work of the mind abstractly modulates the sympathetic propulsions, and responding to the apportioning and measuring project, it establishes principles of classification and derives therefrom the telic guidelines that we call "ideals."

With this highest differentiating, categorizing, and synthesizing working of the fully operational human mind, we hit directly upon the core of the moral directives that spread their rays into all the corners of human life, humanizing it. This core is the issue of the entire network thus outlined. Here the "human predicament" is discovered. Drawing on all the fibers of our moral beingness, the human predicament is by no means an abstract intellectual notion, an idea, or a concept; it is, on the contrary, an experiential vortex around which, *nolens volens,* all our inner, conscious, life vibrates.

In fact, we do not in our experience halt at any one single transactional instance; we do not haphazardly negotiate each of the obstacles and challenges that our life enactment encounters in its path. On the contrary, given the great range of possibilities offered, and the great flexibility we have in our adjustment to life situations, we compare, discriminate, retain, reflect upon, and form synthetic views on one or more recent situations encountered, the ways in which we met them, and the successful or unsuccessful outcomes. Concurrently, the synthesizing, abstracting, and categorizing work of the mind plays an essential role in the process of the symbiotic sharing that is sympatheia. As a matter of fact, we do not enter into this more complex sympathetic trans-action "blindly"; on the contrary, our intellectual circuits, *in a first move,* release a universalizing, organizing endeavor that reviews innumerable instances, considers results, and produces a synthetic valuative view of the overall life situation of the person adopting it. Partly overlapping this endeavor, however, there is the development of a varied set of synthetic valuative views on the situation on the part of the partner or partners who participate in the symbiotic sharing. In this whirl of numerous, varied overall views that complement, discount, supercede, enhance each other, we project an imaginative picture of their many-sided and plurimodal interconnectedness within the vaster scope of the societal world with its hazy contours.

In a second move, one still within the continuing empathic cloak of the psychic/intellectual acts that synthesize the estimations of our partners' interests, concerns, again in a whirl of numberless partial glimpses, we situate our and their places and roles within the common, overall, fragmentarily and hazily seen universal framework.

Lastly, on the wings of imagination as prompted by the sympathic spirit and as lifted by the intentional network woven in these two moves, we fly further to the crowning point of life's sympathic synthesis: with all our faculties operative within the creative orchestration of our functioning, we lift our selves from within our innermost above our own or others' or anybody's narrow specific frame, and extrapolating on the syntheses made in those previous moves, we make a deeply experienced intuitive commitment to life as such. We reach thus the point in all our experiencing of realizing that there is "something" about human life in its origin, course, unfolding, and extinction, about life as such.

We discover, as a matter of fact, we feel it in our very limbs, that life is a fierce race with time, each instant of which has to be wrought from adverse forces-in-turmoil; that at every instant we stand knowingly or not in a fierce struggle to perdure. We feel in our beingness that the natural life energies that carry us, unfurling ecstatically toward a peak, are heading toward an inescapable, insurmountable, finishing line in extinction. We suffer, bleed, suffocate from anguish, thirst for respite, hunger for justice. All in vain. We run, run, run — without respite and mercy other than the illusory — and arrive at the finale to vanish into oblivion. We are led to wonder what is the sense of this course, of this individual life striving to perdure for a short time. Why do we undertake this race for which there is no reward but death, and no recognition but just occasions for spending our forces? Why do we pursue it, submitting to innumerable obstacles, oftentimes insurmountable ones, to traps and exertions beyond our natural forces, attempting most hazardous enterprises on a road that does not lead anywhere? And yet we do not give up. We partly submit, partly address even the biggest challenges of life. And we do this not by willy-nilly passively following a predelineated, instinctive route. We actively carry on and undertake the task of life to the last breath. Why do we undertake? Why do we relentlessly roll the rock up hill over innumerable obstacles and when it falls back, pick it up with the next breath and push it with released forces, expectations, and determination uphill again? Why do we do it? Why do the first obstacles and failures not discourage us? They circumscribe the human predicament. We carry it with us; it has penetrated our entire beingness to the marrow of our bones. Why do we undertake it all? We never stop asking ourselves these questions. And we find no immediate answers.

No doubt, we are prompted in the first place by our vital forces and energies, the spontaneities and dynamisms that burst forth, asking to be engaged. These energies, flowing into our functional system, intergenerating, amplifying their impetus, keep that system "alive" and fit. With their weakening impetus, our readiness to undertake tasks vanishes; our resistance to discouragement weakens. When these forces are ascendant, we glow in spirit and go to encounter the enemy, to strike back, to meet the challenge, and are ready to endure whatever comes along and to propose new goals.

Secondly, there is the sustaining network of all the others who throughout our life course are most intimately linked with our entire functional system. We enter into this course already initiated and nurtured in the womb of nature, which was none other than the innermost vital organs of another living being. And this and other beings must continue the nurturing process as we progress in our self-individualizing course. In this course, our individualizing growth is throughout enmeshed with the life courses of others. These are parents, family, friends, colleagues, comrades in arms, competitors, challengers, etc., etc., with whom our lives are intertwined in all the aspects of the human predicament. We are enmeshed in a fabric of others who sustain us at the decisive moments. Human symbiotic sharing is what carries us along, is what sustains us and stimulates our undertakings. The human predicament, we share it all. It is our common inheritance. We work at it whether as friends or enemies, whether as supporters or competitors, whether as challengers or pacifiers, whether we inflict wounds or heal them; we push others over the precipice or save them from disaster...we bear the task of the life struggle together. Thus, in innumerable trans-actional networks, we endow our undertakings, our life course, with a moral significance.

I have elsewhere repeatedly analyzed the various ways in which Imaginatio Creatrix works at taming the forces that pose insurmountable obstacles and contend with the weaknesses, deficiencies, and congenital limitations of the human being in the struggle for life. It is with respect to the human predicament embodied in them all within the orbit of the moral significance of existence that the life struggle turns into the "taming of primitive forces" so that they assume the subliminal force of the "elements."

It is also with reference to the human predicament that Imaginatio Creatrix, impelled by the sympathic drive embedded within the entire creative orchestration, releases as its pivot the function of fabulation.

The work of fabulation functions as the pivot through which the subliminal work proceeds. Fabulation spins our images, feelings, fragments of representations, emotions, thoughts, ideas in germ or at any stage of development, all deeply felt responses to the common doom. These are spun out in infinitely varying circles of transformation, fusion, coalescence; their spinning is sustained by our innermost personal sentiments.

Even as these develop into an attitude, an appraisal, a view of the concrete life situations with which we are presented, there emerge the subliminal qualities to be conferred upon moral/aesthetic feelings, emotions, etc, which will further on be objectified into ideals, values; in the end they become objectified and categorized as prescriptions of human conduct, that is, norms, rules and laws. We will soon turn our attention to these.

At this point let us emphasize that these subliminal moral/aesthetic qualities constitute the springboard for moral/societal existence. With their qualitative coloring they imbue human reactions, valuations, judgements, tendencies, etc. The moral qualities, as they spread into all the lines of human/societal life and there obtain further corroboration as well as specific directions according to the life conditions of specific societies, gather into emotional streams that give character to societal life itself; furthermore, through conduct they penetrate individual attitudes toward the natural inclinations, bents, and even the individual's personality. Most significantly, however, for our present inquiry, as qualitative complexes and trends, they first become directives for personal orientation and then — interchangeably — they are hypostasized into the ideals of culture and hence directives for conduct.

Recapitulation. Our creative orchestration having taken the lead within the natural workings of things, and our subliminal soul having released the yearning and the call for an understanding of life — an understanding higher than the narrow span of life acceptance proper to the primary phase of psychic existence, that is, awareness of the vast horizons of life with a focus on the human predicament — there having, in short, been released a specifically human quest to discover the meaning of life, we find ourselves in the center of the moral orbit. The moral forces, radiating from this center, permeate our life struggles, endow them with a lasting quality. Our struggles are fought and then pass and vanish without a visible trace. Their moral significance perdures in our memory, in the innermost fabric of our existence, in its meaningfulness, and beyond us in the struggles of others. Even if we were to win all victories in our battles on the field of life, we cannot but lose all to the impending doom or to change. But we may endow these battles with a moral significance and so enhance them into moments of fulfillment. This significance extends beyond our concrete real perishable life and death. In it our life is launched into the open and infinitely expanding sky of the ideal.

We join with all other human beings in a communion in the spirit. The societal spirit of human community takes off.

V. THE JUDGMENTAL APPARATUS, INSTRUMENT OF PERSONAL/SOCIETAL EQUIPOISE

The encounter between the elemental significance of our brute forces, strivings, instinctual drives and our subliminal morally informed virtualities creates a chaotic turmoil of contrary tendencies pulling the soul apart. Concurrently, the moral sensibilia that surge with the benevolent/malevolent sentiment inform the originating experience of the soul.

Now beyond the control of vitally organized functioning, these brute as well as originating subliminal impulses and sensibilities meet and find a measure for equipoise in the intellective apparatus balancing them.

The crucial point of this equipoise to be established afresh with each scanning/spacing of the soul's pulsations is, however, the sympathic sharing-in-life among human beings. This is a sustained empathic sensitivity to the needs, feelings, joys, and sufferings of others. We may have to master attitudes, modes of reaction, appreciation so that we are ready to react negatively, and yet sensitively to other's joy, sorrow, needs, their drawbacks and shortcomings so that we remain capable of entering into a constructive trans-action with them.

That is to say, the crux of social equipoise is to establish in the innermost human being the equipoise allowing constructive trans-action with the other and sustaining profound sensitivity toward the other's felt beingness. Although we usually maintain a focus on laws and their enforcement, on the perfection of the institutions of government and of private, educational, professional, medical, and social welfare to secure for every citizen an adequate sharing-in-life with the others, yet no laws, regulations, enforcement, model provisions, and institutions will in themselves have any tangible effect on individual adequacy in sharing-in-life; this is always in the hands of individual persons. If the person who is expected to perform such or other function in the care and distribution of services or goods to people is not adequately balanced and sensitively attuned to the needs, joys, sufferings of others while at the same time capable of overcoming reactions of personal revulsion, antipathy, real misgivings, discouragement, apathy, loss of heart, or just laziness, no laws and institutions will be able to secure an honest sharing-in-life for their tutees.

While the "truth of life" resides in the human heart, the dignity of a personal life lies in the heart's fortitude.

VI. THE SUBLIMINAL PASSIONS AND THE STANDARDS FOR HUMAN MORAL CONSCIENCE (NORMS, LAWS, IDEALS)

For societal sharing of life among individuals in the ever advancing flux of the world-life as well as of the individual cyclic unfolding in its foreseeable as well as unforeseeable variations and amid strictly personal moral developments in flux, there is still need for the project of ordering a universal set of principles for this sharing in the intimately personal as well as social domain of all humans — and to some degree of all living beings with their due life territory. There is need for some universally valid prescriptions for the inclinations, attitudes, judgements, and conduct, that is, for an ethic obliging all humans or citizens. These prescriptions we find in what are usually called "moral norms," "rules," "laws," etc.

We may say that already in the subliminal sentiments, inclinations, etc., that is, in the subliminal workshop of the twilight of consciousness, we find the groundwork done for these prescriptions. In fact, although moral norms possess a claim to universal validity — in all cases and with respect to all individuals — nevertheless they only appeal to the innermost personal ground of the person for recognition and respect. Obviously, then, it is from this very ground that they emerge.

The qualitative modulations of our feelings, emotions, sensing, instinctive reactions, pulsations, etc. are themselves, as mentioned above, not only carried by functional forces but are part and parcel of sentiments. This means that they are not, as they were considered in classical philosophy, abstract, ideal qualities; they are in their original state sentient forces. Only when hypostasized by the operations of intellectual intentional consciousness into a presentationally posited form do they acquire abstract existential status.

In the processes in which the quality imbued, sentient forces coalesce into complexes, they intergenerate and gather novel energies. Intensifying through fusions and gathering unto themselves varied strivings as well as pulsating vital energies, they become a powerful torrent. Thus, the elemental passions are born. The powerful subliminal passions — on the wings of the prompting forces of the moral sense, under the intrinsic propulsions of sentiment, and drawing freely on the forces and energies of the entire functional system that carries and distributes them, but most of all underpinned by the silent brute vital forces of life from which in the twilight of consciousness their bearers first emerged — drive the person who experiences them to spread and promote the values, ideals, dreams, principles, criteria, etc. that they distill from the multifarious qualitative web in which they are enacted. The subliminal passions are the new driving forces of the living being whose life significance is henceforth tuned in a novel key. Self-interest and primitive greed, which are the main vital drives of living beingness, cede place within the subliminal horizon of life significance to other tendencies and strivings with different coloring.

As Aristotle and Milton foresaw, it is indeed in the depth of the human soul that the entire moral apparatus is worked out. It has still to be shown, however, how it emerges and gets off the ground. In plain words, what we have attempted here, remained to be shown — how the human soul acquires a unique tuning that opens a special zone within which this apparatus may emerge.

There is still a decisive moment among others to be brought out in discussing the origination or generation of this new zone of life significance in the midst of the animal sentient soul. Descartes already raised the question of whether there is not a passion that is the soul's very own. We have, within our thorough investigation of the powers of the soul, to pinpoint the Cartesian question. Indeed, we might say that the soul in its subliminal transformation — or generation — finds within its own resources a prompting and fateful set of wings: the passions of generosity and of infinity.

The passion of generosity springs forth in response to the moral sense's setting to work. It carries in a germinal state all the virtualities of benevolence. The passion of infinity propels our entire functional schema at the level of its intellective/sentient circuits toward the infinite horizons of life, toward perspectives far beyond concrete life, beyond life as such.

Between these two passions, the soul in its subliminal work radiates an entire spectrum of moral passions: sincerity, honesty, solicitude, faithfulness, loyalty, heroic courage, forgiveness, as well as their malevolent opposites — hatred, vengefulness, resentment, etc.

These are the forces of the societal spirit. Individuals are interlinked in societal networks through the forces that these passions promote. But these passions that qualitatively distinguish human conduct give, within the trans-actional apparatus, directions for the establishment of norms, rules, and laws of individual conduct within a human community. In fact, people connect with groups and societies by sharing the societal spirit through a commitment to a set of moral norms, rules, and laws. All the laws from Hammurabi to the present's most elaborate civil codes, the Decalogue and all moral-religious rules refer back to this deepest personal font where each and every member of a community, all citizens spontaneously find themselves in a living unity with all others. A living, *felt* unity is at its primogenital source. It is to this subliminal personal font of the soul that each of us has to return in order to revive his/her commitment to the practice of the moral prescriptions upholding the moral/societal order.

But what about the nature of these prescriptions? How can they concretely direct human conduct in such a way that they not only bring about the order that subtends the communal network but also cause it to expand, differentiating into further circuits of the expanding Logos of Life? In point of fact, on closer analysis, we find that they contain, first, an intensely colored — maximally intense — qualitative moral moment as a background. Being the fruit of the whole series of trans-actional moments that gave rise to the given norm or criterion, this moment partakes of its decisive dynamism. This qualitative moment then confers to the norm the stringency of an imperative appeal (e.g., in the case of the negative precept "Do not steal," the trans-actional background leading to this norm is carried in it visibly). A prescription contains, in its qualitative moment, a "force of attraction." This is seen in positive prescriptions (e.g., "Love thy neighbor") that stimulate "appetite" for experiencing their enactment. Thirdly, morally qualitative norms and prescriptions "fulgurate" in their qualitative moments, creating a "propulsion," bringing them into societal linkage in enactment.

Accompanying these moments, which are of a dynamic energizing nature, is the universalizing of moral norms, precepts, criteria, principles, etc.; their aspect of normativeness, principle, etc., provides validation for each and every concrete trans-actional situation. Their validation is nothing other than a universalizing of the essentially valuative nature of their "imperative" appeal; a valuative moment is an intrinsic, constitutive factor in every moral qualification of sentient acts as well as of their extrapolation in norms and principles for conduct.

These three moments of a dynamic qualitative change peculiar to morally qualitative norms affect our natural tendencies, habits, and inclinations. In fact, in their molding of the primitive natural bent, they orient them in a moral direction.

These bents have, however, in conjunction with the passion of the infinite (and with the cooperation of the intellectual faculty) a still further contribution to make to the moral apparatus, its crowning point: ideals.

As a matter of fact, the moral qualities that now suffuse the workings of the entire creative orchestration of human faculties know attractions and appetitive moments of their own. The morally imbued forces and strivings, inclinations, and passional streams intensify in the "ideal" moment that lies latent as a moment, drawing them, the moment of an "example to follow," a "model to imitate," a "model to set forth," a model of conduct "worth striving for." In the model-like aspect of an "ideal," ignited by imagination and prompted by the passions (from the benevolent sentiment through the entire qualitative passional spectrum up to the passion of the infinite), the ideal moment penetrates the entire functioning of the person, stirs within it an elan to follow, to imitate, to emulate, to become worthy of, to measure up to, to find one's accomplishment in. In short, the thus extrapolated qualitative clusters become what we call "our ideals."

Ideals are the beacons of the specifically human expansion of the Logos of Life. Between the ideals that glimmer at the furthest horizon of our human striving to undertake, to conquer, to create, to accomplish, and that promote the societal unfolding and the platform of the norms, regulations, and laws keeping societal order in the midst of the stormy fluctuations of life's progress, there vibrates the vast and infinitely colorful and varied trans-action of human sharing in the spirit, a sharing that occurs in our circuits ranging from the affective intimate sphere of personal experience to the spirit as it is objectified in abstract values and ideals.

CHAPTER FOUR

MORAL AND CIVIC VIRTUE AS THE BEDROCK OF THE MANIFEST GAME OF LIFE, THE CORNERSTONE OF DYNAMIC SOCIAL EQUIPOISE

I. PLACING THE ISSUE OF VIRTUE WITHIN THE FRAMEWORK OF THE PHENOMENOLOGY OF LIFE

To conceive of virtue as the bedrock foundation of communal life and the social edifice is nothing new. Just as Leibniz in presenting his metaphysics of the monad insisted that his view of the monad was nothing new, but merely a "reviving of the individual substances of the Scholastics," so I will refer in my proposal to Aristotle. However, I have to point out from the start that my approach to questions concerning human conduct is situated within a special — non-Aristotelian — perspective, namely, that of the phenomenology and metaphysics of life.

To situate human conduct within the context of the metaphysics of life and to carry on the inquiry from within that field entails at the outset adopting some decisive points of orientation. We should, to begin with, declare at what points we diverge radically from the presuppositions that played a decisive role in Aristotle's conception of virtue.

But let us state a point of agreement first. Aristotelian philosophy as well as my phenomenology of life conceive of human conduct as being connatural with the societal sphere; that means that individual decisions have as their moral criteria the laws and rules of the society and community in which they will be enacted and, conversely, that the moral tone of a society depends on the individual's exercise of his will. Yet in the

Aristotelian schema the system of a given society and its governing laws — those of the Greek society of his time — seems to be considered fixed. Hence, the criteria of human conduct seem also to be stable. However, since Milton we are perfectly aware and even preoccupied with the fact that societal systems with their ordering of individual conduct are in the process of constant change; this change, as Milton insists in the *Aeropagita,* is the condition of a society's survival. Because it is necessary to reconsider, to readjust, to adapt the attitudes of individuals to cataclysms, revolutions, economic dislocations, epidemics, and the like, and to the progress in thinking and scientific research, individuals have to exercise their imagination and intelligence and go beyond a strict reading of rules posited for situations that preceded such changes. Society has to allow freedom to submit to doubt and scrutiny what has been accepted as absolutely valid.

Phenomenology of life, going to the primogenital conditions of life's unfolding, reveals it to be a striving of forces but precariously kept in an ordering in flux. Human life in its societal dimension appears caught in constant change, progressing from one type of turmoil to another. Thus, the laws and rules governing a community and a society cannot be seen as being absolutely valid but, as has been the case throughout the historical unfolding of human cultures, as always subject to individual re-appreciation. This amounts to saying that, within the perspective of the phenomenology of life, the tension between the individual's decision making concerning his/her conduct and societal laws and regulations is seen as a living and essential subject for scrutiny and re-evaluation, with societal laws standing in vibrant tension with personal moral convictions and with the emphasis falling on the individual's personal life. We shall see here a sharp shift away from the Aristotelian views on virtue. Would this view entail that ethical norms are relative to societal changes or, at the other extreme, to fluctuating personal convictions? Does this mean social and historical relativism? Nothing is further from our mind, as we will show.

It would be incorrect to minimize Aristotle's concern with personal human experience. To the contrary, for him virtue consists not in obedience to law alone but, preeminently, in an inner state. Here, however, we encounter the second point of the phenomenology of life's concurrence with as well as sharp divergence from Aristotelian suppositions. In its conception of the human being, the Aristotelian ethical view refers essentially to what is called our "human nature," as if there were fixed basic essential features defining the human being. *We, in a deliberate move, simply shift our investigation to the plane at which the human being emerges.* Within the framework of the phenomenology of life, the human being is envisaged not in an anthropocentric fashion but as one of the types of beingness that emerges within the evolutionary progress of life as such — not as a crystalized essence, but as a being in the process of unfolding himself. This is in line with classical phenomenology, but this unfolding process occurs at a much deeper level than that reached by the Husserlian probing of constitution, which centers on intellectual consciousness, which is but the last crowning step of this unfolding.

In our phenomenology of life we, in contrast, circumscribe the specifically human phase of the evolving life progress by singling out the specific Human Condition. It is the set of virtualities circumscribing the specifically human condition that as an emergent phase of life in its evolution determines what human beingness as such might be and is. The virtualities of the human condition are the constants as long as human being is a constant.

Thus, we are going to investigate the individual source of morality, focusing on virtue, not only in the personal depths of the human soul but within the innermost depths of the human being caught in the genetic crystalization of his personal beingness. We differentiate this innermost reality in the genesis of life, in the "twilight of consciousness"; here is the "creative forge" where the specifically human significance of life is worked out in the crystalization of the constants offered by the human condition.

On this point we again depart most significantly from the Aristotelian schema of things. In that schema, all the specific individual inclinations that are decisive for the nature of the virtues, all the qualities of emotion, etc. that give to virtue its definition are taken to be naturally given, and their significance to be naturally accepted in individual and societal life as being such and such. In contrast, the great question that the phenomenology of life addresses first and foremost is precisely: "How does the specifically human significance of life originate? How does the human being come on the aesthetic, moral, spiritual significance of his experiences, significance that transforms the primitive meaningfulness of Nature into the myriad novel avenues of life that man invents and crafts for himself as his own human universe?" After this preliminary establishment of bearings for a renewed investigation of virtue as the vehicle of moral conduct and as the bedrock of human society and culture, we must yet indicate the crucial point of our inquiry.

Instead of placing virtue decidedly in the column of intellect — as is the tendency of Kant — and so regarding it as a victory over natural inclinations and emotions as well as the primitive forces of Nature, and instead of placing it in the column of Nature — as is the tendency of the emotive interpretation of ethics of the British (Shaftesbury, Hume) as well as of Rousseau — we situate the question of morality and virtue at the primogenital human plane where reason with its faculties, on the one hand, and the vital forces, on the other, emerge as partners in the creative orchestration of human functioning that forms the crucibles of intelligibility that is specifically human and that accounts for the emergence of the human universe, that is, the human expansion of the schema of Nature.

In short, the question of virtue lies at the heart of the life strategies of the Logos.

1. *From the Subliminal Passions of the Soul to the Moral Virtues in Human Conduct*

In order to delve into the heart of our winding argument let us now recapitulate those of its sequences already exfoliated. Let us recall that our investigations seek the origin, modalities, and cornerstones of the societal order as it emerges in the human phase of life.

Within the field of the phenomenology of life this order is the harnessing through which the primeval Logos attempts to enlist the wild, turbulent flux of primal life into its service. We have seen how the humanizing trans-action surging within the orbit of the Human Condition animates its virtualities and sets in motion the entire apparatus of specifically human linkage-in-sympatheia. We have seen how the subliminal soul of the human being thus emerging is the workshop in which the moral significance of life is worked out together with the apparatus to promote and carry on the linkage of sharing. Moreover we have seen how, in particular, first, the moral agency that deliberates and selects, second, the criteria by which to measure — the ideal guidelines — and, lastly, the driving forces emerge from the subliminal soul.

But what about the judgement to make contractual commitments, to carry them out, to stay on course despite external obstacles, changes of situations, etc., as well as despite internal fluctuations, contrary moves, doubts, etc., and so meet those commitments?

In our inquiry it has clearly come to light how the moral order is suspended between the deep shadow zone (twilight) of the living individual and the luminous circuits of the universal, that is, intellectual norms, precepts, and laws that guide our symbiotic sharing-in-life and inform societal linkage. We have found the model humanizing trans-action to be the intermediary instrumentality between these. The significance of projecting the moral/societal order for the living individual cannot be stressed enough. It is the various humanizing trans-actions that certainly institute and keep alive the societal system. Moral principles, norms, and laws of behavior form the abstract reference system for societal life and seem to hover "above" the individual. And yet all comes down to the concrete societal trans-action improvised and enacted by concrete individuals, that is, to the foundational role that the concrete human individual plays.

The impetus to differentiate and discriminate, to valuate and to select, to decide, and to enter into the trans-actional relationship is an initiative that only the individual may take. And it is still up to him to project and simultaneously agree on a contractual plan. We may well attribute the projection of the contract to the moral agency of conscience, which has already been discussed (Part VI, Chapter 2, Section 5). But the agreement contains in itself the essential moment of the indispensable commitment of the partners involved to fulfill the obligations that it entails according to the terms established. To assume this commitment, to project a series of actions in adherence to this commitment under the moral norms that ultimately govern it, and finally to perform the actions necessary — regardless of circumstances, propitious or adverse — is a different matter.

The measuring, deliberating, and selecting circuits of the moral agency are not an abstract inert schema, but vibrant forces. They call to be made operative. In fact, as we have been attempting to show throughout, the progress of the moral sense as it works its way up from the twilight of consciousness prompts, generates, intergenerates, and promotes ever new instances of force and impetus. Hence, its vocation to be operative in human moral sympatheia and in sharing prompts the elaboration of appropriate conductive circuits; and hence the moral decision to enter into a complex of human acts positing firm steps and thus delineating a consistent course in the midst of adversities within the flux of things. This being granted, the solution to the problem we have raised is still far off.

The symbiotic sharing-in-sympatheia is itself an intense play of forces. Although this play of forces is a process binding two or more "partners," each of whom is caught within a personal life-process-in-progress, it takes place within the turmoil of a still larger scene, within family life, communal life, national life, even within the relentless life advance of mankind. Each of these circuits of living forces is itself composed of innumerable singular lines, which rush relentlessly onwards and exercise some closer or more remote effect on individual lives.

The question of the societal order of life is the question of how within the turmoil of life there may be established among a number of individual beings the sharing not merely of principles and criteria but of their concrete enactment: the behavior of people toward each other in a trans-actional interlinkage.

It boils down to this question: How may the human individuals caught up in the fluctuations of their inner forces and states of feeling, emotion, desires, instinct, etc. and subject to external forces as well develop steadfast conduct? How can they maintain it?

It is in just this possibility of the crystalization of the moral vector of life in steady conduct on the part of individuals that societal sharing in the unity-of-life resides.

We have submitted that righteousness in the measuring up of moral issues and making judgement is the axis of the moral/societal order whatever this would be. However, the probity of judgement of the moral agent does not seem to guarantee the appropriate responsibility or faithful and steadfast continuance in the direction of the fulfillment of obligations. Neither does it have the sufficient assortment of forces to spur the human being to engage, maintain, and complete the series of actions to be performed in the fulfillment of obligations.

With these considerations we have circumscribed a most sensitive area: the area of the passage from the subliminal movements of the soul to the concrete and deliberate active movements of human beings within the world. We can phrase the concern here in a simplified way: How do moral norms so-called find their way into "application" in the chain of personal acts? To put it even more precisely: How may this very passage to "application" be effected, established, and guaranteed not only in single instances, but with constancy whatever the inner/outer fluctuations and the propitious/adverse forces that may be at play inclining the heart in one way or an other in the midst of unique conflicts and situations?

Let us not forget that in the line of life's vision there is such turmoil and unpredictability in the flow of forces that some bit of order, of stability, of security, of trust and confidence must be thrown up to meet them. This is the basic task given people as their life courses simultaneously unroll over a period of time. No societal order necessitating mutual cooperation, constructive exchange, and mutual reliance, that is, no societal symbiotic sharing, could be projected without this stabilizing trust. Without this bit of order, the human creative elan would be completely overwhelmed and the human world of life as expanded by culture would be stifled in germ and never take off.

Thus to unlock the gate to the above-postulated passage we have to find a key. I propose that this key lies in what Western civilization has called the "moral virtues."

Naturally, we are not inventing anything new here. Our analysis of the virtues will have many traditional Aristotelian overtones. But in showing how the moral virtues spring from the twilight circuits of the subliminal soul, we will illuminate them in their fullness. Furthermore, it will become clear that virtues stem from and necessarily maintain always a direct line of communication with the subliminal passions.

To recapitulate, in our new perspective we take into account the underlying onto-cosmological fact of the turmoil of life. Thus, our query boils down to the question: How in the turmoil of life can we outline any consistent course of conduct and then follow through with its concrete life enactment in circumstances that inwardly and outwardly are different, often unpredictable and unforeseeable at the moment at which we projected it? How can we steer a consistent course at all? And if we do not — if no one can — how could we enter into, promulgate, and carry out any of the trans-actional agreements by which human cooperation projects the networks of societal life? Lastly, how could we then constructively cooperate in the development of the common human universe of culture?

2. *The Enigma at the Heart of the Moral Life*

Our life experience oscillates between two extremes in conduct. In principle we live with a conviction of being honest, just, responsible, and reliable, devoted to our friends and ready to help them in need, ready to defend the oppressed, etc. And yet when the occasion occurs to take an open stand on these convictions, we will seek excuses for not acting on them. There are any number of excuses (e.g., we are not sure, we cannot run the risk of recrimination, we have other commitments or responsibilities, etc.). We hide our heads in the sand letting things take their course. Put on the spot, we would forcefully deny having abandoned our convictions or moral ideals and would defend the quality of our active conscience, yet the net of sophisms that we spin in our own defense takes us even farther from implementing ideals and convictions in concrete behavior. If we could for an instant withdraw from the cloud of self-delusion in which we dwell, seeing the situation with all sincerity, we could not deny the strange gap here.

We find this surprising gap at the other extreme as well. Even if we find ourselves under attack, say the victim of an intrigue plotted by a good friend whom we find seeking in a surreptitious way to defraud us in callous disregard of our rights, our due, and his moral obligations toward us and we find ourselves having to defend our property, professional standing, or even our good name, though we vow then to answer this challenge on all fronts and though we are anxious for our very being, feeling endangered, in distress over the personal nature of the deceit, scandalized by the contempt for moral principles, wounded by the ingratitude, and think we are ready to go to the end for the sake of justice no matter what, we will in fact go only so far and no further. We will decide to not expose the other's wrong, to not denounce our adversary, to not compromise his good name though it is undeserved, we just will not.

We do not know why; yet pausing with surprise, we approve of this hesitation in our innermost self. We "know" that "we could not do it."

There is a most fine line up to which we will tenaciously carry our self-defense but that despite our fury and good reasons, and despite whatever pressure inner and outer may be on us, we will not be able to cross. We simply cannot bring ourselves to betray a onetime friend. There seems to be deposited in the deepest reaches of our being a moral "code" that permeates all our functioning imposing a limit on our conduct that we cannot transgress.

In these two cases, that of failure to come to the defense of others out of self-interest, and that of failure to come to the defense of our own interests because restrained by an inner imperative, we enter into the heart of the moral life as lived in our inner self, as projected from there onto the horizon of ideals and models for our conduct, and lastly as enacted in conduct itself. Moral life worthy of the name, in its full-fledged expression in exemplary instances (since it cannot be said that all human beings achieve it) manifests itself when we are put to a crucial test. We have, first, the morally molded propensities of character to interpret events, qualities of feeling, bents, attitudes toward others with whom we enter into trans-actions, a sense of one's responsibility to meet contractual obligations; then we have the moral modalities of valuation to which we refer in setting aims — modalities, criteria, and norms inherent in our moral conscience as an orienting agency. At the far-off horizon we refer to the ideals that illuminate our prompting elan. Within this theoretical moral set up we may investigate exemplary instances illustrating these issues such as that found in *Lord Jim,* Joseph Conrad's most perspicacious psychodrama.

3. *Moral Failure and the Conquest of Virtue*

Jim is presented as a fine specimen of humanity. An educated man and one with fine moral dispositions, he made himself popular through his extraordinary performance of his varied and very demanding duties. As a water clerk he was "To the captain faithful like a friend and attentive like a son, with the patience of Job, the unselfish devotion of a woman, and the jollity of a boon companion." These, even taken singly, are exceptional, rare qualities, but Jim topped them with a many-sided courage braving the winds and waves in order to be serviceable to the ships approaching the port. And in these trying performances he has not displayed bravery but manifested an exultant certitude. The everyday tasks of his profession, one could surmise, made him ready in the moral quality of his feelings, dispositions, and character to acquit himself well when greater or great occasions for measuring his forces, such as he dreamed of, might come up. Jim, indeed, owing to his being the son of a parson and owing to his education had high ideals. He went to sea animated by the propulsion of and attraction to noble, great deeds for which opportunities are so rare. In fact, it is rare to encounter one such opportunity in a lifetime.

And yet when unexpectedly, such an opportunity presented itself to him, he failed. Here is the tragedy of Jim as well as the drama. The conflict between societal morality and his inner self is its engine. As it unfolds the drama centers on the discrepancy between the inwardly grounded ideal and its enactment in concrete conduct.

In point of fact, out of a spirit of and avidity for adventure, Jim enlisted as the second mate on the ship *Patna* and took thereby a share of responsibility for eight hundred Eastern pilgrims. The ship having run into a critical situation, Jim had a unique opportunity to "show a stiff upper lip," to measure his forces by the yardstick of his ideals and follow through with appropriate conduct. Instead, he finds himself a few weeks later in court as a culprit, "a shameful coward" who followed the skipper and two engineers in deserting the ship, leaving all of the passengers entrusted to their care in mortal danger. In the eyes of the court and according to the moral laws of the worldwide human community, this act is a matter of fact pure and simple; there is no justification for this breach of trust.

Sailors are trusted, and this is trust — which is assumed in special trans-actions — in a dedication that says, "Stay at your post." This solidarity with the people who entrust their lives and possessions to them makes a corps of the sailing community (as is the case now in aviation too). It confers on them a dignity before the world. Sailors commit themselves and their lives to the safeguarding of other lives.

Thus, Jim was irremediably condemned by the universally valid laws and rules of the community. But to him the case is anything but clear. Instead of avoiding the trial — as the skipper and the engineers, people of no moral caliber, did — he assumes full responsibility for his actions and submits himself willingly to standing trial all alone. And yet shaken in his entire being, he is absolutely revolted by the significance that the court, the public, and the code that judges by brute facts attribute to his act of jumping from the ship. His aim is to clarify the *meaning* of the events that led to his seeming desertion. Nevertheless, the trial only confuses him more. What the magistrates want to establish and consider in the light of the law are only the facts and nothing more, "as if facts could explain anything," comments Conrad.

We have here, first, acceptance of the validity of the social code of the moral trans-action; second, the individual's full responsibility for overt conduct and his voluntary submission to the judgement of a societal authority. And yet, the supposed culprit is struck to the marrow of his bones — and so are we — by the discrepancy between the quality of his moral inclinations, bents, and attitudes and the current failure to apply them in overt conduct, between his attitude toward life, his noble ideals and the act of desertion that he committed.

Jim's sense of responsibility goes deeper than the laws of the established human order; it touches his innermost self.

This act seems so foreign to his inner self, on the one hand, and to his ideals and his dreams of meeting once in life an occasion for heroism, on the other, that he is determined to find its real causes.

He attempts to translate the facts into a most accurate account of the events that in succession led to the desertion. The facts fail to yield an answer. The felt radical discrepancy in all of it makes him aware that "behind" the empirical facts was the terror that he experienced, a terror that paralyzed his being and clouded his usual self-awareness and threw him into such a hebetated state that he was prompted by a blind impulse to follow the escaping others and jump into the boat. Indeed, as he recalls, while the others, men of a primitive kind, jumped fearing for their lives and with all their attention on how to save themselves, his reflectively inclined imagination was struck by the striking aspects of the situation: a ship doomed to sink at any instant, with no chance of reprieve, nothing to hold on to. A vivid picture of his own death together with the death of hundreds of pilgrims spread a dread throughout his entire being. His horror is at the thought of hundreds of human beings unaware of the peril that will momentarily plunge them into the sea. This horror playing on him subliminally overtook him. He would have needed time to sort things out. There was no time. The succession of events and an ominous darkening of the sky allow no time. An urgent call comes up from the lifeboat, and Jim, without willing or desiring it but prompted by sheer instinct, follows the call.

We would ask like Jim: What could I have done? There is no answer to this question. There is no direct passage or bridge between our subliminal moral self with its heightened moral consciousness and lofty ideals, its reflection and imagination, on the one hand, and primary nervous/physiological reactivity, on the other. The question of moral conduct, conduct consisting in a succession of steps leading from the moral conscience with its subliminal prompting through the entire moral apparatus that progressively comes to grips with the positive, adverse, or neutral moments of the situation, up to the definitive enactment that translates our subliminal self into factual moments, achieves focus, as we have shown, depending on the state and degree to which our subliminal circuits master the primal, elementary circuits of our functioning.

The tests, however, to which life may put our self-esteem go much deeper than just confirming us in or disabusing us of our surmises about our capacity to effect in action the elevated ideals we believe in. If we are put to the test by social norms only, we may withdraw from the contest with an excuse like Jim's. And we may still then be genuinely committed to the ideals we hold, not only in our imagination but actually by the subliminal threads of our innermost being. Then we will not accept our failure as weakness or a revelation of self-delusion; we will instead try by every means to trace its causes, in order to redeem it.

The failure will be taken as a challenge to our innermost self, and it will release the most powerful passion that presides over our entire moral apparatus: the passion to be true to ourselves.

We realize that there is a gap between what we are convinced is our genuine moral stand on matters of life, a stand that involves the complete complex of our inclinations, bents, and tendencies as well as our aspirations, ambitions, and ideals and is sustained by the very fiber of our innermost subliminal forces, promptings, and passions, and our elementary vital functioning with its energies, instinctual drives, nervous reactivity, etc. The latter is meant to carry out the moral mandate in real life enactment or to refuse to carry out morally repugnant expectations. And yet, it might, as in the case of Jim, prevent even the formulation of a mandate. Furthermore, obviously indispensable for the mandate's fulfillment is a consistent and steadfast prompting of the will, a will that has to go to work within the circuits: it has to lead to the formulation of the mandate — the will to take a stand on the matter — then, it has to issue the call for the crystalization of the mandate.

It has, then, to unify all the circuits of our functioning that lead to and perform our acts. That is to say, it has to proceed from the subliminal down the ladder of our functioning to the elementary level of physical acting.

Realizing that there is a gap between our deepest moral self and our natural concrete functioning, we refuse to identify with the latter.

To be true to oneself does not mean to reject the verdict of a society that condemns us for transgressing the rules on which the trust, security, and stability of the human trans-action rely. Even if society in imposing penalties for transgressions might consider the nuances of culpability with regard to the inner state of the individual, it cannot free him from responsibility on this account. Often friends forgive us the various moral failures by which we fall short of their legitimate expectations, but we do not forgive ourselves.

To be true to oneself means seeking the hidden reasons that prevented us from acting morally. We do not thereby seek to excuse ourselves by disclaiming or diminishing our responsibility. On the contrary, we assume it fully in order to redeem ourselves at the next occasion.

This redemption cannot take place just in our explaining to ourselves and our friends the incidental weakness that overcame us. We have to redeem ourselves before the law that we have transgressed: before the human community that this law represents. This entails forging within ourselves a measure of determination, of sustained moral purpose — determination that will penetrate all the decisive joints of our beingness, the entire chain of circuits that leads from the subliminal forge, through the apparatus of the conscious moral agency that issues the moral mandate on the way, to the primogenital functioning that carries the signals, energies for the enactment of those mandates.

In brief, to be true to our moral self we have to tighten the knots of our beingness, to introduce into the functional system a high moral sensibility, an articulated line by which at an instant's notice a moral alert may be sounded and all our forces and spiritual energies mobilized to enact step by step the mandate of morality.

This schema puts the entire ladder of our functioning at the service of the moral agency and under the banner of those ideals shimmering at the horizon and onto which we project the meaning of life; it serves to conduct all our circuits into the fulfillment of the moral decision.

Societal rules and laws that guide and give points of reference for the humanizing trans-action would be inoperative and extraneous were there not in place this innermost individual line for conduct for enacting our personally experienced moral code, a line linking our subliminal self to our moral conscience, the moral decision making apparatus, and the entire psychic and vital operational system of our functioning — a life linkage system animated by will that prompts all our faculties to work out the moral mandate in deeds.

Going a bit further in discussing the above example, we may say that it is the passion to be true to oneself that sets in motion the entire human functioning, orchestrating the entire process of the moral forging of our line of behavior.

The line of moral behavior has to be — as comes out in the above analysis — elaborated within our entire system through the establishment of a conductor of behavior that, first, sensitizes all our reactions and simultaneously informs them out of our behavior's subliminal moral quality, and that, second, tightens the transitional knots and sustains the will, energizing it at every step with the new forces that it gathers at every stage, so that it becomes unwavering although flexible, until moral mandates are fulfilled.

To bring our study into contact with postmodern culture and societal life at large, we have to reject the prevalent lamentations over cultural decline owing to the abandonment of the high ideals and virtues said to be upheld in humanistic times. Those ideals and virtues, attributed to people "larger than life," have indeed fallen into oblivion in the postmodern period. But does that mean that our societies and our contemporary cultures have no moral standards by which they may order humanizing trans-actions? Does that mean that they have no virtues to rely on in radical cases in which norms have to be upheld unconditionally? Quite to the contrary. In contrast to classical times, ideals and virtue are not reserved for exceptional people today, but are held and practiced by average, ordinary folks. Dedication, responsibility, reliability, faithfulness, loyalty, helpfulness, etc., are the common virtues that modern society builds and relies on.

The case of Lord Jim discussed here may apply to innumerable situations in which the safety, well being, possessions, and health is entrusted to professionals. It is on their performance that the order, the life, and the progress of society depend.

As we have seen Jim had to work out within himself this virtue of dedication, loyalty, faithfulness in order to face terror and dread of impending death and accept its challenge.

Within the strictly philosophical perspective, we have seen that the virtues as revealed in their origin and role dissipate rationalistic as well as emotionalistic preconceptions. Indeed, virtues do not consist in accepting the dictates of reason over against the inclinations of desires; neither do they flow spontaneously from our deepest emotional self.

Virtues situate themselves naturally within the critique of reason. The achievement of this orchestration of our inner-outer faculties is what I call 'virtue.'

II. CONCLUSION

Summarizing our inquiry we will submit that virtues such as heroism honesty, justice, selflessness, equanimity, magnanimity — and even generosity, which as I have pointed out above, is the fountain of them all — are neither an inner principle (*arete*), as Aristotelian philosophy maintains, nor a deeply rooted habit, disposition or quality of our psyche (as psychology would have it). We have seen, first, that our moral principles develop from the workings of our subliminal passions operative in the societal — intersubjective — sharing of the humanizing transaction and that the propulsion to implement them in enactment stems from the passions themselves. Second, we have seen that the subliminal passions are also the promulgators of the ideals that orient our moral life giving it aspirations. Third, and most significantly, it appeared that the capacity to act unconditionally according to our moral principles — a capacity that defines virtue as such — consists in the orchestrated subliminal/passional informing and shaping of our entire experiential and functional life system. Thus virtue, as the exercise of this capacity, consists essentially in a dynamic thread uniting the subliminal moral ideal with the deliberative and prompting forces of the will and with all those forces of our functioning that shape and carry conduct forward in its enactment.

To traverse this distance, full of obstacles, conflicts, doubts, etc., between subliminal feeling, at the one pole, and our psycho-physically incarnated acts, at the other, that is, to practice any of the virtues we have to be prompted and conducted by the crucial passion that surges within us at the crossroads of wavering deliberations: *the innermost passion to be true to oneself.*

TYING POINT SIX

THE GOLDEN MEASURE: TOWARD A NEW ENLIGHTENMENT

TOWARD A NEW ENLIGHTENMENT

I. THE UNIVERSAL CALL FOR MEASURE

Today there have come to light previously unthought virtualities of life and vistas for their application as well as new realizations of the limitations of life. Bewildered humanity is challenged to seek by a need to find new ways to assess gains and losses and make socioeconomic and cultural adjustments and resolve political and religious conflicts.

Our bewilderment over life's new enigmas may issue in a New Enlightenment, a new awareness of all of the forces carrying life and with that ever widening horizons. We are challenged to enter into our depths in order to achieve a new understanding of our place in the cosmos and the web of life, to find new wisdom for charting our paths together and fresh inspiration to animate our personal conduct.

I will here submit that the key issue for this New Enlightenment is that of *measure,* the measure of all things concerning life. Just as important is the discovery of *motivation,* the force needed for the commitment to apply measure. We have to realize that ethics is rooted in the two factors of measure and motivation. Without these ethics cannot address the demands of situations.

In what follows, my exfoliation of the ontopoiesis of life as a protophilosophy will serve as the field on which the argument will be carried out.[1] We will sketch an "anatomy of bewilderment" — of the disarray humanity now finds itself in. First of all, we have to situate our problems, indicating first the context of today's challenge and then the context of its answer.

II. THE CONTEXT OF THE UNIVERSAL HUMAN ON-THE-BRINK ORIENTATION

Since Spengler's *Der Untergang des Abendlandes* we have been aware of the profound ferment corroding the relatively stable system of the traditional worldview. Husserl's *Die Krisis der europäischen Wissenschaften und des europäischen Menschentums* explored the overrationalization of the world of life brought about by a hypertrophy of science.[2] The sciences, especially the sciences of life, have evolved in startling new directions since that diagnosis of our civilization's problems, giving us new awareness of what is occurring in the lifeworld. These changes in the world come from within, that is, from the upheavals caused by new discoveries in technology, the life sciences, medicine, etc. as well as by the dismantling of the old patterns for handling world affairs. The connections by which individuals knit a worldview have been disrupted. What in the middle of the century was called a "crisis of culture"[3] has now gone beyond that. The ties that upheld culture are now dissociated and its elements dispersed. The sources of our ideals, moral sentiments, taboos are either tarnished, disavowed, or emptied of significance. The commonly held worldview is now one of total disarray. A new compass for orientation with fresh points of reference is our urgent need. For the present, society is simply throwing provisory patterns of ordering over its life. The key to orientation is lost, namely, *a principle of measure.* A principle of measure needs to be either discovered or invented, a criterion for appreciating a situation and taking a stand on matters.

At the same time this indispensable principle of measure needs to be matched by a source for personal convictions, for innermost personal *commitment* to the chosen course of action. This commitment should accompany the principle of measure from its foundation in the commerce between living human beings and their vital necessities and higher aspirations. If a course is to be charted from the conditions of Nature through our social entanglements to the realization of ideals, commitment that is clear about what should fulfill and what should distress is necessary. We need not only to measure but also to decide.

Measure and commitment are ever the key issues in life, but today they must be investigated anew in full awareness of the burgeoning ethical demands/claims being raised in our institutions, professions, courts, governments, world order. There is turbulence in both the social and private domains. But careful observation shows that the insights behind these demands/claims are fragmentary. The evidence on which they are based is piecemeal. There is lacking a principle by which to estimate matters globally, and consequently in detail. "All things hold together," says Parmenides, but today's partial perspectives hamper realization of this. It is for this reason that I propose a new vision of the world of life as such in order that there may be distilled from it the measure and commitment now needed, for today the very survival of life hangs on our finding and maintaining balance in that world.

Our age, however, has deliberately rejected speculative systems and has nurtured a distrust of apodictic principles and reason. How can a unifying and not unwarrantedly coercive vision of unity be achieved? As I have often voiced, it is the uncovering of the ontopoietic unfolding of life itself that allows us to assess all of the sectors of life's human concern without projecting unwarranted ties or other mind imposed articulations. Just this is the project of my phenomenology of life and the Human Condition. As the givens of existence emerge spontaneously throughout the vast expanse of life and reveal their autochthonic relevancies to a new mesmerized philosophical vision of the all, they offer us a measuring stick as well as the key to its use in striving for the common good of life.

It is within the framework of the phenomenology of life and of the Human Condition that I will treat my topic. First, however, to justify this query I will offer a succinct discussion of the questions being raised and calls being made in contemporary civilization. Today's situation is much more complex and far-reaching than any "crisis." First of all, it does not concern just one group or society, but all human groupings are being pulled into it. It is not the result of just one revolution but of a host of upheavals. The result is a deep-seated loss of bearings in social reality.

III. THE ANATOMY OF OUR MORAL DISARRAY

We might say that the advance of humanity proceeds by an ever recurring return of "crisis," that is, if by "crisis" we mean a break in hitherto established ways of conducting public affairs and the ensuing social and individual disorientation. Such a crisis is a natural outcome of the dynamism of social life, of its outgrowing a certain pattern of life and inventing a new one, with individuals then developing new skills, approaches, orientations as they pursue their aims. A crisis may reach the intensity of a revolution or may involve a slower but inexorable change. In a revolutionary situation these changes are made in the name of some ideal. But under its aegis the previous order will in the end be reinstalled being only somewhat transformed. It may even happen that the old order will withstand the assault.

The "crisis of culture" over which Husserl sounded the alarm was the outcome of a slow fermentation of ideas undermining an order of beliefs and custom, a transformation of the old flesh of experience and attitudes toward the world. But in our day there seems to be no flesh, in which a new culture could inhere. Humanity seems to have stripped itself of all flesh — to the point that not even a skeleton remains. Today the flesh, of humanity is so disjointed that it cannot maintain any posture or direction. This is our disarray.

A diagnosis of our situation often made — by Simone Weil, to name just one — sees our lack of a foothold in this world as coming principally from the uprootedness of human beings as they have abandoned family, community, natural environment, in short, their native soil, their "place," for foreign places, the cities, other countries, other continents. Our innermost ties with the vitally significant factors that sustain us have been snapped. There is a reaction to this situation in ethnic revivals all over the world. But simultaneously it is the trend that human beings now come into the world in such a way that they cannot sink roots in a communal setting.

Uprootedness is generally an urban reality. Although the state has assumed a preponderate role in the regulation of life, the game of life is played out in the cities. That is the place of the commercial, political, and cultural institutions where the strings are pulled. The city in its size, diversity of lifestyles, and ethnic mix cannot serve as a "native soil." Still, human beings crave a place they can call their own. If only by a pot of geraniums set on the balcony, the dweller of an apartment building seeks to put a personal mark on the area in which he or she dwells. Rootedness involves, on the hand, endowing a ground of one's own with meanings that are one's own and, on the other, sharing these meanings with others. These are two contraposed but intimately interplaying factors of social life. It is in their conundrum, even in the empty spaces between them, that social life is played out.

It is a staple of urban sociology that there is incoherence, even a vacuum, in urban social life. Either there is a lack of significant interpersonal communal ties, or there is an intensification of separate group identities with hostility cropping up between groups. Any harmonization is difficult to achieve given the freedom of city life. Authorities must concentrate on seeing to it that the city's vaunted opportunities for the satisfaction of material needs actually materialize lest violence and crime flourish.

Beneath this disorientation, this human uncertainty about personal standing in the surrounding "world," the need for an unbiased ultimate orientation persists. The human universe has seemed to be credibly, if provisionally, charted through most of human history. We sought and found our bearings according to the measure of our ability to discern. Now with the explosion of new science and technology and its worldwide dissemination, those natural bearings have been lost. We have been made ever more acutely aware of our contingent and perishable condition. And we have become familiar enough with technology's advances to know that to seek fulfillment in them is a wild goose chase, that indeed they but transform our bondage to nature in loosening our ties to it.

The rapidity of the pace of technological change means that we cannot keep pace with cause and effect. All action precipitates unexpected reactions. Hence we find ourselves living on the slopes of a volcano that may erupt at any time. The nomads of old at least had a rhythm and cadence to their lives, a measure. Often it was Nature that provided the rhythm. But for us the firmament is not firm. We know that the polestar drifts. It was only because of the slowness of our perception that the world seemed stable. Now we truly need an Archimedean point from which to discern order. We have instruments fast enough to discern the drift of the polestar, but not instruments fast enough to see where our technological prowess is taking us. Amid competing interests, aims, projects, we do not know what standards will meet ever changing situations in an ever changing network of interactions with the natural circumambient world and within the societal sphere. A criterion, a set of scales, a common denominator is indispensable for our apportioning our habitable space and its resources. In one word, the measure of all things, a point from which to gain purchase on reality, is needed. And with that a prompting force, a readiness to apply that measure is needed. Measure and commitment are the crucial requirements of humanity in its present disarray.

Our modern cultural disarray is caused not only by the uprootedness that came with the exodus from the countryside to the cities and the great migrations to new countries but also by the development of means of communication that inform us of what is happening all over the world. This stimulates human hopes and striving while undermining any sense of security and stability. And then the new ideas of science open new horizons, awakening us from "dogmatic slumber." New questions need answers.

With the deepening of research into the "workings" of nature, science has become so compartmentalized that even scientists in one field do not have access to other realms of inquiry. The sciences of life especially are subdivided. How can researchers in one field project a vision of the All? The life sciences need first of all to find a common denominator at the depth of life's functioning. It is only with the help of such a pivotal insight that these sciences may establish a measure, a valuating principle, allowing them to find their way in the great expanse of life. To find this criterion it is necessary to descend into the font of life's unfolding forces and find the tie that binds them.

There is in principle nothing new in this need for measure and commitment. In all of the past civilizations known to us the individual as well as social circuits of existence were guided by "ethical principles and norms" that were in turn expressed in laws, either secular (it is enough to mention the codes of Hammurabi, Confucius, Solon) or religious (the Ten Commandments, the Tao, the Koran). While these laws reflect the state of enlightenment of a particular society, its concerns, they all express the innermost, the gist, of the human predicament of common life — Thou shalt not kill, usurp others' property, slander, etc. They retain their basic validity because they proceed from the very nervous system of this coexistence. Because they are so "universal," these principles require specification to meet individual cases. The specification and explanation of the rules derived from these basic orienting principles is a work taken up by social groups and jurisprudence in the practical domain and by philosophy in the theoretical domain — probing, corroborating, justifying, founding.

Our times stand between two different civilizations. The old one was tied to the earth below and the fixed but moving stars above, and the new one is breaking our ties with these old allegiances to these "basic realities," having entered into the play of forces governing their stations and become enlightened as to the possibilities of human freedom, be they dreams or whims.

But mankind also knows more turmoil. With the common soil stripped of meaning, with habits, creeds, attachments no longer shared, parricidal passions or economic concerns dominate human relations. The above-mentioned universal principles go unused as a measure, and we are no longer impelled to apply them.

But with the new self-awareness of human beings, there comes also a new awareness of the right to respect and dignity, to fair treatment. Society now resounds with calls for ethical comportment in all sectors of life. And there is new awareness of human dependency on the soil. It is no longer "native" soil, but it has become indispensable in its natural characteristics, now threatened by ruthless human exploitation. Calls are now raised for the extension of ethical principles to the previously neutral realm of animal and plant life. In the human realm advances in medical research that increase our ability to control and manipulate have made bioethics the most debated subject of the day.

Ethics as that philosophical discipline dealing with principles of deliberation and choice is being asked to provide the appropriate discernment in these situations. There is no need to recite just how many different conceptions of ethics now compete for society's hearing. None of them compel action, however. We are left to chart our supposedly inventive course with newly deliberate awareness of our ties with the earth below and the skies above. It is between these that the game of sharing in life is being played. All now culminates in the societal circuit; all individual and local actions feed into a vast network of societal interaction, and societal directives in turn reach us where we are. The great game of life has become a gigantic one. More and more of humanity are active players in the game, with each one seeking his or her "due."

The societal circuit is the stage for the play of the strongest forces, passional pulls, cunning, individual and national self-interest, ideology. We will attempt to show the place reserved for ethics in this play of interests. But we will also bring into play entirely new expectations. Only a completely rehauled ethics can expect compliance.

IV. A MORALITY OF NEGOTIATION: THE FALLIBILITY OF THEORETICAL REASON

At the outset it is necessary to make the traditional distinction between morality and ethics. While ethics consists of a set of principles for conduct and norms to be applied in deliberation and choice of direction for action, it is the intention to realize the implementation of a chosen principle, the attitude toward the other or others toward whom the act is directed, its quality, that imbues acts with significance. Indeed, actual conduct may be only very remotely related to some or other ethical precept. While by ethics we understand a prescriptive and directional theory, we seek morality in the actual life of a person, social group, cultural epoch, as the expression of a style in human relations. What is to be considered then are the ways in which ethical principles may or may not be easily applicable, given the already established tenor of morality in a society. It is my contention that today we have not only to contend with vastly divergent, even contradictory ethical theories, but also, and first of all, with a basic tenor in human interaction that does not allow the immediate application of any ethical norm whatsoever. This does not mean that actions never meet ethical norms, but that that can be neither assumed nor expected. Indeed, in matters of merit and punishment, claims, demands, recognition, *it is negotiation not an appeal to norms that now settles things.* We have now a raw clash of interests that has little or even nothing to do with the nature of the claim or the misdeed.

Accommodation, that is the tenor of social relations today. Pressure, influence, societal power to mete out retribution, prestige, networking — this is the gauntlet through which a claim must run to get a hearing. Ethical principles play the role of a far-off echo to which vague reference is made. The most striking examples of this situation are to be found precisely in the place where we would expect exemplary instances of the application of ethics, in the courts. But here procedural justice has come to overshadow substantive justice in the matter of admissible evidence. Similarly, in the negotiations with accused criminals by which they may lessen their punishment by betraying their accomplices, substantive justice is lost sight of. And justice can be a complete stranger in out of court settlements made in order to avoid mounting legal costs.

Nothing evidences our contemporary disorientation more than this ominous shift away from seeing recognition of the true facts of the matter as the crux of any ethical system. It shakes us to the point where we begin to wonder whether what we thought was justice was always a surface phenomenon and is only now known to be such.

These issues are not new. They have captivated the interest of several contemporary philosophers, who either enter into the concrete data and contemplate them in their context, seeking then merely expedient, piecemeal solutions, or turn to classical precepts of philosophy to resolve them — prudence, charity — as if they could awaken them in individuals and society simply through rational argumentation. Even if it be backed by fear and desire, can reason awaken these moral virtues? Are they not to be ignited rather at their source? There are philosophers who seek to uncover that basic principle that reaches to the source and encompasses the entire territory of human concerns. The source experience is the experience of the irreplaceable status of the human being. But whence does this experience come? How is it to be so nurtured that it inspires care for the existence of every living thing and is seized on as the measuring stick to be applied across the span of human concerns?

I agree with Peter Kemp that the ethical concern of the present age must be "anthropocentrifugal" because responsibility for the whole world lies with the human being.[4] But how can man be the source of the measure he adopts and the commitment to apply it? Heidegger and his epigones take death as the measuring stick for life forgetting that it is precisely in the interval between birth and death that a measure is applied; here unfolding, reaching out, pulling back, pains and pleasures, crises, defeats and victories are played out. If we are to take our fragility itself as the informing principle of our ethics, we have first to find the reason why we should be solicitous for the vulnerable. How do we experience vulnerability in such a way that it awakens solicitude? What, indeed, is the source of moral motivation and commitment?

We attempt this succinct anatomy of contemporary moral disarray because our personal weighing and decision making are directed not at inert flesh, but at palpitating, vibrant humanity.

The greater energy of humanity today is behind the call for a novel reassessment of the world and of life's contingency. Life's innermost core, the flux of reason/unreason, order/disorder, chance/necessity, luck/fatum that we cannot escape, has to be unraveled in order to reveal the principles by which its expanse holds together. This would allow us who move with its advance to discover the extent of our prerogatives and commitments.

In the following phenomenology of life I will attempt to expose some of the pivots of life at its core. First and foremost, *the self-individualizing character of life, whether it be at the organic, vital, psychic, social, or cultural level, will serve as our measuring stick in the vast expanse of life.*

Given how the sciences have made us more aware of the fragile contingency of not only ourselves but of the web of life as we extend our power over it, it is to the same sciences that we look for clarification of this contingency in order to better grasp both our prerogatives and dependencies. But the sciences themselves, as I pointed out some decades ago, are in no position to assess their own prerogatives, fathom the reach of their possibilities, or estimate the upheavals caused by the introduction of new technologies. Just given their specialization, they are in no position to deliver any plausible overview of the global situation. Each of them in its own limited dominion, following its own peculiar rationale, misses the underlying common factor in the coordination and harmonization of the whole of life. They but grope for the connective link, the essential principle unifying life, the pattern in the constructive flux of life's becoming.

The upheavals throwing the world and us into disarray have their source at the level of science's excavations into the resources and virtualities of the cosmos-nature-life system, into the radical interdependence of every circuit of existence and its fragile equilibrium. It follows that the measure of the equilibrium to be maintained has to be sought, found, and legitimated elsewhere.

V. SELF-INDIVIDUALIZATION IN THE STREAM OF LIFE AS THE UNIVERSAL MEASURING STICK

1. *Outlining the Territory of the Phenomenology of Life and the Human Condition*

Our sketch of the anatomy of humanity's profound disarray sought to reveal the different circuits in which it has its roots. It became clear that invoking ethics to solve our enigmatic problems would be futile unless the residua of the problems are clarified. These problems are not only moral and economic but also existential in nature. The existence of human beings is conditioned by the situation of nature-bios (at its foundations in the cosmos), and life on earth depends on the measuring and creative wisdom of human beings.

This anatomy is not the dissection of a dead, inert body, but draws on the palpitating pulp of becoming: stars disintegrate and others are born, the heavenly bodies run their course affecting changes in nature-life, the cycles of life's unfolding continue, and human beings pursue the struggle for existence unfolding social and cultural schemata by which they may progress like spiders moving across their own webs. In becoming, all the spheres of existence remain reactive to changes in the other spheres: cosmos, bios, society, culture. The whole is not chaos but is a fabric woven by innumerable threads that form a nervous system, one sensitizing each new loop. This nervous system is our focus. Analyzing it shows how disrupting the interplay of the circuits of existence menaces the entire network. This we have already shown in our phenomenology of life and of the Human Condition. *The specific metaphysical context of the Phenomenology of Life is, as I have developed it in numerous writings, circumscribed by three pivotal factors: the self-individualizing principle of life, the entelechial design of life's unfolding, and that creative virtuality of life that brings about the Human Condition.*

Two points have to be emphasized. First, in order to discover the measuring stick with which to chart the entire field of cosmos/bios/world as the carrier of life, we have to disclaim any anthropocentric privilege for the human being in virtue of which considerations and decisions should be based. The Copernican revolution of our day is the realization that it is with the thriving of the entire field of cosmic/biotic/social circuits that we thrive and with its destruction that we vanish.

Still, as I have been insisting, the creative act as the uniquely human prerogative does give the human being a unique position in that field such that, from his or her own creative participation in the workings of nature-life, the human being has exceptional access to the entire ontopoietic unfolding. But conversely, it is by virtue of creative virtualities that the human being may unfold the entire apparatus of the mind to grasp intricacies, to reflect on and judge, draw conclusions for planning and enacting within nature's own inner workings.

Thus, although we emphatically have to forego the traditional anthropomorphic and anthropocentric perspectives, it is still from a privileged position that we come to know our place in Nature and discover that we are at a loss without it. It is the human being alone among living beings that is capable of mentally encompassing reality's vast horizon so as to feel lost and in need of reorientation.

The phenomenology of life and the human condition, which means the uncovering of the inner workings of nature as they emerge in the human creative act, offers an appropriate philosophical ground for envisaging life's vast expanse and attempting to chart it anew.

2. *The Entelechial Ontopoietic Design of Self-Individualization in Existence As the Axis of a Universal Measure*

Whoever says "life" says "self-individualization-in-existence." Life in its turbulent spontaneity does not proceed blindly and haphazardly, as Bergson seems to have it. No matter at what cost of trial and error, life proceeds by linking its energies, synergies, forces into self-individualizing complexes that we distinguish as individual beings. Invoking Aristotelian intuition, we will characterize each being as projecting an "entelechial" design. I want to say that the individualization of living beings — and only living beings manifest the characteristics of individualization — is a constructive progress that follows a design. This design is "entelechial" because it projects a line of articulations that the process follows from within. No exterior a priori factor, law, or direction fashions its constructive progress. (Enriching this intuition is the Leibnizian idea of the monad or "individual substance.")

The entelechial design is seen to emerge together with the first motions toward life, toward the individualization of forces that had been lying there if not idle then neutral, general, unattached, amorphous. Entelechy springs forth together with synergies that carry in themselves appropriately attuned propensities — germinal virtualities — and come together in a life initiating articulation. It emerges in virtue of this extraordinarily proportioned and measured articulation, which, in turn, occurs in virtue of the entelechial virtuality as if lying-there-in-wait. The entelechial principle of design springs forth within the field of generating life while carrying the ordering of the virtual articulations of the constructive line that processes, operations, formations, the functioning of organs, etc. then follow. It projects, as if horizontally, its articulations to follow a constructive line. While doing so, it differentiates between "sameness" and "otherness," before and after, and interval — in short, the order of succession, simultaneity, motion, and rest. The entelechial design is implemented, then, in a life constructive stream of processes, which themselves carry an intrinsic entelechial pattern, a pattern for the order and relevancies of operations. As we have seen throughout this work, there is indeed an intrinsic axis upon which all the operations and processes are suspended, an axis that introduces the differentiation between "inward" and "outward" and passage in the congenital binding of indissociable opposites.

Why, the coherence of the constructive advance of self-individualizing life ("self-" because from "within" but directed toward and drawing on all "without") is suspended not only upon the propensities of materials and the measure of proportions that they carry, but also upon the inward/outward axis directing the coming together of all their operations, performing a function and leaving the scene. The inward/outward axis operates sameness and otherness, implies the differentiation in the "whereabouts." The "inward" orientation of the influx of materials into a "core" — their coming together — and the "outward" orientation of their "intrinsic" process of fusion and operation, having brought about the appropriate result, throws "out" unused elements in a reverse movement, marking an "outside." The self-individualizing progress of beingness proceeds thus, moving on in an inward-outward directed spiral. Its constructive progress is dependent on the indispensable influx of outside, substances, energies, synergies into its domain, a domain that emerges together with the advance of the life process. The genesis and unfolding of life resides in this innermost functional core of being. It is within this inward/outward oriented system of the constructive advance that nature's metabolic exchange of material and energies, while differentiating and transforming them, finds its foyer. This ongoing cycle right through to the rejection of unused material and energy continues on in all the circuits of unfolding life.

We may first distinguish the nurturing phase of the life promoting inward/outward oriented advance. Photosynthetic sensitivity/receptivity/reaction, assimilation, absorption, symbiotic attunement are the base for the generative phase of the entelechial design's fulfillment. The numerous processes and organs devoted to nourishment, the whole schema we call metabolism, are spontaneously subsumed under the self-individualizing, telos oriented schema, are subsumed spontaneously under the social and cultural circuits of creativity too, becoming its "substantial pipeline." Speaking metaphysically, I call this telos oriented schema the *ontopoiesis of life.*

"Onto-" refers here to the "firstness" of this process with respect to the scale of existential formation. "Before" this schema's articulations there is no beingness that we may ascertain with the criteria of our minds, which are attuned precisely to this and no other reality. "Onto-" here also means the indispensable and universal character of whatever there could be in the "objective" form proper to human reality, in the sense of the classical metaphysics of "onto-" logos, that is, ontology.

However, and this is of crucial significance for the understanding of our vision, this indispensable essential factor of all beingness does not concern beingness in its finished, formed, established, or stabilized state; it is the intrinsic factor of the constructive process of individual becoming. The individual remains always in the process of becoming. It acquires form and transforms it. "Becoming" is "becoming something that is not yet." Becoming is a process in its own advance, in qualification. Hence I call it, going back to Aristotle's *Poetics,* a "poietic" process: ontopoietic.

In brief, *the self-individualization of life is an ontopoietic process.*

It remains now to first show how this self-individualizing progress of life serves as an organizer and distributor of all energies, forces, synergies, etc. in all the circuits of our earthly reality and then to show how from the very principle of measure and its source in self-individualization-in-existence there flows the indispensable principles for valuation. With these latter we will reach the source of motivation and commitment in the human being as the custodian of life.

We will pass to that now. Let me first, however, reiterate that the nurturing function of the inward/outward axis of the inner workings of life is itself immersed in an extensive attention-reception-reaction network that participates in the above-discussed inner workings of the ontopoiesis of life. Let us again emphasize that *these workings extend to other realms of significance, that is, they encompass not only the germinal preorganic and organic circuits but also the sentient, sympathetic, psychic, social and cultural circuits of life as well.*

VI. THE HUMAN CREATIVE CONDITION AS THE ARCHIMEDEAN POINT OF LIFE'S MULTIPLE RATIONALITIES

1. *Measure and Valuation within the Ontopoietic Progress*

The next focus for our inquiry has to be the question of just how the self-individualizing prompting pursues its course. Do the articulations of the inward/outward oriented operations follow only a constructive entelechial design? When one speaks of measure, one implies some order of things to be measured, a measure as such, that is, a measuring stick, a criterion according to which the order intrinsic to the measured object will be visualized, and a valuating criterion. This means that we have to consider *that there is a primordial ordering lying prior to any becoming or process* so that these latter may occur as well as the primogenital ordering of the ontopoietic progress as it spontaneously unfolds. But this ordering function in the progress of becoming has to be matched by a valuating function. In fact, the primogenital material may come together and be measured, proportioned according to intrinsic propensities. In order to portion it out and measure it, a valuation of these propensities is indispensable. That is to say, the primogenital measuring has to be "fitting" to accomplish a transforming symbiosis with the material at hand. Seeing to this is the primogenital valuating function of life. Only order and valuation together may accomplish the appropriate constructive articulations necessary for self-individualizing progress. With the function of valuation we enter directly into the second main issue at stake in this essay, that of the motivation of moral commitment.

2. *The Rational/Valuative Network of the Dynamic Web of Life Stretching between Cosmic Relevancies and Human Creative Invention*

This essential inward/outward exchange in which all the circuits of forces, synergies, types of assimilation, synthesis, and amalgamation take place actuates the surging of articulations, that is, rational pointers that carry significance. With the advancing complexity of unfolding constructive progress, the field of inward/outward exchange expands, as do the modalities of exchange and interchange. Just as much as inward organs differentiate and become more complex, so do outward modalities of interchange differentiate. Just as the preorganic and organic interchange projects articulations that proceed according to a principle of fittedness and the generic operations follow the primogenital rationality of the primogenital significance of life, so does the advancing complexity of functional operations project novel schemata of exchange. These are not limited only to material substances and to the primary generative formation of the initial nucleus of life but carry progressively more sophisticated lines of these same two functions; exchange with the outside forces means also a differentiation from other individualizing processes and beings. The span of this interchange in which the modalities of reaction-action involve modifications with each new type of complexity that sets in and emerges according to the systems of attunement, adaptation, assimilation (and their reverse) covers the entire expansion of life on earth from its "womb" in nature-life[5] to its highest, most sophisticated network of modalities in the Human Condition. The crux of the matter is that the criteria of valuation progress along with the progress of this differentiating rationale.

As I have developed elsewhere, the living being through a series of steps loosens itself and breaks out of the bands that seclude it in a self-interested orbit. There is that sharing-in-life of the simple animal species that spontaneously brings certain species together in a territory that favors their needs.

They share this space but also establish the existential frontiers of their own "dominion," e.g., various kinds of birds. "Sympathy" is a sentient factor that accounts for a gregarious precommunal existence. A "leader" is followed in a herd that shares available resources. While metabolic organic exchange — nourishment and elimination — is directed toward maintaining existence, this inward central place is extended toward the circumambient surroundings into which a living beingness extends its dominion. Significant ties are thrown out toward materials that become shelter from the elements, protection against aggressors. The living being in circumscribing its own domain goes from the inward womb to a world. With the expansion of this inward/outward operative life system, we witness types of living beings that throw out ties toward other individuals of the same species, ties of gregarious significance and that then, and what is more, throw out ties of solidarity by "sharing-in-life." With a growth in self-awareness and flexibility in discernment an animal social significance of life surges forth.[6]

In coming together, generating, processing, transforming, departing, newly emerging, that is, in the span of life's genesis, growth, fruition, decline, extinction, envelopment, and transformation, in what we call 'temporality,' as well as in succession and dispersion, in coming together here, and coming apart there, in processing substances "in here," and seeking them "out there," life establishes proximities and distances that are the points marking its progress. Within this incommensurable network of all interchanges past and possible, of all the equilibria that spontaneously present themselves for possible melting or synthesis according to the very system of life's conditions in the cosmos, according to, first of all, the self-individualizing ontopoietic schema with its pregiven germinal virtualities and directional entelechial and yet at the same time "to some degree" adaptable with its own equilibrium, measure, and criteria, there is an ontopoietic skeleton. All of this makes up the inextricable unity-of-everything-there-is-alive.

We have uncovered in the congenital web of life around this axis of self-individualization the territory of phenomenologico-philosophical inquiry. We find right there where to look for the measure and its criteria for our existential condition, criteria today revealed to be urgently needed but valid ever. Not only does each segment of organic intake have its own intrinsic measure and principle of valuation, not only does each substance in a specific relation to an other entail its own criterion of proportion, but in the organic/vital circuits of operation the setup, patterning, and performance of each organ is proportionate to a specific criterion that matches it to all the others. In short, *Nature-Life bears a measure for the intrinsic/extrinsic life network itself and for each participant throughout the web of the unity-of-everything-there-is-alive.*

Let us turn now to the human condition and the moral measure for life.

3. *The Human Condition As the Unique Station in the Evolution of Types, One Bursting with Novel Creative Rationalities and Establishing the Central Valuative Agency of the Mind*

We have already hinted at the double constructive role of the poietic entelechial design. Let us repeat that it is not only from within the line of unfolding of the singular individualization of life but also with reference to the adjusting and attuning as well the intergenerating proficiencies of the inward/outward network of functioning in life's unfolding, that this design projects the intrinsic transformability of life and its apportioned steps as well as modalities of adjustment to outward conditions. That is to say, *there is already within the self-individualizing devices a provision made for the evolution of types of self-individualizing entities.* Needless to say, this evolutionary progress or regress is manifested also in a twofold way: inward complexity and novel modalities of functioning, and outward novel interactive modalities and the spread of interactive links and networks. Thus, we witness in this manifestation of life within the lifeworld a ladder along which less and more developed types of individualizing life are ranged, with the human being placed at its peak. The emergence of the human type of individual life is not a sudden event; there is a developmental expansion of what I have called the Human Condition within the spans of life to be considered in order to reach understanding of the human individual in its specific prerogatives as well as dependencies. We may define it as a specific "station" in life's dynamically uninterrupted stream, a station because the human being alone among all the existing/known types invents rational networks of such relative stability that outlast his or her own turn on the unfolding cycle of life.

The Human Condition establishes the human living being in a most particular situation with respect to the total life expanse, the entire existential schema of living beingness. Simultaneously it gives the human being an outstanding position — a *knot* position — with respect to the spheres within which living being is suspended. Lastly, it gives to man a unique responsibility toward all.

The Human Condition is a station in life's dynamic stream owing to the virtualities that the progress of life deposited as its foothold on life: the creative virtuality with *three absolutely novel valuative factors of sense — the aesthetic, intellective, and moral senses.* In virtue of their employment the human living being devises new significant avenues for its self-individualizing progress: the societal and cultural circuits of the spirit. On the wings of the creative imagination, the living being has been in a position to stepwise unfold a complete valuating and sense-giving set of faculties orchestrated into a sovereign organization called 'mind.' While the faculties of the mind process the data of the inflow-outflow traffic that establishes and keeps the station afloat and operates as a factor in the network of the common world of life, the three factors of sense are the carriers and distributors of reason.[7]

As I have been unfolding all along, the work and propensities of all the three sense giving/valuating factors are indissociable since they complement each other in the projection of individualizing circuits forging them within the inward networks of an "intimate zone" projected into a vast network of "outward" interlinkage with other living beings within a common world of life.

Beginning with *the aesthetic/poetic sense* we see the human experience prompted in psychic acts expanding into all the directions that life's functioning takes. Each psychic act in its fullness draws on a selection from all of the vital functions that life carries. It is "receptive" and assimilates in a proportioned way all the movements of natural forces in sensing, feeling, hearing, tasting, touching, feeling temperature, etc., movements that without this receptivity and assimilation would remain mute, but that experience endows with a voice. Through this receptivity, assimilation, and endowment with a "voice," the human condition allows man to participate in the innermost movements of nature's life. Through the specific voice that man utters in the assimilation of these forces, the movements of these forces of the innermost natural life reverberate in the conscious sphere that experience causes to emerge. This "conscious" sphere prolongs itself into a typically human "aesthetic" sphere of life. In fact, the essential function of the poetic/aesthetic sense is not only to draw all the life movements into the orbit of psychic experience but, by reaching in its elan far beyond the border of the natural reservoir of life and prompted by the stirrings of imaginatio creatrix, to endow them with significance that also goes beyond the threshold of the performance of life preserving functions. In the movements of the wind, man not only experiences the life of the natural sphere and may see in it the presage of a tempest or of a change of weather but also feels drawn, without his thinking of it, into the aesthetic expanse of the movement of the celestial spheres, which provides a pause in everyday sorrows in a dreamlike repose. Thus, through the poetic sense, the human condition makes the human living being not only partake in a mute fashion in the process of universal life, but makes man's participation in it full through a conscious receptivity by which it has a voice that reverberates throughout his entire frame. Lastly, it prolongs this sphere of the natural life into all of the spheres that man himself creates by bringing it in full, and enlarged a thousandfold, through the aesthetic expansion of experience.

The intellective sense, of which the main instrument is the human intellect, is particularly employed in the construction of the — entirely new in the living universe — sphere of objectivity. The intellect, as the architect of this marvelous edifice, uses the intellective sense in finding meaningful complexes in all of natural life's functions as a base for constructing marvelous avenues of life and inventing projects for life's unlimited expansion.

And so it turns the primitive natural groundwork of the otherwise blind, mute movements of nature into the stupendous dwelling of an objective lifeworld — "objective" because it stands in front of us and brings everything in front of us to behold, to test, to appreciate, to work on, to enjoy, to transform to our wish in order that we may fashion ourselves and be transformed, and to, not least, find appropriate. We dwell and grow within the objective universe assuming it to be our birthright instead of realizing its miraculous appearance within the play of natural forces that, without being expanded in objectivity, are consumed in the game of life.

But through the use of intellect, which measures, observes, judges, and establishes proportions, we human beings come also to have a distance from the rest of living beings, from our surroundings, and from others, as well as from ourselves. We may envisage all mirrored in our experience, and through the knowledge thus obtained — knowledge that goes beyond the receptivity of the sense by which we frame various levels of observation and construct/constitute objects, higher objects, ideas, etc.— we realize the situations of other living beings on earth, the trends of life, and the positive and negative conditions for their growth. In short, owing to the intellect with the intellective sense at work, we may bring everything that surrounds us and relates to our human condition within the unity-of-everything-there-is-alive (which keeps us at the center of all the life forces and processes) into an intelligible form. And then, as from a gazebo, we may view the entire expanse of life in its workings within as well as its manifestations without.

Indeed, through the intellective sense the human condition brings us into a double centrality with respect to the schema of life: it is not only that we tie the knot of its processes, but that we may from within, and as if impartially, encompass in a glance the whole of life. We may in this way, first, gain all the necessary insights for our further progress, avoiding obstacles and the dangers, and inventing means to use the opportunities and the forces at our hand. But, second, this extraordinary position that the human condition offers us gives us also a uniquely human role to assume. With this we pass to the function of the moral sense.

The moral sense is another unique factor that surges as a virtuality of the human condition and is operative particularly in the emergence and expansion of the specifically human sphere of existence. While the intellective sense installs the "objectifying" axis of the specifically human significance through which the human being expands within his self-projected orbit, it is through the moral sense that there is introduced into the otherwise "neutral" but self-centered (driven by egotistic survival interests) individual line of behavior, which is purely "utilitarian" and empirically pragmatic, the Sentiment of Benevolence toward other living creatures, toward oneself, and toward life in general. *The benevolent sentiment introduced into the living arena a new and unprecedented system of appreciation, that of moral valuation.* With it the objective neutrality of self-interest finds a balancing counterpart in the acting agent's subjective attitude toward the other person, so that not only one's own but also his or her or its vital interests are considered.[8] Valuation under the prompting of the benevolent sentiment takes in an entirely new sphere of the objective universe, the societal sphere in which the participants take into consideration each other's interests, inclinations, attitudes, and feelings about things. Now objective neutrality is balanced by subjective concern for the Other. Given this concern the benevolent sentiment will in principle be able to calibrate the balance for the benefit of the other against one's own interests so that he may receive his due according to his own merits. On extending this valuation, subjective concern for the Other, the measuring of "what everyone is due," beyond strictly individual situations, we reach to establish societal organizations and institutions of justice in the name of the interests of the group as opposed to those of the individual, and even international laws that regulate the balance of interests among nations, etc. In short, it may be said that within the human condition's gathering myriads of vitally/socially/sensitively significant forces there surges on the wings of creative virtualities/creative imagination a dynamic, novel synergy, a prompting force that is none other than the benevolent sentiment with its novel life significant factor, the moral sense. Springing up at the heart of all of the crisscrossing arteries of the self-individualizing inward-outward directed articulations of life and drawing on them, this synergy is a unique type of confluence, unique because simultaneously it brings them together in a valuating multisentient attunement while encompassing them in a *sui generis* move of a higher order of significance: the benevolent attitude. Since it is this novel synergy that allows the human being to become "human," that is, to share and unfold trans-actions with others, it is the benevolent sentiment that needs to be called back from forgetfulness and indifference, to be revived and nurtured. Here is the motivation for the common good of life, solidarity, and responsibility.

VII. RECAPITULATION

Toward Tying the Knot of Multiple Natural As Well As Creative Rationalities and the Invigoration of Moral Commitment

Summarizing, we may say that at the point at which the phase of the

Human Condition within the unity-of-everything-there-is-alive offers an outlet for life's unfolded synergies following their own ontopoietic course, there emerges first the entire apparatus of the human mind, which is capable of following this course back down to the inner workings of nature and of making this unique synergy of the benevolent sentiment and its sense giving factor of the moral sense surge forth. Both the mind's vertiginous flexibility in attunement to the real and its judgmental powers as well as the indispensable valuative factor of the moral sense working in tandem with it, if properly recognized and deployed, promise to tie the loose ends of seemingly chaotic reality and to right the sides falling out of balance.

With this succinct presentation I have attempted to show how the self-individualizing entelechial principle of the ontopoietic unfolding of life serves as an axis around which all the prelife and life forces articulate themselves into a vital artery and then entails — vis-à-vis those forces, proficiencies, and proportions in their exercise with respect to innumerable relevant factors, circuits, rules of interaction, etc. — their existential, or vital, or gregarious, or sympathetic, or social significance.

It is this axis of the self-individualizing entelechial code, the axis of life's articulations according to various rationales, that runs through the entire edifice of Nature-Life with all its relevancies in the cosmos that is uniquely fitted to serve as our measuring stick in the search for proportions, measures, criteria. But how can we human beings, while woven into and participating in its entanglements form any picture of this axis?

A picture of the self-individualizing axis would, first of all, be of unique importance for the sciences of life, allowing a view from the depths of their restricted fields of the underlying articulating principle of life throughout its expanse and in its every sector. The biological sciences are becoming aware, first, of a need to find an underlying ontopoietic skeleton, mechanism, design in the innumerable life processes that they distinguish in their investigations so that the articulations of the singular may become meaningful when seen within the vast network. Furthermore, sciences such as those of the environment or ecology are built on the assumption of an underlying all-coordinating network of interarticulation.[9] But they cannot formulate our being if lost in the concrete. Only a phenomenology of life can. They will find the vantage point for orientation and the measuring stick for the equilibria they seek in the self-individualizing principle of life within the unity-of-everything-there-is-alive.[10]

Secondly, the moral factor will find its source in the encounter of the individualizing arteries of life's synergies and the creative orchestration of the mind, whose virtuality of the benevolent sentiment it personifies. Here is the source of human societal life, its action/reaction or interaction, or trans-action. The moral factor is also revealed as we consider the valuative function of human judgement. Hence, it is on its vigor and persuasive linking of individual personal inclinations that human judgement in social affairs relies. Our social turmoil and our current morality of negotiation seem to us an entrapping web hindering us from making proper reference to ethics of any sort. The flow of moral functioning is weakened and disoriented. Could not its genuine spontaneous flow be revived?

With the regaining of the primeval spontaneities stifled in the overcalculative functioning of the mind that is the fruit of a civilization that seems to have overexploited its premises, we could find that beyond the singular self-interest of individuals there lies the common good of life in its marvelous myriads of routes, tendencies, pulls. By striking the essential chord of a situation now understood, the human being would be enabled to get an overview of all these routes, tendencies, pulls and devise new ones and even select the means to counteract all the obstacles to our becoming the custodian of life's equilibrium.

THE META-ONTOPOIETIC CLOSURE

THE META-ONTOPOIETIC CLOSURE

I. RECOVERING THE GREAT VISION OF THE ALL

We have pursued, by fits and bounds, in longer and shorter excursions, the three major lines of the logos of life — the vital, the Dionysian, and the Apollonian. I have metaphorically bestowed on two of these modes of the logos of life the terms of a traditional contrast, namely, Dionysian and Apollonian. Distinctive as these two are, they wondrously promote each other in the logos' radiation through the entire network of life's forces and so come into their own.

These two logoic forces work like two arms in their radiation. In our vision they are grounded in the vital logos that constitutes the primeval bedrock of beingness and bring about the entire choral dance of life, leading it to its manifestation.

Hence we go back to this crucial beginnings.

The bedrock of the logoic spread of life lies, indeed, in its primogenital generative forces, promptings, endowments.

The major intuitions by which we decipher this primal field accord wondrously with those deposited by the Ancients. But in contrast to those of the Ancients, our exploratory insights, sparks, hints, formulations do not come out of speculative thinking about principles and final causes. They are worked out in a concrete investigation, in the scrutiny of the real — first, in its subterranean workings, and later on, in the further manifest expansion of its circuits. *The logoic matrix of generation, which we hold to be the nucleus of the logoic powers, extends, as shown above, in two directions,* each leading toward the unknown, the unknowable, and the infinite, insofar as we can gain access to it from our microcosm.

First, new clues show us that the grand project, the generative matrix of life — the "womb," encircling the spheres of life, its existential conditions — stretches far beyond the biosphere into the stratosphere, the cosmos. The spheres of the innumerable stellar entities: planets, moons, comets, suns, galaxies, in their own revolutions set up by the cosmic logos are the necessary framework of our tiny planet earth with its life. The current revolutions of the celestial bodies indicate an order and a great project that certainly includes the furthest limits of the orbit of life.

Second, we discern another generative matrix of life — that of sharing-in-life with its tandem fusion of the Promethean vibration incarnated in the human creative forge and the Dionysian/Apollonian exchange, which sends forth an ever expanding wealth of vibrating rays of life in our human creative promptings.

Working the primal, unworked ground with painstaking effort, we have opened the doors to the treasures of the logos, to the secrets of its prompting synergies and transformational powers in the generative courses and unfolding sequences of self-individualizing life within its cosmic framework. Entering life's corridors, which crisscross, connect, run parallel to each other, we have pursued them to the limits of our creative ontopoietic intuition. Our yield has been only fragments of the gigantic constructive network displayed in the different logoic spheres. These fragments, if pursued toward their limit, often encroach upon each other's realms in active dynamic streaks; oftentimes they leave behind open intermediary spaces that our minds, whether philosophical or scientific, cannot traverse or encompass with their tentacles! It appears that our mind can not trace the complete map of the trajectories of the logoic project, even if there be one, and I emphasize, "even if there be one."

After our descent into this subterranean forge of becoming and beingness, are we then left with only a fragmentary, half-disjointed puzzle? But even if this game can be envisaged by us only from a limited and truncated perspective, the mind should not be disparaged despite its unsatisfactory situation for it expresses our own, human, living participation in the sublime logoic game.

Furthermore, the vertiginous transformative fits of the logos in the attunement of the generative-evolutionary schema to the ever changeable available conditions seem to indicate *a spontaneous virtuosity, a constant improvisation.* No mechanical stereotyping could grasp it. The *artistry* displayed has to be and can be taken into account.

Lastly, but foremost, we cannot fail to recognize that our limited transcendental, intellective mind, which urges us to question, also postulates the archetype of an ultimate status of reality, thus prompting and guiding our questioning toward that ever elusive goal. We cannot fail to recognize that the transcendental mind bears archetypal postulates of the enduring, of fullness, unity, harmony. Having set up these postulates goading our quest to understand, would the logos not provide means by which to consummate that questioning, allowing a response attuned to, corresponding to the incommensurable spread of the logoic performance in life and the cosmos, even though that spread be so vast as to always elude our human intellective logos? Would the logos have not prepared an answer to its own quest?

Our meta-ontopoietic speculation will attempt to elucidate these questions.

Having framed the two specifically life promoting powers working in tandem with the primogenital vital logos as being Dionysian and Apollonian, we have touched and brought out into light the hidden sources, the logoic springs themselves. And pursuing patiently their radiation in the works of the logos, we are about to revive the Ancients' great vision of the All.

We have lifted the lid of the purely intellectual jar in which modernity locked the logoic treasures away from our minds. We have freed from their chains the radiating wealth of life enjoyments. The Ancients did not sever Apollonian intellect from Dionysian enjoyment but, recognizing their peculiar roles, treated them as partners in the great game of constituting, experiencing, and imaging reality, life, the sidereal spheres. As we will see, our ontopoietic investigation has brought their inherent harmony to light again.

Grounded in the primal and then in the primogenital logos of life, these two complementary powers of the logos of life do not for nothing bear the names we assign them.

One is named after the god Dionysius: the "joy of life for mortals" (Hesiod, *Theogony* 94), whose gifts bring joy, ecstasy, and sorrow, struggle, tragedy, burdens as well, along with the search for the meaning of life, the interpretation of the tragic and the comic. The Dionysian logos receives a lift away from the crude and bawdy from the Apollonian streak of the logos. Yet as crucial as this Apollonian lift may be, the nerve of life enjoyment is Dionysian. In the theater play, this search after the meaning of life, the ciphering of human nature in comedy or tragedy is suffused with the Dionysian thrill and the intrinsic pathos of the human predicament so that the joy of life reverberates and in its exuberance extends through the entire span of life's progress extending to death and the hereafter.

The other, its indispensable partner, the logos of the philosophers, is named after Apollo, the leader of the muses — "musagetes" — and protector of epic singers and cithara players (Hesiod, *Theogony* 94), who with the muses represents all the higher functions of the mind. He is the master musician, who plays the lyre with artistic subtlety — and not the Dionysian drums. Apollo is the master of the intellectual elevations of the human mind and of its intelligibly ordered constitution as well as of its world of life. The Apollonian logos is the logos of the "higher" spheres of human beingness.

What is more, in the ancient Greek vision, Apollo evokes the universal harmony of the choral "dance of the spheres."

Shall we not, in a parallel way, cap the fragmentarily sketched grand logoic project with a harmonizing vision taking these lines of the logos as inspiration?

II. BEGINNINGS AND ENDINGS ARE OUT OF SIGHT, YET HERE WE ARE AT THE FULCRUM OF ALL

The thirst to understand the presently ungraspable intricacies of things throughout the entire stretch of life's becoming is an unquenchable obsession for us humans. We seek explanation of the present by searching out its motivations, inheritances, influences in all directions. We seek explanation of the historical past in evolutionary origins. We ask after the beginnings and seek to extrapolate the endings from the available data.

Our passion for storytelling serves this search. On the one hand, we probe into our historical development, seeking in lost civilizations, dying cultures, the threads leading to our beginnings. On the other hand, are we not equally obsessed with foreseeing the future by conjecture and extrapolating from our estimation of the present? In each of these opposite directions, we reach for explanations that are "definitive," that is, "complete," taking all into account. This is an obsession with perdurance. And so the immediate, palpitating actuality recedes to second place in our appreciation for it always falls short of this completeness, which can be attained only in relation to a whole, a whole, however, that without fail escapes our grip.

There is indeed present in our individual setup vestiges of the previous ontopoietic individuations that are our inheritance. Progressively and in a meta-ontopoietic metamorphosis of type, our innermost endowment of ontopoietic design was passed down to us by innumerable generations. Something like an onto-biologico-vital memory of the inner workings of nature-life in its generative sphere is indeed present, but it does not come directly to our mental recollection. Within our circuits there is a lurking, glimmering awareness of continuity that feeds a longing to situate ourselves within the long, long line of generations and prompts a yearning to attempt to discover in evolution, history, culture, something of human destiny.

The human being just cannot put out of mind these great quests for explanations, for the beginnings and the endings. These questions well up and call for answers of one or another kind, whether scientific or metaphysical. However, the great drumbeat of the logos runs through our innermost arteries, and Dionysian enjoyment of our inward pulsation keeps in step with the pulsation of all nature-life, and with the dance of the sidereal cosmos, with the melodies of the spheres, which in turn lead our most personal dance, our solo performance. To these rhythms modern humanity has become deaf. We are all too dazzled by the purity and clarity of the intellect and its benefices for practical life. However, nothing ingrained into our ontopoietic design and its course may ever be forgotten. What remains in the abyss of our generations will surface in some way or an other. Our ontopoietic, existential sensibility may be deafened and disfigured by the competition between the receding Dionysian logos of life enjoyment and its emerging Apollonian partner. For long stretches their cooperation in the forging of the human destiny may be marred by the latter's pure, detached, impersonal, objective schemata and the projects that these inspire for gaining power over the material universe. The pursuit of such clear objectives may be drained of all the life juices. Indeed, even the sensory image retreats before the impact of the intellect. A rational-intellectual logoic infrastructure is not a home. Thus can human beings become blind to the full inner imaginal. This is not the imaginary, the concocting of images; the imaginal retrieves in images the inward visionary essence of ontopoietic life with its vibrating pulp, with its tentacles thrown out and reaching even to the sidereal spheres.

Our microcosmic route of life points to macrocosmic infinity for its beginnings and its ending. The whole of our very own inner revolutions, and of our world, which makes up but a fraction of the entire cosmos and its "life," may only be conjectured from our perspective. This life of the cosmic sphere — or spheres — which from our microcosmic perspective appears only as the "outskirts" of our nucleus, the planet earth, belongs as such to the Great Plan of Life. It is the existential matrix of the planet earth with its extraordinary conditions accounting for the origin of life and its unfolding; but it stretches into infinitely vast and to us unknowable spheres that the cosmic logos projects for its own reasons, for its own sake. We may not conjecture from the revolutions of the stellar systems their inner reason. Likewise, in investigating the reason of life, of the life course, life's "origin" or ends, we may not project from the spiral path of its inner transformations its "wholeness." We may not — as the myths and sagas have brought to light — consider its beginnings otherwise than in relation to the ordering of the macrocosm. Yet the infinity of the macrocosmic revolutions forbids us to consider it as a "totality," to ask after its "entire cycle," or its *aion.*

On the other side of the panorama, we see the logos of life spiraling through ever new spheres from the vegetal to human sharing-in-life, and extending from the Dionysian logos of enjoyment to the intellective logos of the human mind and to the creative Apollonian logos, and then, on attaining its summit, receding and dismantling the accomplishments of these under the pressure of the innermost moves of the soul toward transcendence of it all, and then, with the advance of the transnatural current of the sacral spirit, vanishing from the horizon. Does the logos of life then vanish into oblivion? Or, does it unfold a new course, a novel logoic sphere in its "other" infinity?

Artistry and creative imagination cry out to occupy their proper place within the experimental and cognitive system of our mind that is entailed by life as such. Incapable of encompassing the logos of life, the projects of the logos as such, are we doomed to ever lack knowledge and reflection, any springboard for a vision of all? Did we not forget at the height of Apollonian speculation and creativity that the intellective logos and reflective intellect is *not* the only and not the principal logoic thrust reigning over the spheres of life? Did not the Greeks, who in the ancient world first discovered the concept of the cosmos, see that the cosmos not only means order but also creative skills, artistic skills?

Artists do not envision reality in the terms of the concepts of the Apollonian mind. They draw on the wealth of life enjoyment: feelings, moods, passions, foreboding, inward striving, yearning, love, and hatred, harmonizing them all in a creative vision. How could we — in following the modern obsession with translucid intellect as the only valid conveyor of knowledge — have forgotten Dionysian life enjoyment, the logos of fecundity and growth, of companionship, of the struggle for existence, of victory and defeat, of thirst after beauty, of the enjoyment of our senses, and of the stirrings and pulsation of our soul?

Have we not, throughout our itinerary through life's inner workings sensed at every step that the juices of life pulse through the radical leaps, disrupted links, intermediary realms and that a harmony pervades all, that when one element of life vanishes another seeks its place and assumes its role?

III. REAL LIFE AS THE PROTOTYPE OF THE ARTISTIC THEATER PLAY

The spectacle of the world of life with all its beings and concatenations of events, things employed in the furtherance of human progress, has been with good reason compared to the theater. The world itself is a stage upon which human beings with their intricate dynamic schemas of existence play the roles of protagonists. And real-life actors, once their role is completed, vanish from the scene. We are both players and spectators in these real-life dramas. But unlike in the theater play, which is scripted ahead of time by the playwright and conducted accordingly by the director, our real life drama is spontaneous, haphazardly planned, uncontrollable in its unfolding. Here it is the players in their very limited frame that have the responsibility of the stage director.

But all along the impervious striving of the logos of life lurks in the shifts of the struggling forces that subtend the enactment of the drama that human beings assume they carry out. As the intellective logos of the human mind — with its underpinnings in nature-life and its soaring toward the discovery of ever more complex, intricate patterns in the world — plays out and appropriates knowledge, allowing further inventive and creative advance, it ultimately soars toward the unbound and infinite. In this its plays its ultimate role.

Let us be carried by the heroic swing of the logos, its quest for the never graspable pleroma, for the unity that is never entirely brought to light, for the perduring — just a wish of the soul — and for harmony, even though the games of the logos of life, its crises, its defeats, its unexplainable disruptions, natural cataclysms, and human greed and recklessness, seem to purport its absence.

Let us then on our own turf conjure the vision of the All-in-becoming, one analogous to the Platonic and Neo-Platonic vision but proceeding from different assumptions, a vision of the gigantic spectacle put up by the Logos, in particular of the dance of the logos of life, the main performer in the logoic spheres.

Who are the dancers? Who leads them? Whose choreography do they perform? Is it anybody who beats a drum, or is all rhymed from within? Is this one gigantic performance, or the work of innumerable orchestrations? Who is the leading soloist?

Before we enter into a reflection drawing on the preceding concrete analyses/investigations, we have to legitimate our procedure within the classical phenomenological tradition. We find this legitimization in the unique role of creative imaging, which crowns the functioning of the devices of the logos of life.

IV. RETRIEVING THE FULL ROLE OF IMAGINATION IN EXPANDING OUR COGNITION-KNOWLEDGE

Recognizing the Archetypal Postulates of the Mind

We find in all spheres of the human mind reference to the transcendental logoic archetypes of imagination: its ideals of harmony and beauty, which are not concepts but are a vision of things, beings, and the cosmic spheres.

The creative imagination of the human mind comprehends several logoic spheres of its own. First of all, imagination as such is the fulgurating spark that the creative logos sends out to kindle the human mind, the human intellective logos per se. In its leaps creative imagination is guided by two archetypal models: the archetype of harmony, and that of beauty. The creative imagination is, as I have exfoliated elsewhere,[1] the conveyor of the aesthetic sense, one of the three essential virtualities of the human condition. At the furthest frontier of the aesthetic-imaginative swing, the ideal of harmony gathers the dispersed elements of the aesthetically significant picture, which picture conversely draws on the aesthetic sense as it shows forth the beauty within and above the pedestrian approach to life. It lifts all of life's elements to a higher level of experiential significance.

The archetype of beauty receives and arrests the swing of our natural valuation and its appreciative tendencies by imbuing them with an aesthetic ideal: delectation in color, sonority, rhythmic cadences, sublimity, charm, gracefulness. Harmony and beauty are intertwined, and both work in tandem with our human setup. They are the indispensable completions of the aesthetic sense, which, albeit to various degrees, pervades the entire network of the workings of the human mind. They lift our intellective-cognitive and aesthetically sober neutral attitude towards life and its course to a visionary height.

Within the setup of the human mind, imaginatio creatrix is a logoic device that a) lifts the presencing in the manifestation of life from its womb of operations; b) provides the variation of constitutive forms within the representational level of manifestation; c) in its guise of creative/aesthetic imagination serves as the source of the beautiful; d) as creative-reflective imagination, is the motor of the heroic turn of the creative/constructive logos, whose reflection engenders constructive work, art, poetry, scientific research, and the ontopoietic investigation reaching even to the cosmic relevancies of life; and above all, e) in its crowning role mirrors the fullness, the All.

This mirroring is a unique operation of the intellective, creative, and reflective logos; it not only brings the performance to the stage, but also gives it a mirror in the arena of spectators, who will seek to find themselves among the dancers.

Should imagination be mistrusted and pushed aside in the cognitive life enterprise of our entire beingness as it participates in the stream of life, in the whole work of the logoic manifestation of life within the world with its ever advancing currents of metamorphosis? Our vision of the beauty of nature, of the world, of the human inner life, of courage, of the heroism of the life struggle, of the heavens so far away and yet ingrained in our very bones, is the carrier of the fullness of our beingness-in-process. Without a vision of the beauty around and within us, the modern attempt to squeeze reality into the schemata of our intellect would make us "bloodless monsters."

The entire investigation we carried out above palpitates with the work of imaginative penetration into the concrete workings of nature-life, of the human sphere of the world and culture, with imagination postulating causes that can not be directly assessed but only inferred, throwing bridges out into murky realms, making Dionysian leaps into the otherwise impenetrable tracts.

In brief, it is through imagination's gathering and reflecting, as in a mirror, all of pulsating life that the ungraspable totality may be envisioned as a harmony orchestrated by the entire albeit infinite logoic project, a harmony of the entire possible spread of life and of the cosmic spheres. This artistic vision of the harmony and beauty pervading is now in retreat before the impact of intellectually concocted schemata. We witness in the fine arts an abandoning of the profound tie between art and the real in its fullness as well of the representational expression of the intuitively felt core of beauty. We see instead the filling out of abstract mental contexts and schemata. We have become blind to even works of art in their power to be received by us in our entire being with delight. Delectation of the imaginal content of beauty gives way to appreciation of the skill with which mental constructs are given flesh. The "viewer" is simply unaware of loosing thereby the essential core of his/her existential enjoyment.

V. THE UNIVERSAL HARMONY IN THE SPECTACLE

Our participation in the universal harmony, the imaginal harmony of the All, corresponds as much to the logoic ideal of the human mind — which seeks to encompass all — as it responds to the archetype of the creative imagination. It ultimately corresponds to the logoic Great Plan in progress, of which the human mind constitutes a microcosmic moment.

Science makes us aware of the harmony in the most intimate interplay and interchange that we maintain with the forces of the universe, harmony essential for our carrying on our life. This interplay maintains our subsistence. We most intimately partake in the energy and the matter of the universe of which we are a living segment. The atoms that constitute us are engaged in a constant process of exchange with the total reservoir of atoms in the universe. To quote the astronomer George V. Coyne: "Each year 98% of the atoms in our bodies are thus renewed. Each time we breathe we take in billions and billions of atoms recycled by the rest of breathing organisms during the past few weeks. Nothing in my genes was present a year ago. It is all new, regenerated from the available energy and matter in the universe. My skin is renewed each month and my liver each six weeks."[2] All beings in the universe are thus recycled.

Amid the revolutions of the heavens, which we envisage through our (mute) participation in them, in their rhythm and tacks, which pervade the very play of our own lives and the whole evolution of life with the succession of its vivid periods, we human beings are privileged by the life-enjoying Dionysian logos, which takes it all in from within, and by the Apollonian reflective logos, which with its creative imagination envisions it all. How are we to envisage this harmony that we realize and toward which we aspire if not as that of the grand theater of life upon the stage of the world?

This is a theater in which we are players along with all the dramatic troupes that appeared before us, the grand performance of the logos in a choral dance in which each living beingness has a part in common with others while the soloists stand out.

The beat of the drum, the rhythmic cadences of the moves, the choreography of the logos of generation, the logos of the immense cycles of the spheres (see the closing of Sec. 1 and Sec. 2 above) — to these we dance in our fated tragedies, in our victorious or failing struggles, in sufferings and in joys, in frustrations and accomplishments. The beat by which we measure each of our moves in order to remain in step with others in this choral performance resounds within. It follows the ever sounding music of the spheres, of our flesh, our soul, our mind. This dance performed in a great spectacle on the scene of life, of the world, is a project directed and performed by the logos in its fulguration. It is performed in order to be beheld. It calls for spectators, and for a Dionysian awakening of all its participants within the immense harmony of the logoic spheres. It is life itself with its innermost pulsation that carries the imaginal vision of the harmony of the All, one that is open-ended but having a fixed albeit movable fulcrum.

Upon this stage, which it sets for itself, the logos in the vertiginous sweeps of its dance brings in the spheres of life, with the great cosmic spheres being carried onwards as much as our individual roles. In this choreography each tiny step is measured with respect to the entire improvised plan of the performance. And the circuits of our individualizing roles carry the significance of our personal destinies, our intraworldly destinies, reaching to the infinite of the transnatural destiny of the soul as much as to the revolutions of the heavenly orbs.

How could we endure our worldly existence, how could we play our role in the choral arrangement, how could we keep the rhythm of the dance, in its struggles with uncertain results, its tragedies, its upheavals, the suffering pervading our earthly doom, the menacing abyss ever threatening to open under our feet, the cataclysms of mother earth — our foothold in beingness — were we not attuned to the entire immense performance of the logos in the harmony of its spheres, did we not envision our existential revolutions as an integral part of the grand dance and count time with the overall spectacle played on the stage of the world of life so that in our own choreography the logos of life brings the revolutions of the heavenly bodies onto the floor? How could we carry on our tortuous patterns often in torment, anguish, and isolation, missing the turns, forgetting the steps, losing the reassuring grip on the choral chain had we not carried in the bosom of our soul the reassuring, resounding beat of the harmony of the spheres, the harmony of the All? Does not the contemporary scientist be it in biology, genetics, physics, astrophysics, in marveling at the hidden joints of the numerical, mathematical structuring of forces seemingly separate but in correspondence with the game of life of the cosmos — finding in fact sooner or later further corresponding segments, fragments of a great play — and at the underlying harmony of their manifestations caught in the cosmic revolutions, nature-life evolutions, sense, as Albert Einstein did in 1929, that "Human beings, vegetables, cosmic dust, we all dance to a mysterious tune, intoned in the distance by an invisible player"?[3] In marveling will the scientist not hear with his/her artistic imagination the music of the spheres that the Pythagoreans and Newton after them heard in the harmony of the All?

At this point it is appropriate to emphasize that the metaphoric vision of the "cosmic" dance that was shared in Antiquity from Plato through Plotinus receives within this new context that posits life on earth to be the fulcrum, the measure of things, a different interpretation. The dance of the logos involves all the logoic spheres, the great universe included, but it is performed upon the stage of the world of life and not upon that of the universe because the logoic spheres of the cosmos cannot in our perspective be encompassed; they have no limits. They can only be inferentially postulated with respect to our own earthly life standards.

VI. THE DANCE OF THE SPHERES

Our reflective, Apollonian intellect, in probing ever further into causes, origins, teloi, falls short of grasping the total choreography that the logos projects in the dance of its spheres, a choreography that has the dance expanding into infinity. By its music, ever sending out new waves in the ever renewed course of the sidereal and earthly spheres, the logos sets into our mind — as a mirror of coherence — the ideal of the harmonious and ever varying choral dance in the gigantic spectacle of the All that intones the music of the spheres. That ideal resides in the logoic setup-device of the creative imaginal imagination as an archetypal harmony that gathers all in an imaginal spectacle to be seen and to resound within us.

The logoic device of an archetypal universal harmony dominates our creative imagination and finds correspondence in the imaginal spectacle of the logoic macrocosmic and earthly dance, in its choreography, its rhythms, its steps, and its accompanying music, which sounds within each living being but is reflectively mirrored in the visualizing vision.

This choreography is, as I have pointed out in a different key, differentiated according to the nature of the logoic spheres, beginning with the vertiginous generative labor that establishes life on earth as a transformative sphere gathering the threads of all the cosmic spheres of the logos as well as its fragmentary peculiarity on planet earth. This choreography of the generative poietic logos of life in the wild onward rush of the forces harnessed for life's unfurling, on the one side, and in the struggle for life of the human creative endeavor and its self-perfecting strivings, on the other, is immersed in the conspiring astral motions — some linear, going from east to west, some rotary, still others perpendicular as in meteoric "falls", and others of an incalculable inner nature, in their leaps forward and retreats backward — within which spheres the tiny motions of our own existence are enveloped. All are numberless partners in the dance, in the innumerable circuits orchestrated into the coherence of the universal logos, which corresponds to an unquenchable thirst in our soul for an "image of it All," an "image" within which we can situate our destiny and thus give it a hold on stability in the becoming of our limitless course, the restless rush of all, as well as a promise, even if one infinitely delayed in its realization, of a full explanation providing the sense of it all.

Yet the human being as creator, as I mentioned above, has become deaf to the beat of the macrocosmic drum, to the beat of the logos. Transported by the Apollonian inspiration to penetrate the inner workings of nature and made drunk by the powers gained by bringing some of it under his control by taking extraordinary turns in our choreography, turns invented by our human creative logos, we go overboard. In this age humanity is intoxicated with the powers, skills, concepts, formulas, calculations, speculations, expectations *ad infinitum,* overwhelmed by the thrilling pace of technological progress.

We stand confused, being at the still unripe, intermediary stage of interpreting the hectic speed of our individual lives within the social sharing of existence, artificially accelerated by the improvement of our practical lives through technology. Such is technology's further promise that we have become blind to the beauty of the All, insensitive to the delicate moves of our own being, and deaf to the inward beating and pulsing of our blood. We have became deaf to the connatural rhyming of our steps with those of the other dancers on the stage, the steps of those who occupied it before and have left and of those who will follow us in the incessant turnover of plays.

We are now lost amid the intrigues, pursuits, tragedies/comedies, and exuberant joys that manifest in the actual pulp of existence the vertiginous dancing proficiencies of the logos. Within the narrow corridor of the Apollonian logos, we have indeed become deaf to the tunes of the grand dance, to the choral music of the logoic spheres. May the discovery of the ontopoietic underpinnings of life, existence, reality, now shake the artifices of philosophical concepts, and ideas accumulated through millennia and liberate and clear our pristine genuine feeling, sight, sensibilities, and intuitions in order to retrieve the rhythm, the cadences, the moves of our originary partaking of the great choral dance.

VII. THE IMAGINAL RESPONSE TO THE POSTULATES OF THE CONSTITUTIVE MIND: HYPOSTASIS

We have reached now the point of rounding up our ontopoietic vision. Using the metaphor of the "dance of the spheres," and seeing our innermost participation as being its fulcrum, we have grounded our vision of the All foremost within the workings of the logos of life as they spread into the relevancies of life in the cosmos, and into the further specifically human, creative spheres. We have focused on direct "poietic" logoic endeavors and strategies, of which only a skeleton could be given in this pioneering effort. We know how significant the last touches of a brush are in completing a painting, here a picture of reality. The vision that we have attempted to conjure would not be complete unless we briefly enhanced and framed it with a succinct treatment of some of the devices of the reflective logos that in real life express transcendentally and cognitively the synthesized, simplified workings of its inner arteries, furthering the smooth unfurling of human reason in life practice.

We began our meta-ontopoietic reflection by pointing out the cognitive quest of the transcendental logos, of the logos of our mind set up in its unique way, for origins and endings, which quest signifies a thirst for unity, fullness, totality, harmony. Having clarified that the quest of the reflective mind for harmony finds a response in the way in which the creative imagination manifests a special logoic device in its archetypal ideals of harmony and beauty, we may now turn to the already enumerated logoic quests to grasp reality, among which the probes into origins and endings has already been somewhat outlined.

It is our contention that the answers that the logoic imagination offers to these puzzling queries of the transcendental cognitive mind are to be looked on as marginal and not yet significant workings of the cognitive apparatus. That is to say, the prompting of a query after origins, endings, unity appears to stem from imaginal reflection on the generation, unfolding, decline, and disappearance from the scene of the dancing troupes seen together as wild and yet rhythmically punctuated unfurling of the logos of life. At the peak of its prerogatives, an archetype of harmony informs the creative imagination, which in response to the logos' quest places a magic — artistic — mirror before that quest, before its grand evolution; even so, the transcendental logos itself sets up postulates occurring within the range of its archetypal reservoir and this in response to the needs of the cognitive-constitutive transcendental system of the human mind. In its striving to offer to the mind the most complete grasp of reality, the logos in its transcendental strategies sets up transcendental postulates to be tested.

In fact, the impetus of the transcendental constitutive logos entails discrimination, selection, error, its correction, etc. That is to say, it entails a questioning streak running through the entire system of the constitutive bulwark. That amounts to saying that in its constitutive impetus the logos of the mind alertly throws a searchlight on and questions the givenness that is indispensable to carrying on its work. Where does this questioning impetus stop? The questioning impetus of the logos runs through the minutest moves of the generative and unfolding poiesis of individualization, through the spheres of sharing-in-life, up to the reflective sphere of the transcendental mind, differentiating its attention in accord with its objective, with differentiated claims to a reaction, a response. It is at the height of this transcendental reflection on its own work of manifesting reality that the questioning logos installs the crucial measure of equipoise: it postulates a completion of its search in an overarching of the particular, the ever moving, endless, and ungraspable within the mirror of its reflective constitutive molds. *It is again the creative imagination that supplies this equipoise and a period for the logoic drive by projecting hypostatic visions of unity, of the fullness of becoming itself, and of the two innermost prime devices of the logos: timing and spacing.*

In fact creative imagination here mediates between the representational and the visionary by bringing in intuitions that transcend the evidences of manifestation and of the ontopoietic reach, creating images and figures in its own imaginal mode. *The totality or unity of becoming and being, of the fleeting processions of generations can never be grasped.* Attempting to do so, *ancient philosophers sought to organize these processions in cycles, in periods,* in "full cycles." The fullness of becoming marked an *aion*. It should have come clearly to light in our analysis that the course of the logos in concrete does not justify the distinction of such significant periods. *These distinctions come in the imagination's response to the postulations of the reflective-constitutive human mind. No concrete reality corresponds to them. They are logoic devices for the final rounding up of our grasp of reality, in a word, hypostases.*

We may venture to say that the hypostasis of unity, fullness, origins and endings may be reduced to the main hypostasis of time and space, with which we will deal henceforth.

VIII. THE IMAGINAL AND THE HYPOSTASIS OF TIME AND SPACE: MOVEMENT AS THE PRIME DEVICE OF THE LOGOIC IMPETUS

Throughout the entire course of this investigation, I have attempted to show how the notion of time, which has had such an important place in Occidental philosophy, is most inherent to the becoming of life in all its spheres. It should be obvious now that life "times" itself in its steps, phases, the unfurling of its virtualities, its struggles to maintain a course, as well as in the specifically human sharing-in-life, in the interactions of individuals, in our striving for accomplishment, and in our seeking a meaning for existence. We have distinguished the chronos of becoming — the unfurling of the ontopoietic route — and the kairos of accomplishment proper to the human logos.

This timing of life is matched by the structure of the specifically human experience of the inner timing of our transcendental consciousness. This is the Bergsonian duration carried on in the inner sphere of our psychic and conscious pulsation, emotions, feelings, whether these single themselves out, presence themselves qualitatively, and so come into our lucid awareness, or not; these tacitly form a thick and opaque flesh of the psychic realm nurtured incessantly by the vital/psychic energies of life's individualizing onward progress, giving us the unique experience of substantial selfhood-in-continuity. It is owing to this experience of our duration that we hypostasize the timing of life, which is life's essential device, into a substantialized *quasi*-entity of time.

Aristotle claimed that measuring time is a prerogative of the human being. Life times itself in various ways in a vast gamut of proportions, successions, coalescences, fusions, filterings, etc. of the "outward" and "inward." It is only in the human being that these various ways, that is, the organic, vital, social, but also psychic and intellectual, creative and spiritual — oftentimes fused together — comes to reflective, intellectual intelligibility.

Only by grasping these proportions and relationships among life's stages and segments of becoming may we raise the question of the nature of timing itself. This question is overlooked, hence the universal acceptance of the hypostasis.

What has been universally overlooked and what the revealing of the hidden, inner working ontopoietic process *in concreto* has brought to light is that life's innermost generative and metamorphic processes consist in measuring/apportioning. The timing of the processes assumes the choreographic precepts of the logos of life. Within the human sphere, it is the transcendental logoic mind that allows the human being to measure and apportion his/her vital, societal, and creative endeavors.

Becoming, proceeding ever onward maintains the continuity of the design and its ontopoietic actualization (crystalization). It is owing to this advancing progress or regress in the constructive course of life's becoming that we surmise the "'continuity' or 'discontinuity' of time." Seen in the light of the manifestation of life, the relative lasting "for a while" of the generative or developmental phases or events, their relative albeit short perdurance with relation to other events and phases, appears as a temporal continuity, as the duration of a reality, a duration that parallels our conscious experience of duration. Yet, this continuity is merely the imaginative completion of our sensing-feeling-constitutive acts of life.

What is the measuring and timing of the stellar revolutions? Otherwise, are there not different measures and proportions specific to each galaxy according to its spacing with respect to the others? Since our own timing devices depend on the conditions of our galaxy, should we not assume that there is another system of proportions, measures, etc. for each galaxy and planet? That implies the diversification of "time," if we want to hypostasize it.

Thus, contrary to the whole of philosophical tradition, "time" as seen within in our horizons appears to be the main vehicle of becoming. It is nothing "in" itself, but reveals itself nevertheless to be the main logoic device for becoming and beingness.

In the direction of revalidating time's hypostatic significance, we will acknowledge that it is by harnessing it through the instruments of measure and proportion that the universal logos pursues the plans of its grand manifestation.

Doing that requires that the entire logoic bulwark so far established in its manifestation maintain the stage and the play, just as the Ancients surmised. It may require as much force to maintain the performance, as it took to set it up and bring it to enactment. However, we would dispute such an assumption on the ground of our ontopoietics of becoming. Each and every logoic move, in their myriads, that sustains and carries the progress of becoming appears as a logoic "advent." It appears as if the logos of life be in incessant "enactment." The advent of its singular moves outlines a pretertemporal present in as much as each and every one of them partakes of the ontopoietic unfurling in its becoming, its manifesting itself. *Each of the logoic moves might be an act in the play, but not in the sense of being an advent of a beingness' manifestation.* All of the logoic generative and developmental as well as transformative moves in their singular and limited "actuality" partake of the culminating point of this unfurling, which is the actual advent of the beingness they carry. This advent means the fullness of the ontopoietic disclosure of the logoic project as actualized in the particular self-individualizing process: the advent of a singular, unprecedented, and self-identical beingness. This advent needs to be carried by the logoic acts of the myriads of logoic sparks through which the logos maintains itself in constant alertness, never resting, never slacking: a nameless, perfect actuality *before* a name, a concept, an idea, in the mode of a constant improvisation. This improvisation is the specific of the logoic poietic drive.

The situation with spacing seems to be different. Space seems to our transcendental experience and reflection to be "out there," ever present and ever in readiness to receive the constructive or existential moves and motivations of the logos — so much so that Plato considered time to be an image of eternity, but space itself to be eternal. Eternal, that means pregiven before any moves of life.

And yet we observe the variability of space in relation to motion; distances measured by the rod, the objective standard, shrink enormously or expand according to the spread of the motions that traverse them. Furthermore, we have seen how spacing with its intrinsic directives, in a way comparable to that found with timing, emerges and unfolds with the generative and evolutionary progress of the self-individualization of life. Moreover, we ponder how different spatial relations seem to be according to the strength of the pull of gravitation. What may we surmise about the incommensurable distances and regulation of motion in the sphere of the galaxies? Should we hypostasize the experience of spacing as we do that of timing? Doing both would call for a novel approach that takes its lead not from the human experience of the cosmos, but rather from the so far ungraspable constructive system of the logos between the two infinities.

The dynamics of timing and spacing are indeed the basic axes of the choreography of the cosmos and life. They are the axes of its unifying system, articulations, coordination, networks of causes, of the immense transformative systems of forces, energies, of the intermediary turmoil and fulguration.

The logoic dance of the spheres on the stage of the world of life is the artistic vision of the fulgurating progress of the logos that brings harmony to its orbit.

From this perspective of the hypostasized spacing/timing, we ask, "Did this dance have a beginning?" The evolutionary phases of life on earth prompt the question. One thing seems certain, namely, that, as some scholastic philosophers thought, it takes as much to maintain the cosmic dance in its course as it did to initiate it. Does this mean that the cosmos and the world of life are in a perpetual coming to be, in a perpetual advent?

Should we speculate about the measures taken in the preparation of the stage of the world so that the spectacle of the logos — and the cosmic logos and the logos of life — could appear on it?

Timaeus stated that the muses are silent on these matters and they remain silent. Still, is there anything that could be revealed?

IX. THE UNCONDITIONED AND THE UNTIED

What, however, is the status of the universal logos, the choreographer, the driving force, the multifaceted colorful performer with ever new invigorating dynamisms, the incalculably entwined soloist and simultaneous spectator that holds a mirror up before our human eyes, a mirror reflecting this bedazzling and never graspable play?

The logos does not imitate; it carries its models with itself, it ever anew acts the *choragus*; it improvises each figure in the dance, and its rhythm and tempo; it is the leader of the dance and the drummer.

Did it originate before the dance? But before and after, beginning and end are temporal notions drawn from the evolving nature of life. They cannot apply outside of life's course. Does the master of the dance have cause to bring this spectacle forth? Causes are ties that condition. An agent carrying its laws within him cannot be tied to any cause. An agent who brings forth its own arsenal of dynamisms and fulgurating forces cannot be held captive by any condition.

Is the logos then eternal? But with eternity we associate immobility and unchangeability, while the logos appears to be infinitely transformable and to maintain merely the nucleus of directive constructive forces.

With these questions we have reached the limit of our transcendental, that is, human logos of interrogation. We remain encircled by the spectacle and its mirror, participating in both, and in infinite perspectives.

This meta-ontopoietic marveling at the enigmas of the Logos, answer to All, might call for a still different key in which to unravel its mysteries.

NOTES

INTRODUCTION

[1] Anna-Teresa Tymieniecka, "The First Principles of the Metaphysics of Life. Charting the Human Condition: Man's Creative Act and the Origin of Rationalities," in Anna-Teresa Tymieniecka (ed.), *Phenomenology of Man and of the Human Condition,* Part II: *The Meeting Point between Occidental and Oriental Philosophies,* Analecta Husserliana 21 (Dordrecht: Kluwer Academic Publishers, 1986), pp. 3-73.

FOREGROUND

[1] C. S. Peirce, "Evolutionary Love," in Morris R. Cohen (ed.), *Chance, Love, and Logic: Philosophical Papers, by the Late Charles S. Peirce* (London: Kegan Paul, Trench, Trubner & Co., 1923).
[2] Cornelis A. van Peursen, "Creativity and the Method of the Sciences: A Problematic Issue in Husserl's Phenomenology," in Anna-Teresa Tymieniecka (ed.), *The Phenomenology of Man and of the Human Condition,* Part I: *Plotting the Territory for Interdisciplinary Communication,* Analecta Husserliana 14 (Dordrecht: D. Reidel, 1983), pp. 375-385.

PART I, CHAPTER 2

[1] Anna-Teresa Tymieniecka, "The Moral Sense: A Discourse on the Phenomenological Foundation of the Social World and of Ethics" in Anna-Teresa Tymieniecka (ed.), *Foundations of Morality, Human Rights, and the Human Sciences,* Analecta Husserliana 15 (Dordrecht: D. Reidel, 1983), pp. 3-78.
[2] Ibid.
[3] Anna-Teresa Tymieniecka, *Logos and Life,* Book 1: *Creative Experience and the Critique of Reason,* Analecta Husserliana 24 (Dordrecht: Kluwer Academic Publishers, 1988).
[4] Husserlian ontologico-intentional assumptions do not allow for Ingarden's ontological conception of process, which does not reveal any of these essential features. Besides, it is obscured by his distinction of two sides in the "essence" of the process: the aggregate of its phases and the process-object. No doubt there is a series of phases in every process, but they cannot be seen as an adjunct aggregate; on the contrary, they install what Husserl called the living present, that is, each of the phases inaugurates the next while it passes. Neither is there in any objective form that can be

abstractly seen in the unfolding of the process through all its phases and that can be called its essence. It is in the process' intrinsic operational constructivism, its specific line of step by step advance that binds the sequence of operational phases together as it delineates the constructive profile of its efficiency. See Roman Ingarden, *Time and Modes of Being*, American Lecture Series, ed. Marvin Farber (Springfield, Ill.: Charles Thomas Publ., 1964), p. 123.

PART I, CHAPTER 3

1 By "regeneration" is meant not only an organism's ability to repair a damaged organ but a reserve capacity of the organism to replace an organ that has been destroyed. This involves a radical constructive reaction of the entire organism to a dysfunctional or missing part. Hans Driesch describes "embryonic restitution," in which an amputated portion of an embryo is replaced: "Es gibt auch eine embryonale *Restitution,...*" See Hans Driesch, "Das Wesen des Organismus," in Hans Driesch and Heinz Woltereck (eds.), *Das Lebensproblem, im Lichte der Modernen Forschungen* (Leipzig: Verlag von Quelle und Meyer, 1931), pp. 394-418.

2 The great question of the macrocosmic organization of life has preoccupied the natural sciences. There are at least two approaches to this organization. One is the scalar approach exfoliated by Stanley N. Salthe in his *Development and Evolution — Complexity and Change in Biology* (Cambridge, Mass.: The MIT Press, 1993), pp. 35-52. The other is the hierarchical conception of the species found already in Plato's *Republic* — the history of which was covered by J. van der Meer in an unpublished 1989 dissertation cited by Salthe — which approach differentiates species according to the features they exhibit above those common to all species. Aristotle with his concept of epigenesis already outlined the developmental approach. It has been taken by William Harvey, who stressed that the developmental stages proceed in, what I call, a "translacing order." It has been corroborated by concepts such as that of the "chain of being" (see Arthur Oncken Lovejoy, *The Great Chain of Being: A Study of the History of an Idea* [Cambridge, Mass.: Harvard University Press, 1933]), levels of organization (see L. R. Aronson, "Levels of Integration and Organization: a reevaluation of the evolutionary scale" in Gary Greenberg and Ethel Tobach [eds.], *Behavioral Evolution and Integrative Levels* [Hillsdale, N.J.: Erlbaum, 1988]; and in the same collection Stanley N. Salthe, "Notes toward a Formal History of the Levels Concept"). The attempt has been made, particularly by Bonnet to establish that this chain is a seamless web. Lamarck was aware of the impossibility of establishing such a continuum; his was the radical concept of "discursive formation." All of these led to the concept of an unfolding speciation in the ontogenetic process, a progression from fewer, more generally shared specifics to more particular sets of features.

3 Driesch, "Das Wesen des Organismus," op. cit., p. 418.

4 I take particular satisfaction in seeing that my conception of the self-individualization of the ontopoietic constructive unfolding of life that I have been voicing since my *Why Is There Something Rather Than Nothing?* (Assen: Royal Van Gorcum 1966) has found varied corroboration in the work of natural scientists, of biologists in particular. One book that gives more substantial analytic details than previous ones is *Development and Evolution — Complexity and Change in Biology* by Stanley N. Salthe (Cambridge, Mass.: The MIT Press, 1993). Salthe places stress similar to ours on individuation in the life process. He writes that the fundamental process of development is self-organization, which is one of my leading ideas. He takes the individuation of life as the key to understanding "construction in development and evolution." Salthe emphasizes "treatment of change as the result of individuation" — which harks back to the Romantics such as Coleridge. This idea is perhaps the most unusual feature of my own approach. "Central to this approach is a distinction between development and individuation" (p. 29). He writes further in a summary pre-

sentation of his book: "I have undertaken to introduce a way of thinking about nature (and some basic tools for that) quite different from Baconian/Cartesian/Newtonian/Darwinian/Comtean (BCNDC) science. I will be concerned with the self-organization of complex systems...." He enumerates several other, specifically biologico-scientific clues to substantiate his project. Further, he emphasizes that "the process of individuation as such seems not only to have been examined in BCNDC science in structuralism; indeed, perhaps only dialectical discourses have been concerned with it at all" (p. 34).

We can say that Salthe names evolution "individuation." He considers its mechanism to be "an ineliminable margin of error." He also emphasizes that organic evolution/individuation is the "general theoretical grounding of biology" (p. 140). In contrast, we distinguish several essential streaks of self-individualization — its intrinsic "organicity," that is, its tightly coordinated functional articulation of operations and processes in progress and its participation in a conundrum of the environmental forces while unfolding its very own individual beingness, on the one hand, and the course of life, on the other.

It is only in terms of these factors of the individualizing process that we distinguish the seminal and generic factors that offer the basis for the individual's reaching over into the evolutionary process of the species. He rightly identifies organic change with development that can be seen as a "trajectory" or an "entity." But according to Salthe evolution cannot be envisaged as a "constructive" process; it "is not any thing at all, it is not a process of any kind" (p. 147). As we agree with the first statement about development, we might also agree with the other. However, what we emphasize is that evolution is not merely, as Salthe says, "an accumulation of informational imprints generated in a system upon recovery from contingent collisions and environmental fluctuations / or as a result of internal excursions, such as mutations/scars which mark its experience in the world" (p. 148) that are transmitted automatically from one individuation to another. Evolution is in itself the crucial strategy of self-individuation as such. Namely, it belongs to its essential ontopoietic generative function to account simultaneously for the life maintainence of the individual and the reproduction of the type.

This entire inquiry into development and evolution is carried out in scientific terms around the clue concept of autoindividualization, or of the ontogenesis that is substantiated by it.

This is the best example of the interaction of philosophy and scientific research in projecting an inquiry. My treatise *Logos and Life*: Book 1 (1988), Book 2 (1988), and Book 3 (1989), and Book 4 (the present volume) marks the steps of my unfolding thought. It all began with a graduate course that I gave at Duquesne University in 1967 surveying "The Origin of Forms in Nature," which was actually a second step in my pursuit since it followed my writing *Why Is There Something Rather Than Nothing* (Assen: Royal Van Gorcum, 1966) as well as *Leibniz' Cosmological Synthesis* (Assen: Royal Van Gorcum, 1964), a book in which for my own satisfaction I compared my ideas with those of Leibniz.

In the Duquesne course I followed the common origin of philosophical and scientific inquiry. Readers may see for themselves that just as much as philosophical intuitions seek corroboration and substantiation in scientific research, the scientists need philosophical intuitive insights as a clue for their projects as well as for the articulation of their results.

In this present study I elaborate much further my key conception of the self-individualization of life as life's promoting and ordering factor, of the ontopoietic devices and principles as well, and I hope that in it there might be further inspiration for science.

5 Considering the most intimate functional integration of the life generating, processing/nourishing, and advancing principles, which come together ultimately in the ontopoietic "entelechial" principle and "design" as we have called it (see *Why Is There Something Rather Than Nothing*), we find strong support from the side of the medical philosophy of Erich Kurt Ledermann, M.D. as set

out in his *Philosophy of Medicine* (London: Tavistock Publications, J. P. Lippincott Co., 1970); Ledermann with reference to Driesch's assuming the Aristotelian notion of "entelechy" elaborates the need for a "holistic principle," some such being indispensable for the understanding of the functions of the body and for the practice of medicine (see pp. 17-34).

Ledermann discusses the difference between living organisms and non-life first by stating that "living organisms are composed of the same molecules which are the elements of inert matter." In both there occur chemical processes (e.g., oxidation plays a vital part in both). "The secrets of life can be unraveled by the isolation of material chemical substances. An example is the discovery of the carrier of hereditary characteristics, the gene: deoxyribonucleic acid (DNA)...." If we accept this fact as essential, we are prone to hold a mechanistic-materialistic approach. And yet, argues Ledermann, more is needed to distinguish the living and the non-living — "...the former have a marked tendency to maintain themselves as wholes by means of *regulatory processes* (emphasis added), whereas the latter lack such a tendency" (p. 4). Ledermann calls a theological explanation necessary for the "maintenance of the wholeness," a "holistic principle" (p. 4). He emphasizes that the holistic principle calls for a "theological" explanation, which mechanistic materialism considers a different emphasis and perspective.

6 See Heraclitus, Fragments 11, 8, 51, 126 (Diels numbering).

TYING POINT 1

1 But according to the metaphysical macrocosmic set up of Plotinus, within which the issues of the origin of the cosmos/world are viewed, this stirring does not come from the "material" but needs an "active principle," that is, comes from the World-Soul, which itself is eternal but which in order to bring the cosmos into being steps out from eternity and assumes the forms of change, motion, temporality. We have to ask just what is the "active principle" within the microcosmic perspective of our investigation. To a large degree this energizing of the ready-in-waiting material, of certain so far inactive logoic synergies accounts for a preparatory degree of activation that will — when an appropriate level of complexity is reached in the matrix in preparation for individualization and when the appropriate concomitant conditions are present — trigger the "first stirring" by itself. By the "first stirring" I mean here the incipient phase of the individualizing process.

2 Plato, *Timaeus and Critias*, trans. Desmond Lee (Harmondsworth: Penguin Books, 1977). Timaeus identifies the *chora*, receptacle, as having space, which he considers eternal and indestructible; as such the receptacle stands apart from the generated, changeable, visible reality that it brings forth. The receptacle also stands apart from the eternal, timeless models that it uses for the modeling of changeable realities. The time that is attributed to the receptacle in this context also has to be formless and undetermined, as well as completely "plastic," to receive all the forms that it is supposed to reproduce. Within the context of our constructive logos, timing and spacing are "connatural" and mutually determining partners.

3 See Sec. VIII of the Closure of the present volume, which is devoted to time and space. On ontopoietic grounds, we cannot on the basis of either the cyclical rotation of the seasons or the cycle of the individual's life envisage a specific mode of "eternal return" peculiar to time "itself." As we concluded above, in Sec. 5 of this chapter, there cannot be in the cycle of life's transformation any return of "the Same." This is as evident as the singularity of living beings. As Leibniz said, and we would agree, no two leaves on a birch are identical. What we witness is all only variation on a similar theme — and even not the "same" theme — a variation that can never be a "return." As Lucretius said, this variation is an effect of the generative regulations of life itself, which in its transformational rotation of cycles always mixes the "same" with the "different" obtaining a novel

shape. Ever advancing — even while retracing its steps, often — in its cyclic unfolding of individuals, life just moves on in its transformational course in ever changeable conditions. Nothing ever returns the same, that is, nothing returns.

We must share Aristotle's view, which measures events "in temporal terms." The "measure of time" is the prerogative of the human being. In fact, measuring is the inherent skill of the logos of life at all its levels, yet it is the human being alone that is intellectively expressive. The human mind operates around the axis of measure. Hence we seek to establish special occasions on which certain meaningful events are "commemorated" at certain moments in the web of time that we project through the course of human history. Yet if we talk about "centuries" of culture, about anniversaries and millennia, that is our human way of organizing life. There is no way to talk about the nature of time as such; no "fullness" of time, no *aion* other than what the human conception of history considers significant moments.

PART II, CHAPTER 1

1 The celebrated notion of "chiasm" proposed by Merleau-Ponty denotes neither a special temporal nor a spatial "distance" between the acts being processed in interwoven sequences, but merely the "invisibility" to the observer's eye of the mute processes of life. There is no chiasm between them unless they are disrupted.

PART II, CHAPTER 2

1 There are in biological research numerous indications of the necessity of admitting the agency as well as selfhood of the individualizing life. In the first place it is unthinkable to deny the identity of a living being despite the continuous replacement of the all the material forces that comprise and sustain it. And then besides this maintenance there is the line of its informed unfolding as it grows. A biologist like Robert E. Ulanowicz (*Growth and Development: Ecosystems Phenomenology* [New York: Springer-Verlag, 1986]) sees this "agent" as exhibiting "integrity in ordering" (or self-interpretation-in-existence, as I call it), which means the sequential unfolding of development. The "agent" is more and more visible with the expanding functions of development. It is directly related to the "autonomy" of the developmental entitity.

2 See Richard B. Carter, *Descartes' Medical Philosophy* (Baltimore: The Johns Hopkins University Press, 1983), pp.108-109.

3 My intuition of the body is kindred to that of Chinese medicine, though there are some divergences. See Tu Wei-Ming's presentation of Confucian views on the body, or as I would say on the "embodiment of the logoic game of forces." He points out first that body is conceived as a modality of *qi* in accord with the view that sees the "lived experience" of the body as "a flowing stream constantly responsive to the vicissitudes of the changing environment." *Qi* is a "self-generative internal dynamism," and the body that it animates is "an open system, encountering, discovering, collecting, and enduring." In this perspective Chinese medicine refers to the "original vital energy [*Yuangi*] of the person." Since Chinese medicine considers the body to be a "gift" coming simultaneously from one's parents and from Heaven, the particular qualitative and quantitative endowment of the individual human body is codetermined by both sources. This body is in the process of self-realization in becoming. It belongs to the cosmic vision, a vision of the "great transformation"

(*dasha*) that "encompasses all modalities of being in the universe." See Tu Wei-Ming, "A Confucian Perspective on Embodiment," in Drew Leder (ed.), *The Body in Medical Thought and Practice* (Dordrecht: Kluwer Academic Publishers, 1993), p. 91.
[4]See Part I, Chapter 2, "Life and Non-Life," *supra.*

PART II, CHAPTER 4

[1] Abandoning the structural conception of the body that sees it in terms of "dimensions" for one that that emphasizes the functional coordination of life processes and their *constructive* character, we bypass an otherwise ingenious distinction made by Drew Leder in his insightful book *The Absent Body?* (Chicago: University of Chicago Press, 1990). Leder emphasizes the mute functioning of the body. But this perspective is misleading for it seems to draw an artificial line precisely at the most strategic point of the intimate cooperation between the visceral, inner, functions and those functions by which they sustain themselves on the "surface," that is, the sensorial functions that bring the body into immediate contact with the surrounding world.
[2] See Part I, Chapter 2, "Life and Non-Life," *supra.*

PART II, CHAPTER 5

[1] Autoeroticism occurs in situations in which intentions directed at others do not for some or other reason meet a response, a situation of isolating frustration, solitude, and abandonment. On the rebound the flesh prompts acts supplying relative fulfillment of carnal desire.

PART III, CHAPTER 1

[1] See Hedwig Conrad-Martius, *Die Seele der Pflanze,* Bücher der neuen Biologie und Anthropologie, Bd. 11 (Breslau: Franke, 1934).

PART III, CHAPTER 2

[1] See Anna-Teresa Tymieniecka, "Life's Primogenital Timing," Inaugural Lecture in Anna-Teresa Tymieniecka (ed.), *Life. Phenomenology as the Starting Point of Philosophy,* Analecta Husserliana 50 (Dordrecht: Kluwer Academic Publishers, 1997), pp. 3-22.
[2] *Ibid.*
[3] See Anna-Teresa Tymieniecka, "Metaphysics of Manifestation: Reason in the Individualization of Life, Sociability, Culture," in Anna-Teresa Tymieniecka (ed.), *Reason, Life, Culture,* Part I: *Phenomenology in the Baltics,* Analecta Husserliana 39 (Dordrecht: Kluwer Academic Publishers, 1993), pp. 3-10.
[4] See Jacques Monod, *Le hasard et la necessité, Essai sur la philosophie naturelle de la biologie moderne* (Paris: Editions du Seul, 1970).
[5] See Anna-Teresa Tymieniecka, "The First Principles of the Phenomenology of Life," in Anna-Teresa Tymieniecka (ed.), *Phenomenology of Man and of the Human Condition,* Part II: *The Meet-*

ing Point of Occidental and Oriental Philosophies, Analecta Husserliana 21 (Dordrecht: Kluwer Academic Publishers, 1986).

[6] See Anna-Teresa Tymieniecka, *Why Is There Something Rather Than Nothing? Prolegomena to the Phenomenology of Cosmic Creation* (Assen, The Netherlands: Royal van Gorcum, 1968).

[7] Quoted from Rudolf Virchow, "Atom and Individual," in *Disease, Life and Man. Selected Essays,* trans. Lelland J. Rather (Palo Alto: Stanford University Press, 1958), p. 39.

[8] Ibid., p. 124.

[9] Ibid., p. 129.

[10] See Roberto Canullo, "Plants and the Problem of the Individual," in Anna-Teresa Tymieniecka (ed.), *Life: Differentiation and Harmony—Vegetal, Animal, Human,* Analecta Husserliana 57 (Dordrecht: Kluwer Academic Publishers, 1998), pp. 39-46.

[11] Ibid.

[12] See Francesco Moiso, "De Candolle et Goethe" in S &TP, II series, 1996.

[13] Hedwig Conrad-Martius, *Die Seele der Pflanze,* Biicher der neuen Biologie und Anthropologie, Bd. 11 (Breslau: Franke, 1934).

[14] See Kathleen A. Akins, "A Bat without Qualities?," in Marc Bekoff and Dale Jamieson (eds.), *Readings in Animal Cognition* (Cambridge, MA: The MIT Press, 1996), pp. 346-358.

[15] See ibid. for a discussion of bat behavior.

[16] See Orazio Ciancio and Susanna Nocentini, "The New Silviculture: Epistemological Considerations," in Anna-Teresa Tymieniecka (ed.), *Life: Scientific Philosophy/Phenomenology of Life and of the Sciences of Life,* Analecta Husserliana 59 (Dordrecht: Kluwer Academic Publishers, 1999), pp. 343-357. The "autonomy" of the forest is built upon all of its vegetation. The forest is conceived of as an "autopoietic system," that is, as a system that "can subordinate structural changes to the conservation of its own organization."

[17] In view of the need to reconcile the autonomy of the lowest, that is, simplest form of beingness or organism, with the autonomy of the more complex forms that they build up, the concept of the "superorganism" or "complex organism" was proposed long back by Frederic Clements in his *The Development and Structure of Vegetation* (Lincoln, Neb.: The Botanical Seminar, 1904); quoted by Ciancio and Nocentini, op. cit.

[18] Anna-Teresa Tymieniecka, *Logos and Life,* Book 1: *Creative Experience and the Critique of Reason,* Analecta Husserliana 24 (Dordrecht: Kluwer Academic Publishers, 1988).

[19] With this life oriented conception of the constructive entelechial *nucleus* and *design* we have struck at the very heart of Ingarden's ontological conception, in which the individual is seen as a transcendental object. That conception is analytically unfolded in Roman Ingarden, *Spor o Istnienie Swiata,* Vol. I (Warsaw: PAN, 1960), pp. 361-390.

PART III, CHAPTER 3

[1] A part of this study was presented at the first Chinese Phenomenology Conference in Nanking, October 25-30, 1994.

[2] Edmund Husserl, *Erfahrung und Urteil, Untersuchungen zur Genealogie der Logik,* comp. and ed. Ludwig Landgrebe (Hamburg: Felix Meiner Verlag, 1972), henceforth cited as *EU.* English translation: Edmund Husserl, *Experience and Judgment,* trans. James S. Churchill (Evanston: Northwestern University Press, 1973), p. 74. Henceforth this text will be cited as "Churchill trans."

[3] *EU,* p. 24. "Wir können auch so sagen: vor jeder Erkenntnis Bewegnung liegt schon der Gegenstand der Erkenntnis als Dynamis, die zur Entelechie werden soll. Mit dem Voranliegen ist

gemeint: er affiziert als im Hintergrund in unser Bewußtseinsfeld tretender, oder auch: er ist schon im Vordergrund, er ist sogar schon erfasst, weckt aber erst dann das gegenüber allen anderen Interessen der Lebenspraxis ausgezeichnete 'Erkenntnisinteresse'. Der Erfassen aber liegt immer voran die Affektion, die nicht ein Affizieren eines isolierten einzelnen Gegenstandes ist. Affizieren heißt Sicherherauscheben aus der Umgebung, die immer mit da ist, das Interesse, eventuell das Erkenntnisinteresse auf sich ziehen."

4 *EU,* p. 21. The definition of experience is principally the evidence of individual objects.

5 *EU,* p. 25. "Die Umgebung ist mit da als ein Bereich der Vorgegebenheit, einer passiven Vorgegebenheit, daß heißt einer solchen, die ohne jedes Zutun, ohne Hinwendung des erfassenden Blickes ohne eines Erwachen des Interesses immer bereits da ist. Diesen Bereich setzt alle Erkenntnisbetätigung...er affiziert aus seinem Felde heraus...er ist ein Seiendes unter anderem schon vorgegeben in einer passiver doxa, in einem Feld, daß selbst eine Einheit passiver Doxa darstellt ...aller Erkenntnisbetätigung voran liegt als universaler Boden eine jeweilige Welt; ...ein Boden universalen passiven Seinsglaubens, den jede einzelne Erkenntnis handlung schon voraussetzt...Alles, was als seiender gegenstand Ziel der der Erkenntnis ist, ist seiendes auf dem Boden der selbstverständlich als seiend geltenden Welt. Einzelnes vermeintlich Seiendes in ihr mag sich als nicht seiend herausstellen, Erkenntnis mag im Einzelnen Korrektur von Seinsmeinungen bringe; aber daß heißt nur, daß statt so anders ist, anders auf dem Boden der im Ganzen seienden Welt."

6 *EU,* p. 39. "Diese Welt ist uns immer vorgegeben 'mit dem Niederschlag logischer Leistungen' in welcher wir schon mit den anderen eine Tradition übernehmen, mit dem Erfahrungsgewerb der anderen die sich schon daran ..." durch Mitteilung, Lernen,..."sich schon logisch urteilend, erkennend betätigt haben."

7 *EU,* p. 24. "Vor jedem Eingang der Erkenntnis sind schon immer Gegenstände für uns da, in schlichter Gewißheit "vorgegeben", das heißt es vermeintlich seiend und so seiend als uns vor der Erkenntnis schon geltende, und das in verschiedener Weise. Da sind sie als schlicht vorgegebene Ansatz und Anreiz für die Erkenntnisbetätigung."

8 *EU,* p. 24. "Vor dem Einsatz der Erkenntnisbewegung haben wir 'vermeinte Gegenstände, schlicht in der Glaubensgewißheit vermeint' solange der Verlauf der Erfahrung oder eine kritische Einstellung diese Glaubensgewißheit erschüttert und sie in 'nur vermutliche', oder nicht so aber anders modifiziert."

9 *EU,* p. 25. "...Dieser universale Boden des Weltglaubens ist es, den jede Praxis voraussetzt, sowohl die Praxis des Lebens als auch die theoretische Praxis des Erkennens. Das Sein der Welt ist die Selbstverständlichkeit, die nie angezweifelt und nicht selbst durch die urteilende Tätigkeit erworben ist, sondern schon die Vorasussetzung für alles Urteilenbildet. Weltbewußtsein is Bewußtsein im *Modus der Glaubensgewißheit,* nicht durch einen im Lebenszusammenhang eigens auftretenden Akt der Seinssetzung, der Erfassung als das seiend oder gar des prädikativen Existenzurteils erworben. All das setzt schon Weltbewußtsein in Glaubensgewißheit voraus."

10 *EU,* pp. 29-30. "Existenz eines Realen hat sonach nie und nimmer einen anderen Sinn als Inexistenz, als Sein in Universum, im offenen Horizont der Raum-Zeitlichkeit, dem Horizont schon bekannter und nicht bloß jetzt aktual bewußten, aber auch unbekannter, möglicherweise zur Erfahrung und künftigen Bekannntheit kommender Realen. Die Einzelapperzeptionen machen einzelnes Reales bewußt, aber unvergeblich mit einem, wenn auch nicht thematisch werdenden Sinnbestand, der über sie, über den gesamten Bestand an einzelnen Apperzepierten hinausreichet."

11 *EU,* p. 29. "... daß heißt auch für menschliche und tieriesche Subjekte als Subjekte der Welt, für Kulturgüter, Gebrauchsdinge, Kunstwerke, usw. Alles Weltliche hat Anteil an der Natur. Die Naturaliesierung des Geistes ist nicht eine Erfindung der Philosophen — sie ist, wenn sie falsch gedeutet und verwertet wird, ein Grundfehler, aber eben nur dann. Aber sie hat einen Grund und ihr

Recht darin, dass mittelbar oder unmittelbar in der raumlich-zeitlichen Sphäre, alles was weltlich real ist, seine Stelle hat: alles ist hier oder dort, und der Ort ist bestimmbar, wie Orte überhaupt durch physikalische Instrumente, mögen es Sanduhren oder Pendeluhren oder sonstige Chronometer sein. Damit hat auch alles Unsinnliche an der Sinnlichkeit Anteil; es ist Seiendes aus der Welt, in dem einen raum-zeit-lichen Horizont Seiendes."

12 See the discussion by Tadashi Ogawa, "The Proto-Synthesis in the Perceptual Dimension According to Husserl: A Reconstructive Reflection," in Philip Blosser *et al.* (eds.), *Japanese and Western Phenomenology* (Dordrecht: Kluwer Academic Publishers, 1993), pp. 25-36.

TYING POINT III, CHAPTER 1

1 See Anna-Teresa Tymieniecka, *Logos and Life*, Book II, *The Three Movements of the Soul*, Analecta Husserliana 25 (Dordrecht: Kluwer Academic Publishers, 1988).

TYING POINT III, CHAPTER 2

1 See Anna-Teresa Tymieniecka, "The First Principles of Metaphysics of Life; Charting the Human Condition," in A-T. Tymieniecka (ed.), *Phenomenology of Man and of the Human Condition*, Part II: *The Meeting Point between Occidental and Oriental Philosophies*, Analecta Husserliana 21 (Dordrecht: Kluwer Academic Publishers, 1986) and, *inter alia*, *Logos and Life*, Book 1: *Creative Experience and the Critique of Reason*, Analecta Husserliana 24 (Dordrecht: Kluwer Academic Publishers, 1988); *Logos and life*, Book 2: *The Three Movements of the Soul*, Analecta Husserliana 25 (Dordrecht: Kluwer Academic Publishers, 1988); *Logos and Life*, Book 3: *The Passions of the Soul and the Elements in the Ontopoiesis of Culture. The Life Significance of Literature* (Dordrecht: Kluwer Academic Publishers, 1990).

2 I refer here, above all, to Fragments 1 and 2, of course:
Fragment 1: But of this account, which holds forever, people forever prove un-comprehending, both before they have heard it and when once they have heard it. For, although all things happen in accordance with this account, they are like people *without* experience when they experience words and deeds such as I set forth, distinguishing (as I do) each thing according to (its) real constitution, i.e. pointing out how it *is*. The rest of mankind, however, fail to be aware of what they do after they wake up just as they forget what they do while asleep. (From T. M. Robinson, *Heraclitus. Fragments: a Text and Translation with a Commentary* [Toronto: University of Toronto Press, 1987]). I use Robinson's translations here finding them the most subtle and accurate in comparison with others. However, the major translations in French, German, and Italian have also been consulted. The Diels numbering of the Fragments is used.)

Fragment 2: That is why one must follow that which is (common) [i.e., universal. For "common" means "universal"]. Though the account is common, the many live, however, as though they had a private understanding.

Fragment 1 might have been the opening statement of Heraclitus' presumed book. But even if it finds itself up front in an arbitrary way, it offers nevertheless a scope, a synthesis, a leading idea of logos (*tou de logou*), of an "everlasting plan" according to which all things happen. This idea of logos has received numerous interpretations in line with the widespread philosophical use of this term in Greek philosophy, as well as in response to numerous intimations coming from the texts of the fragments themselves. Our interpreting it as a "great plan" seems not only to synthesize the various side meanings that the term logos displays and to which it is related but also to do justice to the issue of rationality as an order of things, on the one hand, and the form of cognition, on the other.

First, it is almost unanimously accepted by interpreters that in Fragment 1 a radical distinction is made between the prephilosophical knowledge of the ordinary human being and the knowledge of the philosopher. Already Sextus Empiricus pointed out that according to Heraclitus, man is furnished with two organs for gaining knowledge of truth, namely, "sensation and reason" (*Against the Mathematicians* 7: 126-134). Of these organs he considers sensation to be untrustworthy and posits reason as the standard of judgement. However, although it is reason that he declares to be "the judge of truth," it is not "any sort of reason" but "that reason which is 'common' and divine." As human cognition, reason has to have a specific and unique access to all things to which it is common. This is a special insight into reality, an insight that allows one to penetrate into the hidden laws of reality, laws inaccessible to the ordinary mind. In this sense the logos of the human mind can be said to be, as Euripedes says quoting Homer, "like the day brought to them by the sire of gods and men" (*Odyssey* 18: 136-137), to be "divine" then. By "divine" the Greeks tend to mean that which is above the power of humans, what eludes direct experience and explanation. Empiricus quotes to this effect Euripedes, who addresses Zeus by saying that it is "hard to puzzle out" whether he be "necessity of Nature or mankind's intelligence." Heraclitus speaks about "insight" in Fragment 114. He emphasizes its universality: "those who speak with insight must base themselves firmly on that which is common to all" (all things?). Here he brings the meaning of the "common" to the level of significance of the law, and of a law "higher" than the laws of a society, city laws. Insight refers to the one law, the "divine law," the law of All. Common to all, the logos bespeaks the everlasting law. (However, it remains to be seen in virtue of what human beings that are otherwise "asleep" in sensuous common sense may attain the "private" knowledge that is insight. Heraclitus does not answer this question.) Thus Heraclitean logos is simultaneously insight into the hidden spheres of reality that reaches the level of universal laws. Robinson, in his presentation of Heraclitus' fragments emphasizes that the logos of which Heraclitus speaks in Fragments 1, 2, and 50 means essentially "statement." "This statement is uttered by the cosmos itself, which descriptively tells of how things are, and prescriptively lays down patterns of cosmic activity that serve as the basis for human laws."

Second, Logos understood as a statement of the inner law and measure of cosmic (and human) reality corresponds with the understanding of it as the factor of coincidence (*homologen*) among the insights penetrating reality, binding them together then while bringing forth the statement, the law, the world order.

Third, since Schleiermacher the fundamental meaning of the Heraclitean logos has often derived from the verb *legein* understood as gathering, putting together, the gathering of the similar in order to give an account of it in a statement. (See Edwin L. Minar, Jr., "The Logos of Heraclitus," *Classical Philology* 34 [1939], p. 323.)

I understand the complete meaning of the Heraclitean logos to be that of *a statement, an account of the universal, the "great" plan of reality, of the All.*

3 See Fragment 1.

4 Ibid.

5 Although the innermost flux of things underlies the majority of the Heraclitean statements, the most striking and discussed of these are Fragments 91a (91b).

Fragments 9la? [91b]: (a) [For, according to Heraclitus, it is not possible to step twice into the same river, nor is it possible to touch a mortal substance twice in so far as its state (*hexis*) is concerned. But, thanks to (the) swiftness and speed of change] (b) it scatters (things?) and brings (them?) together again, [(or, rather, it brings together and lets go neither "again" nor "later" but *simultaneously*)], (it) forms and (it) dissolves, and (it) approaches and departs.

6 Fragment 126: Cold things become warm, a warm thing becomes cold; a moist thing becomes dry, a parched thing becomes moist.

7 Fragment 88: And, (?) as (one and) the same thing, there is present (in us?) living and dead and the walking and the sleeping and young and old. For the latter, having changed around, are the former, and the former, having changed around, are (back) again (to being) the latter.

8 Fragments 10 and 51.

9 Fragment 94: The sun (god) will not overstep (his) measures. Otherwise (the) avenging Furies, ministers of Justice, will find him out.

Fragment 100: [The sun ... shares with the chief and primal god the job of setting bounds to ... (the) changes and] seasons that bring all things, [according to Heraclitus].

10 Fragment [115]: Soul possesses a logos (measure, proportion) which increases itself.

11 Anna-Teresa Tymieniecka, "La Fenomenologia in quanto nuova Critica della Ragione," in Enrico Berti and Marcelo Sánchez Sorondo (eds.), *L'Atto aristotelico e le sue ermeneutiche*, Collana "Dialogo di Filosofia" 7 (Rome: Herder, Università Lateranense, 1990), pp. 232-255.

12 See Anna-Teresa Tymieniecka, "La Condizione Umana all'Interno dell'Unità-di-tuttocio-che-e-vivante," in Marcelo Sánchez Sorondo (ed.), *Peri psyches, de homine: antropologia, nuovi approcci*, Collana "Dialogo di Filosofia" 11 (Rome: Herder, Università Lateranense, 1994) and *Logos and Life*, Book 2: *The Three Movements of the Soul*, Analecta Husserliana 25 (Dordrecht: Kluwer Academic Publishers, 1988).

13 Commenting on Fragment 115, Diels writes: "Die Seele is mit ihrem wesen, ihrem Gesetz (logos) in dem Urprinzip am tiefsten gewurzelt. Ihre grenzen reichen also an die grenzen des Alls," Hermann Diels, *Die Fragmente der Vorsokratiker* (Berlin: Weidmann, 3rd ed., 1910).

14 Anna-Teresa Tymieniecka, "Dal Sacro al Divino," in Enrico Berti and Marcelo Sánchez Sorondo (eds.), *Teologia razionale, filosofia della religione, linguaggio su Dio*, Collana "Dialogo di Filosofia" 9 (Rome: Herder, Università Lateranense, 1992), pp. 337-357.

15 Discovering the creative act of the human being as the access to the hidden inner workings of life that exhibit through their ramifications and constructive relevancies the inner order/law of being/becoming, phenomenology of life seems to meet Heraclitus' intuitions as interpreted by Wilhelm Capelle: "Die Natur der Welte enthülle sich ihm als er in die Tiefen seiner eigenen Natur Hinabsteigt" (*Die Vorsokratiker; die Fragmente und Quellenberichte übersetzt und eingeleitet* [Leipzig: A. Kroner, 1935], p. 148).

PART IV, CHAPTER 1

1 See Anna-Teresa Tymieniecka, "Poetica Nova" in Anna-Teresa Tymieniecka (ed.), *The Philosophical Reflection of Man in Literature*, Analecta Husserliana 12 (Dordrecht: , D. Reidel, 1983), pp. 1-93.

2 See my *Logos and Life*, Book 1, *Creative Experience and the Critique of Reason*, Analecta Husserliana 24 (Dordrecht: Kluwer Academic Publishers, 1988).

PART IV, CHAPTER 2

[1] Anna-Teresa Tymieniecka, *Logos and Life*, Book 1: *Creative Experience and the Critique of Reason*, Analecta Husserliana 24 (Dordrecht: Kluwer Academic Publishers, 1988).
[2] See by the present author, "The First Principles of the Metaphysics of Life," a monograph in Anna-Teresa Tymieniecka (ed.), *Phenomenology of Man and of the Human Condition*. Part II, *The Meeting Point of Occidental and Oriental Philosophies*, Analecta Husserliana 21 (Dordrecht: Kluwer Academic Publishers, 1986), pp. 3-73.
[3] Ibid.
[4] Ibid.
[5] Ann-Teresa Tymieniecka, *Leibniz' Cosmological Synthesis* (Assen: Royal Van Gorcum, 1967).
[6] By "human condition" I understand something completely different from the current usage of this term in existential thought as well as in religious perspectives. For this new conception of the human being see my "The First Principles of the Metaphysics of Life; Charting the Human Condition," cited above.
[7] See by the present author, "Harmony in Becoming: The Spontaneity of Life and Self-Individualization," in Anna-Teresa Tymieniecka (ed.), *Phenomenology of Life in a Dialogue between Chinese and Occidental Philosophies*, Analecta Husserliana 17 (Dordrecht: Kluwer Academic Publishers, 1984), pp. 3-19, and the present author's most recent "La Fenomenologia in Quanto Nuova Critica della Ragione," in Enrico Berti and Marcelo Sánchez Sorondo (eds.), *L'Atto Aristotelico e le sue ermeneutiche*, Collana "Dialogo di Filosofia" 7 (Rome: Herder, Università Lateranense, 1990).
[8] See by the present author "The First Principles...," op. cit.

PART IV, CHAPTER 3

[1] In Anna-Teresa Tymieniecka (ed.), *Foundations of Morality, Human Rights and the Human Sciences*, Analecta Husserliana 15 (Dordrecht: D. Reidel, 1985), pp. 3-78.
[2] See ibid.
[3] See my *Logos and Life*, Book 1, *Creative Experience and the Critique of Reason*, Analecta Husserliana 24 (Dordrecht: Kluwer Academic Publishers, 1988), pp. 330-341.
[4] Ibid., pp. 202-217.

TYING PART IV, CHAPTER 1

[1] See Anna-Teresa Tymieniecka, *Logos and Life:* Book 1, *Phenomenology of Creative Experience and the Triple Critique of Reason;* Book 2, *The Three Movements of the Soul*, Book 3, *The Passions of the Soul and the Elements in the Ontopoiesis of Culture* (Dordrecht: Kluwer Academic Publishers, 1988, 1988, 1989).

PART V, CHAPTER 2

[1] Edmund Husserl, *Logische Untersuchungen. Untersuchungen zur Phänomenologie und Theorie der Erkenntnis*, II/1, 5th ed. (Tübingen: Max Niemeyer Verlag, 1968).

[2] Edmund Husserl, *Ideen zu einer reinen Phänomenologie und Phänomenologischen Philosophie*, erstes Buch, ed. Walter Biemel and Marly Biemel, (The Hague: Martinus Nijhoff, 1969-).
[3] See Edmund Husserl, *Ideen zu einer reinen Phänomenologie und Phänomenologischen Philosophie*, zweites Buch, ed. Walter Biemel and Marly Biemel, (The Hague: Martinus Nijhoff, 1969-), p. 8.
[4] "Die Ethischen MS," F121, p. 4, quoted after Roth; see Alois Roth, *Edmund Husserl's ethische Untersuchungen, dargestellt anhand seiner Vorlesungsmanuskripte* (The Hague: Martinus Nijhoff, 1960).
[5] Ibid., F128, p. 247.
[6] Anna-Teresa Tymieniecka, "Imaginatio Creatrix," in Anna-Teresa Tymieniecka (ed.), *The Phenomenological Realism of the Possible Worlds*, Analecta Husserliana 3 (Dordrecht: D. Reidel, 1974), pp. 3-41.

PART V, CHAPTER 3

[1] See Anna-Teresa Tymieniecka, *Leibniz' Cosmological Synthesis* (Assen: Royal Van Gorcum, 1964).
[2] See Anna-Teresa Tymieniecka, "Beyond Ingarden's Idealism/Realism Controversy with Husserl; the New Contextual Phase of Phenomenology," in Anna-Teresa Tymieniecka (ed.), *Ingardeniana*, Analecta Husserliana 4 (Dordrecht: D. Reidel, 1976), pp. 241-418. This is my interpretation of Ingarden's analytic work.
[3] Roman Ingarden, *Der Streit um die Existenz der Welt*, Vol. III (Tübingen: Max Niemeyer Verlag, 1966).
[4] See Tymieniecka, "Beyond Ingarden's Idealism/Realism Controversy...," op. cit., pp. 322-341.
[5] Ibid.
[6] See Anna-Teresa Tymieniecka, *Why Is There Something Rather Than Nothing?* (Assen: Royal Van Gorcum, 1966), pp. 13-25.
[7] Ibid., pp. 77-90.
[8] I have proposed the distinction between 'construction' and 'reconstruction' in "Self and the Other in Man's Self-Interpretation," in Anna-Teresa Tymieniecka (ed.), *The Self and the Other. The Irreducible Element in Man.* Part I: *The "Crisis of Man,"* Analecta Husserliana 6 (Dordrecht: D. Reidel, 1977), pp. 156-161.
[9] See Tymieniecka, *Why Is There Something Rather Than Nothing?*, op. cit., pp. 23-71, 95.
[10] Ibid., pp. 33-34.
[11] Ibid., pp. 29-39.
[12] Ibid., p. 33.
[13] Ibid., pp. 77-158.
[14] See Anna-Teresa Tymieniecka, "The Initial Spontaneity," in Anna-Teresa Tymieniecka (ed.), *The Crisis of Culture*, Analecta Husserliana 5 (Dordrecht: D. Reidel, 1976), pp. 17-21.
[15] See Anna-Teresa Tymieniecka, "Imaginatio Creatrix," in Anna-Teresa Tymieniecka (ed.), *The Phenomenological Realism of Possible Worlds*, Analecta Husserliana 3 (Dordrecht: D. Reidel, 1974), pp. 3-41.
[16] See Tymieniecka, "The Initial Spontaneity," op. cit., pp. 21-26.
[17] See Tymieniecka, "Imaginatio Creatrix," op. cit., pp. 25-35.
[18] See Tymieniecka, "The Initial Spontaneity," op. cit., pp. 20-25.
[19] See Tymieniecka, "The Self and the Other...," op. cit., pp. 156-168.
[20] I have developed these ideas in a treatise written between 1968 and 1970 and published as

Logos and Life, Book 2: *The Three Movements of the Soul*, Analecta Husserliana 25 (Dordrecht: Kluwer Academic Publishers, 1988).

[21] See Anna-Teresa Tymieniecka, "Hope and the Present Instant," in Sebastian A. Matczak (ed.), *God in Contemporary Thought* (New York: Learned Publications; Louvain: Nauwelaerts: 1977), pp. 1082-1087.

[22] See Tymieniecka, "The Self and the Other...," op. cit.

[23] Ibid., pp. 175-186.

[24] See Tymieniecka, "Hope and the Present Instant," op. cit.

TYING POINT V, CHAPTER 1

[1] See my "The First Principles of the Metaphysics of Life," in Anna-Teresa Tymieniecka (ed.), *Phenomenology of Man and of the Human Condition*, Part II: *The Meeting Point of Occidental and Oriental Philosophies*, Analecta Husserliana 21 (Dordrecht: Kluwer Academic Press, 1986), pp. 3-73.

PART VI, CHAPTER 1

[1] On the "communal sense of life," see my monograph "The Moral Sense and the Human Person within the the Fabric of Communal Life," in Anna-Teresa Tymieniecka (ed.), *The Moral Sense in the Communal Sense of Life*, Analecta Husserliana 20 (Dordrecht: Kluwer Academic Publishers, 1986), pp. 3-100; and *Morality within the Life- and Social World*, Analecta Husserliana 22, pp. ix-xii (Dordrecht: Kluwer Academic Publishers, 1986; 1987).

PART VI, CHAPTER 2

[1] I established the term "fitness" in this sense in "The Moral Sense: A Discourse on the Phenomenological Foundation of the Social World and of Ethics," in Anna-Teresa Tymieniecka (ed.), *Foundations of Morality, Human Rights, and the Human Sciences*, Analecta Husserliana 15 (Dordrecht: D. Reidel, 1983), pp. 3-78.

[2] See the present author's monograph, "The Moral Sense and the Human Person within the the Fabric of Communal Life," in Anna-Teresa Tymieniecka (ed.), *The Moral Sense in the Communal Significance of Life. Investigations in Phenomenological Praxeology: Psychiatric Therapeutics, Medical Ethics, and Social Praxis within the Life- and Communal World*, Analecta Husserliana 20 (Dordrecht: Kluwer Academic Publishers, 1986), pp. 3-100.

[3] See my "Morality and the Life-World or the 'Moral Sense' within the World of Life," in Anna-Teresa Tymieniecka (ed.), *Morality within the Life- and Social World*, Analecta Husserliana 22 (Dordrecht: Kluwer Academic Publishers, 1987), pp. ix-xii.

[4] See my "Measure and the Ontopoietic Self-Individualization of Life," *Phenomenological Inquiry* 19 (1995), pp. 26-51.

[5] See Anna-Teresa Tymieniecka, *Logos and Life*, Book II: *The Three Movements of the Soul*, Analecta Husserliana 25 (Dordrecht: Kluwer Academic Publishers, 1988).

[6] See ibid.

[7] See by the present writer, *Phenomenology and Science in Contemporary European Thought* (New York: Farrar, Straus and Cudahy, 1962).
[8] See by the present writer, *Logos and Life,* Book 1: *Creative Experience and the Critique of Reason,* Analecta Husserliana 24 (Dordrecht: Kluwer Academic Publishers, 1988) and "Imaginatio Creatrix, the *Creative* versus the *Constitutive* Function of Man and the *Possible Worlds,*" in Anna-Teresa Tymieniecka (ed.), *The Phenomenological Realism of the Possible Worlds,* Analecta Husserliana 3 (Dordrecht: D. Reidel, 1974), pp. pp. 3-41.
[9] See by the present writer. "Beyond Ingarden's Idealism/Realism Controversy with Husserl; the New Contextual Phase of Phenomenology," in Anna-Teresa Tymieniecka (ed.), *Ingardeniana,* Analecta Husserliana 4 (Dordrecht: D. Reidel, 1976), pp. 241-418.
[10] Edmund Husserl, *Zur Phänomenologie der Intersubjektivität,* ed. and introd. Iso Kern, Husserliana, Vols. XIV and XV (The Hague: Nijhoff, 1973).
[11] See Tymieniecka, "Beyond Ingarden's Idealism/Realism Controversy with Husserl," op. cit.
[12] See Anna-Teresa Tymieniecka, "Dem Wendepunkt der Phänomenologie entgegen," *Philosophische Rundschau* (1967).
[13] Elmar Holenstein, *Phänomenologie der Assoziation; zur Struktur und Funktion eines Grundprinzips der passiven Genesis bei E. Husserl* (The Hague: Nijhoff, 1972), pp. 25-40.
[14] Edmund Husserl, *Analysen zur passiven Synthesis* (The Hague: Nijhoff, 1966), pp, 140, 148-172.
[15] Ibid., pp, 151, 163, 272-275, 286, and *passim.*
[16] See Holenstein, op. cit., p. 37.
[17] See Tymieniecka, "Imaginatio Creatrix," op. cit.
[18] Roman Ingarden, *Der Streit um die Existenz der Welt.* Vol. II, *Formal ontologie,* Parts I and II (Tübingen: Niemeyer, 1964). See also Tymieniecka, "Beyond Ingarden's Idealism/Realism Controversy with Husserl," op. cit., Part I.
[19] Roman Ingarden, *Der Streit um die Existenz der Welt,* Vol. III, *Über die Kausale Struktur der realen Welt* (Tübingen: Niemeyer, 1974) and by the same author, *Über die Verantwortung: ihre ontische Fundamente* (Stuttgart: Reclam, 1970). See also Tymieniecka "Beyond Ingarden's Idealism/Realism Controversy with Husserl," op. cit., Parts II and III.
[20] See Anna-Teresa Tymieniecka, *Why Is There Something Rather Than Nothing? Prolegomena to the Phenomenology of Cosmic Creation* (Assen: Royal Van Gorcum, 1966).
[21] See Tymieniecka, *Logos and Life,* Book I, op. cit. and Tymieniecka, "Imaginatio Creatrix," op. cit.
[22] See Tymieniecka, "Beyond Ingarden's Idealism/Realism Controversy with Husserl," op. cit., Part IV, "The Contextual Framework of Phenomenological Inquiry."
[23] Ibid.
[24] Cardinal Karol Wojtyla, with some reference to Max Scheler, developed in his book *The Acting Person,* Analecta Husserliana 10 (Dordrecht: D. Reidel, 1979) — a new edition of his *Osoba i Czyn* (Krakow: PTT, 1969), revised and expanded in collaboration with Anna-Teresa Tymieniecka — what we could call the "operative structure, of consciousness in the human act." His analysis of human "dynamisms" as centered in the structure of consciousness first attempts to span the entire scale of man's functioning (from the biophysical through the emotional and up to speculative thinking). Second, his analysis emphasizes the constitutive crisscrossing of natural tendencies, spiritual strivings, and intelligence under the determining agency of the ethical deliberation and choice entering into the fully "human act." Emmanuel Levinas in his book *Autrement qu'Être* (The Hague: Nijhoff, 1974) also forcefully situates the irreducible ethical element at the preconstitutive level of man's functioning.
[25] Husserl, *Analysen zur passiven Synthesis,* op. cit., pp. 148ff.

26 Holenstein, following Iso Kern, refers in this respect to original manuscripts of Husserl (AV 21, p. 102b). See Holenstein, op. cit., pp. 111-113.

27 Edmund Husserl in his *Ideen zu einer reinen Phänomenologie und phänomenologischen Philosophie,* Buch II (The Hague: Nijhoff, 1976), Beilage XII, pp. 332-340, draws a distinction between the activity of the ego and its "affectivity."

28 Holenstein also maintains this view; see Holenstein, op. cit., p. 37 and *passim.*

29 Ibid., p. 12.

30 Edmund Husserl, *Phänomenologische Psychologie* (The Hague: Nijhoff, 1968), p. 131.

31 Ibid.

32 Emmanuel Levinas, *Totalité et Infini, essai sur l'exteriorité,* Phænomenologica 8 (The Hague: Nijhoff, 1961).

33 By the term 'enjoyment' we do not understand the same thing that Levinas means by 'jouissance.' It seems that 'jouissance' means for him a fundamentally vital but specific type of experience, whereas by 'enjoyment' we mean a concrete and yet preexperiential type of spontaneity that makes experience qua experience actual. Compare the previously quoted *Totalité et Infini.*

34 Our term "subliminal" — which we introduced already in our previously cited "Beyond Ingarden's Idealism/Realism Controversy with Husserl" to denote *the preconscious realm of spontaneities* that are actually no longer at the "initial" stage of spontaneity but are already endowed with constructive virtualities ready to enter into play within the origin of the specifically human being — is not without reference to Kant's notion of the "sublime," which in the *Critique of Judgment,* as Milton Nahm has since pointed out, seems to contain the intuition that the morality signifying man's independence from nature stems from the preconscious, which is, however not bare of determination but is virtually structured.

TYING POINT 6

1 See Anna-Teresa Tymieniecka, "The First Principles of the Metaphysics of Life," in Anna-Teresa Tymieniecka (ed.), *Phenomenology of Man and of the Human Condition,* Part II: *The Meeting Point of Occidental and Oriental Philosophies,* Analecta Husserliana 21 (Dordrecht: Kluwer Academic Publishers, 1986), pp. 3-73; and Anna-Teresa Tymieniecka *Logos and Life,* Book 1: *Creative Experience and the Critique of Reason* (Dordrecht: Kluwer Academic Publishers, 1988).

2 Husserl's book has been widely discussed. See Anna-Teresa Tymieniecka (ed.), *The Crisis of Culture,* Analecta Husserliana 5 (Dordrecht: D. Reidel, 1976).

3 See ibid.; in this turn of discussion the focus was on the "crisis" of the human being.

4 Peter Kemp, "La Bio-ethique Anthropo-Centrifuge," *Acta Institutionis Philosophiae et Aestheticae,* Vol. 11 (Tokyo: Centre International pour Etude Comparée de Philosophie et d'Esthetique, 1993), pp.119-129.

5 See Anna-Teresa Tymieniecka, "Nature-Life and Culture in the Unity-of-Everything-There-Is-Alive," in Marcelo Sánchez Sorondo (ed.), *Physica, Cosmologia, Naturphilosophie, Nuovi approcci,* Collana "Dialogo di Filosofia" 10 (Rome: Herder; Università Lateranense, 1993). See also Tying Point I, Section 2 above: "The Womb of Life."

6 See Anna-Teresa Tymieniecka, "The Moral Sense and the Human Person within the Fabric of Communal Life," in Anna-Teresa Tymieniecka (ed.), *The Moral Sense in the Communal Significance of Life,* Analecta Husserliana 20 (Dordrecht: Kluwer Academic Publishers, 1986), pp. 3-100.

7 See Anna-Teresa Tymieniecka, *Logos and Life,* Book 1, op. cit.

8 See Tymieniecka, "The Moral Sense...," op. cit.

[9] See Anna-Teresa Tymieniecka, "The Ontopoiesis of Life as a New Philosophical Paradigm," *Phenomenological Inquiry* 22 (1998), pp. 12-59.

[10] See Anna-Teresa Tymieniecka, "The Ontopoietic Design of Life and Medicine's Search of a Norm," in Anna-Teresa Tymieniecka and Zbigniew Zalewski (eds.), *Life — The Human Being between Life and Death, A Dialogue between Medicine and Philosophy,* Analecta Husserliana 64 (Dordrecht: Kluwer Academic Publishers, 2000), pp. 13-36.

THE META-ONTOPOIETIC CLOSURE

[1] See Anna-Teresa Tymieniecka, "The Aesthetics of Enchantment," in Anna-Teresa Tymieniecka (ed.), *The Aesthetics of Enchantment,* Analecta Husserliana 65 (Dordrecht: Kluwer Academic Publishers, forthcoming).

[2] George V. Coyne, "Modern Research in Astronomy," Proceedings of Vatican conference of Nov. 15-18, 1999 (Città del Vaticano: Pontificia Academia Scientiarum, forthcoming).

[3] Quoted by James Miller in his *Measures of Wisdom: the Cosmic Dance in Classical and Christian Antiquity* (Toronto: University of Toronto Press, 1986), p. 5, from an interview of Einstein by George Sylvester Viereck, "What Life Means to Einstein," *Saturday Evening Post* 202: 17 (26 October 1929), p. 117.

INDEX OF NAMES

GPSR Compliance

The European Union's (EU) General Product Safety Regulation (GPSR) is a set of rules that requires consumer products to be safe and our obligations to ensure this.

If you have any concerns about our products, you can contact us on ProductSafety@springernature.com

In case Publisher is established outside the EU, the EU authorized representative is:

Springer Nature Customer Service Center GmbH
Europaplatz 3
69115 Heidelberg, Germany

Batch number: 10815964

Printed by Printforce, the Netherlands